CODE

SECOND BASE OF TRIPLET CODE

ADENINE	GUANINE	THIRD BASE OF TRIPLET CODE
Tyrosine UAU ⎫ Tyrosine UAC ⎭ Tyr	Cysteine UGU ⎫ Cysteine UGC ⎭ Cys	URACIL CYTOSINE
UAA ⎫ UAG ⎭ End Chain	UGA End Chain Tryptophan UGG ⎭ Try	ADENINE GUANINE
Histidine CAU ⎫ Histidine CAC ⎭ His	Arginine CGU ⎫ Arginine CGC	URACIL CYTOSINE
Glutamine CAA ⎫ Glutamine CAG ⎭ Gln	Arginine CGA ⎫ Arginine CGG ⎭ Arg	ADENINE GUANINE
Asparagine AAU ⎫ Asparagine AAC ⎭ Asn	Serine AGU ⎫ Serine AGC ⎭ Ser	URACIL CYTOSINE
Lysine AAA ⎫ Lysine AAG ⎭ Lys	Arginine AGA ⎫ Arginine AGG ⎭ Arg	ADENINE GUANINE
Aspartic Acid GAU ⎫ Aspartic Acid GAC ⎭ Asp	Glycine GGU ⎫ Glycine GGC	URACIL CYTOSINE
Glutamic Acid GAA ⎫ Glutamic Acid GAG ⎭ Glu	Glycine GGA ⎫ Glycine GGG ⎭ Gly	ADENINE GUANINE

The Science of Genetics

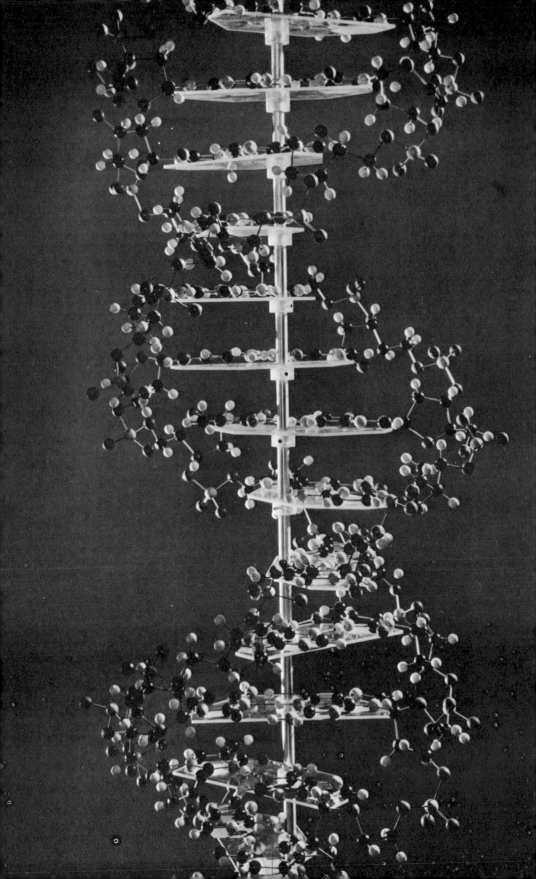

The Science of GENETICS

An Introduction to HEREDITY

THIRD EDITION

George W. Burns
Ohio Wesleyan University

Macmillan Publishing Co., Inc.
New York

Collier Macmillan Publishers
London

MACMILLAN PUBLISHING CO., INC.
866 Third Avenue, New York, New York 10022

COLLIER MACMILLAN CANADA, LTD.

Library of Congress Cataloging in Publication Data

Burns, George W   (data)
    The science of genetics.

    Bibliography: p.
    Includes index.
    1. Genetics.   I. Title. [DNLM: 1.   Genetics.
QH430 B967s]
QH430.B87   1976   575   75-4878
ISBN 0-02-317170-7

Printing: 1 2 3 4 5 6 7 8   Year: 6 7 8 9 0 1

Men love to wonder, and that is the seed of our science.

—EMERSON

Preface

IT has been pointed out that genetic knowledge now doubles every two years. In the four years that have elapsed since publication of the second edition of *The Science of Genetics,* then, the problem has increasingly become one of selecting material so that an up-to-date treatment can be maintained within the bounds of a one-term course.

Because of the growing information about human genetics, and because of interest in oneself as a genetic organism, emphasis on human genetic patterns has been increased in this edition, and examples are drawn from this field wherever possible. New or expanded treatments of sex determination, sex anomalies, chromosome aberrations, polygenic traits, and genetic polymorphism of mankind have been included. More information has been added on the human karyotype, both normal and abnormal, with emphasis on contributions from fluorescence microscopy and other recent studies. Legal implications of blood group genetics have been updated in the light of new laws and court cases. Treatment of possible genetic damage in man by a variety of drugs has been expanded.

Fruits of current research on DNA and RNA structure and behavior, ribosome structure, transcription, translation, and gene regulation have been included. Mutation and mutagenic agents have received greater coverage. The topic of population genetics has been expanded to include examination of the combined effect of mutation and selection on gene frequencies; at the same time the mathematics has been kept understandable for the general student. Implications of the "new genetics" (e.g., genetic engineering) for the future of the human race have been broadened in the light of current and ongoing research and controversy. Problems have been revised and augmented, and many new references have been added. Most of the latter are referred to directly within the chapters themselves, but a few significant, more general references have also been included.

Although most of the book has been completely rewritten for this edition, the problem approach has been retained, using work of the men and women who have contributed most to the development of the science. Topics are developed inductively, from observation to explanation to principle. The historical theme—starting with "classical" genetics and progressing through molecular genetics, wherein the story of our increasing knowledge about the field unfolds for the student as it has over the years for mankind—has been employed again in this edition. However, no loss of coherence will result if, for example, one wishes to proceed to DNA and molecular genetics early, for example, after considering monohybrid and dihybrid genetics.

This book is intended for undergraduates who have had a previous college course in one of the life sciences. However, salient points of cell structure and behavior are reviewed where needed. Appendixes contain answers to problems, life cycles of animals and plants used in genetic research, structural formulas of the biologically important amino acids, a summary of mathematical formulas, ratios and statistical tests that have been developed in the text, a table of metric values, and a list of journals and reviews. The glossary has been expanded.

The help of many persons who have made this book possible is gratefully acknowledged. Special thanks for their invaluable help are due to Jeffrey and Susan Carpenter Laycock for assistance on legal implications of blood group genetics and to Dawn DeLozier, John Derr, Gary Bock, and Chris Arn for providing or helping to secure photographic material. Mrs. Laycock, Miss DeLozier, Mr. Derr, Mr. Bock, and Miss Arn are former students of mine, and their interest is therefore particularly gratifying. In addition, many persons active in genetic research have most graciously provided other photographs of their own; these have added immeasurably to accounts of several topics. These latter persons include Drs. Theodore A. Baramki, Mihaly Bartalos, E. J. DuPraw, Richard C. Juberg, C. C. Lin, and Herbert A Lubs. Dr. S.-H. Kim kindly permitted use of a three-dimensional diagram of transfer RNA, and the National Foundation–March of Dimes generously gave permission to reproduce certain drawings and tables from *Paris Conference (1971): Standardization in Human Cytogenetics.* Full acknowledgment is given with each figure or table. I am also indebted to the Literary Executor of the late Sir Ronald A. Fisher, F.R.S., and to Oliver and Boyd, Ltd., Edinburgh, for their permission to reprint Table 3 from their book *Statistical Methods for Research Workers.* Sincere thanks are due also to the many professional associates, users of previous editions of *The Science of Genetics,* who responded so helpfully to requests for comments and suggestions. No expression of gratitude would be complete without recognition of the wisdom and continuing helpful counsel of Mr. Charles E. Stewart, Jr., Biology Editor, and Mr. Ronald Harris, Production Supervisor, of the Macmillan staff. Finally, to my students who have always made the teaching of genetics an exciting and rewarding experience, this book is dedicated with real affection.

G. W. B.

Contents

The Science of Genetics

CHAPTER 1
Introduction

ERTAINLY one of the most exciting fields of biological science, if not of all science, is genetics. This is the study of the mechanisms of heredity by which traits or characteristics are passed from generation to generation. Not only has modern genetics had a compact history, being essentially a product of the twentieth century, but it has made almost explosive progress from the rediscovery in 1900 of Mendel's basic observations of the 1860s to a fairly full comprehension of underlying principles at the molecular level. As our knowledge of these operating mechanisms developed, it became apparent that they are remarkably similar in their fundamental behavior for all kinds of organisms, whether man or mouse, bacterium or corn. But geneticists' quest for truth and understanding is far from completed; as in other sciences, the answer to one question raises new ones and opens whole new avenues of inquiry.

Genetics is personally relevant to everyone. Man is a genetic animal; each of us is the product of a long series of matings. People differ among themselves with regard to the expression of many traits; one has some inherited characteristics of his father and certain ones of his mother, but often, as well, some not exhibited by either parent. Familiar examples abound in persons of your own acquaintance—hair or eye color, curly or straight hair, height, intelligence, and baldness, to list but a few. Less obvious genetic traits include such diverse ones as form of ear lobes (Fig. 1-1), ability to roll the tongue (Fig. 1-2), ability to taste the chemical phenylthiocarbamide (PTC), red-green color blindness, hemophilia or "bleeder's disease," extra fingers or toes, or ability to produce insulin (lack of which results in diabetes). Note that some of these characteristics seem purely morphological, being concerned primarily with form and structure, whereas others are clearly physiological. Look about you at your friends and family for points of difference or similarity. Most of these characteristics have genetic bases. You are undoubtedly aware also of mentally retarded persons in our population, and of infants so grossly deformed that they are born dead or die within a very short time. Many of these unfortunate cases have underlying genetic and/or cytological causes. The determination of one's sex and the occurrence of sex intermediates (hermaphrodites and pseudohermaphrodites) also have genetic and cytological causes.

Although a survey of man's long interest in heredity is outside the scope of this book, it is well established that as much as 6,000 years ago he kept records of pedigrees of such domestic animals as the horse or of crop plants like rice. Because certain animals and plants were necessary for his survival and culture, man has, since the beginning of recorded history at least, attempted to develop improved varieties. But the story of man's concern with heredity during his

1

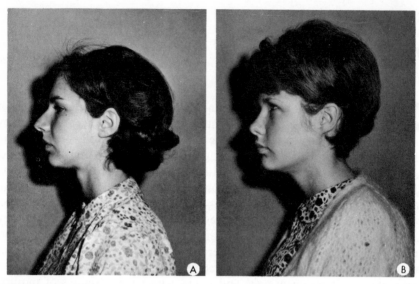

FIGURE 1-1. *Inherited difference in form of ear lobe. (A) Free ear lobe, the result of a dominant gene. (B) Attached ear lobe, caused by a recessive allele.*

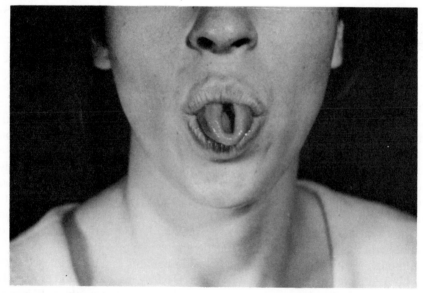

FIGURE 1-2. *Genetic or learned trait? Ability to roll the tongue has been ascribed to action of a dominant gene. A less commonly encountered trait, ability to fold the tip of the tongue back toward its base, is often described as resulting from another dominant gene. However, Martin (1975) finds no evidence for a genetic basis to tongue rolling.*

lifetime on this planet has been, until recently, one of interest largely in results rather than in fundamental understanding of the mechanisms involved.

As one examines the development of ideas relating to these mechanisms, he finds the way replete with misconceptions, many of them naïve in the light of modern knowledge. These theories may be divided roughly into three categories: (1) *"vapors and fluids,"* (2) *preformation,* and (3) *particulate inheritance.* Such early Greek philosophers as Pythagoras (500 B.C.) proposed that "vapors" derived from various organs unite to form a new individual. Then Aristotle assigned a "vitalizing" effect to semen, which, he suggested, was highly purified blood, a notion that was to influence thinking for almost 2,000 years.

By the seventeenth century sperm and egg had been discovered, and the Dutch scientist Swammerdam theorized that sex cells contained miniatures of the adult. Literature of that time contains drawings of models or manikins within sperm heads which imaginative workers reported seeing (Fig. 1-3). Such theories of preformation persisted well into the eighteenth century, by which time the German investigator Wolff offered experimental evidence that no preformed embryo existed in the egg of the chicken.

But Maupertuis in France, recognizing that preformation could not easily account for transmission of traits to the offspring from both parents, had pro-

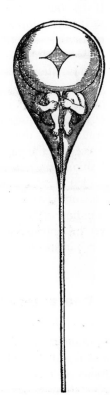

FIGURE 1-3. *Homunculus, "little man in a sperm cell."* (Drawing by N. Hartsoecker, *Journal des Scavans,* Feb. 7, 1695.)

posed in the early 1800s that minute particles, one from each body part, united in sexual reproduction to form a new individual. In some instances, he reasoned, particles from the male parent might dominate those from the female, and in other cases the reverse might be true. Thus the notion of particulate inheritance came into consideration. Maupertuis was actually closer to the truth, in general terms, than anyone realized for more than a century.

Charles Darwin suggested in the nineteenth century essentially the same basic mechanism in his theory of pangenesis, the central idea of which had first been put forward by Hippocrates (400 B.C.). Under this concept, each part of the body produced minute particles ("gemmules") that were contained in the blood of the entire body but eventually concentrated in the reproductive organs. Thus an individual would represent a "blending" of both parents. Moreover, acquired characters would be inherited because as parts of the body changed, so did the pangenes they produced. A champion weight lifter, therefore, should produce children with strong arm muscles; such transmission of acquired traits we know does not occur.

Pangenesis was disproved later in the same century by the German biologist Weismann. In a well-known experiment he cut off the tails of mice for 22 generations, yet each new lot of offspring consisted only of animals with tails. If the source of pangenes for tails was removed, how, he reasoned, could the next generation have tails? Yet, in spite of these early problems with the idea of particulate inheritance, its basic concept is the central core of our modern understanding.

Most attempts to explain observed breeding results failed because investigators generally tried to encompass simultaneously *all* variations, whether heritable or not. Nor was the progress of scientific thought or the development of suitable equipment and techniques ready to help point the way. It was the Augustinian monk Gregor Mendel who laid the groundwork for our modern concept of the particulate theory. He did so by attacking the problem in logical fashion, concentrating on one or a few observable, contrasting traits in a controlled breeding program. Both by his method and by his suggestion of causal "factors" (which we now call *genes*), Mendel came closer to a real understanding of heredity than had anyone in the preceding 5,000 years or more, yet he only opened the door for others. An understanding of the cellular mechanisms was still to be developed.

Characteristics of Useful Experimental Organisms

Even though Mendel's approach was probably more the result of luck than of thoughtful planning, it was superb in its simplicity and logic. First, Mendel was fortunate to have in the garden pea (*Pisum sativum*) what we recognize today as a good subject for genetic study. There are six important considerations for choosing a plant or animal for genetic experiments:

1. Variation. The organism chosen should show a number of detectable differences. Nothing could be learned of the inheritance of skin color in man, for example, if all human beings were alike in this respect. In general, the larger the number of discontinuous traits and the more clearly marked they are, the greater the usefulness of the species for genetic study.

2. Recombination. Genetic analysis of a species is greatly expedited if it has some effective means of combining, in one individual, traits of two parents. Such *recombination* permits comparison of one expression of a character with another expression of the same trait (e.g., tall versus dwarf *size,* brown versus blue *eye color*) through several generations. In many organisms recombination occurs as the result of *sexual reproduction,* in which two sex cells (**gametes**), generally from two different parents, combine as a fertilized egg (**zygote**). The sexual process is characteristic of higher animals and plants, and occurs in many lower forms as well. In bacteria and viruses there occur such processes as conjugation and transduction which also bring about recombination. Life cycles of several genetically important organisms are reviewed in Appendix B.

On the other hand, many organisms reproduce *asexually* or *vegetatively* and cannot furnish recombinational information. Basically, asexual reproduction may involve specialized cells (often called **spores**), daughter cells (in unicellular forms), parts of a single parent (cuttings, grafts, fragmentation, etc.), or **parthenogenesis,** in which an individual develops from an unfertilized egg, as in the male honeybee. By and large, a means of recombination is required for genetic study.

3. Controlled Matings. Systematic study of an organism's genetics is far easier if we can make controlled matings, choosing parental lines with particular purposes in mind, and keep careful records of offspring through several generations. The mouse, the fruit fly (Fig. 1-4), corn, and the red bread mold (*Neurospora*), for instance, make better genetic subjects in this respect than does man. In human genetics we are dependent largely on pedigree analysis, or studies of traits as they have appeared in a given family line for several past generations. Although such analyses are certainly useful, the human geneticist must rely largely on lines of progeny that *have been* established and cannot devise desired crosses of his own.

4. Short Life Cycle. Acquisition of genetic knowledge is facilitated if the organism chosen requires only a short time between generations. Mice, which are sexually mature at five or six weeks of age and have a gestation period of about 19 to 31 days, are much more useful, for instance, than elephants, which mature in eight to 16 years and have a gestation period of nearly two years. Likewise, the fruit fly, *Drosophila,* is much used in investigations because it may provide as many as five or six generations in a season. But first place for short life cycle goes to bacteria and to bacteriophages (or simply phages, viruses that infect bacteria), which, under optimum conditions, have a generation time of only 20 minutes! Both bacteria and phages also offer a number of other advantages for genetic study, as we shall subsequently see.

FIGURE 1-4. *The fruit fly,* Drosophila melanogaster, *an animal highly useful in genetic research.*

5. Large Number of Offspring.   Genetic studies are greatly speeded if the organism chosen produces fairly sizable lots of progeny per mating. Cattle, with generally one calf per breeding, do not provide nearly as much information in a given time as many lower forms of life where offspring may number many thousands.

6. Convenience of Handling.   For practical reasons, an experimental species should be of a type that can be raised and maintained conveniently and relatively inexpensively. Whales are obviously less useful in this context than are bacteria.

Methods of Genetic Study

Mendel's pea studies illustrate an approach useful both in classical, descriptive genetics and in modern molecular studies. This is the *planned breeding experiment,* in which parents exhibiting contrasting expressions of the same trait or traits are mated or crossed and careful records of results kept through several generations. We shall examine a number of such experiments in the next and succeeding chapters.

In addition to experimental breeding, we have already noted *pedigree analysis* in cases where controlled breeding programs are impossible. Pedigrees of three different conditions are shown in Figures 1-5 to 1-7; Figure 1-5 illustrates a pedigree of polydactyly, the occurrence of extra fingers, shaded symbols indicating this condition and unshaded ones representing individuals with the usual five fingers. As is customary in pedigree diagrams, squares represent males, circles females. Here we have a marriage between a polydactylous man and woman (generation I). They have three children, a polydactylous girl, a polydactylous son, and a "normal" son (generation II). The first and third individuals of generation II each marry "normal" persons; their children are shown in generation III. Note in this case that affected individuals appear

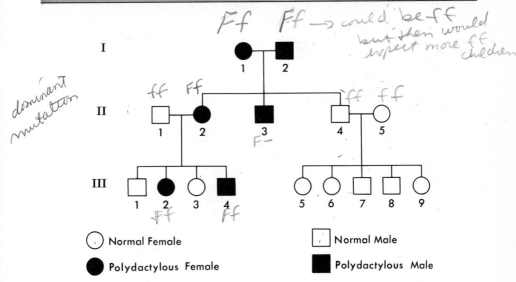

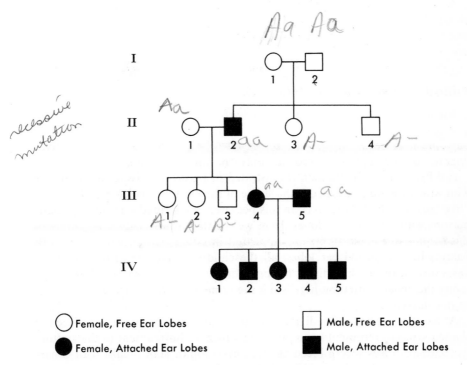

FIGURE 1-5. *Pedigree of polydactyly, occurrence of extra fingers, in human beings.*

FIGURE 1-6. *Pedigree of ear-lobe shape in human beings.*

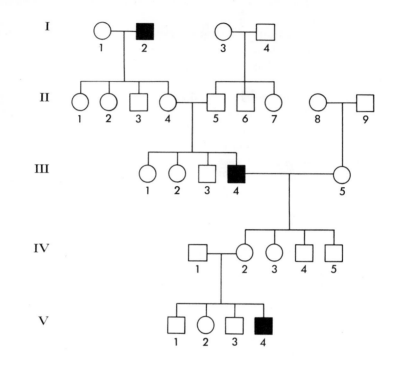

sex-linked

Normal Female   Normal Male   Hemophilic Male

FIGURE 1-7. *Pedigree of hemophilia in man.*

the mutation is thus is a dominant allele (will show up in at least 1 parent)

only when at least one parent is polydactylous; marriages between five-fingered persons appear to produce only "normal" offspring.

A different kind of situation is illustrated in Figure 1-6. Here the condition with which we are concerned is the shape of the ear lobe, which may be either "free" or "attached" (Fig. 1-1). In this pedigree, the shaded symbols represent persons with attached ear lobes. Here we see that an individual may display this condition without its having appeared in either parent, whereas if both parents have attached ear lobes, all the children have the same condition. Does this mean that heredity is not operating in this case? Not at all. But certainly the genetic situation here is not the same as in the previous illustration of polydactyly.

Thus, this mutation is recessive allele (may not show in either parent)

As another example of pedigree, note the case shown in Figure 1-7. Here the shaded symbols represent hemophilia, in which an impairment of the clotting mechanism causes severe bleeding from even minor skin breaks with consequent danger to life. In this instance the trait appears to be confined to the males but transmitted from an affected male through his daughters to some of his grandsons. This pattern, repeated in many pedigrees of hemophilia, certainly suggests a genetic basis, but one that is quite different from either of the first two.

Finally, *statistical analyses* of several kinds are used (1) to predict the probability of certain results in untried crosses and (2) to provide degrees of confidence in a theory regarding the specific genetic mechanism operating in a given case. Geneticists employ primarily several statistics of probability, which may be applied either to the results of an experimental breeding program or to a particular pedigree.

Fields of Study Useful in Genetics

After an initial and appropriate preoccupation with descriptive genetics, scientists turned naturally to problems of the mechanics of the processes they observed. The "what" of the earliest twentieth century rapidly gave way to a concern with "how." Parallels between inheritance patterns and the structure and behavior of cells were noted by a number of pioneer investigators. Thus *cytology* rapidly became an important adjunct to genetics. In fact, a pair of papers by Sutton as early as 1902 and 1903 clearly pointed the way to a physical basis for the burgeoning science of heredity. Sutton concluded his 1902 paper with a bold prediction: "I may finally call attention to the probability that (the behavior of chromosomes) may constitute the physical basis of the Mendelian law of heredity." Truly the door was thereby opened to an objective examination of the physical mechanisms of the genetic processes.

As the science of genetics developed rapidly during the first quarter of this century, a considerable body of knowledge was built up for such organisms as the fruit fly (*Drosophila*), corn (*Zea*), the laboratory mouse (*Mus*), and the tomato (*Lycopersicon*) concerning *what* traits are inherited and how different expressions of these are related to each other. Genetic *maps,* based on breeding experiments, were constructed for these and other species showing which genes were located on which chromosomes and the distances between *linked* genes. (We shall examine such maps and the methods used to acquire information needed to construct them in Chapter 6.) Geneticists next began to turn from concern with inheritance patterns of such traits as eye color in fruit flies to problems of *how* the observable trait is produced. Especially in the period since the beginning of World War II, a central question has been the *structure* of the gene and the mode of its operation. As the search for answers has proceeded ever more deeply into molecular levels, an increasingly important part in genetic study has been played by chemistry and physics. Contributions of these sciences have enabled geneticists to gain a clear concept of the molecular nature of the gene and its operation.

Practical Applications of Genetics

Genetics appeals to many of us not only because we are part of an ongoing genetic stream, but also because it has had such an exciting history in which theory has evolved out of observation and led, in turn, to experimental proof of fundamental operating mechanisms. Of course, any science may make the same claim, but the history of man's knowledge and understanding of genetics is to other sciences as a time-lapse movie of a growth process is to a normal-speed film. A fraction of a century ago the scientific community at large knew nothing of genetic mechanisms. Now, however, we can, with considerable accuracy, construct molecular models of genes, atom by atom. In fact, one relatively simple gene has recently been synthesized in the laboratory of Khorana and his associates (Agarwal et al., 1970). But besides being a fascinating intellectual discipline intimately related to ourselves, genetics has many important practical applications. Some of these are fairly familiar; others may be less so.

The history of improvement of food crops and domestic animals by selective breeding is too well known to warrant detailed description here (Figs. 1-8, 1-9). Increases in yield of crops such as corn and rice, improvement in flavor and size, as well as the production of seedless varieties of fruits, and advances in meat production of cattle and swine have markedly benefited mankind. As the population of the world continues to increase, this practical utilization of genetics is likely to assume even greater significance. Appropriately, the 1970 Nobel Peace Prize was awarded to a scientist, Norman Borlaug, for more than a quarter century of successful work in breeding high-yield, stiff-stemmed varieties of Mexican wheat. These new varieties, incorporating genes from American, Japanese, Australian, and Colombian stocks, not only have much-improved yield, but also wide geographic, photoperiodic, and climatic adaptability. They are successfully grown in such varied parts of the world as Mexico, Turkey, Afghanistan, Pakistan, and India. In the five-year period ending in 1970, introduction of the new varieties into India resulted in raising the wheat crop from 12 to 21 million tons, a rate of increase greater than that of India's population. Borlaug's efforts are also buying precious time in the race to control population growth. Similarly, the effort to breed disease-resistant plants must be never-ending. The 1970 epidemic of southern corn blight in the United States is a case in point.

Applications of genetics in the general field of medicine are numerous and growing. Many diseases and abnormalities are now known to have genetic bases. Hemophilia, some types of diabetes, an anemia known as hemolytic icterus, some forms of deafness and of blindness, several hemoglobin abnormalities, and Rh incompatibility are a few conditions that fall into this category. Recognition of their inherited nature is important in anticipating their possible future occurrence in a given family, so that appropriate preventive steps may be taken.

FIGURE 1-8. *Effects of breeding programs in cattle.* (*A*) *Texas Longhorn.* (USDA photo). (*B*) *Hereford, bred for meat production.* (Photo courtesy National Hereford Association.)

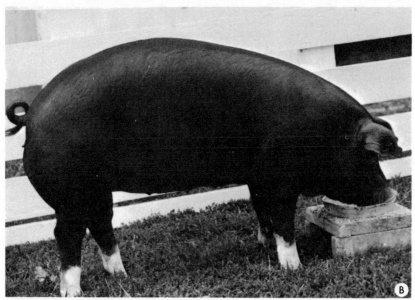

FIGURE 1-9. *Effects of breeding programs in swine.* (*A*) *Wild boar.* (Photo courtesy of Tennessee Game and Fish Commission. By permission.) (*B*) *Modern Poland China gilt.* (Photo courtesy Poland China Record Association. By permission.)

Closely related is the whole field of genetic counseling. Some estimate of the likelihood of a particular desirable or undesirable trait appearing in the children of a given couple can be provided by one who has sound genetic training and some information on the ancestors of the prospective parents. Questions encountered might range from the probability of a couple's having any red-haired children to the chance of muscular dystrophy appearing in the offspring.

Genetics has its legal applications, too. Analysis of blood type, a genetically determined character, may be used to solve problems of disputed parentage. Questions of baby mixups in hospitals, illegitimate children, and estate claims can often be clarified by genetics.

So in our study of genetics we shall pursue the same general route followed by other scientists in their search for an understanding of the mechanisms of inheritance. Beginning with simple observations that could be made by anyone, in many cases without a laboratory or special equipment, we shall continually raise questions and explore possibilities concerning the causes of phenomena we observe. From these we shall attempt to arrive, by inductive reasoning, at the specific principles that form the groundwork of modern genetics. These we will then test deductively by application to still other cases. Our quest for the "why" and "how" of genetics will grow ever more specific until we are able to answer on the deepest molecular levels yet penetrated by science. In the process we should not only acquire some fundamental genetic knowledge about ourselves, but also sharpen our powers of critical, analytical, skeptical thinking.

REFERENCES

AGARWAL, K. L., H. BÜCHI, M. H. CARUTHERS, N. GUPTA, H. G. KHORANA, K. KLEPPE, A. KUMAR, E. OHTSUKA, U. L. RAJBHANDARY, J. H. VAN DE SANDE, V. SGARAMELLA, H. WEBER, and T. YAMADA, 1970. Total Synthesis of the Gene for an Alanine Transfer Ribonucleic Acid from Yeast. *Nature*, **227:** 27–34.

ILTIS, H., 1932. *Life of Mendel.* New York, W. W. Norton.

MARTIN, N. G., 1975. No Evidence for a Genetic Basis of Tongue Rolling or Hand Clasping. *Jour. Hered.*, **66:** 179–180.

MENDEL, G., 1865. *Experiments in Plant Hybridization.* Reprinted in J. A. Peters, ed., 1959. *Classic Papers in Genetics.* Englewood Cliffs, N.J., Prentice-Hall.

STURTEVANT, A. H., 1965. *A History of Genetics.* New York, Harper & Row.

PROBLEMS

1-1. Suggest a number of ways in which bacteria make good genetic subjects.

1-2. Why are organisms that have only a single set of chromosomes (monoploid) often more favorable for genetic experiments than those having two sets (diploid)?

1-3. On the basis of the criteria of useful genetic organisms listed in this chapter, how do Mendel's peas measure up?

1-4. List as many reasons as possible why man is not a good experimental genetic organism.

1-5. Why is the inheritance of acquired characters no longer accepted as fact?

1-6. Devise an experiment of your own to test the question of inheritance of acquired characters.

1-7. How could you, at this stage of your study of genetics, justify Sutton's prediction (page 9) that the chromosomes may serve as a physical basis of inheritance?

1-8. Some races of corn produce red kernels, some white, and others white kernels that turn red only if they are exposed to light during maturation. Does kernel color in corn appear to depend on heredity, environment, or both? Explain.

1-9. Poliomyelitis is a disease known to be caused by a specific virus. It has sometimes been observed to be more frequent in some families than in others, even where children appear to have the same degree of exposure to the disease. Suggest a possible explanation.

1-10. Blue-green algae are primitive plants that reproduce only by cell division or fragmentation. They appear to have changed little over a very long period of geologic time. Moreover, all individuals of a given species appear to be morphologically and physiologically alike. Suggest a rational explanation for these latter two observations.

1-11. Huntington's chorea is a genetically based disorder of the central nervous system leading to physical and mental deterioration and culminating in death. It most often begins to appear in middle age. Would this condition be easier or more difficult than ear lobes to study genetically? Why?

1-12. Considering variation in height in members of your family and in other families, would you say that height in human beings is determined by heredity, environment, or both? Why do you say so?

CHAPTER 2
Monohybrid Inheritance

I N the introduction we noted that early attempts to determine fundamental genetic mechanisms frequently failed because investigators tried to examine simultaneously all discernible traits. We saw that Mendel's success in preparing the groundwork of our modern understanding lay in (1) concentrating on one or a few characters at a time, (2) making controlled crosses and keeping careful records of the results, and (3) suggesting "factors" as the particulate causes of various genetic patterns. If we wished, for instance, to learn something of the inheritance of *vestigial* wing in the fruit fly, *Drosophila* (Fig. 2-1), we would cross an individual having normal, full-sized wings with one having vestigial wings, ignoring all other traits. The appearance of the offspring through several generations and the relative numbers of "normal" and "vestigial" individuals produced would be carefully recorded and evaluated.

Such a cross, involving contrasting expressions of the same trait, is referred to as a **monohybrid cross.** Although this term is restricted by some geneticists to cases in which the parents *differ* in one trait (as in our suggested cross of the preceding paragraph), it may also be used to refer to any cross in which only a single character is being considered, whether or not the parents differ *detectably.* We shall use it in this latter, wider sense.

Let us begin our study of genetics along the general lines of Mendel, but using a different organism for the sake of variety.

The Standard Monohybrid Cross

COMPLETE DOMINANCE

The common house plant variegated coleus (*Coleus blumei,* of the mint family) is a frequently encountered flowering plant that lends itself readily to simple genetic experimentation. It has many of the desirable characteristics, listed in Chapter 1, of an organism useful for genetic experimentation. Its reproductive cycle is relatively short, crosses between different individuals are easily made, yet self-pollination is also possible, and many of its genetically determined characteristics are vegetative, appearing even in the seedling stage.

One such character is seen in the shape of the leaf margins. Some plants have shallowly crenate edges; others have rather deeply incised leaves (Figs. 2-2 and 2-3). Plants displaying one or the other of these two traits may be referred to as "shallow" and "deep," respectively. Imagine two such plants, each the product of a long series of generations obtained by self-pollination,

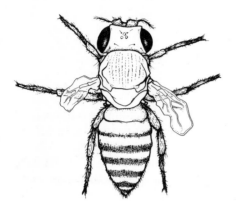

FIGURE 2-1. *Vestigial wing, resulting from the action of a recessive gene, in* Drosophila melanogaster.

in which the same individual serves as both maternal and paternal parent.[1] If, in such repeated breeding, deep always gave rise to deep offspring only, and shallow to shallow only, each such line of descent would be established as "pure-breeding" for this trait. The geneticist customarily uses the term ***homozygous*** in such cases. The precise genetic implication of this latter term will be seen shortly.

Suppose now a cross were to be made between two such homozygous individuals, one deep and the other shallow. In theory, the outcome might be any one of the following: (1) all the offspring may resemble one or the other parent only; (2) some of the progeny may resemble one parent, whereas the remainder may look like the other parent; (3) all the offspring, though like each other, may be intermediate in appearance between the two parents; (4) the offspring may look like neither parent and yet not be clearly intermediate between them; or (5) there may be a considerable range of types in the progeny, some being about as deeply lobed as one parent, some about as shallowly cut as the other, with the remainder forming a continuum of variation between the two parental types. When the offspring of this cross are examined, however, all of them are seen to be *deep*, like one parent. The *shallow* trait does not appear at all. Since the days of Mendel, such a characteristic that thus expresses itself in all the offspring (as *deep* does in this case) has been termed **dominant,** and the trait that fails to be expressed (here, *shallow*) has been referred to as **recessive.**

If we now either cross various individuals of this progeny or allow them to self-pollinate, what will be the results? From similar work of Mendel (Table 2-1) we can predict that about three fourths of the second generation should be deep, and about one fourth shallow. This is what we actually observe. (It should be pointed out that only with quite large samples would we expect to approach a 3:1 ratio closely.)

---

[1] See Appendix B for a review of life cycles which demonstrates how this can occur in many plants.

FIGURE 2-2. *The common house plant,* Coleus, *showing shallow-lobed leaves.*

FIGURE 2-3. *Deep-lobed leaves in* Coleus. *Compare with Figure 2-2.*

TABLE 2-1. Mendel's Earliest Experiments on Peas

Parents	First Generation	Second Generation		Ratio
Round × wrinkled seed	round	5,474 round	1,850 wrinkled	2.96:1
Yellow × green cotyledons	yellow	6,022 yellow	2,001 green	3.01:1
Gray-brown × white seed coats	gray-brown	705 gray-brown	224 white	3.15:1
Inflated × constricted pods	inflated	882 inflated	299 constricted	2.95:1
Green × yellow pods	green	428 green	152 yellow	2.82:1
Axial × terminal flowers	axial	651 axial	207 terminal	3.14:1
Long × short stems	long	787 long	277 short	2.84:1
TOTALS		14,949	5,010	Av. 2.98:1

According to standard genetic terminology, our results in *Coleus* may be summarized as follows:

(*Parental generation*) P deep × shallow

(*First filial generation*) F_1 all deep

(*Second filial generation*) F_2 $\frac{3}{4}$ deep + $\frac{1}{4}$ shallow

F_2 progeny are obtained either by "selfing" or by interbreeding members of the preceding F_1.

MECHANISM OF THE MONOHYBRID CROSS

Genes and Their Location. It now becomes our task to determine what mechanism might exist that will account for these results. Obviously something is transmitted from parent to offspring; just as obviously it must not be the trait itself, but rather something that *determines* the later development of that trait at an appropriate time and place in a particular environment. These determiners are called *genes;* we shall leave for later discussions their actual structure and behavior.

Because the sex cells or gametes constitute the only link in sexually reproducing organisms between parent and offspring, it is clear that genes must be transmitted from generation to generation via this gametic bridge. Where, then, are the genes located in the gametes? Cytological examination of the gametes of a great many sexually reproducing organisms, both plant and animal, discloses that, although the egg cell has a large amount of cytoplasm, the sperm is largely nucleus, having relatively little cytoplasm. This relationship breaks down, of course, in such isogamous forms as some of the algae where no size difference exists between fusing gametes. Though this observation does not furnish conclusive *proof* of the location of the genes within the nuclei of the sex cells, it does offer a *probability* that the genes with which we are dealing here are more likely to be nuclear than cytoplasmic in location. The possibility of other genes being located in the cytoplasm is discussed in Chapter 20. For

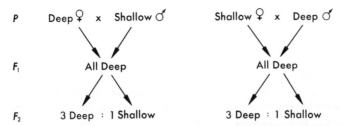

FIGURE 2-4. *Diagram of reciprocal cross in* Coleus. *Note that, in this case, results in the* F$_1$ *and* F$_2$ *are the same regardless of which plant is used as the pistillate parent.*

the present, until and unless contrary evidence becomes available, we shall therefore pursue the thesis that these genes are nuclear in location.

That sperm and egg in this particular kind of cross do make equal genetic contribution to the next generation is strongly indicated by the fact that, whether *deep* serves as the pistillate (egg-contributing) parent or as the staminate (sperm-contributing) parent, the result is the same: all the F$_1$ are deep lobed (Fig. 2-4). It should be emphasized, however, that such **reciprocal crosses** do not, in every case, produce identical results. In fact, whether reciprocal crosses give the same or different results provides certain important additional information concerning the genes involved. (See Chapter 12.)

If these genes are situated in the nucleus, it would be helpful to be able to fix their location within the nucleus more exactly. We shall have to look, therefore, for something that is (1) quantitatively distributed with complete exactness during nuclear division and (2) contributed equally by sperm and egg at gametic union, or syngamy. Although we shall look more closely in the next chapter at the cellular process involved, previous work that you as a student have had in the life sciences surely suggests, on these two bases, implication of the chromosomes as the vehicles of gene transmission in the kind of case under consideration. Pending a test of that hypothesis, this is the assumption on which we shall base our present attempts to explain our observations in *Coleus*. Regardless of the attractiveness of a theory, however, our acceptance of it must rest on testing, and we must always be ready to revise our theory in the light of new experimental evidence.

For the present, then, we shall assume genes to be situated on the chromosomes in the cell nuclei, transmitted via the gametes from one generation to the next, and contributed equally by both parents. If we designate the dominant gene for deep lobes as *D* and the recessive as *d,* we can represent the gametic contribution of the P generation as

$$\textcircled{D} \quad \text{and} \quad \textcircled{d}$$

where the two genes, one for deep (*D*) and one for shallow (*d*), are said to be **alleles** of each other, or to form an allelic pair.

Based on (1) the observation that chromosomes of both fusing gametes become incorporated within the nucleus of the zygote, maintaining their separate identities, and (2) the assumption that chromosomes serve as likely candidates for gene location, it then follows that the zygotes (which will become F_1 individuals) can be represented as *Dd*. So, also, the P individuals must be represented as *DD* and *dd*, respectively. Thus each of the parental individuals possesses two identical genes in each of its somatic (body, or vegetative) cells; that is, each is **homozygous.** In the same way, the F_1 plants must have in each of their somatic cells one gene for deep (*D*) *and* one for shallow (*d*); they are thus **heterozygous** with respect to this pair of alleles. Because in the heterozygous F_1, the effect of the dominant gene appears to mask completely the presence of the recessive allele, this case illustrates **complete dominance.**

The assumption is, therefore, that there are two genes for a given trait in the somatic cells and only one in the gametes. Some sort of nuclear division, prior to sex cell formation, which will reduce the gene number from two per trait to one—i.e., separate the alleles—is thus required. The next chapter explores this possibility further. Our original *Coleus* cross may now be written as follows:

$$
\begin{array}{ccc}
\text{P} \quad \text{deep} & \times & \text{shallow} \\
DD & & dd \\[4pt]
\textit{P gametes} \quad \textcircled{D} & \times & \textcircled{d} \\[4pt]
F_1 & \text{deep} & \\
& Dd & \\[4pt]
\textit{F}_1 \textit{ gametes} \quad \textcircled{D} & + & \textcircled{d}
\end{array}
$$

Phenotype and Genotype.   The *appearance* of the organism, with regard to the character or characters under consideration, constitutes its **phenotype,** which is generally designated by a descriptive word or phrase. On the other hand, an individual's *genetic makeup* is its **genotype,** and is customarily given by letters of the alphabet or other convenient symbols. In this case, *deep* and *shallow* represent *phenotypes,* whereas *DD, Dd, dd, D,* and *d* represent *genotypes* (somatic and gametic).

An individual's phenotype is not always a rigidly expressed, "either-or" condition but is often modified by environmental influences. For example, such a quantitative character as height is genetically determined in many plants (and in man), yet even in those cases in plants where only a single pair of genes can be shown to be operating, height variation in both the tall and short individuals often occurs. These latter variations, usually clustering fairly closely about a mean, can be shown, in suitable, controlled experiments, to be due to environment. In other words, genotype determines the phenotypic *range* within which an individual will fall; environment determines where in that range the individual will occur. In certain environments, too, it is conceivable that a particular genotype will not express itself at all. In corn, for

FIGURE 2-5. *"Sun-red" corn. Action of the sun-red gene is to produce red kernels when these are exposed to light; in the absence of light, kernels remain white. In this instance, the husks had been replaced by the black mask shown below.* (Courtesy Department of Plant Breeding, Cornell University.)

example, one gene ("sun-red") produces red grains if the ear is exposed to light (Fig. 2-5), but as long as the husks are intact the grains remain white, so that both sun-red and white genotypes remain phenotypically indistinguishable. Furthermore, as we shall soon see, phenotype may often be physiological and therefore "observable" only in the biochemical sense.

Probability Method of Calculating Ratios. If a nuclear division, which segregates the alleles D and d, occurs in the F_1 prior to gamete formation, then *one half* of the F_1 gametes should carry the dominant D and *one half* the recessive d. If, furthermore, syngamy is random, so that a D egg has an equal chance to be fertilized by either a D or a d sperm, for example, we can represent mating of the F_1 to produce the F_2 in this way:

$$\text{eggs from } ♀\ F_1 \quad \tfrac{1}{2}\,Ⓓ \ + \ \tfrac{1}{2}\,Ⓓ$$
$$\text{sperms from } ♂\ F_1 \quad \tfrac{1}{2}\,Ⓓ \ + \ \tfrac{1}{2}\,Ⓓ$$
$$\overline{\qquad F_2 \quad \tfrac{1}{4}DD + \tfrac{1}{4}Dd + \tfrac{1}{4}dD + \tfrac{1}{4}dd \qquad}$$

$(\tfrac{1}{2}D : \tfrac{1}{2}d)(\tfrac{1}{2}D : \tfrac{1}{2}d)$
$\tfrac{1}{4}DD : \tfrac{1}{2}Dd : \tfrac{1}{4}dd$
$1 : 2 : 1$

or $\tfrac{1}{4}DD + \tfrac{2}{4}Dd + \tfrac{1}{4}dd$ for a **1:2:1 F_2 monohybrid genotypic ratio.**

The basis for this kind of calculation is the **product law of probability.** Briefly stated, this law holds that the probability of the simultaneous occurrence of two independent events equals the product of the probabilities of their separate occurrences. Thus if one half the eggs are of genotype D and one half of the sperms have the same genotype, and if fertilization is a completely random event, then the probability of getting a DD zygote is $\frac{1}{2} \times \frac{1}{2} = \frac{1}{4}$.

As to phenotypes, we see that DD and Dd organisms appear indistinguishable on visual bases, so that the $1:2:1$ genotypic ratio gives our observed **3:1 F$_2$ monohybrid phenotypic ratio.** A $3:1$ ratio is shown in Figure 2-6. (As we shall see, these ratios may occur in the F$_1$, given certain P individuals, and are, of course, not the only possible monohybrid ratios.) Thus our assumptions have led to an explanation completely compatible with our earlier observations. These assumptions remain to be tested further, both by cytological examination and by additional breeding experiments. It should be noted that we have not used the so-called checkerboard or Punnett square method but rather a probability method of calculating genotypic and phenotypic ratios. This latter system becomes extremely advantageous when calculating ratios involving several pairs of genes.

THE TESTCROSS

As we have indicated, DD and Dd individuals in *Coleus* cannot be distinguished visually from each other. This raises two questions. First, can homozygous dominant and heterozygous individuals be distinguished in any manner? The answer is yes, and the **testcross** can supply the answer. In the testcross the individual having the dominant phenotype (which can be represented as having genotype $D-$) is crossed with one having a recessive phenotype that is, of course, homozygous. In our *Coleus* case the testcross of a homozygous dominant is as follows:

$$\text{P} \quad \text{deep} \times \text{shallow}$$
$$DD \qquad dd$$

$$\textit{P gametes} \quad \textit{eggs} \quad (1)\; \textcircled{D}$$

$$\textit{sperms} \quad (1)\; \textcircled{d}$$

$$\text{F}_1 \quad \text{all} \quad Dd\,(\text{deep})$$

If the P dominant phenotype had been heterozygous, the result would be the classic **1:1 monohybrid testcross ratio:**

$$\text{P} \quad \text{deep} \quad \times \quad \text{shallow}$$
$$Dd \qquad\qquad dd$$

$$\textit{P gametes} \quad \textit{eggs} \quad \tfrac{1}{2}\,\textcircled{D} + \tfrac{1}{2}\,\textcircled{d}$$

$$\textit{sperms} \quad 1\;\textcircled{d}$$

$$\text{F}_1 \qquad\qquad \tfrac{1}{2}\,Dd + \tfrac{1}{2}\,dd$$
$$\text{deep} \quad \text{shallow}$$

FIGURE 2-6. *A 3:1 purple:white ratio in corn. This is the* F_2 *of homozygous purple* × *white.*

Note that $DD \times dd$ produces an F_1 all of the dominant phenotype, whereas $Dd \times dd$ gives rise to a 1:1 ratio in the progeny. A 1:1 testcross ratio in corn is shown in Figure 2-7.

At this point it is important to emphasize our usage of the expression F_1. We shall use it to designate the first generation resulting from *any* given mating regardless of the parental genotypes. Therefore the term F_1 does not necessarily imply heterozygosity. In the two testcrosses just outlined, note that the F_1 genotype depends on the P genotypes and may, of course, be either heterozygous (as in the first instance) or homozygous (as in the *dd* individuals of the second case).

The second question suggested by the visual similarity of homozygous dominants and heterozygotes is the matter of how genes operate to exert their phenotypic effect. Although a complete answer must be deferred to later considerations of the molecular nature of the gene and the chemistry of its action, a genetic trait in Mendel's peas offers a tempting suggestion. It also furnishes an introduction to a variation in the classic 3:1 phenotypic ratio we have seen in *Coleus*. We shall examine this case in the following section.

Modifications of the 3:1 Phenotypic Ratio

INCOMPLETE DOMINANCE

Peas. The data of Table 2-1 were first reported by Mendel in a pair of originally widely ignored papers read before the Natural History Society at Brünn on February 8 and March 8, 1865. Mendel's results led him to designate round seeds as (completely) dominant to wrinkled seeds; indeed, homozygous and heterozygous round seeds cannot be differentiated macroscopically. However, when cells of the cotyledons of the three genotypes (*WW, Ww* round, and *ww* wrinkled) are examined *microscopically,* an abundance of well-formed starch grains is seen in the *WW* plants, and very few in *ww* individuals. Heterozygotes show an intermediate number of grains, many of them being imperfect or eroded. Moreover, *ww* embryos test higher in reducing sugars (the raw material from which starch molecules are constructed) than do those of *Ww* genotype. *WW* embryos test lowest of all for sugar. In plant cells starch is

FIGURE 2-7. *A 1:1 testcross ratio in corn. This ear resulted from the cross white* × *heterozygous purple.*

synthesized from glucose-1-phosphate under the influence of an enzyme system. Although a discussion of the chemistry of the sugar $\rightleftharpoons$ starch interconversion is outside the scope of a beginning course in genetics, it appears plausible that gene *W* is responsible for the production of one of the enzymes required for the reaction glucose-1-phosphate $\rightarrow$ starch. Likewise, it may be surmised that gene *w* produces either a smaller quantity of this enzyme or, perhaps, a molecule differing sufficiently in structure so that it functions only imperfectly, producing a less efficient enzyme molecule. The larger amount of starch in *WW* plants is associated with higher water retention (starch is a hydrophilic colloid) and therefore with plump, distended, spherical seeds. Homozygous recessive embryos, on the other hand, retain considerably less water upon reaching maturity and hence appear shriveled. The starch content of heterozygotes, however, is apparently sufficient to produce embryos visually indistinguishable from those of *WW* genotype. Consistent with these differences in starch content, it is found that heterozygotes have an intermediate amount of functional enzyme. The relationship between genes and enzymes suggested here is a fundamental concern of modern genetics and offers much help in elucidating the nature and action of the gene. It will receive considerable attention in later chapters of this book.

For our present purposes, however, it is clear that on the gross, macroscopic level, we have a case of complete dominance, yielding the familiar 3:1 phenotypic and 1:2:1 genotypic ratios. But on the microscopic, chemical, or molecular level we are faced with the realization that gene *W* must be considered only **incompletely dominant** to its allele. Therefore at the *physiological* level the cross *Ww* × *Ww* yields identical phenotypic and genotypic ratios of 1:2:1. Whether round versus wrinkled seeds is considered as complete dominance or as incomplete dominance depends entirely on the level of analysis. Because the physiological level more accurately reflects the actual operation of the genes involved, most geneticists would prefer to regard this as a case of incomplete dominance. It is important to note that in incomplete dominance the heterozygotes are phenotypically *intermediate* between the two homozygous types.

Other Organisms. Examples of incomplete dominance at the macroscopic level occur in both plants and animals. For example, radishes may be long, oval, or round. Crosses of long × round produce an F_1 of wholly oval phenotype:

$$P \quad \underset{l_1 l_1}{\text{long}} \quad \times \quad \underset{l_2 l_2}{\text{round}}$$

$$F_1 \qquad \underset{l_1 l_2}{\text{all oval}}$$

$$F_1 \text{ gametes} \quad eggs \quad \tfrac{1}{2}\,\boxed{l_1}\, + \tfrac{1}{2}\,\boxed{l_2}$$

$$sperms \quad \tfrac{1}{2}\,\boxed{l_1}\, + \tfrac{1}{2}\,\boxed{l_2}$$

$$F_2 \qquad \underset{\text{long} \quad \text{oval} \quad \text{round}}{\tfrac{1}{4}\,l_1 l_1 + \tfrac{2}{4}\,l_1 l_2 + \tfrac{1}{4}\,l_2 l_2}$$

It will be our practice to use lower-case letters with numerical subscripts to designate incompletely dominant alleles.

A classical case of incomplete dominance, and one of the earliest to be described, is that of the common ornamental flowering plant snapdragon (*Antirrhinum majus*). Homozygotes are either red or white, whereas heterozygotes are pink (Fig. 2-8).

Many morphological and physiological traits in a wide variety of animals and plants give evidence of being the result of a single pair of genes, and new ones are continually being reported in many different organisms. These include resistance to blister rust in the sugar pine (dominant; Parks and Fowler, 1970), coloboma in fowl "with profound effects on all body parts through effects on cartilage formation" (recessive lethal; Abbott et al., 1970), early seed stalk development, or "bolting," in celery (dominant; Bouwkamp and Honma, 1970), height in cultivated geranium, *Pelargonium* (incomplete dominance; Hennault and Craig, 1970), lack of anthocyanin pigment in cotton (recessive; Konoplia et al., 1973), resistance to downy mildew in grape (dominant; Filippenko and Shtin, 1973), resistance to a mottling virus in pepper plants (recessive; Zitter and Cook, 1973), resistance to stem rust in certain species of wheat (dominant; Kerber and Dyck, 1973), and lactose tolerance (lactase production) in man (dominant; Kretchmer, 1972), to list only a few.

CODOMINANCE

Cattle. In shorthorn cattle genes for red and white coat occur. Crosses between red ($r_1 r_1$) and ($r_2 r_2$) produce offspring ($r_1 r_2$) whose coat appears at a distance to be reddish gray, or *roan*. Superficially this would seem to be a case of incomplete dominance, but close examination of roan animals discloses the coat to be composed of a *mixture of red hairs and white hairs,* rather than hairs all of a color intermediate between red and white. Instances such as this, where

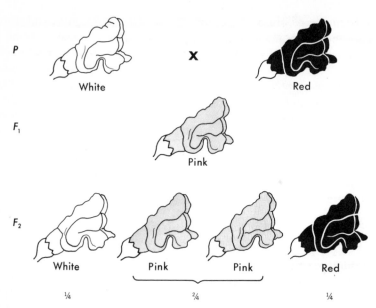

FIGURE 2-8. *A case of incomplete dominance in snapdragon. The pink* F_1 *hybrid is truly intermediate between the red and white parents, and the* F_2 *segregates in a characteristic 1:2:1 phenotypic ratio.*

the heterozygote exhibits a mixture of the phenotypic characters of both homozygotes, instead of a single intermediate expression, illustrate **codominance.** Genotypic and phenotypic ratios are identical in incomplete dominance and codominance, but the distinction reflects something of the way in which genes operate, a topic to which we shall devote considerable attention in later chapters.

Man. Several illustrations of codominance are to be found in human genetics. One of these involves a pair of codominant alleles, M and N, responsible for production of antigenic substances M and N, respectively, on the surfaces of the red blood cells. Although the genetics of blood antigens is explored more fully in Chapter 7, it is interesting to note that, whereas persons of genotype MM produce antigen M and NN individuals produce the somewhat different antigen N, heterozygotes (MN) produce *both* M and N antigens, not a single intermediate substance. All persons belong to one of three phenotypic classes, M, MN, or N. However, because antibodies to these antigens rarely occur, they do not need to be considered in transfusion, and most people do not know their M–N blood group. Accumulated data from families where both parents have been identified as MN show a close approximation to the expected 1:2:1 ratio in the children.

Another and more important case is found in the sickle-cell phenotype, which is particularly prevalent in blacks in this country and in certain African peoples. The hemoglobin of most persons is of a particular chemical structure and is known as hemoglobin A (for adult hemoglobin). Many chemical variants of hemoglobin A are found in relatively small numbers of people; one of these, hemoglobin S, is involved in the sickle-cell disorder. Genes responsible for hemoglobin types here are Hb^A and Hb^S. Most persons belong to genotype Hb^AHb^A. Their erythrocytes contain only hemoglobin A and are biconcave disk-shaped (Fig. 2-9A). Persons with *sickle-cell anemia* are of the genotype Hb^SHb^S and are characterized by a collection of symptoms, chiefly a chronic hemolytic anemia. In the blood of such persons the erythrocytes become distorted, many being essentially sickle-shaped (Fig. 2-9B). These cells not only impede circulation by blocking capillaries, but also cannot properly perform their function of carrying oxygen and carbon dioxide to and from the tissues.

In heterozygotes, Hb^AHb^S, some red cells contain hemoglobin A, others hemoglobin S. Because both types of hemoglobin, rather than a single intermediate form, are produced, this is another case of codominance. Microscopic examination of heterozygotes' blood under low oxygen tension discloses both normal and sickled erythrocytes (Fig. 2-10). Under normal conditions heterozygotes manifest none of the severe symptoms of Hb^SHb^S persons, though they may suffer some periodic discomfort and even develop anemia after a time at high altitudes. Genotypes and phenotypes in this disorder are as follows:

Hb^AHb^A normal (hemoglobin A only; no sickling of red cells)

Hb^AHb^S sickle-cell trait (hemoglobins A and S; sickling under reduced oxygen tension)

Hb^SHb^S sickle-cell anemia (hemoglobin S only; sickling under normal oxygen tension)

We shall examine in more detail the chemistry of these and other hemoglobins as well as the operation of the genes involved in Chapter 18.

LETHAL GENES

A second major modification of the classic monohybrid ratios is produced by genes whose effect is sufficiently drastic to kill the bearers of certain genotypes. Here both the effect itself and the developmental stage at which the lethal effect is exerted are important.

Man. In persons with sickle-cell anemia, blocking of the capillaries may produce infarction in almost any organ, leading to painful episodes, tissue destruction, and death, usually before attainment of reproductive age. Because homozygosity for Hb^S causes premature death it is a lethal gene, but its lethality is recessive inasmuch as heterozygotes are of generally good health. Marriages between heterozygotes may be expected to produce children in the ratio of 1 sickle-cell anemia : 2 sickle-cell trait : 1 normal at birth, but this phenotypic ratio becomes 2:1 after the sickle-cell anemia persons have died. A 2:1 phenotypic ratio, either at birth or later in life, is indicative of a pair of

#3

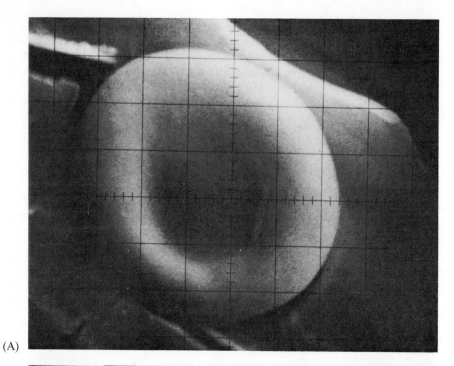

(A)

(B)

1) Regular dominance & cross   2) incomplete or codominance   3) incom or codominance with
F₂ geno 1:2:1   geno 1:2:1   recessive lethality of 1 allele
   pheno 3:1   pheno 1:2:1   at birth or later in life

                                                                   so pheno   { geno 1:2:1 dies recessive
                                                                   & geno   pheno 1:2   (homoz recessive)
                                                                   ratios
                                                                   both
                                                                   become 1:2

FIGURE 2-9. *(A) Normal red blood cell and (B) sickled red blood cells.* [Samples courtesy Dr. Patricia Farnsworth, Barnard College.] *The photomicrographs were taken at magnifications of (A) 10,000× and (B) 5,000× by Irene Piscopo of Philips Electronic Instruments on a Philips EM 300 Electron Microscope with Scanning Attachment.* (Photos courtesy Philips Electronic Instruments.)

codominant or incompletely dominant genes in which one allele shows recessive lethality.

Corn. Most plants with which a nonbotanist is familiar are characteristically autotrophic, being able to manufacture their own food from carbon dioxide and water in the process of photosynthesis. For this process, in all but the few autotrophic bacteria, the presence of a green, light-absorbing pigment, chlorophyll, is required. In corn (*Zea mays*) several pairs of genes affecting chlorophyll production have been described in the literature. One such gene, which we shall designate *G*, for normal chlorophyll production, is completely dominant to its allele *g*, so that *G*− plants contain chlorophyll and are photosynthetic. On the other hand, *gg* plants produce no chlorophyll and are yellowish white (because they still produce the yellow carotenoid pigments). (See Fig. 2-11.)

On the average, about one fourth of the progeny of two heterozygous parents are thus without chlorophyll, and seedlings show the classic 3:1 ratio. (F_2)
(pheno)

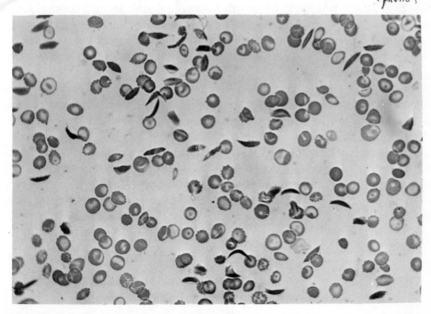

FIGURE 2-10. *Blood smear showing normal and sickle erythrocytes.* (Courtesy Carolina Biological Supply Co.)

4) recessive lethal gene with regular dominance (ie complete)
geno 1:2
pheno 1:0 (or actually 3:0)

5) Also get geno & pheno ratios
of 2:1 with normal (complete)
dominance when its the
necessary lethal (not homog recessive)
homog dominant which dies

In corn, germination and early seedling development take place at the expense of a food storage tissue, the endosperm, in the grain. In normal green plants, by the time this food reserve has been exhausted (in about 10 to 14 days' time), the seedling has developed a sufficient root system and amount of green tissue to be physiologically independent. Therefore in the cross of two heterozygotes, the initial 3:1 phenotypic ratio becomes what might be termed a 3:0 or a "1:0" ratio after some two weeks:

$$P \qquad \text{green} \times \text{green}$$
$$\qquad\qquad Gg \qquad\quad Gg$$

$$F_1 \qquad \tfrac{1}{4}\text{green} + \tfrac{2}{4}\text{green} + \tfrac{1}{4}\text{nongreen (die)}$$
$$\qquad\qquad GG \qquad\quad Gg \qquad\qquad gg$$

Note that after the lethal gene g has exerted its effect, the genotypic ratio has been converted from 1:2:1 to 2:1, as in the case of sickle-cell in man. So whereas homozygous green plants, for example, initially make up one fourth of the F_1, they later come to comprise *one third* of the surviving progeny because of the death of *gg* individuals. This 2:1 genotypic ratio, or the occurrence of only one phenotypic class where two would be expected in a 3:1 ratio, again is a clear indication of a (recessive) lethal gene.

In the instance just described, the occurrence and action of the lethal gene is easily discerned because the death of the homozygous recessives takes place only after nearly two weeks of growth following germination. Yet other lethals might conceivably produce their effect at almost any time between syngamy (i.e., in the zygote stage) on through embryogeny to a very late point in life. Obviously, lethals killing very late in life might well be hard to separate from other causes of death.

Mouse.   A classic case of a recessive lethal that kills early in embryo development, and one of the first to be reported in the literature, was "yellow" in mice. Early in this century Cuénot (1904, 1905) noted that black × black always produced black offspring, but that yellow × black produced yellow and black in a 1:1 ratio. He correctly concluded that yellow is heterozygous. Yet crosses of yellow × yellow always produced yellow and black in a ratio of 2:1, with litters of such matings being about one-fourth smaller than those from other crosses. Letting A^Y represent a gene for yellow and a one for black, we have:

Testcross:

$$P \qquad \text{yellow} \times \text{black}$$
$$\qquad\qquad A^Ya \qquad\quad aa$$

$$F_1 \qquad \tfrac{1}{2}\text{yellow} + \tfrac{1}{2}\text{black}$$
$$\qquad\qquad A^Ya \qquad\qquad aa$$

or,

FIGURE 2-11. *A lethal gene in corn. The chlorophyll-less plants are unable to manu-facture their own food and will die as soon as food stored in the grain has been consumed. Photo shows progeny plants of the cross heterozygous green × heterozygous green; there are 36 green and 12 "albino" seedlings in the flat, a perfect 3:1 ratio.*

simple monohybrid:

$$P \qquad \text{yellow} \times \text{yellow}$$
$$A^Y a \qquad A^Y a$$
$$F_1 \qquad \tfrac{1}{4} A^Y A^Y + \tfrac{2}{4} A^Y a + \tfrac{1}{4} aa$$
$$\text{die} \qquad \text{yellow} \qquad \text{black}$$

#5

A is lethal when recessive

Thus gene A^Y appears to be dominant with respect to coat color, but <u>recessive as to lethality</u>. For some time the nature of the action of $A^Y A^Y$ was unknown. Although selective fertilization was considered a possible factor, Castle and

since only the homozygote dies

Little (1910) suggested that A^YA^Y animals are conceived but die soon after. Robertson (1942) and later Eaton and Green (1962) were able to demonstrate that about one fourth of the embryos of pregnant yellow (A^Ya) females that had been mated to yellow males did die soon after conception. In this case death usually occurs at gastrulation.

Fowl. The well-known "creeper" condition in fowl falls into the same category. Creeper birds have much-shortened and deformed legs and wings, giving them a squatty appearance and creeping gait. Creeper $\times$ creeper always produces two creeper to one normal, the homozygous creepers having such gross deformities (greater than in heterozygotes) that they die during incubation, generally about the fourth day:

$$P \qquad \text{creeper} \times \text{creeper}$$
$$\qquad\quad c_1c_2 \qquad\qquad c_1c_2$$
$$F_1 \qquad \tfrac{1}{4}\text{normal} + \tfrac{2}{4}\text{creeper} + \tfrac{1}{4}\text{(die)}$$
$$\qquad\quad c_1c_1 \qquad\quad c_1c_2 \qquad\quad c_2c_2$$

Landauer has shown that the creeper gene produces general retardation of embryo growth, the effect being greatest at the stage of limb bud formation.

A Dominant Lethal in Man. Thus far we have examined only cases where lethality itself is recessive. Reasoning a priori, there is no cause not to expect genes whose lethal effect is dominant, provided death of the affected individual occurs somewhat after reproduction has taken place. Huntington's chorea, a disease in man characterized by involuntary jerking of the body and a progressive degeneration of the nervous system, accompanied by gradual mental and physical deterioration, illustrates just such a situation. The mean age of onset of these symptoms is between 40 and 45 (though it is reported to occur as early as the first decade of life and as late as 60 or 70), by which time, of course, many afflicted persons have produced children. Affected offspring always have at least one parent who, sooner or later, is also choreic, though the variability of the age of onset (which may be due to the action of still other genes) makes this difficult to demonstrate in some cases, and impossible where parents and grandparents die at early ages from other causes. Clearly, then, on available evidence, this disease is due to a dominant gene, both as to lethality and as to its abnormal phenotype.

The lethals discussed in this chapter are summarized in Table 2-2.

REFERENCES

ABBOTT, U. K., R. M. CRAIG, and E. B. BENNETT, 1970. Sex-Linked Coloboma in the Chicken, *Jour. Hered.,* **61:** 95–102.

BOUWKAMP, J. C., and S. HONMA, 1970. Vernalization Response, Pinnae Number, and Leaf Shape in Celery. *Jour. Hered.,* **61:** 115–118.

CUÉNOT, L., 1904. L'Hérédité de la Pigmentation chez les Souris, (3me Note). *Arch. Zool. Exp. et Gén.,* 3me Série, **10,** Notes et Revues, 27–30.

CUÉNOT, L., 1905. Les Races Pures et Leur Combinaisons chez les Souris. *Arch. Zool. Exp. et Gén.,* **3:** 123–132.

TABLE 2-2. Comparison of Certain Lethal Genes

Organism	Phenotype	Dominance		F_1 Phenotypic Ratio of Heterozygote × Heterozygote		Age at Death
		Phenotype	Lethality	Before Lethality Occurs	After Lethality Occurs	
Man	Sickle-cell	Codom.	Recessive	1 normal : 2 sickle-cell trait : 1 sickle-cell anemia*	1 normal : 2 sickle-cell trait	Adolescence
Corn	Albinism	Recessive	Recessive	3 green : 1 albino	All green ("1:0")	10–14 days
Mouse	Yellow	Dominant?	Recessive	Unknown	2 yellow : 1 black	Postzygote
Fowl	Creeper	Inc. dom.	Recessive	Unknown	2 creeper : 1 normal	Early embryo
Man	Huntington's chorea	Dominant	Dominant	3 choreic : 1 normal	All normal ("1:0")	Middle age, but somewhat variable

* The 1:2:1 ratio listed under F_1 phenotype for sickle-cell anemia is based on microscopic examination of blood samples.

EATON, G. J., and M. M. GREEN, 1962. Implantation and Lethality of the Yellow Mouse. *Genetica,* **33:** 106–112.

FILIPPENKO, I. M., and L. T. SHTIN, 1973. Inheritance of Downy Mildew Hardiness (*Plasmopara viticola*) in European-American Grape Hybrids. *Genetika,* **9:** 53–60.

GREEN, E. L., 1967. Shambling, a Neurological Mutant of the Mouse. *Jour. Hered.,* **58:** 65–67.

HENAULT, R. E., and R. CRAIG, 1970. Inheritance of Plant Height in the Geranium. *Jour. Hered.,* **61:** 75–78.

KERBER, E. R., and P. L. DYCK, 1973. Inheritance of Stem Rust Resistance Transferred from Diploid Wheat (*Triticum monococcum*) to Tetraploid and Hexaploid Wheat and Chromosome Location of the Gene Involved. *Canad. Jour. Genet. Cytol.,* **15:** 397–409.

KONOPLIA, S. P., V. N. FURSOV, A. A. DRUZHKOV, and G. N. NURYEVA, 1973. Independent Inheritance of the New Character "Lack of Anthocyanin" and of the Type of Generative Branches in the Cotton Species *Gossypium peruvianum. Genetika,* **9:** 154–156.

KRETCHMER, N., 1972. Lactose and Lactase. *Sci. Amer.,* **227:** 71–78.

PARKS, G. K., and C. W. FOWLER, 1970. White Pine Blister Rust: Inherited Resistance in Sugar Pine. *Science,* **167:** 193–195.

ROBERTSON, G. G., 1942. An Analysis of the Development of Homozygous Yellow Mouse Embryos. *Jour. Exp. Zool.,* **89:** 197–231.

STERN, C., 1973, 3rd ed. *Principles of Human Genetics.* San Francisco, W. H. Freeman.

ZITTER, T. A., and A. A. COOK, 1973. Inheritance of Tolerance to a Pepper Virus in Florida. *Phytopathology,* **63:** 1211–1212.

PROBLEMS

2-1. Make a list of several phenotypic characters in your family for as many generations and individuals as possible. Considering traits singly, try to determine the kind of inheritance involved. Save any that seem not to fit patterns developed in this chapter until somewhat later on.

2-2. In human beings, ability to curl the tongue into a U-shaped trough is often described as a heritable trait. "Curlers" always have at least one curler parent, but "noncurlers" may occur in families where one or both parents are curlers. Using *C* and *c* to symbolize this trait, what is the genotype of a noncurler?

2-3. Some individuals have one whorl of hair on the back of the head whereas others have two. In the following pedigree, solid symbols represent one whorl, open symbols two:

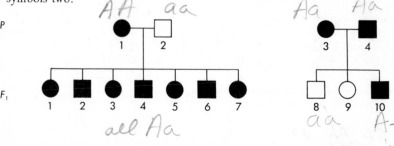

   (a) Using the first letter of the alphabet, give the probable genotype of (1) P 2;
       (2) P 3; (3) F$_1$ 7; (4) F$_1$ 8; (5) F$_1$ 10.
   (b) What should be the phenotypic ratio of an F$_2$ produced by the marriage of
       F$_1$ 7 × F$_1$ 8, assuming a large family? *So both must be Aa since must receive an a from the albino parent*

2-4. Albinism, the total lack of pigment, is due to a recessive gene. A man and a girl
     plan to marry and wish to know the probability of their having any albino chil-
     dren. What could you tell them if (a) both are normally pigmented, but each
     has one albino parent; (b) the man is an albino, the girl is normal, but her father
     is an albino; (c) the man is an albino and the girl's family includes no albinos
     for at least three generations?

2-5. Cystic fibrosis of the pancreas is an inherited condition characterized by faulty
     metabolism of fats. Affected individuals are homozygous for the gene respon-
     sible and ordinarily die in childhood. Such individuals also produce a much
     higher concentration of chlorides in their sweat than homozygous normals,
     whereas heterozygotes, who live a normal life-span, have an intermediate chlo-
     ride concentration. In terms of the types of genes discussed in this chapter, how
     would you designate (a) this pair of alleles as to dominance; (b) the gene for
     cystic fibrosis? *lethal recessive* *incomplete*

2-6. One study has estimated the number of persons in the United States who are
     heterozygous for the cystic gene at about 8 million. A couple planning marriage
     decide to have a sweat test because a brother of the man died in infancy from the
     disorder. The tests disclose the man to be heterozygous and the woman homo-
     zygous normal. (a) What is the chance that any of their children might have
     cystic fibrosis? (b) Could any of the couple's grandchildren have it?

2-7. In a certain plant the cross purple × blue yields purple- and blue-flowered
     progeny in equal proportions, but blue × blue always gives rise only to blue.
     (a) What does this tell you about the genotypes of blue- and purple-flowered
     plants? (b) Which phenotype is dominant? *blue is recessive*

2-8. In cattle, the cross horned × hornless sometimes produce only hornless off-
     spring, and in other crosses horned and hornless appear in equal numbers. A
     cattleman has a large herd of hornless cattle in which horned progeny occasion-
     ally appear. He has red, roan, and white animals and wishes to establish a pure-
     breeding line of red hornless animals. How should he proceed?

2-9. In corn, resistance to a certain fungus is conferred by gene *h*, which is completely
     recessive to its allele *H* for susceptibility. If a resistant plant (♀) is pollinated
     by a homozygous susceptible plant (♂), give the genotypes for (a) pistillate *female*
     parent, (b) the staminate parent, (c) sperm, (d) egg, (e) polar nucleus, (f) F$_1$
     embryo, (g) endosperm surrounding the F$_1$ embryo, (h) epidermis of kernels
     that contain the F$_1$ embryos.

2-10. Two curly-winged fruit flies (*Drosophila*) are mated; the F$_1$ consists of 341 curly
      and 162 normal. Explain. *close to 2:1 instead of 3:1. Reduced # pheno classes recessive lethal*

2-11. Using the sixth letter of the alphabet, give the genotype of each of the following
      persons in Figure 1-5: I-1, I-2, II-1, II-2, II-3, II-4, III-1, and III-2.

2-12. The marriage between II-3 and II-4 in Figure 1-5 represents what sort of ge-
      netic cross? *II-1 & II-2 testcross bet hetero & homo recessive*

2-13. Using the first letter of the alphabet, give the genotype of each of the following
      persons from Figure 1-6: I-1, II-1, III-4.

AA   Aa   aa
↓   curly normal
die

2-14. No ancestry information is given for II-1 in Figure 1-6; how do you justify your designation of her genotype? *dominant*

2-15. Rh negative children (those not producing rhesus antigen D) may be born to either Rh positive or Rh negative parents, but Rh positive children always have at least one Rh positive parent. Which phenotype is due to a dominant gene?

2-16. Thalassemia is a hemoglobin defect in humans; it occurs in two forms, (a) thalassemia minor, in which erythrocytes are small (microcytic) and increased in number (polycythemic), but health is essentially normal, and (b) thalassemia major, characterized by early, severe anemia, enlargement of the spleen, microcytes, and polycythemia, among other symptoms. The latter form usually culminates in death before attainment of reproductive age. From the following hypothetical pedigree, determine the mode of inheritance:

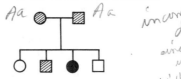

Aa ⊘ — ▨ *Aa*

incomplete dominance, since there is intermediate form with lethal recessive

Clear symbols represent normal persons, hatched symbols thalassemia minor, and the shaded symbol thalassemia major.

2-17. In families where both parents have sickle-cell trait, what is the probability of their having (a) a child also with sickle-cell trait, (b) a normal child?

2-18. A normal individual, $Hb^A Hb^A$, receives a transfusion of blood from a person who has sickle-cell trait. Would this transfusion transmit the sickle-cell trait to the recipient? Explain.

2-19. Phenylketonuria (PKU) is a heritable condition in humans involving inability to metabolize the amino acid phenylalanine because of failure to produce the enzyme phenylalanine hydroxylase. If not diagnosed and treated very soon after birth, PKUs develop such severe mental retardation (among other symptoms) that they almost never reproduce. Almost all PKU children, therefore, are born to parents who are not PKUs. (a) Is the gene responsible for phenylketonuria completely dominant, incompletely dominant, codominant, or recessive? (b) Inasmuch as PKUs so rarely reproduce, why does such a disadvantageous gene persist in the population?

but depends on level what look you at

2-20. The normal brother of a PKU seeks the advice of a genetic counselor before a contemplated marriage. (a) What is the probability that he is heterozygous? (b) If PKU occurs once in 25,000 live births in the United States, and he contemplates marrying a normal woman in whose family no cases of PKU have occurred since her ancestors came over on the *Mayflower,* what is the probability of their having a PKU child? (c) What else might the genetic counselor consider telling this couple?

good

2-21. In juvenile amaurotic idiocy children are normal until about age six. Subsequently there is a progressive decline in mental development, an impairment of vision leading to blindness, and muscular degeneration, culminating in death, usually before age 20. The trait may appear in families where both parents are completely normal. A couple, age 25, planning marriage, are first cousins; siblings of both parties have died of the disorder. (a) Knowing no more than you do

so it recessive

at this point about the genotypes of these two persons, what is the probability of *both* of them being heterozygous? (b) On the basis of your answer to part a, what could you tell them about the chance of their having an affected child? (c) Heterozygotes can be detected by an increase in vacuolization of lymphocytes (a type of white blood cell). If such a test should disclose that both persons are actually heterozygotes, what then could you say about the probability of their having an affected child?

a) prob of each one is:
  Aa Aa

  ¼AA : ½Aa : ¼aa

Note: can't be Aa × aa since aa parent wouldn't live to reproduce

we know they're probably not this

so ⅔ prob Aa

prob both: ⅔ × ⅔ = 4/9

b) ¼ (4/9) = 1/9   since ¼ prob of being aa child + 4/9 prob they are Aa

c) ¼

parents must be
a) Aa , Aa
  → 50% Aa

  ¼AA : ½Aa : ¼aa
  so ⅔% he's Aa   we know he's not this so can eliminate this from our probability

b) if he's AA, then AA × AA
  → 0%

if he's Aa, then Aa × AA → 0%

CHAPTER 3
Cytological Bases of Inheritance

In Chapter 2 we assumed that the genes with which we were dealing might well be located in the nucleus. If this assumption is true, we should be able to find some confirming, objective evidence. Such evidence does exist and can be conveniently designated as of two kinds, cytological and chemical. In the history of the development of the science of genetics the former preceded the latter by many years. Let us now examine the cytological evidence, deferring the chemical to a later and more logical point in the development of our concepts of genetic mechanisms.

The Interphase Nucleus

[handwritten annotation: an organism or cell with a structurally discreet nucleus]

As your previous experience in the life sciences has shown you, living cells of most organisms (*eukaryotes*) are characterized by the presence of a discrete, often spherical body, the **nucleus** (Fig. 3-1). Notable exceptions are such *prokaryotes* as the blue-green algae and the bacteria in which, although "nuclear material" can be shown to occur, the visible, structural organization so typical of the cells of higher organisms is lacking. Prokaryotes, of course, lack the conventional mitotic and meiotic nuclear divisions that characterize eukaryotes. In addition to these two types of genetic systems, we can recognize still a third in the viruses. The structure of an important type of virus, the bacteriophages, is described more fully in Chapter 15.

The interphase (nondividing) nucleus is bounded by an interface, the **nuclear membrane,** which is not clearly visible with the light microscope. Electron micrographs, however, reveal this membrane to be a double layer, provided with numerous "pores" of about 20 to 80 nm in inside diameter.[1] Data summarized by Du Praw (1970) from several sources show an average of about 50 nm. Each is surrounded by a thickened, electron-dense ring in the nuclear membrane, giving an outside diameter of up to 200 nm. Du Praw (1970) prefers the term *annulus* for the entire specialized structure whose function is active control of passage of macromolecules between nucleus and cytoplasm. The nuclear membrane is continuous with a cytoplasmic double membrane system, the **endoplasmic reticulum,** to which dense granular structures some 17 by 22 nm in size, the **ribosomes,** can usually be seen attached in electron micrographs (Fig. 3-2). Ribosomes are rich in **ribonucleic acid** (RNA) and play an important role in protein synthesis (see Chapters 16 and 17).

---

[1] Twenty to 80 nanometers (nm); a nanometer is 1×10^{-9} meter or 10 Angstrom units. In older usage a nanometer was called a millimicron (mμ).

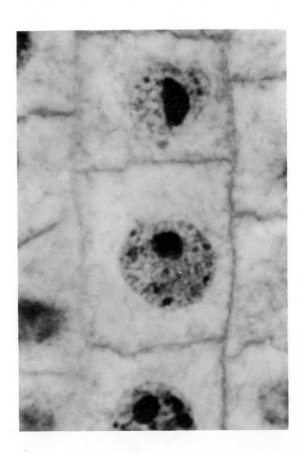

FIGURE 3-1. *Interphase nucleus of onion root tip, as seen with the light microscope. Compare with Figure 3-2. Note the prominent nucleolus.* (Courtesy Carolina Biological Supply Co.)

Within the interphase nucleus three major components can be distinguished. The first of these is nuclear sap, or **karyolymph,** a clear, usually nonstaining, largely proteinaceous, colloidal material. The second intranuclear component is a generally spherical, densely staining body, the **nucleolus.** Many nuclei contain two or more nucleoli. The general size range is 2 to 5 μm, but size varies with the tissue, the degree of protein synthesizing activity of the cell (the nucleolus is larger when this activity is high), and nutritional factors. Nucleoli contain RNA, as well as some DNA and protein. Typically each nucleolus is produced by and is physically associated with a *nucleolus organizing region* (which is the site of synthesis of ribosomal RNA, as described in Chapter 16) of a particular **chromosome,** the third major nuclear component.

The remainder of the nucleus consists of separate, fine, threadlike strands some 10 to 50 nm in diameter, the *chromatin.* Chromatin consists of nucleoprotein, a complex of **deoxyribonucleic acid** (DNA) and basic proteins, either *histones* or *protamines,* the role of which will be explored in Chapter 19. During nuclear division the chromatin strands contract and appear thicker, to become the **chromosomes,** which are visible with an ordinary light microscope. The

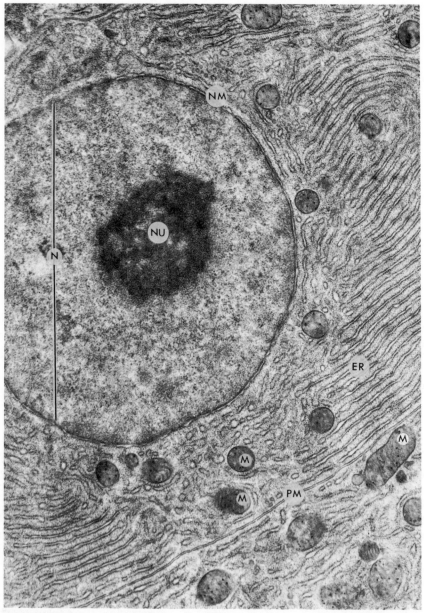

FIGURE 3-2. *Electron micrograph of interphase cell from bat pancreas. N, nucleus; NU, nucleolus; NM, nuclear membrane; ER, endoplasmic reticulum with ribosomes attached; M, mitochondria; PM, plasma membrane delimiting the cell. Note double-ness of the membrane systems.* (From *Cell Ultrastructure* by William Jensen and Roderic Park. © 1967 by Wadsworth Publishing Co., Inc., Belmont, California. Reproduced by permission.)

number and morphology of the chromosomes are specific, distinct, and ordinarily constant for each species, although, especially in plants, subspecific taxonomic categories with multiple sets (polyploids) are not infrequent (Table 3-1; see also Chapter 13). Some details of chromosome morphology are described on pages 55–64.

TABLE 3-1. Chromosome Numbers in Some Plants and Animals*

	Common Name	Scientific Name	Chromosome Number Monoploid†	Diploid†
		PLANTS		
I Chlorophyta	Chlamydomonas	*Chlamydomonas moewussi*	8	
	Chlamydomonas	*C. reinhardi*	16	
II Phycomycota	Bread mold	*Mucor hiemalis*	2	
III Ascomycota	Red bread mold	*Neurospora crassa*	7	
	Penicillium	*Penicillium* spp.	2, 4, 5	
IV Bryophyta	Liverwort	*Sphaerocarpos donnelii*	♂ 7 + Y ♀ 7 + X	
V Pterophyta	Adder's tongue fern	*Ophioglossum vulgatum*		480–520
	Adder's tongue fern	*O. reticulatum*		1,262
VI Coniferophyta	Spruce	*Picea abies*		24‡
	White pine	*Pinus strobus*		24
VII Anthophyta Dicotyledonae	Watermelon	*Citrullus vulgaris*		22§
	Coleus	*Coleus blumei*		24
	Jimson weed	*Datura stramonium*		24
	Tomato	*Lycopersicon esculentum*		24
	Evening primrose	*Oenothera* spp.		14‡§
	Pea	*Pisum sativum*		14
	African violet	*Saintpaulia ionantha*		28
	Swamp saxifrage	*Saxifraga pensylvanica*		56‡§
	Alsike clover	*Trifolium hybridum*		16
	Red clover	*T. pratense*		14
	White clover	*T. repense*		32§
Mono-cotyledonae	Onion	*Allium cepa*		16‡
	Slender oat	*Avena barbata*		28
	Wild oat	*A. brevis*		14
	Cultivated oat	*A. sativa*		42
	Asiatic tree cotton	*Gossypium arboreum*		26
	Upland cotton	*G. hirsutum*		52
	Regal lily	*Lilium regale*		24
	Rice	*Oryza sativa*		24

* An extensive list of chromosome numbers is given in P. L. Altman and D. S. Ditmer, eds., 1962, and 1972. In each plant example here the chromosome number of the *dominant* generation (gametophyte or sporophyte) is cited. See also Appendix B.

† *Diploid,* from the Greek *di* (as a prefix; "two") and *ploid* ("unit"), refers to any nucleus, cell, or organism that has two "units" or sets of chromosomes that, in sexually reproducing organisms, are normally composed of one paternal and one maternal set. The often used but anomalous "haploid" (literally half unit or set) for structures having but one set of chromosomes is here replaced by the more logical *monoploid* (Greek *monos,* "only" or "alone," hence "one," and *ploid*).

‡ Tetraploid (4*n*) forms also known.

§ Triploid (3*n*) forms also known.

TABLE 3-1. *Continued*

	Common Name	Scientific Name	Chromosome Number Monoploid†	Diploid†
		PLANTS		
Mono-	Rye	*Secale cereale*		14
cotyledonae	Smooth cordgrass	*Spartina alterniflora*		62
	Cordgrass	*S. maritima*		60
	Cordgrass	*S. anglica*		124
	Emmer wheat	*Triticum dicoccum*		28
	Wild emmer	*T. dicoccoides*		28
	Durum wheat	*T. durum*		28
	Einkorn	*T. monococcum*		14
	Spelt	*T. spelta*		42
	Common wheat	*T. vulgare*		42
	Corn	*Zea mays*		20
		ANIMALS		
I Protozoa	Paramecium	*Paramecium aurelia* (micronucleus)		30–40
II Aschelminthes	Roundworm	*Ascaris megalocephala*		2
III Arthropoda	Honeybee	*Apis mellifica*	♂ 16	♀ 32
	Crayfish	*Cambarus clarkii*		200
	Fruit fly	*Drosophila affinis*		10
		D. hydei		12
		D. melanogaster		8
		D. prosaltans		6
		D. pseudoobscura		10
		D. virilis		12
		D. willistoni		6
	Grasshopper	*Melanoplus differentialis*		♀ 24, ♂ 23
	Habrobracon	*Habrobracon juglandis*	♂ 10	♀ 20
	Housefly	*Musca domestica*		12
IV Chordata	Cattle	*Bos taurus*		60
	Cat	*Felis domesticus*		38
	Fowl	*Gallus domesticus*		78
	Man	*Homo sapiens*		46
	Rhesus monkey	*Macaca mullatta*		42
	Mouse	*Mus musculus*		40
	Chimpanzee	*Pan troglodytes*		48
	Rat	*Rattus norvegicus*		42
	Rabbit	*Sylvilagus floridanus*		44

Cell Division

Growth and development of every organism depends in large part upon multiplication and enlargement of its cells. In multicellular individuals attainment of adult form depends on a coordinated sequence of increase in cell number, size, and differentiation from zygote to maturity. In unicellular organisms cell division serves also as a form of reproduction, often the only one. Sexually reproducing forms also depend in most instances directly on cell division for the formation of sex cells or gametes.

Division of nucleate cells consists of two distinct but integrated activities, nuclear division (**karyokinesis**) and cytoplasmic division (**cytokinesis**). In general, cytokinesis begins after nuclear division is well underway, but in many instances may be deferred or, indeed, entirely lacking. For example, in the development of the female gametophyte of pine (see Appendix B), about 11 nuclear divisions occur before cytokinesis begins, and in some algae and fungi the plant body is a coenocyte, without any walls separating nuclei except for the reproductive cells.

Two types of nuclear division, **mitosis** and **meiosis,** are characteristic of most plant and animal cells. The first of these is regularly associated with nuclear division of vegetative or somatic cells; the latter occurs in conjunction with formation of reproductive cells (either gametes or meiospores) in sexually reproducing species.

THE INTERPHASE CYCLE

The interphase cycle may be defined as the entire sequence of events transpiring from the close of one cell division to the beginning of the next one. The total time varies with the species, maturation, tissue, and temperature, among other factors. Duration of as little as about three hours to as much as 174 hours has been reported for various organisms. In a number of kinds of human tissue in vitro the interphase cycle typically occupies some 18 to 24 hours. For convenience we may recognize the following stages:
experimentally induced processes outside the organism

G_1 The first *growth* stage of interphase in which nucleus (and cytoplasm) are enlarging toward mature size. Chromosomes are fully extended and not distinguishable as discrete entities with the light microscope. This is a time of active protein and RNA synthesis; it may occupy 30 to 40 per cent of the total time of the interphase cycle, or be entirely lacking in rapidly dividing cells such as those of the early mammalian embryo, as in such lower forms as slime molds (*Physarum*) and yeasts (*Schizosaccharomyces*), but as long as 151 hours in mature cells of corn (*Zea*) roots.

S During the *synthesis* stage synthesis of DNA and histones occurs, the former doubling in amount. This is reflected in the chromosomes becoming longitudinally double so that each consists of two *chromatids*. This phase may occupy roughly 35 to 45 per cent of the interphase cycle; in human cells in culture it extends over some six to eight hours.

G_2 A second *growth* period, less well understood than G_1, in which some protein synthesis continues. It may occupy about 10 to 20 per cent of the interphase cycle.

King (1972) lists duration of the several stages of the interphase cycle in cultured cells from a carcinoma of the cervix (HeLa cells) as follows: G_1, 8.2 hours; S, 6.2 hours; and G_2, 4.6 hours. An idealized diagram of the relative amounts of time occupied by each stage of the interphase cycle and mitosis is

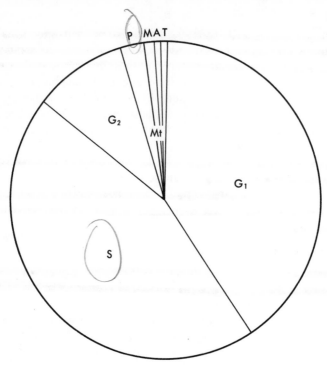

FIGURE 3-3. *Idealized diagram of the relative amounts of time occupied by each stage of interphase and the mitotic cycle. G₁, first growth stage; S, synthesis of DNA and histones; G₂, second growth stage; P, prophase of mitosis; M, metaphase of mitosis; A, anaphase of mitosis; T, telophase of mitosis; Mt, mitotic period.*

shown in Figure 3-3. As the G_2 stage draws to a close, the cell gradually enters division, though governing factors for this transition are unclear.

MITOSIS

As a process, mitosis is remarkably similar in all but relatively small details in both plants and animals, from the least specialized to the most highly evolved forms. Although mitosis is a smoothly continuous process, it is divided arbitrarily into several stages or phases for convenient reference. The following description of the mitotic process in plant cells will adequately serve our purpose of determining whether the mechanism provides a reasonable basis for our genetic assumptions developed in Chapter 2 (Fig. 3-4).

Prophase.   As stage G_2 of interphase gives way to prophase of mitosis, chromosomes progressively shorten and thicken to form individually recognizable, elongate, longitudinally double structures (Fig. 3-5) arranged randomly in the nucleus. This contraction of the chromosomes is one of the most conspicuous features of prophase. The two **sister chromatids** of each chromosome are closely aligned and somewhat coiled on themselves. The tightening

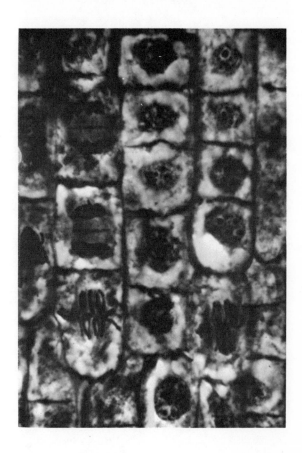

FIGURE 3-4. *Various stages of mitosis in onion root tips cells.* (Courtesy General Biological Supply House, Inc., Chicago.)

of these coils contributes in large measure to the shortening and thickening of the chromosomes.

The two sister chromatids of each chromosome are held together by strands in a specialized region, the **centromere** (Fig. 3-6). In well-prepared stained material under the light microscope the centromere appears as a small, relatively clear, spherical zone, but in electron micrographs it appears dark. The centromere has been shown by electron microscopy to be a complex region of the chromosome. It includes two kinetochores, one for each sister chromatid, to which the microtubules of the spindle (see later) are attached. Kinetochores appear to become organized in prophase only after chromosome contraction has taken place. That the centromere is an essential part is indicated by the fact that chromosome fragments, such as may result from radiation-induced breakage, that lack it (*acentric fragments*) fail to move normally during nuclear division and are generally, therefore, lost from one or both of the reorganizing daughter nuclei.

During prophase the nucleolus gradually disappears in most organisms (Figs. 3-7 and 3-14). In those forms in which it does not disappear (e.g., some

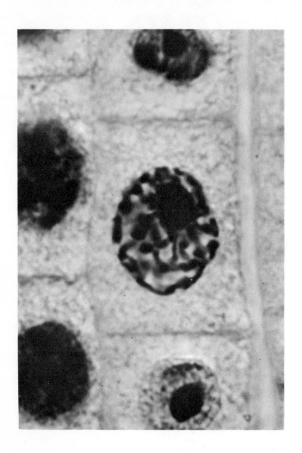

FIGURE 3-5. *Mitosis in onion root tip cell; early prophase. Longitudinal doubleness is evident in a few areas (e.g., lower left portion of the nucleus); the nuclear membrane is becoming indistinct.* (Courtesy Carolina Biological Supply Co.)

grasses, the green alga *Spirogyra,* and in *Euglena*), it apparently ceases to function and is discarded into the cytoplasm.

Another common feature of prophase is the degeneration and disappearance of the nuclear membrane (Figs. 3-7 and 3-14). The mechanism is unknown, but an aggregating of mitochondria around the membrane (especially in animal cells) suggests enzymatic processes. Bajer (1968) theorizes that breakdown of the nuclear membrane results from physical stresses exerted by microtubules (see next paragraph) that are attached to it.

As prophase progresses an important and often conspicuous component, a football-shaped **mitotic apparatus** (*spindle* and associated structures) begins to form (see Figs. 3-11 and 3-14). By the end of prophase this structure occupies a large portion of the cell volume and often extends very nearly from one "end" of the cell to the other. The mitotic apparatus consists of slender, tubular **microtubules,** each some 15 to 30 nm in outside diameter, arranged along the long axis of the spindle. Some of the microtubules are continuous from pole to pole of the mitotic apparatus; these are the *continuous fibers.* Groups of others, the *chromosomal fibers,* extend from one pole or the other

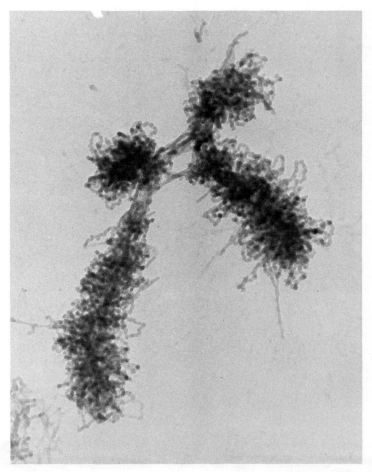

FIGURE 3-6. *Electron micrograph showing two sister chromatids held together by chromatin fibers in the centromere region. This is one of the human chromosomes.* (Photo courtesy Dr. E. J. Du Praw. From E. J. Du Praw, 1970. *DNA and Chromosomes.* New York, Holt, Rinehart and Winston, Inc. Used by permission.)

to each chromosomal kinetochore to which they are attached (Fig. 3-8). Some workers also recognize a third type of microtubule, the *interzonal fibers,* which run between the centromeres of the separating daughter chromosomes later on in anaphase. Microtubules are composed of some 90 per cent protein, with lesser amounts of RNA (about 5 per cent), polysaccharide, and lipid.

As prophase draws to a close, the longitudinally double chromosomes move or are moved in the direction of the midplane, or equator, of the developing spindle, a period of time often designated as **prometaphase** or **metakinesis** (Fig. 3-9). In most organisms prophase comprises the bulk of the time consumed by mitosis (Fig. 3-3 and Table 3-2).

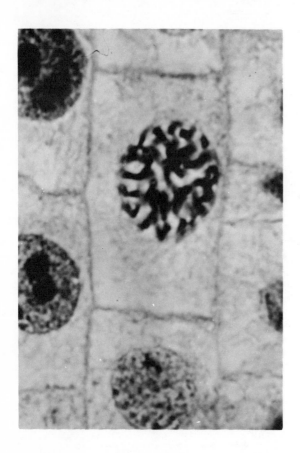

FIGURE 3-7. *Mitosis in onion root tip cell; late prophase. Chromosomes are shorter and thicker than in early prophase; note the chromatids.* (Courtesy Carolina Biological Supply Co.)

Metaphase.   Metaphase (Fig. 3-10) is that period of time in which the centromeres of the longitudinally double chromosomes occupy the plane of the equator of the mitotic apparatus, although the chromosomal arms may extend in any direction. At this stage the sister chromatids are still held together by chromatin fibers connecting their centromere regions (Figs. 3-6 and 3-8). Electron microscopy shows that the kinetochores of the two sister chromatids face opposite poles; this will permit their separation in the next phase (anaphase). During metaphase the chromosomes are at their shortest and thickest. Polar views furnish good material for chromosome counts; both lateral and polar views are useful for studying chromosome morphology.

Anaphase.   Anaphase (Fig. 3-11) is characterized by the separation of the metaphase sister chromatids and their passage as **daughter chromosomes** to the spindle poles. It begins at the moment when the centromeres of each of the sister chromatids become *functionally* double and ends with the arrival of the daughter chromosomes at the poles. For clarity, it is important to distinguish between sister chromatids and daughter chromosomes. A metaphase cell contains a number of sister chromatids, associated in pairs through centromeric

daughter chromosomes in anaphase = # chromatids in metaphase

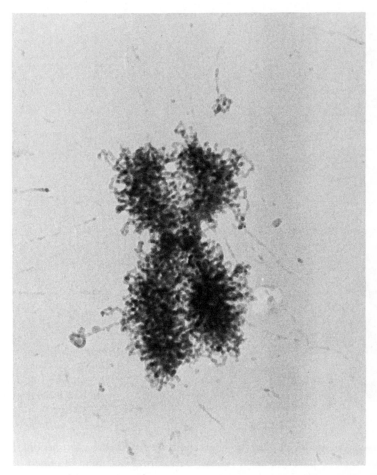

FIGURE 3-8. *Electron micrograph of one of the human chromosomes showing the two kinetochores of the centromere.* (Photo courtesy Dr. E. J. Du Praw. From E. J. Du Praw, 1970, *DNA and Chromosomes.* New York, Holt, Rinehart and Winston, Inc. Used by permission.)

connection, equal to twice the number of chromosomes. On the other hand, the anaphase cell contains no sister chromatids but does, by definition, contain a number of daughter chromosomes twice great as the metaphase or prophase number. Anaphase therefore accomplishes the *quantitatively* equal distribution of chromosomal material to two developing daughter nuclei. That distribution is also *qualitatively* equal if the replication process of the preceding S stage has been exact. We shall examine this replication process at the molecular level in Chapter 15.

The mechanism of the anaphasic movement of chromosomes is not understood in spite of intensive experimentation and study. Du Praw (1970) believes

FIGURE 3-9. *Chromo-
somes nearing equatorial
plane of developing spindle;
prometaphase.* (Courtesy
Carolina Biological Sup-
ply Co.)

the most satisfactory explanation lies in an ability of the chromosomal microtubules to slide past the continuous fibers, dragging their attached daughter
chromosomes with them. Under this concept, Du Praw (1970) envisions that
the "poleward end (of the chromosomal fiber) progressively melts or disag-
gregates, and its protein subunits contribute to the increasing mass of the
aster" in animal cells. Or, as Mazia (1961) describes it, the "contraction (of
the chromosomal microtubules) involves the loss of some molecules from the
contracted region to the background and consolidation of the rest to maintain
continuity. The apparent result is that the fiber is largely 'consumed' as the
wave of local contraction proceeds." In some kinds of cells anaphasic separation is by elongation of the entire mitotic apparatus, which increases the
separation of the poles and, therefore, of the daughter chromosomes. In at
least a few cases (e.g., grasshopper spermatocytes) both mechanisms operate.

   Telophase.   The arrival of the (longitudinally single) daughter chromosomes
at the spindle poles marks the beginning of telophase; it is, in turn, terminated
by the reorganization of two new nuclei and their entry into the G_1 stage of
interphase (Figs. 3-12 and 3-14). In general terms, the events of prophase occur

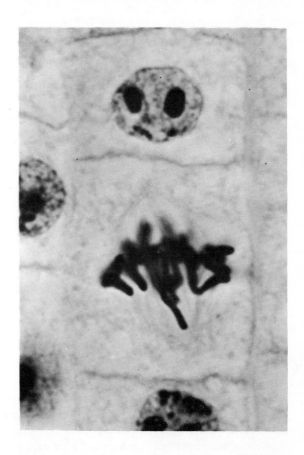

FIGURE 3-10. *Mitosis in onion root tip cell; metaphase. Note the spindle.* (Courtesy Carolina Biological Supply Co.)

in reverse sequence during this phase. New nuclear membranes are constructed from materials that may be remnants of the original membrane, or derived from the endoplasmic reticulum, or newly synthesized from appropriate cellular components. The mitotic apparatus gradually disappears, the nucleoli are reformed at the nucleolar organizing sites of specific chromosomes, and the chromosomes resume their long, slender, extended form as their coils relax. Replication of chromosomal material, by which each chromosome comes again to consist of two sister chromatids, then occurs, as we have seen, in the S stage of the succeeding interphase.

CYTOKINESIS

Cytokinesis, if it is to occur, takes place during telophase, though it may be initiated during anaphase. In cells of higher plants cytokinesis is typically accomplished by formation of a **cell plate** (Figs. 3-12B and 3-14). Early steps in its development include formation of vesicles in the midplane of the mitotic apparatus and their coalescence, starting at the center of the spindle, to form a

FIGURE 3-11. *Mitosis in onion root tip cell; anaphase. The daughter chromosomes are nearing the spindle poles, which are clearly evident.*
(Courtesy Carolina Biological Supply Co.)

phragmoplast. The cell plate forms within the phragmoplast, gradually extending centrifugally, dividing the phragmoplast into two parts before the latter disappears. The cell plate is then converted into a *middle lamella,* with new cross walls between the daughter cells deposited on each side of the middle lamella (Fig. 3-13). In animal cells cytokinesis is accomplished by furrowing. This is a process of progressive constriction of a contractile ring, perhaps of contractile fibers. Figure 3-14 summarizes the entire process of mitosis and cytokinesis.

SIGNIFICANCE OF MITOSIS

The process of mitosis and the subsequent replication of chromosomal material in the succeeding interphase has the inevitable result, if the cell "makes no mistakes," of creating from one cell two new ones that are chromosomally identical. (Actually, "mistakes" do sometimes occur; these, however, shed considerable light on the nature of the gene, as detailed in Chapter 18.) Certainly in mitosis the chromosomal material is distributed in equal quantity to two daughter cells in a strikingly precise manner.

If the replication process during the S stage of interphase is qualitatively equal (and we shall examine this possibility in Chapter 15), then the chromosomes will serve quite adequately as physical bearers of the genes we have considered in Chapter 2. The pattern of inheritance we followed for *Coleus,*

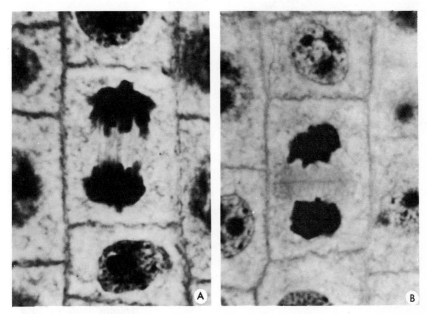

FIGURE 3-12. *Mitosis in onion root tip; telophase. (A) Early telophase; chromosomes have just arrived at the poles. The cell plate has begun to form. (B) Later telophase. The cell plate is prominent, and will later extend from side wall to side wall.* (Courtesy Carolina Biological Supply Co.)

for example, requires that genes be transmitted in cell division from the zygote to every somatic cell of the mature organism. In mitosis there exists a process by which the precise, equal distribution of structures called chromosomes can be carried through cell generation after cell generation. Our assumptions in Chapter 2 regarding genes are well served if the genes are indeed located on the chromosomes; certainly the behavior of these bodies makes them ideal vehicles for genes in terms of our earlier speculations. Before we can be certain of the chromosomal location of genes, however, we need to note additional parallels between the behavior of genes as deduced from breeding experiments and the behavior of chromosomes as seen under the microscope. In the mechanism of mitosis we have begun to accumulate good presumptive evidence that many of our theories regarding gene location are sound. We still require physical and chemical *proof,* but our theory looks sufficiently promising to retain for the present.

DURATION OF MITOSIS

Although not directly germane to our purposes at this point, it is interesting to note that the rather complicated physical and chemical changes in mitotic nuclei often occur in a surprisingly short time. Table 3-2 summarizes a few studies from the literature.

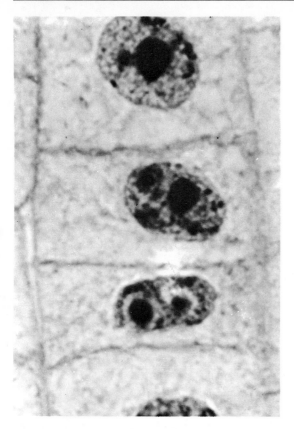

FIGURE 3-13. *After completion of mitosis in an onion root tip cell, two new interphase nuclei have been formed. In plant cells these are typically separated by a wall as a result of cyto-kinesis begun during telophase.* (Courtesy Carolina Biological Supply Co.)

TABLE 3-2. Duration of Mitosis in Living Cells

Name of Organism				Duration, minutes				
Common	Scientific	Tissue	Temp. °C	P	M	A	T	Total
PLANTS (ANGIOSPERMS)								
Onion	*Allium cepa*	Root tip	20	71	6.5	2.4	3,8	83.7
Oatgrass	*Arrhenatherum sp.*	Stigma	19	36–45	7–10	15–20	20–35	78–110
Pea	*Pisum sativum*	Endosperm	–	40	20	12	110	182
Pea	*Pisum sativum*	Root tip	20	78	14.4	4.2	13.2	110
Spiderwort	*Tradescantia sp.*	Stamen hair	20	181	14	15	130	340
Broad bean	*Vicia faba*	Root tip	19	90	31	34	34	155
ANIMALS								
Fowl	*Gallus Sp.*	Fibroblast culture	–	19–25	4–7	3.5–6	7.5–14	34–52
Grasshopper	*Melanoplus differentialis*	Neuroblast	–	102	13	9	57	181
Mouse	*Mus musculus*	Spleen mesenchyme	38	21	13	5	20	59
Salamander	*Salamandra maculosa*	Embryo kidney	20	59	55	6	75	195

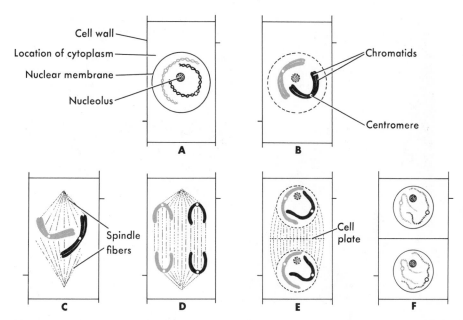

FIGURE 3-14. *Diagrammatic summary of mitosis in onion root tip cell. Only two of the total complement of 16 chromosomes are shown; the black and gray represent a "paternal" and a "maternal" chromosome, respectively, of one of the eight pairs. (A) **Prophase**; nuclear membrane and nucleolus still evident. (B) **Late prophase**; chromosomes becoming shorter and thicker, nucleolus and nuclear membrane disappearing. Note that each chromosome consists of two chromatids. (C) **Metaphase**; centromeres aligned on the midplane of the spindle, chromatids still present. (D) **Anaphase**; daughter chromosomes moving poleward. (E) **Telophase**; chromosomes have reached poles, nucleoli and nuclear membranes reappearing, cytokinesis by cell plate underway. (F) **Interphase**. See text for details.*

CHROMOSOME MORPHOLOGY

General Structure. In stained preparations as seen in the light microscope, chromosomes possess few definitive morphological characteristics by which individual members of a set may be distinguished from each other. The only useful features are (1) length, (2) centromere location and relative arm lengths, and (3) presence of satellites (if any), set off by secondary constrictions. During mitotic metaphase and anaphase chromosomes are at their shortest and thickest, as has been pointed out, varying from as short as a fraction of a micrometer[2] to as long as around 400 μm in exceptional cases, and between about 0.2 and 2 μm in diameter (Fig. 3-15). Each chromosome of a complement has its own characteristic length (within relatively narrow limits) and centromere location. However, in some organisms, especially those with large numbers of

[2] A micrometer (μm) is 1×10^{-6} meter; in older usage, a micron (μ).

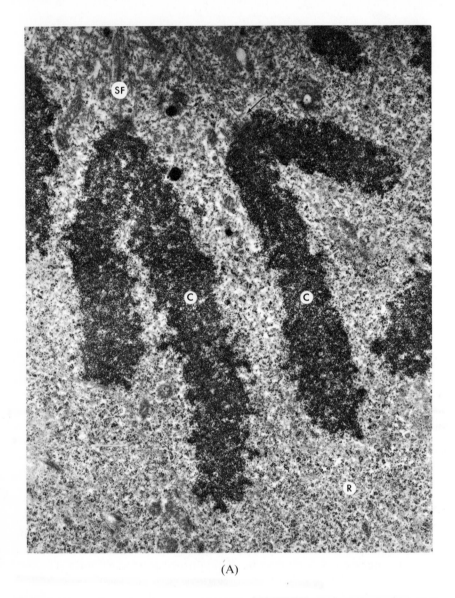

(A)

chromosomes, there is often considerable size similarity among some of the members in a set. This is true for some of the chromosomes of man and, until recently, it has not been possible to distinguish particular members with certainty. This difficulty has been especially troublesome in assigning groups of genes to the proper chromosome, as well as in distinguishing various chromosomal aberrations. Before the development of new techniques, results of which will be described in the next section, it was possible only to photograph smears of metaphase cells (Fig. 3-16), cut the chromosomes apart and arrange them in

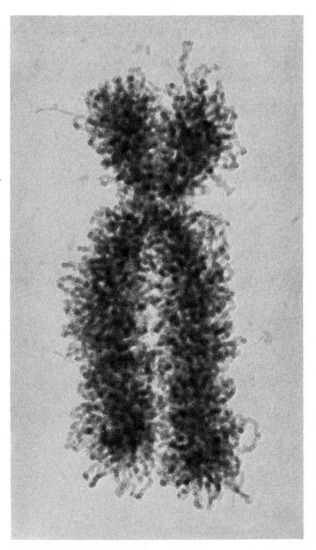

FIGURE 3-15. *(A) Electron micrograph of anaphase chromosomes of the Tasmanian Wallaby. C, chromosome; SF, spindle fibers; R, ribosomes. Note spindle fibers at centromere (arrow). (B) One of the human chromosomes showing centromere and DNA fibrils.* [(A) From W. Jensen and R. Park, 1967, *Cell Ultrastructure.* Copyright 1967 by Wadsworth Publishing Co., Inc., Belmont, California. Reproduced by permission. (B) Courtesy Dr. E. J. Du Praw, who took the electron micrograph; from E. J. Du Praw, 1970. *DNA and Chromosomes.* New York, Holt, Rinehart and Winston, Inc. Used by permission.]

(B)

matching pairs so far as could be determined from length, centromere location, and presence or absence of satellites. Photographs so prepared (or drawings made from them) are referred to as **karyotypes** (Fig. 3-17).

Although no such thing as a "typical" chromosome exists, a composite one is diagramed in Figure 3-18. Many chromosomes, however, show a distressing lack of morphological landmarks such as satellites. Every normal one does possess a centromere; its position and therefore the lengths of the arms are relatively constant. Such properties as these do, of course, permit to some

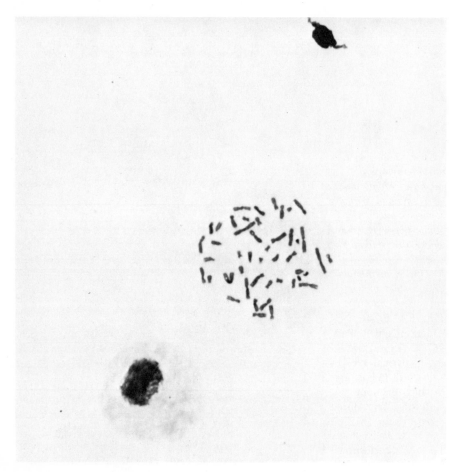

FIGURE 3-16. *Mitotic metaphase of human male chromosomes.* (Photomicrograph by Mr. John Derr.)

degree the identification of some of the chromosomal aberrations that are described in Chapter 13. On the basis of centromere location cytologists recognize four major types of chromosomes (Fig. 3-19):

Metacentric centromere central; arms of equal or essentially equal length.
Submetacentric centromere submedian, giving one longer and one shorter arm.
Acrocentric centromere very near one end; arms very unequal in length.
Telocentric centromere terminal; only one arm.

Heterochromatin and Euchromatin. In prophase (and, to a lesser degree, in interphase) chromatin can be seen to exist in two states: **heterochromatin,** darkly staining and tightly coiled, and **euchromatin,** lightly staining and loosely coiled. Heterochromatic regions do not relax their coils as telophase

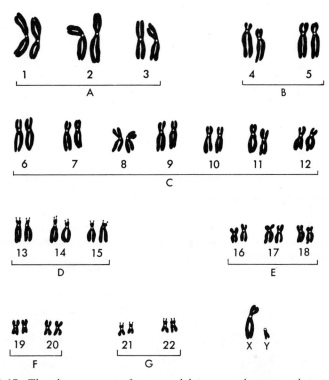

FIGURE 3-17. *The chromosomes of a normal human male arranged as a karyotype. Note the knobs or satellites on several pairs. Because of morphological similarities among several of the human chromosomes, groups are assigned letter designations as shown here.*

gives way to interphase, whereas euchromatic regions uncoil in telophase. Our interest in these regions attaches to the observation that euchromatin is genetically active, but heterochromatin is genetically inert. However, this distinction is not a hard-and-fast one, as we shall see in later chapters. Heterochromatin occurs in particular regions of the chromosome, the chromocenters, but, as Hsu (1973) points out, "identification of heterochromatic regions on metaphase chromosomes (is) difficult because at that stage, though best for

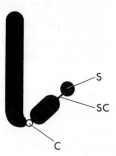

FIGURE 3-18. *Diagram of a composite chromosome as it might appear at high magnification with a light microscope. For simplicity chromatids are not represented. C, centromere; S, satellite or knob; SC, secondary constriction.*

FIGURE 3-19. *Four major morphological chromosome types are recognized on the basis of centromere position and consequent relative arm length, (A) metacentric; (B) submetacentric; (C) acrocentric; (D) telocentric.*

determining chromosome morphology, both euchromatin and heterochromatin are condensed, thus giving no differentiation in staining reaction. In interphase and prophase where differentiation in staining reaction is best, recognition of individual chromosomes is not feasible."

Chromosome Banding. Progress in positive identification of chromosomes came in a fast-breaking series of developments starting in 1968 and 1969 with the work of Caspersson and his colleagues in Sweden. Briefly stated, the basis for this breakthrough lies in the fact that many dyes that have an affinity for DNA fluoresce under ultraviolet light. After suitable treatment with such dyes, each chromosome shows bright and dark zones, or *bands,* which are specific

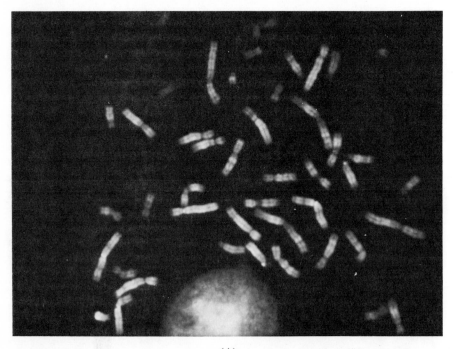

(A)

FIGURE 3-20. *(A) Metaphase spread of normal human male chromosomes showing Q-banding. (B) Karyotype of normal human male, Q-banding. The short bar beside each chromosome pair marks the location of the centromere. Note the brightly fluorescing long arm of the Y chromosome.* (Photos courtesy Dr. C. C. Lin.)

in location and extent for that chromosome. This feature is best seen in metaphase chromosomes. Considerable success has been had using quinacrine mustard, which produces fluorescent bands of various degrees of brightness. Bands produced by quinacrine mustard are referred to as **Q-bands** (Fig. 3-20A). They are unique for each member of the human chromosomal complement, so that positive identification and construction of a karyotype are easy (Fig. 3-20B). There are, however, some disadvantages to the Q-banding technique, particularly the impermanence of the fluorescence. That difficulty can be overcome by staining with the Giemsa dye mixture, which produces its own unique set of bands, the **G-bands.** Figure 3-21 presents both a metaphase spread and a karyotype for the same cell from a human male comparing G-bands (Figs.

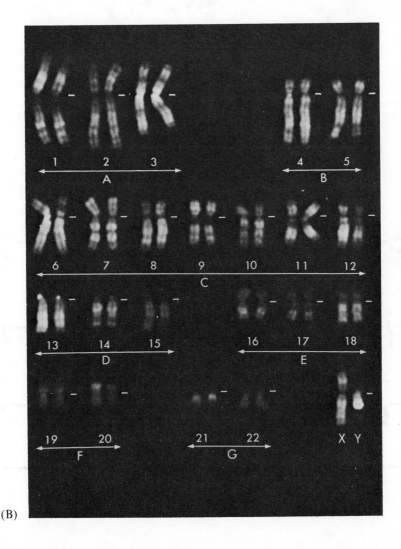

(B)

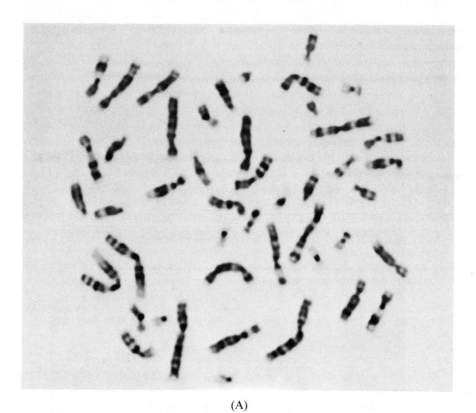

(A)

(B)

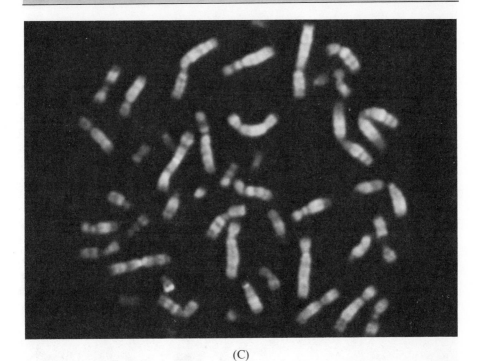

(C)

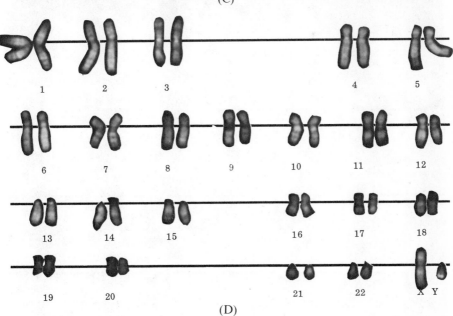

(D)

FIGURE 3-21. (*A*) *Metaphase spread of normal human male chromosomes, G-banding.* (*B*) *Karyotype of normal human male, G-banding.* (*C*) *Metaphase spread of normal human male chromosomes, Q-banding.* (*D*) *Karyotype of normal human male, Q-banding.* (Photos courtesy Dr. Herbert A. Lubs.)

3-21A and B) with Q-bands (Figs. 3-21C and D). In general, darkly stained Giemsa bands correspond to a large degree with the brightly fluorescing Q-bands, though agreement is not complete. However, virtually every chromosome in a complement can be positively identified, and structural alterations of parts of chromosomes clearly determined, as in the case of Q-banding. One variation of the Giemsa method produces band patterns that are the reverse of the G-bands, that is, darkly stained G-bands are lightly stained **R-bands** (Fig. 3-22) and vice versa. Figure 3-23 shows diagrammatically the Q-, G-, and R-band patterns of the human chromosome complement.

MEIOSIS

One of two fundamental cytological and genetic events in the life cycle of sexually reproducing plants and animals is the union of gametes, or sex cells, to form a zygote, a process to which the term **syngamy** is applied. Studies from as long ago as the last century clearly indicate that in gametic union the chromosomes contributed by each gamete retain their separate identities in the zygote nucleus. The zygote thus contains twice as many chromosomes as

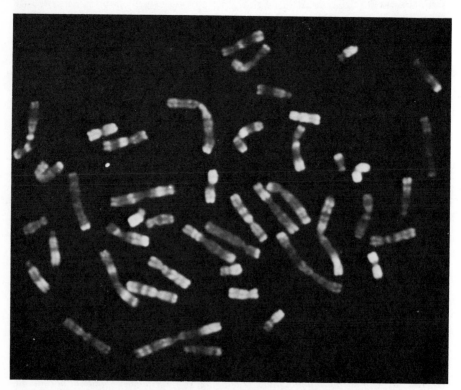

FIGURE 3-22. *Reverse (R-) banding with acridine orange. Metaphase spread of chromosomes of a normal human female.* (Courtesy Dr. C. C. Lin.)

does a gamete or, more accurately, all the chromosomes of each of the two gametes whose union created it.

This fact, of course, is responsible for the occurrence in diploid or $2n$ cells of matching pairs of chromosomes; each member of a given pair is the **homolog** of the other. In each diploid nucleus, then, there occurs the monoploid number of homologous *pairs* of chromosomes, one member of each pair having been contributed by the paternal parent, the other by the maternal parent. Thus, if among the chromosomes of the sperm, there is a metacentric chromosome (Fig. 3-19) of, for example, an average metaphase length of 5 μm, there will be in the chromosomal complement of the egg an identical chromosome. This statement applies to all the **autosomes** (those chromosomes not associated with the sex of the bearer). In the zygote and all cells derived from it by mitosis, *two* metacentric autosomes having an average metaphase length of 5 μm will be found. As we shall see later, these statements will have to be modified for the so-called sex chromosomes where these occur.

The result of syngamy is the incorporation into a zygote nucleus of all the chromosomes of each gamete. Such facts would seem to require a counterbalancing event whereby the doubling of chromosome quantity at syngamy is offset by a nuclear division that halves the amount of chromatin per nucleus at some point prior to gamete formation. Such a "reduction division" does occur in all sexually reproducing organisms that have discrete nuclei. This is **meiosis,** the second of the two fundamental cytological and genetic events in the sexual cycle.

Meiosis as a form of nuclear division differs greatly from mitosis. It consists of two successive divisions, each with its own prophase, metaphase, anaphase, and telophase; it thereby results in *four* daughter nuclei instead of two as in mitosis (although in many species not all the cellular products are functional). Furthermore, because of fundamental "procedural" differences between mitosis and meiosis, not only do the nuclear products of a meiotic division have one set of chromosomes each (monoploid or "haploid") as opposed to the two sets (diploid) of the parent nucleus, but the nuclear products are, if genes and chromosomes have any relationship, also genetically unlike the original diploid nucleus and often genetically unlike each other.

The term **meiocyte** (Fig. 3-24) is a convenient general one to designate any diploid cell whose next act will be to undergo meiosis. In seed plants and other heterosporous forms, for example, meiocytes are represented by the megasporocytes and microsporocytes ("spore mother cells"), in homosporous plants (those producing only one kind of spore by meiosis) by sporocytes or "spore mother cells," and in higher animals by primary spermatocytes and primary oocytes. The position of the meiocyte in the life cycle varies greatly from one major group to another. For instance, in the vascular plants the meiocytes give rise directly to megaspores and microspores (i.e., the **meiospores**), which, by mitosis, produce gamete-bearing plants. In the seed plants these gamete-bearing plants (the gametophytes) are much reduced; refer to Appendix B. In

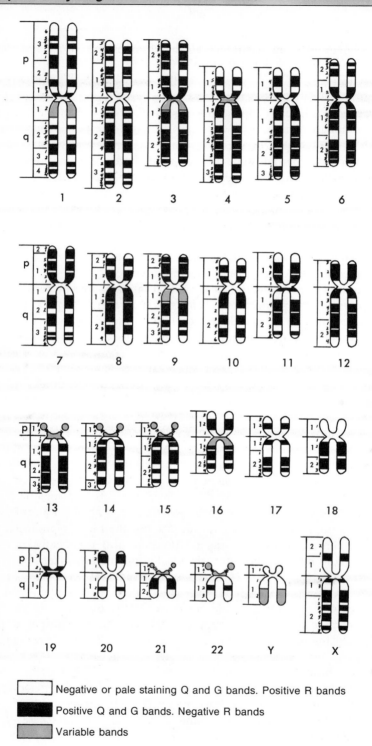

Negative or pale staining Q and G bands. Positive R bands

Positive Q and G bands. Negative R bands

Variable bands

FIGURE 3-23. *Diagrammatic comparison of normal human male karyotype as observed with Q-, G-, and R-banding techniques. The centromere is represented as observed in Q-banding only.* [Redrawn from *Paris Conference (1971): Standardization in Human Cytogenetics.* In *Birth Defects: Orig. Art. Ser.*, D. Bergsma, ed. Published by The National Foundation–March of Dimes, White Plains, New York, Vol. VIII (7), 1972. Used by permission.]

man, on the other hand, meiocytes give rise directly to gametes. In many algae and fungi the zygote nucleus itself functions as a meiocyte, giving rise to monoploid cells that ultimately, by mitosis, produce monoploid, gamete-bearing plants. In summary, meiosis may be sporic (vascular plants and others), or gametic (man and many other animals), or zygotic, and may be performed by (1) sporocytes ("spore mother cells," Fig. 3-18), (2) primary spermatocytes and oocytes, or even (3) by the zygote itself.

First Division: Prophase-I. Although most cytologists recognize and name at least five stages of prophase-I because of its complexities of chromosome

FIGURE 3-24. *Meiocytes (microsporocytes, microspore mother cells, or pollen mother cells) in another of lily stamen. These cells are diploid (2n) and are about to undergo meiosis.* (Courtesy Carolina Biological Supply Co.)

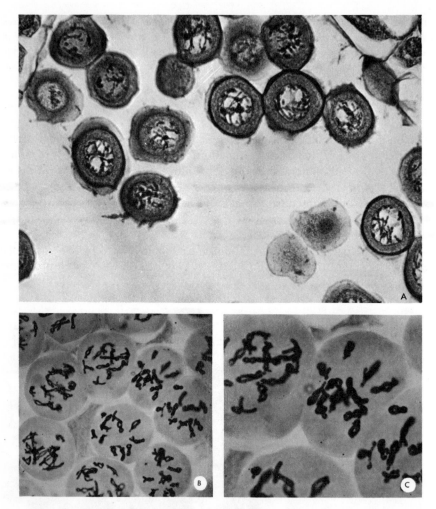

FIGURE 3-25. *Meiosis in microsporocytes of lily* (Lilium); *prophase-I, low power. Synapsis is underway but chromatids cannot easily be distinguished at this magnification.* (Courtesy Carolina Biological Supply Co.)

behavior, it will suffice for our purposes merely to emphasize events in sequence. In very early prophase-I, the diploid number of chromosomes gradually becomes recognizable under the light microscope as long, slender, threadlike structures (Fig. 3-25). In the light microscope the chromosomes *appear* to be longitudinally single in earliest prophase-I (*leptotene*), although chemical and autoradiographic studies show that replication has occurred in the S stage of the preceding interphase. By definition, then, no *visible* chromatids exist at this time in a meiocyte. Chromosomes shorten and thicken progressively, presumably by the same mechanism as for mitosis.

While this contraction is underway, a second event that characterizes meiosis (as opposed to mitosis) takes place. Recall that each diploid nucleus (including meiocytes) contains *pairs* of homologous chromosomes. In early prophase-I (*zygotene*) these homologs begin to pair, or **synapse.** This **synapsis** is remarkably exact and specific, taking place point for point, with the two homologs usually somewhat twined about each other (Fig. 3-26).

Electron microscope studies show at this stage a feature of dimensions too small to be discernible in light microscopy, the **synaptinemal[3] complex.** This complex structure is the physical mechanism by which (1) homologs are held together in synapsis, and (2) crossing-over, to be described shortly, is made possible. The synaptinemal complex is composed of three parallel bands, interconnected by fine strands at right angles to them. The two outer bands are axial components of the homologous chromosomes and are some 30 to 60 nm in diameter; the central one, which Westergaard and Von Wettstein (1972) believe is preformed in the nucleolus, is about 100 nm in width. These elements may be amorphous (both central and lateral in most higher plants and animals, including man), banded (lateral elements in ascomycete fungi and some insects), or latticelike with fine rods spaced about 10 nm apart (central element in insects). The whole structure is some 140 to 170 nm in width, just below the limit of resolution in light microscopy. Formation of the synaptinemal complex is initiated at several points along the pair of homologs. There is good chemical evidence that breakage and repair of DNA molecules of the chromosomes take place within the synaptinemal complex; this can and often does result in exchange of material between nonsister chromatids (*crossing-over*). Furthermore, synaptinemal complexes are not found in organisms in which crossing-over does not occur.

Synapsed chromosomes continue to shorten and thicken and, as they do, their longitudinal doubleness becomes apparent in light microscopy.

The synaptinemal complex is completed at the next stage of prophase-I (*pachytene*), during which the synapsed homologs now are clearly seen to be composed of two chromatids each. The points at which exchange of material occurs between nonsister chromatids is evidenced by more or less X-shaped configurations, the **chiasmata** (singular, **chiasma**), as seen in Figure 3-25. The longer the chromosome pair, the greater likelihood of more than one chiasma, although one chiasma appears to interfere with the formation of another in a closely adjacent region of the chromosomes on the same side of the centromere. The basis for this **interference** is not altogether clear.

Next, in the *diplotene* stage of prophase-I, separation of homologs (except at points where chiasmata occur) is initiated. Chromosomes continue to contract and the nucleolus begins to disappear.

Finally, in the last stage of prophase-I (*diakinesis*), chromosomes reach maximum contraction. At the outset of diakinesis the synapsed homologs be-

---

[3] Also spelled *synaptonemal*.

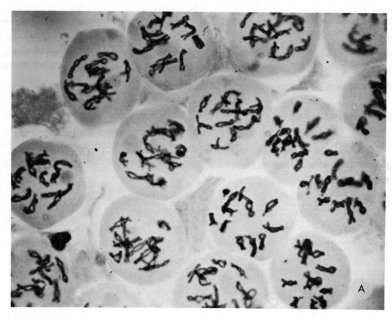

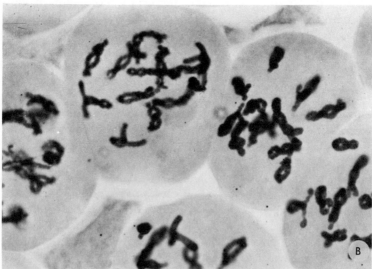

FIGURE 3-26. *Prophase-I in microsporocytes of the fritillary plant* (Fritillaria) *showing synapsis and chiasmata.* (Courtesy Dr. L. F. LaCour, John Innes Institute.)

come well spaced out in the nucleus, often near the nuclear membrane. Chiasmata gradually *terminalize,* that is, they appear to move toward the ends of the arms and finally "slip off," owing to the continued shortening of the chromosomes. Lastly, the nucleolus disappears, the nuclear membrane degenerates, and a spindle is formed.

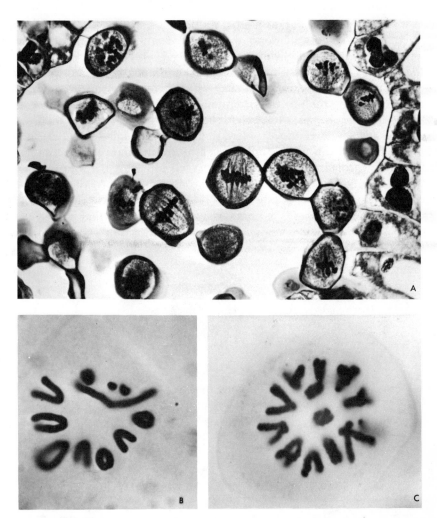

FIGURE 3-27. *Meiosis, metaphase-I.* (*A*) *In lily microsporocytes. Cell in the upper left of the anther cavity is seen in polar view; most of the remainder are seen in side view with spindles clearly evident.* (Courtesy Carolina Biological Co.) (*B*) *Polar view in spermatocytes of the orthopeteran* Mecostethus grossus. (*C*) *Microsporocytes of* Fritillaria, *a plant of the lily family.* [(B) and (C) courtesy Dr. L. F. LaCour, John Innes Institute.]

Metaphase-I. The arrival of synapsed homologous chromosome pairs at the equator of the spindle begins metaphase-I. This phase differs from mitotic metaphase in (1) the arrangement of the monoploid number of chromosome *pairs* on the equatorial plane and (2) the tendency for the centromere of each homolog to be directed somewhat toward one of the poles (Fig. 3-27). An important point to be noted here is the randomness of arrangement of the paired homologs; that is, for a given pair, it is just as likely that the paternal member be directed toward the "north" pole as it is for the maternal member to be so oriented. This will be well worth recalling in considering later the gamete genotypes a polyhybrid individual produces.

Anaphase-I. In anaphase-I actual **disjunction** of synapsed homologs occurs, one longitudinally double chromosome of each pair moving to each pole, thereby completing the process of terminalization. Here is an additional difference from mitosis, mitotic anaphase being marked by separation of sister chromatids which then move poleward as longitudinally single daughter chromosomes. Thus in mitosis one of each of the chromosomes present (i.e., of the entire chromosomal complement) travels to each pole, with the result that each new nucleus has just as many chromosomes as had the parent nucleus, whether monoploid or diploid. In meiosis, however, whole chromosomes of each homologous pair (as modified by any crossing-over that has occurred in prophase-I) separate, so that each pole receives either a paternal or a maternal, longitudinally double chromosome of each pair. This ensures a change in

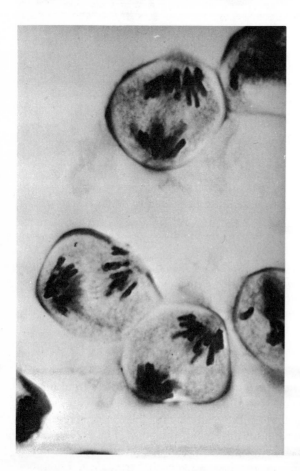

FIGURE 3-28. *Anaphase-I in lily microsporocytes.* (Courtesy General Biological Supply House, Inc., Chicago.)

chromosome number from diploid to monoploid in the resultant reorganized daughter nuclei. In short, whereas mitotic anaphase was marked by the separation of sister chromatids, anaphase-I of meiosis is characterized by separation of homologous entire chromosomes (Fig. 3-28).

Telophase-I.   The arrival of chromosomes at the poles of the spindle signals the end of anaphase-I and the beginning of telophase-I. During this phase the chromosomes may persist for a time in the condensed state, the nucleolus and nuclear membranes may be reconstituted, and cytokinesis may also occur (Fig. 3-29). In some cases, as in the flowering plant genus *Trillium*, meiocytes are reported to progress virtually directly from anaphase-I to prophase-II or even metaphase-II; in other organisms there may be either a short or fairly long interphase between the first and second meiotic divisions. In any event, the first division has thus accomplished the separation of the chromosomal complement into two monoploid nuclei.

Second Division: Prophase-II.   Prophase-II is generally short and superficially resembles mitotic prophase except that sister chromatids of each chromosome are usually widely divergent, exhibiting no relational coiling.

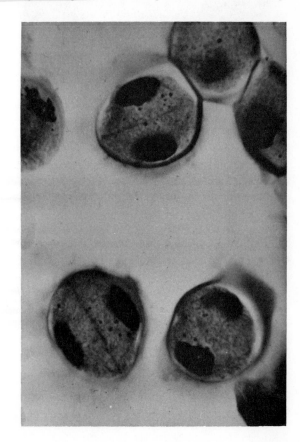

FIGURE 3-29. *Telophase-I in lily microsporocytes. Note the prominent cell plate.* (Courtesy General Biological Supply House, Inc., Chicago.)

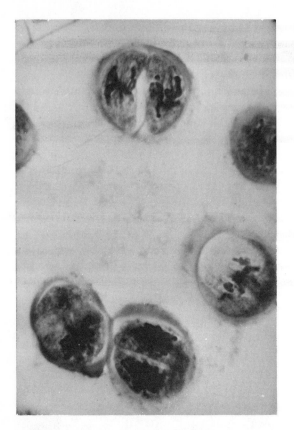

FIGURE 3-30. *Metaphase-II in lily microsporocytes.* (Courtesy General Biological Supply House, Inc., Chicago.)

Metaphase-II.   On two spindles, generally oriented at right angles to the first division spindle and often separated by a membrane or wall, the monoploid numbers of chromosomes, each consisting of two chromatids joined by a common centromere, are arranged in the equatorial plane. This stage is generally brief (Fig. 3-30).

Anaphase-II.   Centromeres now separate and the sister chromatids of metaphase-II now move poleward as daughter chromosomes, much as in mitosis. Their arrival at the poles marks the close of this phase.

Telophase-II.   Following the arrival of the monoploid number of daughter chromosomes at the poles, the chromosomes return to their long, attenuate, reticulate conformation, nuclear membranes are reconstituted, nucleoli reform, and cytokinesis generally separates each nucleus from the others (Fig. 3-31).

The process of meiosis is summarized diagrammatically in Figure 3-32.

CONCLUDING VIEW

Electron microscopy has supplied answers to some questions about meiosis, but others remain unanswered. For instance, it is now clear why in anaphase-I

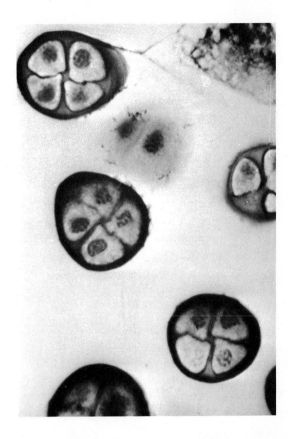

FIGURE 3-31. *Tetrads of lily microspores within the old microsporocyte wall. In some of the nuclei the chromosomes retain the telophase-II morphology, not yet having assumed the long, attenuate form of interphase.* (Courtesy General Biological Supply House, Inc., Chicago.)

homologs separate, whereas in mitotic anaphase it is *sister chromatids* that move apart to the poles. In mitotic metaphase the kinetochores of sister chromatids face in opposite directions, one toward each pole, whereas in metaphase-I the two kinetochores of each pair of synapsed homologs face different poles. Because of the attachment of microtubules to the kinetochores of the centromeres, then, it is homologous whole chromosomes that are separated in anaphase-I, but sister chromatids in mitotic anaphase.

It should also be noted that the second meiotic division is quite different from mitosis, so that the two meiotic divisions are integral parts of one overall process of nuclear division. On the spindles in meiosis-II the chromosomes are always present in the monoploid number; in mitosis they may occur in either monoploid or diploid number, depending upon the tissue in which division is taking place. Chromatids in the second meiotic division are widely separated, exhibiting no relational coiling, and, in addition, they have often been modified by crossing-over.

Although electron microscope observations of the synaptinemal complex have given some information on the mechanism and time of crossing-over,

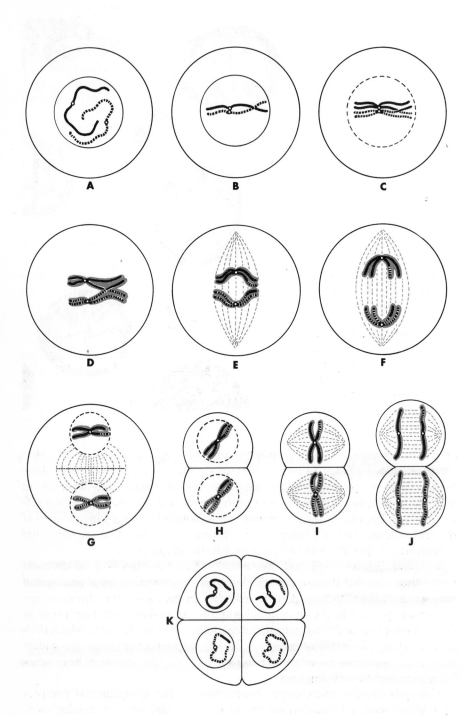

a full explanation of the precise physical and chemical details of synapsis remains to be elucidated. Finally, the triggering stimulus for meiosis is still unknown, although it appears likely to be hormonal or, at least, chemical. Meiocytes excised in interphase often undergo mitosis instead of meiosis in vitro; excision of cells successively later in prophase-I results in progressively larger numbers of cells that do undergo meiosis (Henderson, 1970). It appears likely that commitment to meiosis, as opposed to mitosis, may occur as early as the S or G_2 stage of interphase.

SIGNIFICANCE OF MEIOSIS

Cytologically, the basic significance of meiosis is the formation of four monoploid nuclei from a single diploid one in two successive divisions, thus balancing off, as it were, the doubling of chromosome number that results from syngamy. Note that the first meiotic division accomplished the reduction in chromosome number from diploid to monoploid, whereas the second is equational in distributing equal numbers of daughter chromosomes to developing new nuclei. In higher animals the cellular products of meiosis directly become gametes and/or polar bodies (Appendix B-8), but in vascular plants they are meiospores that give rise to reduced gamete-bearing plants (Appendix B-6).

But how does meiosis relate to the hypotheses we raised concerning probable gene distribution in somatic and gametic cells in the preceding chapter? If genes are located on chromosomes, the process of meiosis generates genetic variability in two important ways: (1) random assortment of paternal and maternal chromosomes and (2) crossing-over. Assume, for example, an organism heterozygous for three pairs of genes (a trihybrid), *AaBbCc* in which *ABC* was derived from its paternal parent and *abc* from its maternal parent. If these three pairs of genes are located on three *different* chromosome pairs (which we might designate as pair 1, pair 2, and pair 3), then in prophase-I the paternal and maternal number 1 chromosomes (bearing genes *A* and *a*, respectively) will synapse, as will the number 2s (bearing genes *B* and *b*) and the number

---

FIGURE 3-32. *Diagrammatic representation of meiosis in a meiocyte having one pair of homologous chromosomes ($2n = 2$). (A) prophase-I, chromosomes long and slender, appearing longitudinally single; (B) prophase-I, homologs synapsing; (C) prophase-I chromatids now evident, with one chiasma; (D) prophase-I, disjunction underway, chiasma still evident; (E) metaphase-I, chromosomes on midplane of spindle with centromeres divergent; (F) anaphase-I, poleward separation of previously synapsed homologs; note that each chromosome is still composed of two chromatids but some of these have been modified by crossing-over; (G) telophase-I, each reorganizing nucleus now contains the monoploid number of chromosomes that are still composed of two chromatids each; (H) prophase-II; (I) metaphase-II; (J) anaphase-II; (K) a postmeiotic tetrad of monoploid cells; note that each in this case contains a genetically unique chromosome because of crossing-over.*

3s (with genes C and c). Arrangement of each pair of synapsed homologs on the metaphase-I spindle is random; i.e., the paternal member of each pair has an equal chance of being oriented toward either pole, as does the maternal member. In anaphase-I each (longitudinally double) chromosome moves toward the nearer pole as it separates from its homolog. Therefore each telophase-I nucleus has an equal chance of receiving a paternal or a maternal chromosome. The same chance exists for each remaining chromosome pair. Hence for three pairs of genes on three different pairs of chromosomes, eight possible telophase-I genotypic combinations are possible:

Daughter Nucleus No. 1	Daughter Nucleus No. 2
ABC	abc
ABc	abC
AbC	aBc
Abc	aBC
aBC	Abc
aBc	AbC
abC	ABc
abc	ABC

$n = $ # pairs of genes of chromosomes

The second division will simply increase the number of each genotype from one to two. The number of possible gamete genotypes occurring after the second meiotic division is here 2^3, or 8. That is, a given gamete in this example has $(\frac{1}{2})^3$ chance of receiving any particular arrangement of paternal and/or maternal chromosomes and genes. Furthermore, in a sample of several hundred gametes from such an individual, each of the eight genotypes would be expected to occur in approximately equal numbers. For organisms with many chromosomes the number of possible combinations becomes very large. In man, for instance, with 23 pairs of chromosomes, the probability that any particular gamete will have a specific combination of chromosomes (excluding crossing-over) is $(\frac{1}{2})^{23}$, or about one in 8 million. Possible zygote genotypes resulting from random fusions of these gametes rise to about 64 trillion. A good deal of the variation in natural populations is due to this kind of **independent segregation** of genes normally occurring in the breeding population.

What would be the situation if these three gene pairs are, instead, located on a single pair of homologous chromosomes (i.e., **linked**)? Assume, for the purpose of illustration, that one member of the pair of homologs bears genes A, B, C on a given arm and that the other homolog bears genes a, b, and c. If no chiasmata are formed, then only two kinds of gametes (ABC and abc) will be produced, and these will occur in equal number. If, on the other hand, as is more often the case, chiasmata do form between genes A and B and between genes B and C in at least some meiocytes, then eight gamete genotypes will

see problem 3-17

again be produced. However, in this case the fraction of each type produced will depend upon the frequency with which crossing-over occurs between *A* and *B* and between *B* and *C*. This has considerable importance in the mapping of genes, which will be considered in detail in Chapter 6.

In this examination of two types of nuclear division so nearly universal in a wide variety of plant and animal forms, in fact always occurring in all organisms having discrete nuclei and sexual reproduction, we have met some impressive similarities between visually observable chromosome behavior and our postulated gene behavior of Chapter 2. Of course, this could conceivably be an unusual case of multiple coincidences, but the very multiplicity of these parallels furnishes a strong presumptive basis for the **chromosomal theory** of genetics. We shall, however, demand not only further parallels but also, more importantly, some positive experimental *proof.* This proof we shall seek not only from the science of genetics itself, but also from cytology, physics, and chemistry. Much of the remainder of this book is devoted, then, to the quest for a sound understanding of what a gene is, even at the molecular level, how it operates, and where in the cell it is located.

REFERENCES

ALTMAN, P. L., and D. S. DITTMER, eds., 1962. *Growth Including Reproduction and Morphological Development.* Washington, Federation of American Societies for Experimental Biology.

ALTMAN, P. L., and D. S. DITTMER, eds., 1972. *Biology Data Book,* 2nd ed., volume I. Washington, Federation of American Societies for Experimental Biology.

BAJER, A., 1968. Behaviour and Fine Structure of Spindle Fibres During Mitosis in Endosperm. *Chromosoma,* **25:** 249–281.

BERGSMA, D., ed., 1972. *Paris Conference (1971): Standardization in Human Cytogenetics.* In *Birth Defects: Orig. Art. Ser.,* volume VIII. White Plains, N. Y., The National Foundation–March of Dimes.

BRACHET, J., and A. E. MIRSKY, eds., 1961. *The Cell,* volume III. New York, Academic Press.

CASPERSSON, T., T. S. FARBER, G. E. FOLEY, J. KUDYNOWSKI, E. J. MODEST, E. SIMONSSON, U. WAGH, and L. ZECH, 1968. Chemical Differentiation Along Metaphase Chromosomes. *Exp. Cell Res.,* **49:** 219–222.

CASPERSSON, T., L. ZECH, E. J. MODEST, G. E. FOLEY, and U. WAGH, 1969a. Chemical Differentiation with Fluorescent Alkylating Agents in *Vicia faba* Metaphase Chromosomes. *Exp. Cell Res.,* **58:** 128–140.

CASPERSSON, T., E. J. MODEST, G. E. FOLEY, and U. WAGH, 1969b. DNA-Binding Fluorochromes for the Study of the Organization of the Metaphase Nucleus. *Exp. Cell Res.,* **58:** 141–152.

CASPERSSON, T., L. ZECH, and C. JOHANSSON, 1970a. Differential Binding of Alkylating Fluorochromes in Human Chromosomes. *Exp. Cell Res.,* **60:** 315–319.

CASPERSSON, T., L. ZECH, and C. JOHANSSON, 1970b. Analysis of the Human Metaphase Chromosome Set by Aid of DNA-Binding Fluorescent Agents. *Exp. Cell Res.,* **62:** 490–492.

CASPERSSON, T., L. ZECH, C. JOHANSSON, and E. J. MODEST, 1970c. Identification of Human Chromosomes by DNA-Binding Fluorescing Agents. *Chromosoma,* **30:** 215–227.

DE ROBERTIS, E. D. P., W. W. NOWINSKI, and F. A. SAEZ, 1970. *Cell Biology,* 5th ed. Philadelphia, W. B. Saunders.

DU PRAW, E. J., 1970. *DNA and Chromosomes.* New York, Holt.

HENDERSON, S. A., 1970. The Time and Place of Meiotic Crossing-over. In H. L. Roman, ed. *Annual Review of Genetics,* volume 4. Palo Alto, Calif., Annual Reviews, Inc.

HSU, T. C., 1973. Longitudinal Differentiation of Chromosomes. In H. L. Roman, ed. *Annual Review of Genetics,* volume 7. Palo Alto, Calif., Annual Reviews, Inc.

JOHN, B., and K. R. LEWIS, 1973. *The Meiotic Mechanism.* In J. J. Head, ed. Oxford Biology Readers. New York, Oxford University Press.

KING, R. C., 1974. *A Dictionary of Genetics,* 2nd ed. revised. New York, Oxford University Press.

MAZIA, D., 1961. Mitosis and the Physiology of Cell Division. In J. Brachet and A. E. Mirsky, eds. *The Cell,* volume III. New York, Academic Press.

RIS, H., and D. F. KUBAI, 1970. Chromosome Structure. In H. L. Roman, ed. *Annual Review of Genetics,* volume 4. Palo Alto, Calif., Annual Reviews, Inc.

SWANSON, C. P., 1969. *The Cell,* 3rd ed. Englewood Cliffs, N.J., Prentice-Hall.

SWANSON, C. P., and W. J. YOUNG, 1965. Chromosome Reproduction in Mitosis and Meiosis. In M. Locke, ed. *Reproduction: Molecular, Subcellular, and Cellular.* New York, Academic Press.

WESTERGAARD, M., and D. VON WETTSTEIN, 1972. The Synaptinemal Complex. In H. L. Roman, ed. *Annual Review of Genetics,* volume 6. Palo Alto, Calif., Annual Reviews, Inc.

PROBLEMS

3-1. In *Coleus* the somatic cells are diploid, having 24 chromosomes. How many of each of the following are present in each cell at the stage of mitosis or meiosis indicated? (Assume cytokinesis to occur in mid-telophase.) (a) Centromeres at anaphase, (b) centromeres at anaphase-I, (c) chromatids at metaphase-I, (d) chromatids at anaphase, (e) chromosomes at anaphase, (f) chromosomes at metaphase-I, (g) chromosomes at the close of telophase-I, (h) chromosomes at telophase-II.

3-2. Corn is a flowering plant whose somatic chromosome number is 20. How many of each of the following will be present in *one* somatic cell at the stage listed: (a) centromeres at prophase, (b) chromatids at prophase, (c) kinetochores at prophase, (d) chromatids in G_1, (e) chromatids in G_2?

3-3. Either from information you already have or based on facts in Appendix B, for a corn plant how many chromosomes are present in each of the following: (a) leaf epidermal cell, (b) antipodal nucleus, (c) endosperm cell, (d) generative nucleus, (e) egg, (f) megaspore, (g) microspore mother cell?

3-4. Based on your present knowledge, or after consulting Appendix B, how many human eggs will be formed from (a) 40 primary oocytes, (b) 40 secondary oocytes, (c) 40 ootids? *40* *40* *40*

3-5. From your present knowledge, or after consulting Appendix B, how many human sperms will be formed from 40 primary spermatocytes? *160*

3-6. From your present knowledge, or from Appendix B, 20 microsporocytes of a flowering plant would be expected to produce how many (a) microspores, (b) sperms? *AB Ab aB ab   AbC Abc $2^n = 2^5 = 32$*

3-7. How many different gamete genotypes will be produced by the following parental genotypes if all genes shown are unlinked: (a) *AA*, (b) *Aa*, (c) *AaBB*, (d) *AaBb*, (e) *AAbbCc*, (f) *AaBbCcDdEe*? *① ②② ①② ②②*

3-8. Consult Table 3-1 in answering the following questions. (a) What is the probability in cattle that a particular egg cell will contain only chromosomes derived from the maternal parent of the cow producing the egg? (b) If this cow is mated to its brother, what is the probability that their calf will receive only chromosomes originally contributed by the calf's grandmother? *$\frac{1}{2^{30}}$* ?

3-9. Triploid watermelons have the advantage of being seedless. (a) What is the somatic chromosome number of such plants? (b) What explanation can you offer for their lack of seeds? *sterile (can't pair right in meiosis) 22 diploid so 11 haploid & thus 33 triploid*

3-10. A cell of genotype *Aa* undergoes mitosis. What will be the genotype(s) of the daughter cells? *Aa*

3-11. A cell of genotype *Aa* undergoes meiosis. What will be the genotype(s) of the daughter cells if all of them are functional? *2 A + 2 a*

3-12. A student examining a number of onion root tips counted 1,000 cells in some phase of mitosis. He noted 692 cells in prophase, 105 in metaphase, 35 in anaphase, and 168 in telophase. From these data what can be concluded about relative duration of the different stages of the process? *prophase longest*

3-13. Garden peas have 14 chromosomes in their somatic cells. How many groups of linked genes occur in this plant? *7*   *7 haploid*

3-14. What is the probability that any ascospore of the red bread mold (*Neurospora crassa*) will have all its chromosomes derived from the + parent? (You may wish to consult Table 3-1 and Appendix B before trying to answer.) *($\frac{1}{2}$)⁷*

3-15. The red bread mold, *Neurospora crassa,* is an ascomycete fungus much used in genetic research. As pointed out more fully in Appendix B, sexual reproduction occurs when the tubular filaments of plants of opposite mating strain come into contact. The fusion nucleus (zygote) that results is the only diploid nucleus in *Neurospora's* life history. It quickly undergoes meiosis into meiospores within a developing saclike structure, the ascus, which is relatively long and narrow, so that the meiospores cannot slip past each other. They are arranged serially and may be removed in sequence for separate germination and study with full knowledge of which spore, with reference to position in the ascus, produces which phenotype. One mitosis follows meiosis, so that eight ascospores occur in each ascus. Prior to sexual reproduction, large numbers of asexual reproductive bodies, called conidia, are produced. These are pink to red (wild type), or yellow in another strain.

    In the case of asci resulting from the cross (+) × yellow (*y*), where + and *y* are alleles, it is observed that in some cases the ascospores, in order from tip of ascus to bottom, were arranged + + + + *y y y y*, whereas less often they

The prob of a particular combination of chromosome in a gamete is ($\frac{1}{2}$)ⁿ where n = haploid #

in a zygote is ($\frac{1}{2}$)ⁿ ($\frac{1}{2}$)ⁿ

see p137 *tricky*

were $+ + y\,y + + y\,y$. In terms of chromosome behavior, account for each of these arrangements.

3-16. In corn recessive gene *hm* determines susceptibility to the fungus *Helminthosporium*, which produces lesions on leaves and in kernels, reducing yield, vigor, and market value of the crop. Its dominant allele determines resistance to the fungus. Another recessive gene, br_1, produces shortened internodes (segments of stem between successive leaves) and stiff, erect leaves (brachytic plants). Its dominant allele is responsible for normal plant form. These two pairs of genes are linked on chromosome 1. If a resistant, normal plant, heterozygous for both pairs of genes (with dominants of each pair on one homologous chromosome) *if tis* is pollinated by a susceptible, brachytic plant, (a) would you expect any resistant, brachytic and susceptible, normal individuals in the progeny? (b) Explain the cytological basis of your answer. *i.e. the dominants be linked*

3-17. Assume an individual heterozygous for three genes, *A*, *B*, and *C*. Assume further that these three dominant genes are linked in that order on one chromosome and *a*, *b*, and *c* linked (in that sequence) on the homologous chromosome. If crossing-over occurs between *A* and *B* with a frequency of 0.05, and between *B* and *C* with a frequency of 0.10, what should be the frequency of *aBc* and *AbC* gametes produced by that individual if there is no interference? *double crossovers*

why? don't understand answer

3-18. Assume a particular metaphase human chromosome has an average length of 5 μm. Give the equivalent in (a) meters, (b) millimeters, (c) nanometers, (d) Angstrom units.

3-19. What possible advantages to a species can you see in crossing-over?

3-20. Can you see any possible disadvantages that might result from crossing-over?

3-16 a)　　　$HM\,hm\,BR_1\,br_1$ × $hm\,hm\,br_1\,br_1$

gametes: $HMBR_1$ $hmbr_1$　　　　$hmbr_1$

progeny: $\frac{1}{2}\,HM\,hm\,BR_1\,br_1$, $\frac{1}{2}\,hm\,hm\,br_1\,br_1$
　　　　　(resist. normal)　　　　(suscept. brach.)

only if crossing over occurs

3-17　do:　$\underset{.05}{A} \quad \underset{.10}{B} \quad C$

　　　　　　　　$a \quad b \quad c$

gametes:
　　parental: ABC }　　　　AbC } double crossover
　　　　　　　　abc }　　　　aBc)
　　recombinant: Abc } .05)
　　　　　　　　aBC }　　) single crossover
　　　　　　　　ABc } .10)
　　　　　　　　abC }

freq of double if no crossovers interference = (.05)(.10)
= .005

CHAPTER 4
Dihybrid Inheritance

We have seen in Chapter 2 the genetic results of the action of a single pair of genes through several generations. In Chapter 3 we examined both mitotic and meiotic nuclear divisions, noting an extensive array of parallels between observable chromosome behavior and our postulates on gene behavior. It will be interesting now to determine whether the ideas developed thus far apply equally well to crosses involving two pairs of genes.

Classic Two-Pair Ratios

COMPLETE DOMINANCE IN TWO PAIRS

In addition to the leaf margin trait of *Coleus* that we considered in Chapter 2, another vegetative character involves the venation pattern of leaves. This is easily observed on their lower surfaces. A commonly encountered vein arrangement is the typically regular one shown in Figure 4-1. Here a single midvein branches in a rather standard pinnate fashion. An alternative expression is the highly irregular arrangement seen in Figure 4-2. For convenience we may attach the terms *regular* and *irregular,* respectively, to these two phenotypes. By a simple monohybrid cross involving these two characters, it can easily be determined that *irregular* is completely *dominant* to *regular.*

Therefore a cross of a doubly homozygous *deep irregular* with a *shallow regular* individual (using *D* and *d* again to represent the deep and shallow genotypes, and *I* and *i* to denote irregular and regular) may be represented as follows:

$$\text{P} \quad \text{deep irregular} \quad \times \quad \text{shallow regular}$$
$$DDII \qquad\qquad ddii$$

P gametes $\widehat{DI}$ $\widehat{di}$

$$\text{F}_1 \qquad\qquad \text{deep irregular}$$
$$DdIi$$

The F_1 individuals in this instance are referred to as **dihybrid** individuals because they are heterozygous for each of the two pairs of genes.

What will be the result of crossing two members of this F_1 to produce an F_2? Let us try to predict the outcome of this cross on the basis of the theses so far developed. Recalling our assumptions and the experimental evidence therefor, it is clear that, *considering one pair of genes at a time,* the phenotypic result will be the usual 3:1 ratio:

FIGURE 4-1. *Regular venation pattern, a trait caused by a recessive gene, in* Coleus.

F_1 $Dd \times Dd$

F_1 gametes $\begin{cases} \text{eggs} & \frac{1}{2}\,\textcircled{D} + \frac{1}{2}\,\textcircled{d} \\ \text{sperms} & \frac{1}{2}\,\textcircled{D} + \frac{1}{2}\,\textcircled{d} \end{cases}$

F_2 genotypes $\underbrace{\frac{1}{4}\,DD + \frac{2}{4}\,Dd + \frac{1}{4}\,dd}$

F_2 phenotypes $\frac{3}{4}$ deep $+ \frac{1}{4}$ shallow

[handwritten: $DdIi \times DdIi$]

Likewise, $Ii \times Ii$ will produce the same $\frac{3}{4}$ irregular (I–): $\frac{1}{4}$ regular (ii) progeny. To determine the *phenotypic ratio* of the dihybrid cross $DdIi \times DdIi$, it should appeal to you as logical and as the *simplest* (though not necessarily therefore correct) expectation, that either the deep or shallow phenotype may be associated *at random* with either the irregular or regular phenotype if the two pairs of genes involved are on different pairs of chromosomes (recall the behavior of chromosomes from Chapter 3). That is, utilizing the product law of probability noted in Chapter 2, we expect the phenotypic ratio here to be the *product* of its two component monohybrid ratios, thus:

$Dd \times Dd$ yields: $\frac{3}{4}$ deep $+ \frac{1}{4}$ shallow
$Ii \times Ii$ yields: $\frac{3}{4}$ irregular $+ \frac{1}{4}$ regular

[handwritten annotations: $(\frac{3}{4}\,\text{deep} : \frac{1}{4}\,\text{shallow})(\frac{3}{4}\,\text{irr} : \frac{1}{4}\,\text{reg.})$]
[handwritten: $\frac{9}{16}\,\text{deep irr} : \frac{3}{16}\,\text{shallow irr} : \frac{3}{16}\,\text{deep reg} : \frac{1}{16}\,\text{shallow reg}$]
[handwritten: $(3:1)(3:1) \rightarrow 9:3:3:1$]
[handwritten: monohybrid   dihybrid]

[handwritten annotation, top left:] Treat dihybrid crosses as 2 monohybrid crosses being done at same time Dd×Dd and Ii×Ii

[handwritten annotation, right:] p.21: Product Law of probability — The probability of the simultaneous occurrence of 2 independent events = the product of the probabilities of their separate occurrence

FIGURE 4-2. *Irregular venation pattern, produced by the dominant allele of the gene for regular venation in* Coleus.

F_2 result, based on random combination of phenotypic classes: $\frac{9}{16}$ deep irregular $+ \frac{3}{16}$ shallow irregular $+ \frac{3}{16}$ deep regular $+ \frac{1}{16}$ shallow regular.

This is what we will actually observe as soon as F_2 seedlings have developed from this cross. It is also the same ratio that Mendel observed in his work with peas, and is one of the classic phenotypic ratios. On this basis, it should also be possible to calculate readily the genotypic ratio of this cross:

$$Dd \times Dd \text{ yields:} \qquad \frac{1}{4}DD + \frac{2}{4}Dd + \frac{1}{4}dd$$
$$Ii \times Ii \text{ yields:} \qquad \frac{1}{4}II + \frac{2}{4}Ii + \frac{1}{4}ii$$

$$F_2 \text{ genotypes:} \qquad \frac{1}{16}DDII + \frac{2}{16}DdII + \frac{1}{16}ddII$$
$$+ \frac{2}{16}DDIi + \frac{4}{16}DdIi + \frac{2}{16}ddIi$$
$$+ \frac{1}{16}DDii + \frac{2}{16}Ddii + \frac{1}{16}ddii$$

[handwritten annotation, right:] 9 different genotypes

[handwritten work, bottom:] $\left(\frac{1}{4}DD : \frac{1}{2}Dd : \frac{1}{4}dd\right)\left(\frac{1}{4}II : \frac{1}{2}Ii : \frac{1}{4}ii\right)$

$\frac{1}{16}DDII + \frac{1}{8}DDIi + \frac{1}{16}DDii + \frac{1}{8}DdII + \frac{1}{4}DdIi + \frac{1}{8}Ddii + \frac{1}{16}IIdd + \frac{1}{8}Ii\,dd + \frac{1}{16}iidd$

or $(1:2:1)(1:2:1) \rightarrow 1:2:1:2:4:2:1:2:1$

9 different genotypes

Closer examination of this $1:2:1:2:4:2:1:2:1$ genotypic ratio reveals the $9:3:3:1$ phenotypic ratio:

$\frac{1}{16}DDII$	$\frac{1}{16}ddII$	$\frac{1}{16}DDii$	$\frac{1}{16}ddii$
$\frac{2}{16}DdII$	$\frac{2}{16}ddIi$	$\frac{2}{16}Ddii$	
$\frac{2}{16}DDIi$			
$\frac{4}{16}DdIi$			

$$\frac{9}{16}D{-}I{-} \; + \; \frac{3}{16}ddI{-} \; + \; \frac{3}{16}D{-}ii \; + \; \frac{1}{16}ddii$$

deep	shallow	deep	shallow
irregular	irregular	regular	regular

Alternatively, we could have calculated this $9:3:3:1$ ratio more quickly in this fashion:

$$Dd \times Dd \text{ yields } \quad \tfrac{3}{4}D{-} + \tfrac{1}{4}dd$$
$$Ii \times Ii \;\; \text{ yields } \quad \tfrac{3}{4}I{-} \; + \tfrac{1}{4}ii$$

to get the same result as we obtained by collecting and summing the nine genotypes immediately above. It will be useful later on to recall this ratio:

9 D–I–

3 ddI–

3 D–ii

1 ddii

A $9:3:3:1$ F_2 phenotypic ratio in corn is shown in Figure 4-3.

Note that in calculating our expected two-pair results, we once more followed the probability method, avoiding the more cumbersome "checkerboard" or Punnett square. The probability approach reflects the random association of different pairs of genes that is operating here and is more direct, especially in cases where one wishes to know what fraction of the progeny in a particular cross will be of a given genotype or phenotype. For instance, in this example with *Coleus,* the probability method permits an almost instant answer to such questions as what fraction of the offspring of the cross $DdIi \times DdIi$ will be (1) shallow irregular or (2) of the genotype *ddII?* In the first case, if we note that both parents are doubly heterozygous and that we seek just those offspring that combine the recessive phenotype of one trait with the dominant expression of the other character, it is clear that the cross $DdIi \times DdIi$ will produce $\frac{1}{4}$ shallow $\times \frac{3}{4}$ irregular, or $\frac{3}{16}$. The second question is calculated in the same manner, the P individuals given producing $\frac{1}{4}$ of their progeny having *each* of the genotypes *dd* and *II,* or $\frac{1}{4} \times \frac{1}{4} = \frac{1}{16}$. Compare these calculations, which, with a little practice and remembering the classic $3:1$ and $1:2:1$ ratios, can be determined mentally, with the results previously arrived at on page 85. The same approach can be used to great advantage with trihybrid and other polyhybrid crosses where, as in the sample just given, only the specific information

[handwritten top margin: can also do it like this: $(\frac{1}{4}DI : \frac{1}{4}Di : \frac{1}{4}di : \frac{1}{4}dI)(\frac{1}{4}DI : \frac{1}{4}Di : \frac{1}{4}di : \frac{1}{4}dI)$ *]*

[handwritten: 4 kinds of gametes are produced by DdIi]

[handwritten: $\frac{1}{16} DDII : \frac{1}{16} DDIi$ etc.]

FIGURE 4-3. *A 9:3:3:1 purple-starchy:purple-sweet:white-starchy:white-sweet* F_2 *phenotypic ratio in corn. Sweet kernels are shriveled, starchy are plump.*

sought is obtained with no need for culling out those facts from a mass of un-needed information about the entire range of progeny types.

It is most important to recognize one tacit assumption involved in our method of calculating these particular dihybrid ratios. Recall particularly the behavior of chromosomes at meiosis and the subsequent gamete genotypes. If genes are indeed located on chromosomes, it is clear that our calculation (page 85) is based on the expectation that, in gametes, association of D with I or i, and of d with I or i, is random. Therefore we are implying *equal* numbers of four possible gamete genotypes: DI, dI, Di, and di from a $DdIi$ individual. We could calculate the results of the cross $DdIi \times DdIi$ either as we did on page 85 or by expected gamete genotypes:

$$DdIi \text{ eggs:} \quad \tfrac{1}{4}\, \textcircled{DI} + \tfrac{1}{4}\, \textcircled{dI} + \tfrac{1}{4}\, \textcircled{Di} + \tfrac{1}{4}\, \textcircled{di}$$
$$DdIi \text{ sperms:} \quad \tfrac{1}{4}\, \textcircled{DI} + \tfrac{1}{4}\, \textcircled{dI} + \tfrac{1}{4}\, \textcircled{Di} + \tfrac{1}{4}\, \textcircled{di}$$

Satisfy yourself that if gamete union is also random, the phenotypic and gen-otypic ratios arrived at in this way will be identical with those calculated on page 85. Production of four kinds of gametes in equal number by a doubly heterozygous individual will occur, of course, if each pair of genes is on a dif-ferent pair of chromsomes. That is, each pair of genes here behaves exactly as it would in a one-pair cross. Therefore *each pair of genes located on a differ-ent pair of chromosomes segregates independently of all other such pairs at meiosis.* Recall the behavior of chromosomes in meiosis, especially their metaphase-I arrangement and anaphase-I movement. *[handwritten: DdIi individual]*

[handwritten: from mother, from father]

GAMETE AND ZYGOTE COMBINATIONS

As seen earlier, a monohybrid such as Dd produces two kinds of gametes (D and d) in equal numbers that can combine by syngamy to form three dif-ferent zygote genotypes (DD, Dd, dd), from which two phenotypes (deep and shallow) will be discernible in the progeny. Likewise, a doubly heterozygous individual ($DdIi$) produces four kinds of gametes (DI, Di, dI, di), again in equal numbers if the two pairs of genes are located on different chromosome pairs. With random syngamy, nine different zygote genotypes are produced:

[handwritten bottom: D ⇄ d / I ⇄ i or D ⇄ d / i ⇄ I 1 chromosome pair]

$$DDII \quad Ddii$$
$$DDIi \quad ddII$$
$$DdII \quad ddIi$$
$$DdIi \quad ddii$$
$$DDii$$

from which four phenotypes (deep irregular, deep regular, shallow irregular, shallow regular) will be apparent in the young seedlings.

A third pair of genes in *Coleus* can be symbolized as follows:

[handwritten: Treat DdIiWi × DdIiWw as 3 separate monoh. crosses like: Dd×Dd   IixIi'   Wwx'Ww]

W no white area at the base of the leaf blade
w white area at the base of the leaf blade

What gamete combinations can be produced by the trihybrid *DdIiWw?* If each of the three pairs of genes is on a different chromosome pair, then either allele of any pair can combine with either allele of any other pair. A simple application of the probability method indicates eight possible gamete genotypes. We have just seen that the dihybrid *DdIi* produces four gamete genotypes, $\frac{1}{4}$ *DI* $+ \frac{1}{4}$ *Di* $+ \frac{1}{4}$ *dI* $+ \frac{1}{4}$ *di*. With the addition of the *W, w* pair, two additional gamete genotypes, *W* and *w*, become possible. These, of course, may combine with any of the dihybrid gamete genotypes:

$$\frac{1}{4}\,\widehat{DI} + \frac{1}{4}\,\widehat{Di} + \frac{1}{4}\,\widehat{dI} + \frac{1}{4}\,\widehat{di}$$
$$\frac{1}{2}\,\widehat{W} + \frac{1}{2}\,\widehat{w}$$

$\frac{1}{8}$ each of *DIW, DiW, dIW, diW, DIw, Diw, dIw, diw*

If we apply again the same probability method used for monohybrid and dihybrid cases, these eight gamete genotypes may be expected to combine randomly into 27 zygote genotypes, producing eight phenotypic classes in the progeny.

[handwritten margin: Note: like formula work only when parents is heterozygous either Dd singly, Ii Dd doubly, Ii Dd Ww or triply IiDd]

Thus, considering the number of possible gamete genotypes per pair of heterozygous alleles in which dominance is complete and which are located on different chromosome pairs, we can detect the emergence of these mathematical relationships (where n = number of pairs of chromosomes with single gene differences):

number of gamete genotypes produced by parents $= 2^n$
number of progeny phenotypic classes $= 2^n$
number of progeny genotypic classes $= 3^n$
[handwritten: " " possible combinations of gametes = 4ⁿ] $= 4^n$

These relationships are summarized and extended in Table 4-1.

TESTCROSS

Just as the testcross is a useful way of determining homozygosity or heterozygosity of a dominant phenotype in one-pair cases, it is equally valuable in determining genotypes in situations with two or more pairs of genes. In a two-

problems
see 4-7a 4-9
Does this equal the
Total # progeny?

TABLE 4-1. Parental Gamete Genotypes, Gamete Combinations, and Progeny Phenotypes and Genotypes (Dominance Complete in All Pairs)

Number of Pairs of Heterozygous Genes	Number of Gamete Genotypes	Number of Progeny Phenotypes	Number of Progeny Genotypes	Number of Possible Combinations of Gametes
1	2	2	3	4
2	4	4	9	16
3	8	8	27	64
4	16	16	81	256
n	2^n	2^n	3^n	$4^n = 2^{2n} = (2^n)^2$

pair cross, where each pair of genes is on a different chromosome pair (i.e., the genes are not linked), the resulting progeny ratio can be seen to be the product of two one-pair ratios. Thus, as we have seen, a monohybrid testcross ratio is 1:1, and a dihybrid testcross produces a 1:1:1:1 ratio when one parent is heterozygous for both gene pairs: *see bottom p 22*

P $DdIi \times ddii$ → *treat like :*

P gametes $\frac{1}{4}DI + \frac{1}{4}dI + \frac{1}{4}Di + \frac{1}{4}di$ $Dd \times dd$ $Ii \times ii$

             $1\,di$ (1:1)   (1:1)

                                               1:1:1:1

F$_1$ $\frac{1}{4}DdIi + \frac{1}{4}ddIi + \frac{1}{4}Ddii + \frac{1}{4}ddii$ *or*
       deep   shallow   deep   shallow $1\,di\,(\frac{1}{4}DI:\frac{1}{4}Di:$
      irregular   irregular   regular   regular $\frac{1}{4}dI:\frac{1}{4}di)$

The testcross is a very useful technique in mapping of gene locations on chromosomes, as we shall see in Chapter 6.

With the information you now have, you might also find it useful to determine testcross ratios in (1) cases where only one of two pairs is heterozygous and (2) trihybrids or other polyhpbrids. Problems at the end of this chapter explore cases such as these.

Modifications of the 9:3:3:1 Ratio

INCOMPLETE DOMINANCE

We have seen the breeding results in dihybrids where dominance is complete. What effect on ratios and on numbers of phenotypic classes does incomplete dominance in one or both pairs produce?

In tomato, two pairs of genes, located on different pairs of chromosomes, are

completely dominant   { $D-$ tall plant h_1h_1 hairless stems
                   dd dwarf plant h_1h_2 scattered short hairs
                               h_2h_2 very hairy stems

incomplete dominance

Crossing two individuals of the genotype Ddh_1h_2 produces progeny as follows:

$\frac{3}{16}$ tall, hairless

$\frac{6}{16}$ tall, scattered hairs

$\frac{3}{16}$ tall, very hairy

$\frac{1}{16}$ dwarf, hairless

$\frac{2}{16}$ dwarf, scattered hairs

$\frac{1}{16}$ dwarf, very hairy

[handwritten margin notes:]
$Ddh_1h_2 \times Ddh_1h_2$
treat like:
$Dd \times Dd$ $h_1h_2 \times h_1h_2$
pheno: $(3:1)$ $(1:2:1)$
    $3:6:3:1:2:1$
geno: $(1:2:1)$ $(1:2:1)$
$1:2:1:2:4:2$
$1:2:1$

Note that this is precisely what one would expect. Tall and dwarf segregate in a 3:1 ratio, and hairless, scattered hairs, and very hairy segregate in a 1:2:1 ratio, producing a dihybrid 3:6:3:1:2:1 phenotypic ratio here. In such a case as this, because $Dd \times Dd$ produces two phenotypic classes in the offspring, and $h_1h_2 \times h_1h_2$ is responsible for three phenotypic classes, the cross $Ddh_1h_2 \times Ddh_1h_2$ produces offspring of $2 \times 3 = 6$ phenotypic classes. Adding to our mathematical expressions developed on page 88, we see that, if dominance is *in*complete, the number of phenotypic classes is 3^n (where again $n = $ the number of chromosome pairs with a single gene difference). In a dihybrid situation where one pair of genes exhibits complete dominance and the other incomplete dominance, as in this example from tomato, the number of F_1 phenotypic classes from two doubly heterozygous parents is $2^n \times 3^n$.

Thus, again, incomplete dominance increases the number of phenotypic classes. Further possibilities of this sort are suggested in some of the problems at the end of this chapter.

[handwritten:] $2^1 \times 3^1 = 6$

EPISTASIS

[handwritten:] since pheno same as geno in incomplete dom.

Mouse. The laboratory mouse occurs in a number of colors and patterns. The wild type,[1] or "agouti," is characterized by color-banded hairs in which the part nearest the skin is gray, then a yellow band, and finally the distal part is either black or brown. The wild type has rather obvious selection value in natural surroundings where it enhances concealment of the individual. Two other colors are albino and solid black. In albinos there is a total lack of pigment, producing white hair and pink eyes (the latter results when blood vessel color shows through unpigmented irises).

The cross *black* $\times$ *albino* produces a uniform F_1 of agouti which, in certain instances, when inbred, results in an F_2 of 9 *agouti*:3 *black*:4 *albino*. The segregation of the F_2 into sixteenths immediately suggests two pairs of genes, and this particular ratio implies a 9:3:3:1 ratio in which the $\frac{1}{16}$ class and one of the $\frac{3}{16}$ classes are indistinguishable.[2] These results would then indicate further that the F_1 individuals in this case are heterozygous for both pairs of *[handwritten:]* epistasis

---

[1] That is, the customary or most frequently encountered phenotype in natural populations, often used as a standard of comparison.

[2] A way to demonstrate the likelihood that this is *not* an approximation of a 1:2:1 ratio is explored in the next chapter.

[handwritten bottom notes:] Note incomplete dominance increases no. of phenotypes from 9:3:3:1 to 3:6:3:1 for example, epistasis decreases no. of phenotypes from 9:3:3:1 to, for example, 9:3:4 types 3

genes. We could assume one of the two pairs of genes to include one allele for color production and another allele for color inhibition (the latter perhaps responsible for either a defective enzyme or the absence of a particular enzyme required for a specific intermediate biochemical step in pigment production). We might further assume the other pair of genes to include one allele for agouti and one for black. Let us try genotypes as follows:

A	agouti	*C*	color
a	black	*c*	color inhibition

Note that dominance of agouti over black is suggested by the $\frac{9}{16}$ agouti class versus the $\frac{3}{16}$ black in the F$_2$ (this is tantamount to a 3:1 segregation). On these assumptions, the cross in mouse may be diagramed in this way:

P   black × albino
      aaCC   *AAcc*

F$_1$ agouti
               AaCc

F$_2$ $\frac{9}{16}$ *A—C—* agouti
        $\frac{3}{16}$ *aaC—* black
        $\frac{3}{16}$ *A—cc* $\Big\}$ albino
        $\frac{1}{16}$ *aacc*

[handwritten margin notes: colorless C → ~~agouti~~ (black) A → agouti. Here, only need 1 dominant C do produce a 3 rd pheno class*]*

In this particular example, gene *c* (which is recessive to its own allele *C*) actually masked the effect of either *A—* or *aa* so that any *—cc* individual is albino. Such a gene which masks the effect of one or both members of a *different* pairs of genes is said to be *epistatic;* the masked gene or genes may be termed *hypostatic*. Here *c* is epistatic to *a* and *A,* and we may refer to this case as one of recessive epistasis, because it is the recessive of one pair that is epistatic to another pair. Note that epistasis is quite different from dominance in that the masking operates between different pairs of alleles rather than between members of one pair.

   Clover.   An interesting example in white clover (*Trifolium repens,* the clover so frequently seen in lawns), reported in 1943 by Atwood and Sullivan, furnishes presumptive evidence that epistasis depends on a gene-enzyme relationship in a series of sequential biochemical steps.

   Some strains of white clover test high in hydrocyanic acid (HCN), whereas others test negatively for this substance. HCN content is associated with more vigorous growth; it does not harm cattle eating such varieties. Often the cross *positive × negative* results in an F$_1$ testing uniformly positive for HCN and an F$_2$ segregating 3 positive: 1 negative, suggesting a single pair of genes with positive dominant.

   One series of crosses reported by Atwood and Sullivan, however, produced unexpected totals:

[handwritten at bottom: 9:7 where both pairs of genes are epistatic as in clover example $\frac{9:7}{2}$ *]*

P   positive × negative
F_1 positive
F_2 351 positive + 256 negative

A 3:1 expectancy in the F_2 would be approximately 455 positive:152 negative; the actual results differ sufficiently from a 3:1 ratio to cast doubt on its relevance here. (Statistical tests, described in Chapter 5, give objective support to the hypothesis that, although a ratio of 351:256 *could* occur by chance alone in a 3:1 expectancy, this result is unlikely enough to cause one to look for a better explanation.) Note that these results are very close to a 9:7 ratio, which would be 342 positive:265 negative. A 9:7 ratio immediately suggests an epistatic expression of the 9:3:3:1, which, in turn, indicates two pairs of genes. One such 9:7 ratio (purple:yellow) in corn is illustrated in Figure 4-4.

It is known that HCN formation follows a path that may be represented thus:

$$\longrightarrow \quad \text{(precursor)} \quad \xrightarrow{\text{enzyme } "\alpha"} \quad \text{cyanogenic glucoside} \quad \xrightarrow{\text{enzyme } "\beta"} \quad \text{HCN}$$

Each of the conversions indicated by the arrows is enzymatically controlled. Tests of F_2 individuals of the cross just described for (1) HCN, (2) enzyme "β," and (3) cyanogenic glucoside revealed four classes of individuals:

Class	HCN	Enzyme "β"	Glucoside
1	+	+	+
2	0	+	0
3	0	0	+
4	0	0	0

Each of these four classes was then tested for HCN after adding either enzyme "β" or glucoside to their leaf extracts, with the following results:

Class	Control Test for HCN	HCN Test After Adding Enzyme "β"	HCN Test After Adding Glucoside
1	+	+	+
2	0	0	+
3	0	+	0
4	0	0	0

Thus it appears that class 1 plants produce both glucoside and enzyme "β" (and, by inference, also enzyme "α"); class 2 plants produce enzyme "β" but no glucoside (hence, by inference, no enzyme "α"); class 3 plants produce glucoside (and, therefore, enzyme "α") but no enzyme "β"; class 4 plants

FIGURE 4-4. *A 9:7 purple:yellow epistatic ratio in corn.*

produce neither enzyme "β" nor glucoside (therefore, by inference, no enzyme "α"). We can summarize our conclusions in tabular form:

Class	Production Enzyme "α"	Glucoside	Enzyme "β"	Substance Accumulating
1	+	+	+	HCN
2	0	0	+	precursor
3	+	+	0	glucoside
4	0	0	0	precursor

It therefore seems highly likely that production of enzymes "α" and "β" is determined by two different pairs of genes:

Gene A   Gene B
↓   ↓
enzyme "α"   enzyme "β"

⟶ (precursor) ⟶ cyanogenic glucoside ⟶ HCN

this isn't true class c doesn't show so 9:7
instead
of 9:3:4
when the
3
shows
as
a
colored
pigment
for
example

So an individual must have at least one dominant of each of two pairs of genes, A and B, in order to carry the process from precursor to HCN. We may now add genotypes to the cross developed on page 92 as follows:

P   positive × negative
     AABB   *aabb*

F₁   positive
            AaBb

F₂ $\frac{9}{16}$ HCN positive *A—B—* ("class 1")   (9)
      $\frac{3}{16}$ HCN negative *aaB—* ("class 2")
      $\frac{3}{16}$ HCN negative *A—bb* ("class 3")   (7)
      $\frac{1}{16}$ HCN negative *aabb* ("class 4")

A little reflection will serve to indicate that this situation may be considered in any of the following ways:

$9:7$ HCN : no HCN

$12:4$ glucoside : no glucoside

$9:3:4$ HCN : glucoside : precursor

$9:3:3:1$ enzymes "α" and "β" : enzyme "β" only :
enzyme "α" only : neither enzyme

The ratio we choose depends, of course, on the level of chemical analysis to which consideration is carried. Thus we are talking about phenotype in terms of chemical reaction and content or of enzyme production. Because enzymes are proteins in whole or in part, this, in turn, suggests again a relationship between gene and enzyme and, therefore, between protein synthesis and phenotype. This will be a promising avenue to explore later in this book.

Incidentally, an explanation for the occurrence of all positive cyanide content progeny from crosses of negative $\times$ negative (which is also reported) works out on these bases quite readily:

P   negative $\times$ negative   *treat like*
       aaBB   *AAbb*   *aax AA   BBxbb*
                                       (1 Aa) (1Bb)
F_1 positive
               AaBb   *AaBb*

The cross (P) positive $\times$ negative which produces an F_2 of 3 positive : 1 negative, referred to on page 91, genotypically would look like this:

P   positive $\times$ negative
       AABB   *aaBB*
F_1 positive   *treat*
               AaBB ———→ *AaBBx AaBB*
                              as
F_2 3 positive : 1 negative   *Aax Aa   BBxBB*
        A—BB   *aaBB*   *4 AABB : 2 AaBB : 1aaBB*
                                      3 : 1

We can recognize epistasis (which can operate in any cross involving two or more pairs of genes) *by the reduction in number of expected phenotypic classes in which two or more of the classes become indistinguishable from each other.*

Many other ratios are, of course, possible and have been reported in the literature. Additional examples are included in the problems at the end of this chapter, as well as in Table 4-2. One of these, however, requires additional explanation. In certain breeds of domestic fowl (e.g., White Leghorn), individuals are white because of a dominant color-inhibiting gene *I*. Even though birds may carry genes for color, such genes cannot be expressed in the presence of *I—*. On the other hand, such breeds as the White Silkie, are white because they are homozygous for the recessive gene *c*, which blocks synthesis of a necessary pigment precursor.

Crosses of White Leghorn (*IICC*) $\times$ White Silkie (*iicc*) produce an F_2 ratio of 13:3 as follows:

*sixteenths suggest 2 pairs of genes   geno suggests epistasis
modification of expected 9:3:3:1 ratio*

TABLE 4-2. Summary of Dihybrid Ratios in F_2 of the Cross $AABB \times aabb$

Phenotypic classes	Interaction	AABB	AABb	AaBB	AaBb	AAbb	Aabb	aaBB	aaBb	aabb
More than four phenotypic classes	A and B both incompletely dominant	1	2	2	4	1	2	1	2	1
	A incompletely dominant; B completely dominant	3		6		1	2	3		1
Four phenotypic classes	A and B both completely dominant (classic ratio)	9				3		3		1
Fewer than four phenotypic classes	aa epistatic to B and b Recessive epistasis	9				3		4		
	A epistatic to B and b Dominant epistasis	12						3		1
	A epistatic to B and b; bb epistatic to A and a Dominant and recessive epistasis	13*						3		
	aa epistatic to B and b; bb epistatic to A and a Duplicate recessive epistasis	9				7				
	A epistatic to B and b; B epistatic to A and a Duplicate dominant epistasis	15								1
	Duplicate interaction	9				6				1

* The 13 is composed of the 12 classes immediately above, plus the 1 $aabb$ from the last column.

9 *I–C–* white (because of color-inhibitor *I*)
3 *iiC–* colored
3 *I–cc* white (because of both *I* and *cc*)
1 *iicc* white (because of *cc*)

The actual color of the *iiC–* individuals depends on the presence of additional genes for particular colors. These are not shown here.

LETHAL GENES

It no doubt occurs to you that lethal genes also reduce the number of expected phenotypic classes in a two-pair cross, but the change is somewhat different from that caused by epistasis. For example, consider the case of corn where tall (*D–*) and dwarf (*dd*) phenotypes are known in addition to the green and "albino" condition described in Chapter 2. Note that the cross *DdGg* × *DdGg* will produce the classic 9:3:3:1 phenotypic ratio of seedlings, but because of the lethal effect of *gg*, this becomes 9 tall green : 3 dwarf green (that is, 3 tall:1 dwarf) after *gg* has exerted its lethal effect. In man, for example, consider the dominant lethal for Huntington's chorea (Chapter 2) together with another pair of genes, free ear lobes (*A–*) versus attached ear lobes (*aa*) (Fig. 1-1). A marriage involving two double heterozygotes, *HhAa* (where *H* represents the dominant lethal for chorea, and *h* its recessive allele for "normal"), would, statistically (or actually in collections of family data), produce a 9:3:3:1 ratio initially, which would ultimately become 3 free (normal):1 attached (normal) after the choreic individuals had died.

REFERENCES

ATWOOD, S. S., and J. T. SULLIVAN, 1943. Inheritance of a Cyanogenetic Glucoside and Its Hydrolyzing Enzyme in *Trifolium repens*. *Jour. Hered.*, **34**: 311–320.
STEWART, R. N., and T. ARISUMI, 1966. Genetic and Histogenic Determination of Pink Bract Color in Poinsettia. *Jour. Hered.*, **57**: 217–220.

PROBLEMS

4-1. How many different matings can be made in a population where only one pair of genes is considered?

4-2. If two *DdIi* Coleus plants are crossed, what fraction of the offspring will be (a) shallow irregular, (b) deep regular, (c) *DDIi*, (d) *ddII*?

4-3. How many progeny phenotypic classes result if two Coleus plants (a) of genotype *DdIiWw* are crossed; (b) heterozygous for one pair of genes (showing complete dominance) on each of its pairs of chromosomes are crossed?

4-4. What is the phenotypic ratio of the testcross (a) *DdII* × *ddii* in Coleus, (b) *DdIiWw* × *ddiiww* in Coleus?

cross between two double heterozygotes
→ 9:3:3:1 pheno ratio

4-5. How many different gamete genotypic classes are produced by the tetrahybrid *AaBbCdDd*? $2^4 = 16$

4-6. If two tetrahybrids like that of the preceding problem are crossed, how many of each of the following can be expected in the progeny: (a) phenotypic classes, $n = 16$ (b) genotypic classes? $3^n = 81$

4-7. In how many ways can gametes of two tetrahybrids (*AaBbCcDd*) be combined to form zygotes?   *see Table 4-1*   *# of possible gamete genotypes = 2^4*   *zygote = $2^4 \times 2^4 = 256$*

4-8. How many phenotypic classes are produced by a testcross where one parent is heterozygous for (a) two pairs of genes, (b) three pairs of genes, (c) four pairs of genes, (d) *n* pairs of genes?   *gametes: AB × aabb*   *Aabb* $→ 2^n$ *as same when two hetero parents are crossed*

4-9. A few of the many known genes (each on a different chromosome pair) in tomato are

P smooth-skinned fruit	*p* "peach" (pubescent fruit)
W yellow flowers	*w* white flowers
C cut leaves *(Pp Ww, Cc, h_2)*	*c* "potato" (entire leaves)
h_1 hairless stems and leaves	h_2 hairy stems and leaves

$h_1 h_2 \times h_1 h_2 = \tfrac{1}{4} h_1 h_1, \tfrac{1}{2} h_1 h_2, \tfrac{1}{4} h_2 h_2$

Recall that the h_1, h_2 pair shows incomplete dominance. A tetrahybrid smooth, yellow, cut, scattered-hair plant is self-pollinated. (a) How many phenotypic classes can occur in the progeny? (b) What fraction of the offspring can be expected to be peach, white, potato, hairless? (c) What fraction of the progeny can be expected to be peach, white, potato, with scattered hairs?   *See Table 4-1*

2×16 since incomplete dominance   *$4 \times 4 \times 4 \times 4 = 256$*   *$\tfrac{1}{4} \times \tfrac{1}{4} \times \tfrac{1}{4} \times \tfrac{1}{2}$*

4-10. You raise 100 tomato plants from seed received from a friend and find 37 red-fruited plants with scattered short hairs on stems and leaves, 19 red hairless, 18 red very hairy, 13 yellow fruited with scattered short hairs, 7 yellow very hairy and 6 yellow hairless. Suggest genotypes and phenotypes for the unknown parent plants from which the 100 seeds were obtained.   *(3:1)(1:2:1)*

4-11. Coat in guinea pigs may be either long or short; matings of short × short may produce long-haired progeny, but long × long gives rise only to long. Additionally, coat color may be yellow, cream, or white. The mating cream × cream produces progeny of each of the three colors. Given the following incomplete pedigree: *hair*

P   long yellow × short white   *ll y_1 y_1*   *LL y_2 y_2*   *So long is recessive*

F_1 all short *cream*   *y_1 y_2*   *so cream is incompletely dominant*

(a) What is the coat color in the F_1? (b) If members of the F_1 were interbred, what fraction of their progeny would be long cream?   *Ll y_1 y_2 × Ll y_1 y_2 → $\tfrac{1}{4}$ ll × $\tfrac{1}{2}$ y_1 y_2 = $\tfrac{1}{8}$*

4-12. What progeny phenotypic ratio results from the cross $AaBbc_1c_2 \times AaBbc_1c_2$ if *bb* individuals die during an early embryo stage?   *in complete dom*

4-13. In addition to the genes for flower color in snapdragon described in Chapter 2, leaves in this plant may be broad, narrow, or intermediate. From the cross red broad × white narrow this F_2 was obtained: 10 red broad, 20 red intermediate, 10 red narrow, 20 pink broad, 40 pink intermediate, 20 pink narrow, 10 white broad, 20 white intermediate, and 10 white narrow. (a) How many pairs of genes are involved and what kind of dominance is demonstrated by this case? (b) Which of these F_2 phenotypic classes can you recognize as homozygous?

4-14. The Christmas poinsettia (*Euphorbia pulcherrima*) produces colored modified leaves (bracts) below the clusters of small flowers. These bracts may be red,

a) 1:2:1:2:4:2:1:2:1 = (1:2:1)(1:2:1)
so 2 pairs of genes involved with incomplete dominance for both
b) red broad, red narrow, white broad, white narrow (in other words, all the 1's)

pink, or white. Work of Stewart and Arisumi (1966) shows that red pigmentation results from a two-step, enzymatically controlled biochemical process from a colorless precursor via an intermediate pink pigment that is converted to a red pigment in the second step. Let W represent the completely dominant gene that is responsible for production of the enzyme catalyzing the first step, from colorless to pink, and P the completely dominant gene for producing the enzyme bringing about conversion of the pink pigment to a red one. If a white-bracted plant is crossed to a doubly homozygous red-bracted one, what phenotypic ratio can be expected in the F_2?

4-15. Normal hearing depends upon the presence of at least one dominant of each of two pairs of genes, D and E. If you examined the collective progeny of a large number of $DdEe \times DdEe$ marriages, what phenotypic ratio would you expect to find?

4-16. In sweet pea the cross white flowers $\times$ white flowers produced an F_1 of all purple flowers. An F_2 of 350 white and 450 purple was then obtained. (a) What is the phenotypic ratio in the F_2? Using the first letter of the alphabet and as many more in sequence as needed, give (b) the genotype of the purple F_2, (c) the genotype of the F_1, (d) the genotypes of the two P individuals.

4-17. The fruit of the weed shepherd's purse (*Capsella bursa-pastoris*) is ordinarily heart-shaped in outline and somewhat flattened, but occasionally individuals with ovoid fruits occur. Crosses between pure-breeding heart and ovoid yield all heart in the F_1. Selfing this F_1 produces an F_2 in which 6 per cent of the individuals are ovoid. Starting with the first letter of the alphabet, and using as many more as necessary, give the genotypes of (a) ovoid, (b) F_1 heart, (c) F_2 heart.

4-18. In the mouse, $C-$ animals are pigmented, cc individuals are albino. Another pair of genes determines the difference between black $(B-)$ and brown (bb). What F_2 will be produced as a result of the cross $CCBB \times ccbb$?

4-19. In some plants cyanidin, a red pigment, is synthesized enzymatically from a colorless precursor; delphinidin, a purple pigment, may be made from cyanidin by the enzymatic addition of one $-OH$ group to the cyanidin molecule. In one cross where these pigments were involved, purple $\times$ purple produced F_1 progeny as follows: 81 purple, 27 red, and 36 white. (a) How many pairs of genes are involved? (b) What is the genotype of the purple parents? (c) What is the genotype of each of the three F_1 phenotypic classes? (Use as many letters of the alphabet as needed, starting with A.)

4-20. In terms of *enzyme production,* instead of flower color, as a phenotypic character in the data of problem 4-19, what is the F_1 phenotypic ratio? The enzyme catalyzing conversion of percursor to cyanidin may be designated enzyme 1, and that controlling the production of delphinidin from cyanidin may be designated enzyme 2.

4-21. In the plants of problem 4-19, one cross of white $\times$ red produced all purple progeny, whereas another white $\times$ red cross gave rise to 1 purple:2 white:1 red. What were the parental genotypes in each of these two crosses?

4-22. In addition to the round, oval, and long radishes mentioned in Chapter 2; radishes may be red, purple, or white in color. Red $\times$ white produces progeny all of which are purple. If purple oval were crossed with purple oval, how many pure-breeding types occur in the progeny?

4-23. In cattle, "short spine" is lethal shortly after birth; it is caused by the homozygous recessive genotype, ss. Heterozygotes are normal. A series of matings between roan animals heterozygous for the short spine gene produces what phenotypic ratio (a) at birth and (b) after several days?

4-24. In the summer squash, fruits may be white, yellow, or green. In one case, the cross of yellow × white produced an F_1 of all white-fruited plants that, when selfed, gave an F_2 segregating 12 white:3 yellow:1 green. (a) Suggest genotypes for the white, yellow, and green phenotypes. (b) Give genotypes of the P, F_1, and F_2 of this cross. Use genotype symbols starting with the first letter of the alphabet.

4-25. Summer squash fruit shape may be disk, sphere, or elongate. The cross of sphere × sphere produced an F_1 with all disk-shaped fruits. Selfing the F_1 gave 9 disk:6 sphere:1 elongate. (a) Suggest genotypes for disk, sphere, and elongate. (b) Give genotypes of the P, F_1, and F_2 of this cross. Use genotype symbols beginning with the first letter of the alphabet *after* those used for problem 4-24.

4-26. Considering the facts suggested by problems 4-24 and 4-25, how many different genotypes are responsible for the (a) yellow sphere, (b) elongate green, (c) disk, white phenotypes?

4-27. A tetrahybrid white disk plant is selfed. (a) How many phenotypic classes could occur in its F_1? (b) What fraction of the F_1 will be white disk?

4-28. In Duroc Jersey pigs, two pairs of interacting genes, R and S, are known. (a) The cross of red × red sometimes produces an F_1 phenotypic ratio of 9 red:6 sandy:1 white. What is the genotype of each of these F_1 phenotypes? (b) For each of the crosses that follow, give the parental genotypes:

		P	F_1	F_2
Case 1.	Red × red	All red	All red
Case 2.	Red × red	3 red:1 sandy	Not reported
Case 3.	Red × white	All red	9 red:6 sandy:1 white
Case 4.	Sandy × sandy	All red	9 red:6 sandy:1 white
Case 5.	Sandy × sandy	1 red:2 sandy:1 white	Not reported.

4-29. Determine the genotypic and phenotypic ratios resulting from each of the following dihybrid crosses (assume lethals to exert their effect during early embryo development):

Parental	Gene Characteristics		Progeny Ratios	
Genotypes	First Pair	Second Pair	Genotypic	Phenotypic
(a) $AaBb \times AaBb$	Complete Dominance	Complete Dominance		
(b) $Aab_1b_2 \times Aab_1b_2$	Complete Dominance	Incomplete Dominance		

(c) $a_1a_2b_1b_2 \times a_1a_2b_1b_2$	Incomplete Dominance	Incomplete Dominance	$(1:2:1)(1:2:1)$ $(1:2:1)(1:2:1)$ $1:2:1:2:4:2$ $\to same$ $1:2:1$
(d) $AaBb \times AaBb$	Complete Dominance	Recessive Lethal	$(1:2:1)(1:2\times)$ $(3:1)(1)=$ $=1:2:1:2:4:2$ $3:1$
(e) $a_1a_2Bb \times a_1a_2Bb$	Incomplete Dominance	Recessive Lethal	$(1:2:1)(1:2:1)$ $(1:2:1)(1)=$ $1:2:1:2:4:2$ $1:2:1$
(f) $AaBb \times AaBb$	Recessive Lethal	Recessive Lethal	$(1:2\times)(1:2\times)$ $(1)(1)=$ $=1:2:2:4$ (1)

assuming bb dies not BB

4-30. In a hypothetical flowering plant assume petal color to be due to a pair of co-dominant genes, a_1 and a_2, heterozygotes being purple, and homozygotes either red (a_1a_1) or blue (a_2a_2). Assume also another pair of genes, completely dominant B for color, and recessive b for color inhibition. This pair is not linked to the a_1, a_2 pair. Two doubly heterozygous purple plants, a_1a_2Bb, are crossed. What phenotypic ratio results in the progeny?

remember the inhibited are white

$$a_1a_2 \, Bb \times a_1a_2 \, Bb$$

$$\left(\tfrac{1}{4} \, a_1a_1 \, \tfrac{1}{2} \, a_1a_2 \, \tfrac{1}{4} \, a_2a_2\right)\left(\tfrac{1}{4} BB : \tfrac{1}{2} Bb : \tfrac{1}{4} bb\right)$$

$$\left(\tfrac{1}{4} red : \tfrac{1}{2} purple : \tfrac{1}{4} blue\right)\left(\tfrac{3}{4} color : \tfrac{1}{4} color\ inhibition\right)$$

$$\tfrac{3}{16} red\ color \qquad \tfrac{3}{8} purple\ color \qquad \tfrac{3}{16} blue\ color$$

$$+ \left(\tfrac{1}{16} + \tfrac{1}{8} + \tfrac{1}{16} inhibited\right)$$

3 red
6 purple
3 blue
4 white

CHAPTER 5
Probability and Goodness of Fit

AN understanding of the laws of probability is of fundamental importance in (1) appreciating the operation of genetic mechanisms, (2) predicting the likelihood of certain results from a given cross, and (3) assessing how well a progeny phenotypic ratio fits a particular postulated genetic mechanism. We have already been applying one of the fundamental laws of probability in our study of dihybrid ratios in the preceding chapter, based on transmission of two pairs of genes that are assumed to be on separate chromosome pairs. That law is the law of the probability of coincident independent events, and states, in essence, that **the chance (or probability) of the simultaneous occurrence of two or more independent events is equal to the product of the probabilities that each will occur separately.** Application of this "product law of probability" to genetics permits elucidation of the three points referred to.

Two Independent, Nongenetic Events

SINGLE-COIN TOSSES

The laws of probability can be applied to *any* chance or random event. For example, a coin tossed into the air and allowed to come to rest is likely to land either "heads" or "tails," if we neglect the highly improbable chance of its landing on edge. We would, therefore, predict (on the basis of certain implicit assumptions to be detailed) that, in the total array of possibilities (heads plus tails), the probability of a "head" is 1 in 2, or $\frac{1}{2}$, and the same for a "tail." But if we toss one coin four successive times, we would not be surprised to get some ratio other than 2 heads : 2 tails. In addition to that possibility, one should expect, on occasion, to get such a combination as four heads, or three heads and one tail, or one head and three tails, or even four tails. If, however, a very large number of tosses were to be made, we would expect to come quite close to a 1 : 1 ratio. Note that we are assuming successive tosses to be independent of each other; that is, the result of one toss has no effect on any succeeding toss.

TWO-COIN TOSSES

What can we expect if we toss two coins simultaneously for, say, 50 tosses? Let us approach this problem through some actual situations. Two students were each asked to toss two coins simultaneously 50 times by shaking the coins in closed, cupped hands and letting them fall lightly on a table. Ignoring the possibility of a coin landing "on edge," only two results can occur: each coin will come to rest flat on the table, with either one side (which we elect to call heads) up or the other side (which we elect to call tails) up. With two coins, then, the outcome of any single toss will be: HH, HT, or TT. The result of the series of tosses by student A was

$$
\begin{array}{ll}
\text{HH} & 12 \\
\text{HT} & 27 \\
\text{TT} & 11
\end{array}
$$

Now one has to ask himself, "Are these the results I could have expected, based on the theory that each coin has an equal chance of landing either heads or tails?" Or, in a genetic experiment, having made a given cross and gotten certain results, "What possible mechanism could be operating in order to produce these results?" In either a coin toss or a genetic breeding experiment, one of the first necessary steps is to formulate certain hypotheses to predict or explain results, then ask oneself whether, *in terms of those hypotheses, deviation from the predicted result is within limits set by chance alone?* If so, sufficient credence can be placed in the hypothesis to use it in predicting outcomes of untried cases; if not, we will have to alter our hypotheses.

In order to make a judgment on this question, one must make several **simplifying assumptions** about the coins and the general conditions surrounding the experiment itself. We would have to ask ourselves: *were the coins themselves unbiased;* that is, were they so constructed physically that each coin had an equal chance of landing heads or tails? Because we have no real evidence to the contrary, let us assume, at least for the time being, that our coins are indeed unbiased. If this assumption should turn out not to be true, then our ultimate evaluation of the results will have to be changed.

We must also examine the question of whether the "headness" or "tailness" of the fall of one of the two coins will have any effect on the fall of the second coin. That is to say, if two coins are tossed simultaneously, does *each* one of them have an *equal* chance of coming to rest heads or tails? Certainly in this experiment there should be no more than a remote chance that the ultimate fall of one coin affects the other. Therefore our *second assumption* will be that *the coins themselves are independent of each other.*

A *third assumption* we will have to make, at least at the outset, is that *successive tosses (events) are also independent of each other.* This would imply, for example, that having gotten two heads the first toss does not in any way affect the outcome of the second or any other toss.

If, then, we assume for the present that the **coins are (1) unbiased and (2) independent of each other and also that (3) the several trials or tosses are likewise independent of each other,** what kind of results would we *expect* to get? Obviously, each coin has one chance in two, or a probability of $\frac{1}{2}$, of coming to rest heads and a probability of $\frac{1}{2}$ of landing tails. By the product law of probability, the chance of *both* coins showing heads in the same toss is $\frac{1}{2} \times \frac{1}{2}$, or $\frac{1}{4}$. We see our student did, indeed, get two heads in almost exactly one fourth of his tosses.

Likewise, the probability of having two tails simultaneously is also $\frac{1}{2} \times \frac{1}{2}$ $= \frac{1}{4}$. This result was not quite obtained in the example under consideration.

What should we expect with regard to one head and one tail? The chance of coin 1 landing heads is $\frac{1}{2}$; the chance of coin 2 coming to rest tails is also $\frac{1}{2}$. Now $\frac{1}{2} \times \frac{1}{2} = \frac{1}{4}$, but the total array of possibilities thus arrived at ($\frac{1}{4}$ HH $+$ $\frac{1}{4}$ TT $+ \frac{1}{4}$ HT) equals only three fourths, leaving one fourth of the possibilities unaccounted for. Note that the chance of one head plus one tail is really the sum of the probability of coin 1 being heads and coin 2 being tails, *plus* the probability of coin 1 being tails and coin 2 being heads; i.e., our HT category is really HT $+$ TH, or 2 HT. Therefore the probability of one head and one tail is $2(\frac{1}{2} \times \frac{1}{2})$, or $\frac{1}{2}$.

The general statement for two independent events of known probability may be written as

$$a^2 + 2ab + b^2$$

where a represents the probability of a head and b the probability of a tail. If, as in this instance, a and b each equal $\frac{1}{2}$, then the value of this expression upon substitution becomes $(\frac{1}{2})^2 + 2(\frac{1}{2} \times \frac{1}{2}) + (\frac{1}{2})^2$ or $\frac{1}{4} + \frac{2}{4} + \frac{1}{4} = 1$. Notice that the total array of probabilities here is 1, and also that $a^2 + 2ab + b^2$ is the expansion of the binomial $(a + b)^2$.

Our observations and expectations for a two-coin toss can therefore be summarized as follows:

Class	Observed	Expected
HH	12	12.5
HT	27	25.0
TT	11	12.5
	50	50.0

Student B's results, however, were a little different:

Class	Observed	Expected
HH	10	12.5
HT	33	25.0
TT	7	12.5
	50	50.0

Obviously the first set of data is closer to the results expected with our assumptions, yet the second set of results was actually obtained. We shall shortly examine each of these sets of data to see if the departure from the expected results is too great for *chance alone* to have operated within the framework of the three assumptions set forth on page 102.

FOUR-COIN TOSSES

But let us first examine an experiment involving four coins tossed together 100 times. The possible combinations of heads and tails, and the results actually obtained in a particular trial, are as follows:

Class	Observed
HHHH	9
HHHT	32
HHTT	29
HTTT	25
TTTT	5

Having found that the expansion of $(a + b)^2$ enables us to determine a set of (ideal) expected results in a two-coin toss, let us try, for the four-coin toss, expanding $(a + b)^4$ to determine the expectancy. Again, let

$$a = \text{probability of a head for any coin} = \tfrac{1}{2}$$

and

$$b = \text{probability of a tail for any coin} = \tfrac{1}{2}$$

Expanding $(a + b)^4$, we get

$$\underset{4H's}{a^4} + \underset{3H's}{4a^3b} + \underset{2H's}{6a^2b^2} + \underset{1H}{4ab^3} + \underset{0H's}{b^4}$$

Substituting the numerical values of a and b, we arrive at

$$(\tfrac{1}{2})^4 + 4[(\tfrac{1}{2})^3 \cdot \tfrac{1}{2}] + 6[(\tfrac{1}{2})^2 \cdot (\tfrac{1}{2})^2] + 4[(\tfrac{1}{2}) \cdot (\tfrac{1}{2})^3] + (\tfrac{1}{2})^4 = 1$$

or

$$\tfrac{1}{16} + \tfrac{4}{16} + \tfrac{6}{16} + \tfrac{4}{16} + \tfrac{1}{16} = 1$$

The first term of the expression, $(\tfrac{1}{2})^4$, gives us the probability of all four coins coming up heads simultaneously; the second term the probability of three heads (a^3) plus one tail (b), and so on. We can now compare our observed results with these calculated results for expectancy in a four-coin toss:

Class	Observed	Calculated	
HHHH	9	6.25	($= \tfrac{1}{16}$ of 100)
HHHT	32	25.00	($= \tfrac{4}{16}$ of 100)
HHTT	29	37.50	($= \tfrac{6}{16}$ of 100)
HTTT	25	25.00	($= \tfrac{4}{16}$ of 100)
TTTT	5	6.25	($= \tfrac{1}{16}$ of 100)
	100	100.00	

Again we are faced with the problem of determining *how well* our observations compare with the results expected on the basis of our three assumptions with only chance deviations—that is, whether we should accept our results as chance expressions of two- and four-coin tosses wherein the coins are unbiased and independent of each other, as are successive tosses. We shall probe this question very shortly.

The Binomial Expression → *The prob of the occurrence simultaneous of n independent events*

Answers to many genetic problems involving "either–or" situations, including those with which members of the medical or legal profession must deal from time to time, are easily provided by the binomial approach. Therefore we should be certain we understand the method of binomial expansion before going on. Rather than multiplying out algebraically, one can remember the following simple rules for expansion of $(a + b)^n$:

1. *The power of the binomial chosen, that is, the value of* n, *is determined by the number of coins, individuals, and so on.* Thus for two coins the expression $(a + b)^2$ is used, even if the coins are tossed 50 times; for a four-coin toss, $(a + b)^4$ is used; and so on. The same principle applies to families of *n* children displaying one or the other of two phenotypic expressions.
2. *The number of terms in the expansion is* n + 1. Thus $(a + b)^2$ expands to $a^2 + 2ab + b^2$, having three terms, $(a + b)^4$ expanded contains five terms, and so on.
3. *Every term of the expansion contains both* a *and* b. The power of *a* in the first term equals *n*, the power of the binomial, and descends in units of 1 to 0 in the last term (and a^0, which equals 1, is not written in); likewise, *b* increases from b^0 (not written) in the first term to b^n in the last. Furthermore, the sum of the powers *in each term* equals *n*. Check this against the expansion of $(a + b)^4$ just used.
4. *The coefficient of the first term is* 1 (not written); the coefficient of the second term is found by multiplying the coefficient of the preceding term by the exponent of *a* and dividing by the number indicating the position of that preceding term in the series:

coefficient of next term =

$$\frac{\text{coefficient of preceding term} \times \text{exponent of preceding term}}{\text{ordinal number of preceding term}}$$

The same principle may be represented graphically in the form of Pascal's triangle, where each coefficient is shown as the sum of two numbers immediately above. Note that the second coefficient in each horizontal line also represents the power of the binomial. Thus the second number of the last line is 6, so that this line represents the series of coefficients in the expansion of $(a + b)^6$.

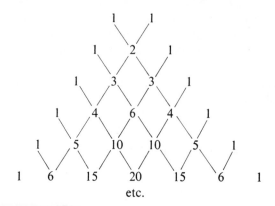

etc.

Thus, in expanding $(a + b)^4$, the first term is a^4 (for $1a^4b^0$). The second is $4a^3b$; i.e., from the preceding term, a^4,

$$\frac{4 \times 1}{1} = 4$$

which is the coefficient of the second term. The third term, which from rule 3 we know contains a^2b^2, becomes $6a^2b^2$ in the same way:

$$\frac{4 \times 3}{2} = 6$$

and so on.

In this way we obtain the binomial expansions listed in Table 5-1.

TABLE 5-1. Expansions of Binomials

$(a + b)^1$	$a + b$
$(a + b)^2$	$a^2 + 2ab + b^2$
$(a + b)^3$	$a^3 + 3a^2b + 3ab^2 + b^3$
$(a + b)^4$	$a^4 + 4a^3b + 6a^2b^2 + 4ab^3 + b^4$
$(a + b)^5$	$a^5 + 5a^4b + 10a^3b^2 + 10a^2b^3 + 5ab^4 + b^5$
$(a + b)^6$	$a^6 + 6a^5b + 15a^4b^2 + 20a^3b^3 + 15a^2b^4 + 6ab^5 + b^6$

Another useful way of thinking of binomial expansions is to remember that the expression $(a + b)^2$ means, of course, $(a + b) \times (a + b)$:

$$\begin{array}{r} a + b \\ \times \quad a + b \\ \hline a^2 + ab + ab + b^2 \end{array}$$

Collecting like terms from this last expression gives us $a^2 + 2ab + b^2$. Similarly, $(a + b)^3$ can be thought of as $(a + b) \times (a + b) \times (a + b)$, or

$$\begin{array}{r} a^2 + 2ab + b^2 \\ \times \quad a + b \\ \hline a^3 + 2a^2b + ab^2 + a^2b + 2ab^2 + b^3 \end{array}$$

which may be rewritten as

$$a^3 + 3a^2b + 3ab^2 + b^3$$

If a and b are each equal to $\frac{1}{2}$, symmetrical *probability curves*, as shown in Figure 5-1, are produced. On the other hand, if $a = \frac{3}{4}$ and $b = \frac{1}{4}$, a skewed curve is obtained as indicated in Figure 5-2.

GENETIC APPLICATIONS OF THE BINOMIAL

Just as we have utilized the binomial to ascertain expectancies in coin tosses, we can also employ the same method to determine the probability of children showing a given heritable trait in a particular family.

For instance, ptosis (drooping eyelids) is an inherited trait in which affected persons are unable to raise the eyelids, so that only a relatively small space between upper and lower lids is available for vision, giving them a "sleepy" appearance. Most pedigrees indicate ptosis to be due to an autosomal dominant. Suppose a young man with ptosis whose father also displayed the trait but whose mother did not wishes to marry a woman with normal eyelids. They consult a physician to determine the likelihood of the occurrence of the defect in their children. If, for example, they plan to have four children, what is the probability of three of those being normal and one having ptosis?

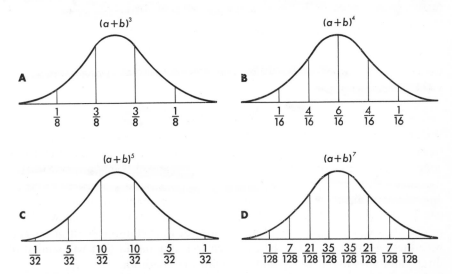

FIGURE 5-1. *Probability curves for binomial expansions where $a = b = \frac{1}{2}$. (A) $(a + b)^3$; (B) $(a + b)^4$; (C) $(a + b)^5$; (D) $(a + b)^7$.*

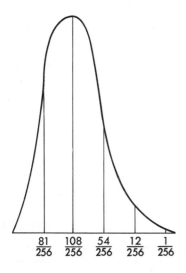

$$\frac{81}{256} \quad \frac{108}{256} \quad \frac{54}{256} \quad \frac{12}{256} \quad \frac{1}{256}$$

FIGURE 5-2. *Skewed curve for expansion of the binomial $(a + b)^4$ where $a = \frac{3}{4}$ and $b = \frac{1}{4}$.*

From the known facts, it is evident that the young man is heterozygous because he has ptosis (letting P represent the trait, his phenotype alone tells us he is P—), although his mother was normal and, therefore, pp. So having the trait and having necessarily received a recessive gene from his mother, he must be Pp. The woman, on the other hand, is pp, as determined by her normal phenotype. So, we are faced with the cross $Pp \times pp$. This is recognized as a testcross, so that the probability of normal children is $\frac{1}{2}$. But what is the chance of a family of *three normal children and one affected child* if they have four? The expansion of the binomial $(a + b)^4$ will provide the answer. If we let

$$a = \text{probability of a normal child} = \tfrac{1}{2}$$
$$b = \text{probability of a child with ptosis} = \tfrac{1}{2}$$

we see that the second term, $4a^3b$, of the expansion will, upon substitution, yield the information:

$$4a^3b = 4[(\tfrac{1}{2})^3 \times \tfrac{1}{2}] = \tfrac{4}{16} = \tfrac{1}{4}$$

We choose the second term ($4a^3b$) in this expansion because it contains a^3 (representing three children of the phenotype denoted by a) and b (for one child of the phenotype represented by b). Therefore there is a probability in this case of 1 in 4, or 0.25, that *if they have four children,* three will be normal and one will have ptosis. Similarly, there is one chance in sixteen (a^4), or 0.0625, that none of the four will exhibit ptosis or (b^4) that all four will have ptosis. Or it can be said that, of all families of four children born of parents with these genotypes, it is to be expected that one out of four will consist of three children without ptosis and one with the condition.

On the other hand, consider a young man and woman, each with ptosis, whose pedigrees indicate that *each* is heterozygous. If such a couple should

Pp × Pp

$\frac{1}{4}PP : \frac{1}{2}Pp : \frac{1}{4}pp$

marry and have four children, what is the probability of three being normal and one having ptosis? We can represent the potential parents as *Pp* × *Pp*. From previous experience with monohybrid crosses of two heterozygous individuals, we know that, in cases of complete dominance, the expected progeny phenotypic ratio is 3:1. That is, we should expect the probability here of an affected child to be $\frac{3}{4}$, and of a normal one to be $\frac{1}{4}$. As in the preceding example, let *since the condition is dominant*

a = probability of a normal child = $\frac{1}{4}$
b = probability of a child with ptosis = $\frac{3}{4}$

Because we are again concerned with a family of four, we once more use the binomial $(a + b)^4$. From the expansion of this binomial we must again substitute in the second term and solve:

$$4a^3b = 4[(\tfrac{1}{4})^3 \times \tfrac{3}{4}] = \tfrac{12}{256} \;=\; \frac{3}{64}$$

So with two heterozygous parents the probability of three normal children and one affected child is $\frac{12}{256}$, or 0.047. This is quite different from the 0.25 proability for the preceding case.

Had we been concerned with, say, two normal and two affected children, we would have utilized the third term, $6a^2b^2$, in the expansion of $(a + b)^4$. In any such problem it is imperative to select both a binomial to the proper power and the correct term within the expansion. The former is determined by the number of events (children, heads, and so on), the latter by the number of individuals of each of the two alternative types (e.g., affected versus normal, heads versus tails). Of course, ptosis, in itself, is not a serious departure from "normal," but this method of calculating probability might be of considerable importance to parents, both of whom are heterozygous for some recessive disabling or lethal gene.

PROBABILITY OF SEPARATE OCCURRENCE OF INDEPENDENT EVENTS

Just as the probability of the simultaneous occurrence of two independent events is the product of their separate probabilities, **the probability of the separate occurrence of either of two independent events equals the square root of their simultaneous occurrence,** provided, of course, the two events are of equal probability. In the preceding section we saw that the probability of tossing two heads simultaneously is $\frac{1}{4}$. The probability of tossing a head in one toss is $\sqrt{\frac{1}{4}}$, or $\frac{1}{2}$. $= \sqrt{a^2} = a = \frac{1}{2}$

This same approach can be applied in genetics. Suppose we wish to determine the frequency of the gene for albinism in a particular human population. Studies of Magnus (1922) and of Froggatt (1960) indicate that about 1 in 10,000 babies in Norway and in Ireland is an albino, a condition due to a recessive autosomal gene. Each albino child represents the simultaneous occurrence of independent events of equal frequency, namely, the union of two gametes

1st term of binomial

each of which carries the recessive gene. Inasmuch as the frequency of albinos in this case is 0.0001, the probability that any given gamete in the population carries the gene for albinism is equal to $\sqrt{0.0001}$, or 0.01. Because this gene is one member of a *pair* of alleles whose total frequency, therefore, must be 1, the frequency of the dominant gene for normal pigmentation is $1 - 0.01 = 0.99$.

Having these two values, we can readily calculate the frequencies of the three possible genotypes. Let

a = frequency of the gene for normal pigmentation (A) = 0.99
b = frequency of the gene for albinism (a) = 0.01

Substituting in the expansion of $(a + b)^2 = 1$ provides genotypic frequencies:

binomial
expansion: $\qquad a^2 \qquad + \qquad 2ab \qquad + \qquad b^2$
genotype: $\qquad AA \qquad\qquad\qquad Aa \qquad\qquad\qquad aa$
frequency: $(0.99)^2 = \mathbf{0.9801} + 2(0.99 \times 0.01) = \mathbf{0.0198} + (0.01)^2 = \mathbf{0.0001}$

Determining "Goodness of Fit"

NONGENETIC EVENTS

In our discussion of calculating expected results of coin tosses, we raised one as yet unanswered question: *In terms of our hypotheses, are the results obtained valid by chance alone?* This question implies that we do not expect to achieve results *exactly* equal to our calculations very often, and raises the problem of how much deviation from our expectations we can accept as likely being due purely to chance. Reference to Table 2-1 shows that even Mendel's results, involving hundreds or thousands of individuals, did not reflect *exactly* the expected ratios (although they are surprisingly close!). In other words, we need a mathematical tool to determine "goodness of fit." We have such a tool in the chi-square (χ^2) test.

To understand how to use this important statistical test, let us re-examine our earlier coin tosses. Recall our two sets of two-coin tosses made by students A and B:

	A's Tosses			B's Tosses	
Class	Observed	Calculated	Class	Observed	Calculated
HH	12	12.5	HH	10	12.5
HT	27	25.0	HT	33	25.0
TT	11	12.5	TT	7	12.5

Obviously, A's tosses are closer to our expected results, but they do not correspond exactly. Do we accept both sets of results, or just A's, or neither, as valid reflections of our assumptions regarding the coins and tosses (page 102)? That is, are the deviations shown in each case to be expected when our assumptions are correct and complete, and only chance is operating? How much deviation from results expected under a given set of assumptions can we tolerate as representing merely chance departures from our calculated expectancy? The chi-square test will enable us to make a judgment in this question.

The formula for calculating chi-square is

$$\chi^2 = \Sigma \left[\frac{(o - c)^2}{c} \right]$$

where o = observed frequencies, c = calculated frequencies, and Σ indicates that the bracketed quantity is to be summed for all classes. Both o and c must be calculated in actual numbers and not in percentages.

To calculate chi-square for student A's coin tosses (1:2:1 expectation), his data may conveniently be set up as follows:

Class	Observed o	Calculated c	Deviation $o - c$	Squared Deviation $(o - c)^2$	$\frac{(o - c)^2}{c}$
HH	12	12.5	−0.5	0.25	0.02
HT	27	25.0	+2.0	4.00	0.16
TT	11	12.5	−1.5	2.25	0.18
TOTALS	50	50.0	0		$\chi^2 = 0.36$

With the value $\chi^2 = 0.36$, our question can now be stated as, "How often, *by chance alone*, will we find a deviation of this much or more when we expect a 1:2:1 ratio?" or, "How often, by chance, can we expect a value of $\chi^2 \geq 0.36$?" If this probability is quite high, then we will be able to accept our original assumptions (page 102) as valid and also the thesis that the observed deviation was produced only by chance.

The answer to our question may be obtained by consulting a table of chi-square (Table 5-2).

To use the table it is necessary only to know the "degrees of freedom" operating in any particular case. This is one less than the number of classes involved and represents the number of *independent* classes that contribute to the calculated value of χ^2. In A's coin tosses, two of the classes may have any value (between 0 and 50), but once we have values for these two, the third is automatically determined as the difference between the total of all classes and the total of all other classes. The values of P across the top in the table indicate the probability of obtaining a value of chi-square (and, therefore, a deviation) as large or larger, *purely by chance.*

In Table 5-2, then, with two degrees of freedom, we read across until we find either our value of χ^2 or two which "bracket" it. In the case at hand, we

TABLE 5-2. Table of Chi-Square

Degrees of Freedom	P = 0.99	0.95	0.80	0.70	0.50	0.30	0.20	0.05	0.01
1	0.00016	0.004	0.064	0.148	0.455	1.074	1.642	3.841	6.635
2	0.0201	0.103	0.446	0.713	1.386	2.408	3.219	5.991	9.210
3	0.115	0.352	1.005	1.424	2.366	3.665	4.642	7.815	11.341
4	0.297	0.711	1.649	2.195	3.357	4.878	5.989	9.488	13.277
5	0.554	1.145	2.343	3.000	4.351	6.064	7.289	11.070	15.086
6	0.872	1.635	3.070	3.828	5.348	7.231	8.558	12.592	16.812
7	1.239	2.167	3.822	4.671	6.346	8.383	9.803	14.067	18.475
8	1.646	2.733	4.594	5.527	7.344	9.524	11.030	15.507	20.090
9	2.088	3.325	5.380	6.393	8.343	10.656	12.242	16.919	21.666
10	2.558	3.940	6.179	7.267	9.342	11.781	13.442	18.307	23.209

Taken from Table 3 of Fisher, *Statistical Methods for Research Workers,* published by Oliver and Boyd, Ltd., Edinburgh, by permission.

find that our χ^2 value of 0.36 does not appear in the table, but for two degrees of freedom we do find two values, 0.103 and 0.446, between which it lies. Reading up to values of P, we see that $\chi^2 = 0.36$ corresponds to a probability value of between 0.95 and 0.80. This means that, for an expectancy of 1:2:1, we can expect a deviation as large as or larger than we experienced in something between 80 and 95 per cent of repeated trials. Such a deviation could, therefore, readily be due to chance, and both our expectancy and the assumptions on which it was based appear good. That is, we have a good fit between observed results and our calculated expectancy.

In the same way, let us calculate χ^2 for B's coin tosses (1:2:1 expectation):

Class	Observed o	Calculated c	Deviation $o - c$	Squared Deviation $(o - c)^2$	$\dfrac{(o - c)^2}{c}$
HH	10	12.5	−2.5	6.25	0.50
HT	33	25.0	+8.0	64.00	2.56
TT	7	12.5	−5.5	30.25	2.42
TOTALS	50	50.0	0		$\chi^2 = 5.48$

In Table 5-2, it is seen that we may expect, by chance, a value of $\chi^2 \geq 5.48$ in between 5 and 20 per cent of such trials. This is obviously not as good a fit between observed and calculated results as obtained by student A, but is it close enough to accept? In answer, statisticians usually choose the 5 per cent probability value as the significance level. In other words, whenever a χ^2 value is equal to or greater than that for a probability value of 0.05, the deviation is considered significant and the likelihood of such a large discrepancy arising by chance too low for acceptance. At this level, only 1 in 20 such trials will produce this large a deviation by chance. A chi-square value equal to or greater than that for a probability of 0.05, that is, large enough to indicate a probability of 0.05 or less, does *not* tell us that the large deviation necessary to produce such a high value of chi-square could *not* occur purely by chance, only that it is highly unlikely to be due to chance alone. In our two-coin tosses, then, values of χ^2 up to (but not including) 5.991 indicate a sufficient probability of chance alone to accept, but larger values would require a careful review of the assumptions that led us to, in this case, a 1:2:1 expectancy.

In the same way we calculate χ^2 for the four-coin toss reported on page 104 to be 5.35. Table 5-2 shows that, for this value of χ^2 and four degrees of freedom (remember, five different combinations are possible), P = 0.30 to 0.20, which shows our value of χ^2 to be well below the level of significance.

Note that χ^2 does not *tell* us that our results do or do not fit our theory, but this test does permit a judgmental answer regarding goodness of fit.

GENETIC APPLICATIONS OF CHI-SQUARE

Recall the case of hydrocyanic acid in clover discussed in Chapter 4. A series of crosses (page 92) of two parental strains, one producing this substance in its leaves, the other not doing so, gave rise to an F_2 of 351 HCN:256 no HCN. Although the 607 F_2 individuals reported by Atwood and Sullivan resulted from some 23 different crosses, it will suffice for our purposes to treat them as though they were all sister progeny of the same cross. Chi-square calculation for a 3:1 expectancy gives the value $\chi^2 = 95.49$:

Class	o	c	$o - c$	$(o - c)^2$	$\dfrac{(o - c)^2}{c}$
HCN positive	351	455.25	− 104.25	10,868	23.87
HCN negative	256	151.75	+ 104.25	10,868	71.62
TOTALS	607	607.00	0.0		$\chi^2 = 95.49$

Reference to Table 5-2 shows that, for one degree of freedom, such a value is highly significant, being likely to occur by chance in far fewer than one in a hundred trials. While the great deviation that produces this extremely high value of chi-square is not impossible when one expects a 3:1 ratio, it is so improbable that we prefer to re-examine our assumptions. For a 3:1 ratio, these would, of course, include the concept of a single pair of genes with one allele completely dominant, or (for a 12:4 ratio) two pairs in which, let us say, *A* and *a* are both epistatic to *B* and *b*, though *A* is dominant to *a*:

$$\left.\begin{array}{l} 9 \ A{-}B{-} \\ 3 \ A{-}bb \end{array}\right\} \quad 12 \text{ if } A \text{ is epistatic to } B \text{ and } b$$

$$\left.\begin{array}{l} 3 \ aaB{-} \\ 1 \ aabb \end{array}\right\} \quad 4 \text{ if } a \text{ is epistatic to } B \text{ and } b \text{ but recessive to } A$$

Since 351:256 is considerably closer to a 9:7 expectancy, it is worth determining χ^2 for this ratio:

Class	o	c	$o - c$	$(o - c)^2$	$\dfrac{(o - c)^2}{c}$
HCN positive	351	341.4	+9.6	92.16	0.27
HCN negative	256	265.6	−9.6	92.16	0.35
TOTALS	607	607.0	0.0		$\chi^2 = 0.62$

With one degree of freedom, Table 5-2 indicates, for this value of χ^2, a probability of between 0.30 and 0.50. This value of χ^2 is well below the level of significance, and interpolation in Table 5-2 indicates that such a value of χ^2 will occur in slightly more than 44 per cent of similar trials when the assumptions underlying a 9:7 ratio are operating. Therefore, until conflicting data may be turned up, these assumptions are acceptable. As described in Chapter 4, these include the concept of two pairs of genes with "duplicate recessive epistasis" where only the $A-B-$ individuals are phenotypically distinguishable from other possible genotypes resulting from the crosses studied.

   The chi-square test is a very useful one for obtaining an objective approximation of goodness of fit, but it is reliable only when the observed or expected frequency in any class is five or more. Its proper use in genetic situations can, as we have seen, shed considerable light on the mechanisms operating in particular crosses.

REFERENCES

DANKS, D. M., J. ALLAN, and C. M. ANDERSON, 1965. A Genetic Study of Fibrocystic Disease of the Pancreas. *Ann. Hum. Genet.*, **28**: 323–356.
FROGGATT, P., 1960. Albinism in Northern Ireland. *Ann. Hum. Genet.*, **24**: 213–238.
MAGNUS, V., 1922. Undersøkelser over albinisme hos mennesket. *Norsk Mag. Laegevidenskaben*, **83**: 509–513.

PROBLEMS

5-1. In tossing three coins simultaneously, what is the probability, in one toss, of (a) three heads, (b) two heads and one tail?

5-2. A couple have two girls and are expecting their third child. They hope it will be a boy. What is the probability that their wish will be realized?

5-3. Another couple have eight children, all boys. What is the chance that their ninth child would be another boy?

[handwritten top margin: 1) Perform cross & calculate probabilities 2) Then, plug in binomial expression]

[handwritten: $p \leq .05 \rightarrow$ deviation is significant (unlikely, due to chance alone)]

5-4. What is the probability of getting (a) a 5 with a single die, (b) a 5 on each of two dice thrown simultaneously, (c) any combination totaling 7 on two dice thrown simultaneously?

5-5. In crossing two heterozygous deep *Coleus* plants of genotype *DdIi*, what is the probability in the F_1 of (a) *DdIi*, (b) *ddII*, (c) deep irregular, (d) shallow irregular? *[handwritten: Simultaneous occurring ↑ I just did this the old way at first]*

5-6. In crossing two *Coleus* plants of genotype *DdIi*, what is the probability in the F_1 of (a) *DdIi*, (b) *ddII*, (c) deep irregular, (d) shallow irregular?

5-7. (a) Give the third term in the expansion of $(a + b)^8$. (b) If $a = b = \frac{1}{2}$, what is the numerical value of this term? *[handwritten: I get $\frac{7}{64}$]*

5-8. Astigmatism is a vision defect produced by unequal curvature of the cornea, causing objects in one plane to be in sharper focus. It results from a dominant gene. Wavy hair appears to be the heterozygous expression of a pair of alleles for straight (h_1) or curly hair (h_2). A wavy-haired woman who has astigmatism, but whose mother did not, marries a wavy-haired man who does not have astigmatism. What is the probability that their first child will be (a) curly-haired and nonastigmatic, (b) wavy-haired and astigmatic? (c) How many different phenotypes could appear in their children with respect to these hair and eye conditions? *[handwritten: I just did this problem the old way]*

5-9. Free ear lobes ($A-$) and clockwise whorl of hair on the back of the head ($C-$) are dominant to attached lobes and counterclockwise whorl, respectively. A husband and wife know their genotypes to be *AaCc* and *aaCc*. They expect to have five children. What is the probability that three will be "free-clockwise" and two "attached-counterclockwise"? *[handwritten: so $a + b$ here $= 5$, doesn't have to $= 1$]*

5-10. Red hair (*rr*) and possession of two whorls of hair (*ww*) on the back of the head both appear to be inherited as recessive traits in most pedigrees. (a) How many times in families of three children, where both parents are *RrWw* (nonred-haired, one whorl), will these consist of one red-haired boy with two whorls and two nonred-haired girls with one whorl? (b) Does $a + b = 1$ in this case? Why?

5-11. Multiple telangiectasia in man is the heterozygous expression of a gene that is lethal when homozygous. Heterozygotes have enlarged blood vessels of face, tongue, lips, nose, and/or fingers and are subject to unusually frequent, serious nose bleeding. Homozygotes for the trait have many fragile and abnormally dilated capillaries; because of severe, multiple hemorrhaging these individuals die within a few months after birth. Two heterozygotes married forty years ago and now have four grown children. What is the probability that two of these are normal and two have multiple telangiectasia? *[handwritten: $6a^2b^2 = 6\left(\frac{2}{3}\right)^2\left(\frac{1}{3}\right)^2 = \frac{24}{81}$]*

5-12. A study by Danks et al. (1965) shows that, in Australia, four in 10,000 live births is an individual who has cystic fibrosis of the pancreas, an inherited recessive metabolic defect in digestion of fats, which is fatal in children homozygous for the gene. This figure is similar to that of other studies in the United States. (a) What is the probability that any given gamete in these populations carries this recessive gene? (b) What is the expected frequency of heterozygotes in these populations? *[handwritten: a) $\sqrt{.0004} = .02$ b) freq of dominant allele is .98 $(a+b)^2 = a^2 + 2ab + b^2$ with $2ab = 2(.98)(.02) = .0392$]*

5-13. A certain cross yields a progeny ratio of 210:90. Chi-square for a 2:1 expectancy is 1.5, and that for a 3:1 expectancy is 4.0. (a) How many degrees of free-

[handwritten bottom: $p \leq .05$ so deviation sig ; recessive lethal since reduced from 1:2:1]

[handwritten bottom left: Don't understand]

dom are there in this case? (b) Is deviation significant in either case? (c) What genetic explanation would you, therefore, prefer?

5-14. A certain cross produces an F_1 ratio of 157:43. By means of the chi-square test, determine the probability of a chance deviation this large or larger on the basis of a 13:3 expectancy.

5-15. Another cross involving different genes gives rise to an F_1 of 110:90. By means of the chi-square test determine the probability of a chance deviation this large or larger on the basis of (a) a 1:1 expectancy and (b) a 9:7 expectancy. (c) Is the deviation to be considered significant in either case? (d) What do you do with these results? *obtain large sample*

5-16. Suppose, with the genes involved in problem 5-15, the F_1 ratio had been 1,100:900. Try a chi-square test with this sample to determine whether there is a significant deviation from (a) 1:1 and (b) 9:7 expectancy. (c) Is the deviation now significant in either case? (d) What is the effect of sample size on the usefulness of the chi-square test?

5-17. Following are listed some of Mendel's reported results with the garden pea. Test each for goodness of fit to the given hypothesis:

Cross	Progeny	Hypothesis
(a) Yellow × green cotyledons	(F_2) 6,022:2,001	3:1
(b) Green × yellow pods	(F_2) 428:152	3:1
(c) Violet red × white flowers	(F_1) 47:40	1:1
(d) Round yellow × wrinkled green seeds	(F_1) 31:26:27:26	1:1:1:1

$df=1$

$df = 3$

CHAPTER 6

Linkage, Crossing-over, and Genetic Mapping of Chromosomes

IN our discussion of dihybrid inheritance (Chapter 4) we saw that the *Coleus* testcross *DdIi* (deep irregular) × *ddii* (shallow regular) produces four progeny phenotypic classes in a 1:1:1:1 ratio. We also noted that this result is to be expected on the basis of the behavior of chromosomes in meiosis, if each of the two pairs of genes is on a different pair of chromosomes. With this *unlinked* arrangement of genes, either member of one pair of genes can combine at random with either member of the other pair, resulting in the production of four kinds of gametes, *DI, Di, dI,* and *di,* in equal number, by the doubly heterozygous individual. Random fusion of these four gamete genotypes with the *di* gametes of the completely recessive parent results in the 1:1:1:1 progeny phenotypic ratio. But as our knowledge of the genetics of various organisms increases, it becomes abundantly clear that the number of genes per species considerably exceeds its number of chromosomes pairs. For example, the fruit fly (*Drosophila melanogaster*) has only four pairs of chromosomes, yet experimental estimates place its number of genes at 10,000 or more. Therefore each chromosome must bear many genes.

All the genes carried on a given chromosome constitute a **linkage group** and would be expected to be inherited as a block were it not for crossing-over (Chapter 3). Therefore we should expect the number of linkage groups for any organism to equal its monoploid chromosome number.[1] Interestingly enough, linkage was anticipated before it was actually demonstrated. Just three years after the rediscovery of Mendel's pioneer papers, Sutton (1903) suggested that each chromosome must bear more than a single gene and that genes "represented by any one chromosome must be inherited together." However, Sutton was unable to support his hypothesis experimentally. Only a few years later, Bateson and Punnett (1905–1908) did have the data with which to do so, but failed to recognize that they were dealing with linked genes.

Linkage and Crossing-over

BATESON AND PUNNETT ON SWEET PEA

In the sweet pea (*Lathyrus odoratus*) two pairs of genes affecting flower color and pollen grain shape occur, each pair exhibiting complete dominance:

---

[1] This expectation will have to be amended when we consider the sex chromosomes (Chapter 11).

R purple flowers Ro long pollen grains
r red flowers ro round pollen grains

Bateson and Punnet crossed a completely homozygous purple long with a red round. The F_1 was, as expected, all purple long, and a $9:3:3:1$ phenotypic ratio was expected in the F_2. Results, however, were quite different, as seen in Table 6-1.

P: $RR\ RoRo \times rr\ ro\ ro$

F₁: $Rr\ Ro\ ro \times Rr\ Ro\ ro$.

TABLE 6-1. Bateson and Punnett Sweet Pea Cross

Phenotype	Observed		Expected (9:3:3:1)	
	Number	Frequency	Number	Frequency
Purple long	296	0.6932	240	0.5625 ($= \frac{9}{16}$)
Purple round	19	0.0445	80	0.1875 ($= \frac{3}{16}$)
Red long	27	0.0632	80	0.1875 ($= \frac{3}{16}$)
Red round	85	0.1991	27	0.0625 ($= \frac{1}{16}$)
	427	1.0000	427	1.0000

Satisfy yourself that chi-square for a $9:3:3:1$ expectancy is 219.28. Inspection of Table 5-2 shows that, for three degrees of freedom, the basis for this expected ratio (independent segregation of the two pairs of alleles) is, therefore, not acceptable. From the arithmetic standpoint Bateson and Punnett recognized that their observed ratio was "explicable on the assumption that . . . the gametes were produced in a series of 16, viz., 7 purple long, 1 purple round, 1 red long, and 7 red round." That is, for example, if each parent produced $r\ ro$ gametes with a frequency of $\frac{7}{16}$ ($= 0.4375$), then $(\frac{7}{16})^2$, or $\frac{49}{256}$ ($= 0.1914$), of the progeny should be "red long." The observed frequency of such plants, 0.1991, is quite close to the calculated 0.1914 frequency. Or, to look at it another way, the "red long" plants represent the fusion of two $r\ ro$ gametes, and, by the product law of probability, the frequency of such gametes must equal the square root of the frequency of the "red long" plants. The square root of 0.1991, the observed frequency of the "red long" individuals, is 0.4462, which is acceptably close to 0.4375, the Bateson and Punnett figure of $\frac{7}{16}$. Expressed as decimals the $7:1:1:7$ array of gamete genotypes, then, would be 0.4375 $R\ Ro$, 0.0625 $R\ ro$, 0.0625 $r\ Ro$, and 0.4375 $r\ ro$. However, Bateson and Punnett were unable to explain why or how this gamete ratio came about because they did not relate it to the behavior of chromosomes in meiosis.

We now know that genes for flower color and shape of pollen grains are **linked.**

ARRANGEMENT OF LINKED GENES

When two pairs of genes are linked, the linkage may be of either of two types in an individual heterozygous for both pairs: (1) the two dominants, R and Ro, may be located on one member of the chromosome pair, with the

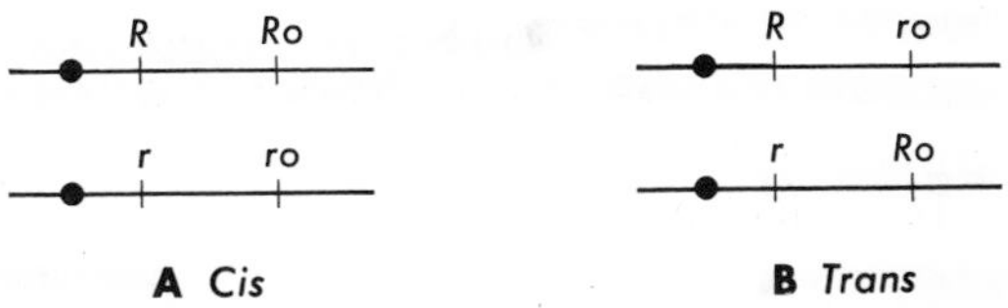

FIGURE 6-1. Cis *and* trans *arrangements for two pairs of linked genes in a diploid cell.*

two recessives, r and ro on the other, or (2) the dominant of one pair and the recessive of the other may be located on one chromosome of the pair, with the recessive of the first gene pair and the dominant of the second gene pair on the other chromosome. The first arrangement, with two dominants on the same chromosome, is referred to as the *cis* arrangement, the second, having one dominant and one recessive on the same chromosome, as the *trans* arrangement. Figure 6-1 illustrates these possibilities.

Thus the genotypes of the parental and first filial generations of the Bateson and Punnett cross may be written in standard fashion to reflect linkage:

P $R\,Ro/R\,Ro \times r\,ro/r\,ro$
F_1 $R\,Ro/r\,ro$

Without crossing-over, the F_1 would produce but two types of gametes, $R\,Ro$ and $r\,ro$. Crossing-over, though, produces two additional gamete genotypes, $R\,ro$ and $r\,Ro$. That these four genotypes are *not* produced in equal frequency could easily be seen from a testcross in which the double heterozygote carries these genes in the *cis* linkage:

P ♀ purple long × red round ♂
$R\,Ro/r\,ro$ $r\,ro/r\,ro$

If the two pairs of genes were not linked, a 1:1:1:1 dihybrid testcross ratio would, of course, result. Based, however, on the gamete frequencies that Bateson and Punnett recognized, testcross progeny should occur in these frequencies:

Phenotype	Genotype	Frequency	Type
1. Purple long	$R\,Ro/r\,ro$	0.4375 } 0.875	{ Noncrossovers or
2. Red round	$r\,ro/r\,ro$	0.4375 }	{ nonrecombinants
3. Purple round	$R\,ro/r\,ro$	0.0625 } 0.125	{ Crossovers or
4. Red long	$r\,Ro/r\,ro$	0.0625 }	{ recombinants

The first two phenotypic classes have received from the pistillate parent an unaltered chromosome carrying either $R\,Ro$ or $r\,ro$. The latter two classes, however, have received a *crossover* chromosome, either $R\,ro$ or $r\,Ro$. Gametes carrying each of these altered chromosomes must occur with a frequency of only 0.0625 ($= \frac{1}{16}$), rather than 0.25 as would be the case with unlinked genes.

With genes *R* and *Ro,* the *crossover frequency* is 12.5 per cent. This frequency is remarkably constant, regardless of the type of cross or the kind of linkage (*cis* or *trans*).

To understand how this happens, recall the occurrence of chiasmata in prophase-I as described on page 69. If a chiasma occurs between these two pairs of genes, *crossing-over* will take place so that the original linkage, *R Ro/r ro,* becomes *R ro/r Ro,* as indicated in Figure 6-2. For each meiocyte in which this happens, the result is four monoploid nuclei of genotypes *R Ro, R ro, r Ro,* and *r ro.* For each meiocyte in which a crossover fails to occur between these two pairs of genes, the result is four monoploid nuclei, two each of genotypes *R Ro* and *r ro.* If the meiocyte carries these genes in the *cis* configuration, as in this case, *R ro* and *r Ro* gametes are referred to as *crossover gametes,* and *R Ro* and *r ro* sex cells as *noncrossover* types. Progeny such as classes 1 and 2, which receive one or the other *intact chromosome* from the heterozygous parent, are referred to as **parental types.** Progeny making up classes 3 and 4 in this testcross incorporate new combinations of linked genes and are referred to as **recombinants.**

Linkage, Linkage Groups, and Mapping

The chromosomal basis of heredity became clearly established in the second decade of this century, verifying Sutton's earlier hypothesis and supplying experimental data to explain and extend such cases as had puzzled Bateson and Punnett. Discovery followed discovery in rapid succession. In 1910 Thomas Hunt Morgan was able to provide evidence for the location of a particular gene of *Drosophila* on a specific chromosome (Morgan 1910a). Within a short time he was able to show clearly that linkage does exist and that linked genes are often inherited together, but may be separated by crossing over (Morgan 1910b, 1911a). Even more exciting than these proofs of earlier surmises was Morgan's conclusion that a definite relation exists between recombination frequency and the linear distance separating genes on their chromosome. As

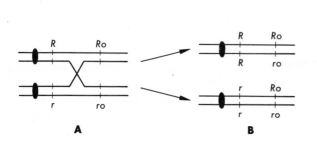

FIGURE 6-2. *Result of crossing-over. A chiasma occurring between linked genes* R *and* Ro, *and involving two nonsister chromatids as shown in (A) may result in "repair" such that the original* cis *linkage is converted to a* trans *linkage in two of the four chromatids, as depicted in (B).*

he wrote (Morgan 1911b), "In consequence, we find coupling in certain characters, and little or no evidence at all of coupling in other characters, the difference *depending on the linear distance apart of the chromosomal materials that represent the factors*" (italics added). Morgan used the term *coupling* in the sense developed by Bateson, Saunders, and Punnett (1905), that is, referring to a greater frequency of gametes carrying two dominants (or two recessives) than would occur by random segregation. A similar association of one dominant and one recessive they called *repulsion.* Largely through cytological work of Morgan, the concept of coupling and repulsion was replaced by that of linkage and crossing-over. The terms *cis* and *trans* came to be employed later (Haldane, 1942).

With these researches, it now became possible not only to identify certain genes with particular chromosomes, but also to begin construction of *chromosome maps* showing linkage groups and relative distances between successive genes. Linkage groups were developed rapidly for *Drosophila* and also for a variety of animals and plants. Perhaps the most remarkable aspect of this burgeoning genetic knowledge was that geneticists could now assign relative positions on chromosomes to genes whose nature and precise function were not to be clarified for another 40 or 50 years and that could not be seen in the microscope.

As this kind of information was accumulated, it became clear that the number of linkage groups in any species ultimately is found to equal its monoploid chromosome number, or, in the sex having dissimilar sex chromosomes, one more than the monoploid number. Thus four linkage groups are known in females of *Drosophila melanogaster,* where $n = 4$, five in males of the same species, and 10 in corn (*Zea mays*), which has 10 pairs of chromosomes. In species whose genetics is less completely worked out, of course, the number of *known* linkage groups is temporarily smaller than its monoploid chromosome number. In mice, for example, only 19 linkage groups are presently known with certainty, although there are 20 chromosomes in a single set. In no event, however, does the number of linkage groups exceed the number of pairs of chromosomes in diploid species except, as we have noted, for one additional linkage group in the sex with dissimilar sex chromosomes. These statements apply only to nuclear genes; the matter of extranuclear genes will be examined later (Chapter 20).

More than a dozen autosomal linkage groups have been established with varying degrees of certainty in man (see Table 6-2), but all are very incomplete. This, of course, leaves many autosomal linkage groups totally unknown.

In addition to these autosomal linkage groups, McKusick (1975) lists 93 genes with confirmed loci on the X chromosome (one of the sex chromosomes), plus 78 additional probable ones, for a total of 171.

Of the 22 autosomes, progress is being made toward a determination of which chromosome bears particular genes. Those listed in Group A of Table 6-2 have been found to be on autosome 1, the longest of the human genome.

TABLE 6-2. Some Probable Autosomal Linkage Groups in
Man (in Alphabetical Order)

Group	Phenotype or Product
A	
	Amylase, pancreatic
	Amylase, salivary
	Auriculo-osteodysplasia (ear and other abnormalities)
	Cataract (zonular pulverulent)
	Duffy blood antigen
	Elliptocytosis (elliptical erythrocytes)
	Peptidase C
	Phosphoglucomutase-1
	6-Phosphogluconate dehydrogenase
	Rhesus blood antigens
	5s RNA
	Uncoiler-1 (uncoiled long arm of autosome 1)
B	
	A and B blood antigens
	Nail-patella syndrome (abnormalities of nails and kneecap)
C	
	Lutheran blood antigen
	Secretor (water-soluble forms of blood antigens A and B)
D	
	β polypeptide chain of hemoglobin
	δ polypeptide chain of hemoglobin
	γ polypeptide chain of hemoglobin
	β thalassemia (a type of anemia)
E	
	Group specific component Gc-1 ⎰ types of
	Group specific component Gc-2 ⎱ alpha-2-
	Serum bisalbumin (types of) globulin
F	
	Serum cholinesterase
	Transferrin (a serum protein)
G	
	Kell blood group
	Phenylthiocarbamide (PTC) taste

The location of some of these has been confirmed only recently, as indicated in the references at the end of this chapter. The probable chromosomal location of genes for certain other traits include chromosome 2, MNSs blood antigens; chromosome 5, hexosaminidase B; chromosome 7, hexosaminidase A, and mannosephosphate isomerase; chromosome 10, glutamate oxalacetic transaminase; chromosome 11, lactate dehydrogenase subunit A; chromosome 12, lactate dehydrogenase subunit B; chromosome 13, ribosomal RNA; chromosome 14, nucleoside phosphorylase, and ribosomal RNA; chromosome 15, ribosomal RNA; chromosome 16, alpha chain of haptoglobins; chromosome 17, thymidine kinase; chromosome 18, peptidase A; chromosome 19, glucosephosphate isomerase, and polio sensitivity; chromosome 21, ribosomal RNA. Progress and problems in mapping human chromosomes are described by McKusick (1971).

CYTOLOGICAL EVIDENCE FOR CROSSING-OVER

Whenever a particular chromosome pair (bearing certain genes) can be clearly identified because of some structural characteristic, it can be shown that when there occurs an interchange of material between two homologs, there is likewise an interchange of genes (i.e., genetic crossing-over). The work to be described in this section also provides further evidence that genes are located on chromosomes. In an analysis of such a situation in corn, Creighton and McClintock (1931) furnished such a convincing correlation between cytological evidence and genetic results that their work has rightly been called a landmark in experimental genetics.

In corn, chromosome 9 (the second shortest in the complement of 10 pairs) ordinarily lacks a small knob (satellite). But in one particular strain investigated by these two workers, a single plant was found to have dissimilar ninth chromosomes. One of these possessed a satellite at the end of its short arm and also an added segment that had been translocated from chromosome 8. The other member of the pair was normal, lacking both the satellite and the added segment of number 8. Chromosome 9 bears, among other genes, the following:

C colored aleurone	*Wx* starchy endosperm
c colorless aleurone	*wx* waxy endosperm

Aleurone and endosperm are parts of the triploid food storage tissue in the grain. The plant having these dissimilar ninth chromosomes was heterozygous for both the aleurone and endosperm genes and, from earlier work, Creighton and McClintock knew which chromosome of the pair carried which gene. Thus the two ninth chromosomes of this plant, which they used as the pistillate parent in their crosses, could be identified visually, and may be diagramed thus:

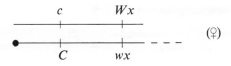

The dashed portion indicates the translocated segment of chromosome 8. An individual possessing such a dissimilar ninth pair of chromosomes was crossed with a *c Wx/c wx* plant possessing two knobless ninth chromosomes which also did not have the added segment of chromosome 8 (i.e., "normal" ninth chromosomes):

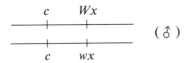

Although only 28 grains resulted from this cross, cytological examination of microsporocytes from adults grown from these grains confirmed the predicted relation between cytology and genetics (except for one class that was not found).

This cross, with the ninth chromosome in each parent and progeny class, is diagramed in Figure 6-3.

The fact that endosperm and aleurone are triploid tissues need not complicate our understanding of this masterful research because, as indicated in Figure 6-3, Creighton and McClintock used chromosomes of the *diploid* microsporocytes to demonstrate the correlation between cytological and genetic crossing-over.

A similar verification that genetic crossing-over is accompanied by a physical exchange between homologous chromosomes was also established in 1931 by Stern for *Drosophila*. He utilized a strain in which the females had a portion of the Y chromosome attached to one of their X chromosomes. Work by Stern and by Creighton and McClintock indicated a clear relationship between the interchange of material between homologs and genetic crossing-over.

We may summarize our progress thus far as follows:

1. Certain genes assort *at random;* these are *unlinked genes* (Chapter 4).
2. *Other genes do not segregate randomly, but are linked* (Chapter 6). These *linkage groups* tend to be transmitted in unitary groups.
3. In diploid cells chromosomes also occur in pairs that tend to be transmitted as units to daughter nuclei (Chapter 3).
4. Linked genes do not always "stay together" but are often exchanged reciprocally (genetic *crossing-over*) (Chapter 6).
5. Chromosomes may be seen to form *chiasmata* and to exchange parts reciprocally (Chapters 3 and 6). Such exchange is reflected in genetic crossing-over. Furthermore, chiasma formation and genetic crossing-over occur with closely similar frequencies.

Thus linkage is an exception to the pattern of random segregation of genes, and crossing-over results in an exception to the consequences of linkage.

Mapping Chromosomes

As Morgan (1911b) predicted, frequency of crossing-over is governed largely by distances between genes. That is, the probability of its occurring between two *particular* genes increases as the distance between them becomes larger, so that crossover frequency appears to be directly proportional to distances

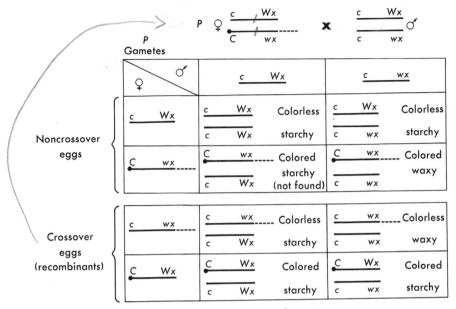

FIGURE 6-3. *A diagrammatic representation of parallelism between cytological and genetic crossing-over, showing the phenotypes of aleurone and endosperm of F_1 grains and the chromosome morphology and genotypes for microsporocytes produced by F_1 plants.* (Based on the work of Creighton and McClintock. See text for full explanation.)

between genes. As we shall see, this relationship is quite valid, though not equally so in all parts of a chromosome, for proximity of one crossover to another decreases probability of another very close by. The centromere has a similar interfering effect; frequency of crossing-over is also reduced near the ends of the chromosome arms. Because of this general relationship between intergene distance and crossover frequency, and because we cannot measure such distances in the customary units employed in microscopy, geneticists use an arbitrary unit of measure, the **map unit,** to describe distances between linked genes. *A map unit is equal to 1 per cent of crossovers* (recombinants); i.e., it represents the linear distance for which 1 per cent crossovers (recombinants) is observed. Thus for sweet peas the distance from *R to Ro* would be described as 12.5 map units.

THE THREE-POINT TESTCROSS

The most commonly employed method in genetic mapping of chromosomes is the trihybrid (or "three-point") testcross. Let us try such a cross in *Drosophila melanogaster,* the little fruit fly whose genetics is so well known. Using a plus sign to denote the so-called wild type, as is customary in mapping problems, we shall examine a three-point cross involving these known genes:

Gene Symbol	Phenotype
+	Normal wing (*dominant*)
cu	Curled wing (*recessive*)
+	Normal thorax (*dominant*)
sr	Striped thorax (*recessive*)
+	Normal bristles (*dominant*)
ss	Spineless bristles (*recessive*)

The actual numbers of each progeny class in the following illustration are hypothetical, but the map distances and the genes are real. For the moment we shall arbitrarily choose the gene sequence that follows, recognizing that the results may either confirm that order or dictate a different one. We shall see how to determine the correct sequence as soon as we have looked at the cross:

[handwritten: Trihybrid testcross]

[handwritten: heterozygous for all 3 traits] *[handwritten: homoz. recessive for all 3 traits]*

P ♀ normal normal normal × curled spineless striped ♂
+ + + /cu ss sr   cu ss sr/cu ss sr

F$_1$ Phenotype	Maternal Chromosome	#	%	Type
Normal normal normal	+ + +	430	*[handwritten: recombinant]* 88.2	parental or
Curled spineless striped	cu ss sr	452	*[handwritten: crossover frequencies]*	noncrossover
Normal spineless striped	+ ss sr	45	8.3	cu-ss single
Curled normal normal	cu + +	38		crossovers
Normal normal striped	+ + sr	16	3.3	ss-sr single
Curled spineless normal	cu ss +	17		crossovers
Normal spineless normal	+ ss +	1	0.2	Double
Curled normal striped	cu + sr	1		crossovers
		1,000	100.0	

[handwritten at left: +++ / cu ss sr]

By grouping F$_1$ data as we have, several important considerations emerge:

1. **The maternal chromosome received by members of each of the two numerically largest classes** (noncrossover flies), determinable from the phenotypes, **discloses whether *cis* or *trans* linkage obtained in the maternal parent.** Here, for example, normal normal normal individuals must have received + + + from the maternal parent and *cu ss sr* from the paternal.

[handwritten: since mother alone determines phenotype (because father is homoz recessive)]

Because noncrossover gametes will be more frequent than crossover sex cells, $+ + +/cu\ ss\ sr$ and $cu\ ss\ sr/cu\ ss\ sr$ flies will occur in the majority if the linkage is in the *cis* configuration.[2]

2. **Double-crossover individuals** (those that would not be noted in a two-pair cross involving only *cu* and *sr*) **show up.** These are individuals that, as the term implies, result from the occurrence of two crossovers between the first and third genes in order on the maternal chromosome. *Even numbers of crossovers between two successive genes will not be picked up:*

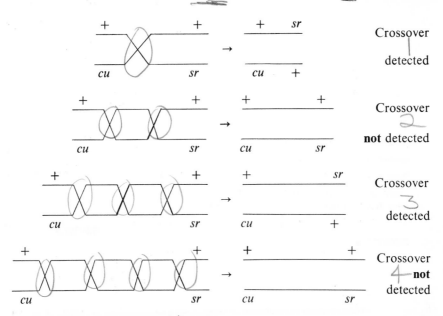

3. **The double crossovers are recognizable as the numerically smallest progeny groups and may be used to determine gene sequence.** Because $+ ss +$ and $cu + sr$ are, therefore, here identifiable as double crossovers, there is only one sequence of genes in the *cis* configuration in the maternal parent that would yield these two gene combinations following a double crossover. This becomes clear if we think of the original *cis* arrangement and how the $+ ss +$ and $cu + sr$ chromosomes can be derived therefrom. This will be only by two crossovers between the first and third genes in the sequence:

---

[2] See also problem 6-20.

If, on the other hand, these genes were arranged in either of the other two possible sequences (*sr cu ss* or *ss sr cu*), the outcome of double crossing-over would be incompatible with the results observed in our example:

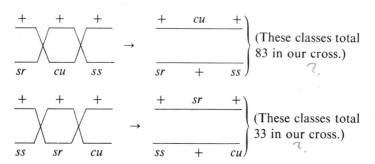

Thus the true sequence here can only be *cu ss sr*.

4. **The true distance** between *cu* and *ss* is, therefore, 8.3 + 0.2 = 8.5 (single crossovers + double crossovers).

5. **The true distance** between *ss* and *sr* is, therefore, 3.3 + 0.2 = 3.5 (single crossovers + double crossovers).

6. **The true distance** between *cu* and *sr* is 8.3 + 0.2 + 3.3 + 0.2 = 12.0 (*cu ss* single crossovers + *ss sr* single crossovers + *twice* the double crossovers). This is so because the double crossovers represent just what their name implies: *two* crossovers, one between *cu* and *ss*, plus a second one between *ss* and *sr*.

The genes here described constitute three members of a linkage group in *Drosophila*. With the cross outlined above, a beginning of genetically mapping one chromosome of *Drosophila* can be made. If these are thought of as the first three to be known in a new linkage group, they can be placed arbitrarily at particular *loci*:

cu	*ss*	*sr*
0.0	8.5	12.0

or, of course, *sr* may be placed at the "left" end and the sequence reversed. If later work shows another gene, *W* (dominant for wrinkled wing), to be located 4.0 map units to the "left" of *cu*, then the map is redrawn and each locus renumbered accordingly:

W	*cu*	*ss*	*sr*
0.0	4.0	12.5	16.0

Figure 6-4 shows the locations of these and other genes as presently determined. Knowledge of which chromosome bears which genes in *Drosophila*

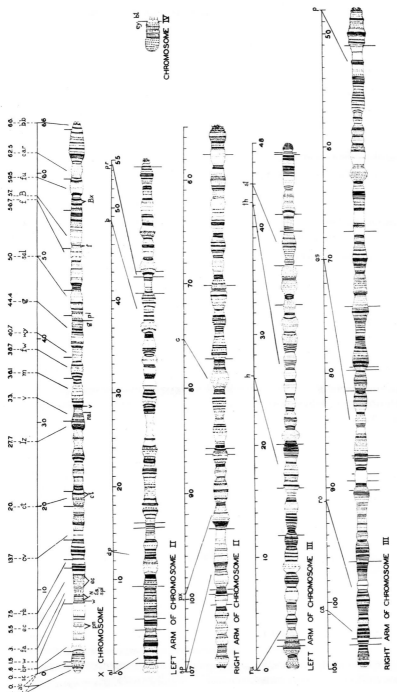

FIGURE 6-4. Comparison of cytologic and genetic maps of the chromosomes of Drosophila melanogaster. For each chromosome, the genetic map is above the cytologic. (From T. S. Painter, 1934, Journal of Heredity, **25**: 465–476. By permission.)

(and other dipterans) is aided by a study of giant chromosomes (Chapter 12). A similar linkage map for corn (*Zea mays*) is shown in Figure 6-5. Some of the problems at the end of this chapter further explore the techniques of genetic mapping.

Interestingly enough, had we performed the reciprocal of the trihybrid testcross given on page 126, namely,

$$♀ \; cu \; ss \; sr/cu \; ss \; sr \; \times \; + + +/cu \; ss \; sr \; ♂$$

only the two parental phenotypes (normal normal normal and curled spineless striped) would have been recovered in the progeny! This is so because crossing-over does not occur in the male flies. Dipterans are unusual in this respect.

INTERFERENCE AND COINCIDENCE

Our discussion of mapping techniques thus far would seem to imply that crossing-over in one part of a chromosome is independent of crossing-over elsewhere in that chromosome. That this is not so was demonstrated as long ago as 1916 by Nobel-Prize-winning geneticist H. J. Muller.

Consider for a moment a chromosome bearing three genes, *a*, *b*, and *c*:

$$\underset{\text{region I}}{\underbrace{\overset{a}{\mid} \qquad \overset{b}{\mid}}} \quad \underset{\text{region II}}{\underbrace{\qquad \overset{c}{\mid}}}$$

If we designate the *a-b* portion as region I and the *b-c* segment as region II, then, recalling the now familiar product law of probability (page 101), the frequency of double crossovers between genes *a* and *c* should equal the crossover frequency of region I times the crossover frequency of region II. Actually, this is seldom true.   *in class notes :* $(10\%)(5\%) = .005 = .5\%$

For example, in our three-point mapping experiment in *Drosophila* we obtained the following crossover frequencies:

"Region"	Genes	Percentage Crossovers	Map Distance (in map units)
I	cu ss	8.3	8.3 + 0.2 = 8.5
II	ss sr	3.3	3.3 + 0.2 = 3.5
Double Crossovers	cu ss sr	0.2	

If crossing-over in regions I and II were independent, we should predict 0.085×0.035, or almost 0.3 per cent, double crossovers, whereas only 0.2 per cent was observed. A disparity of this kind, in which the number of actual double crossovers is less than the number calculated on the basis of independence, is very common. It clearly suggests that, once a crossover occurs,

In class, we would predict $(10\%)(5\%) = .005 = .5\%$
But only 5% *is observed*

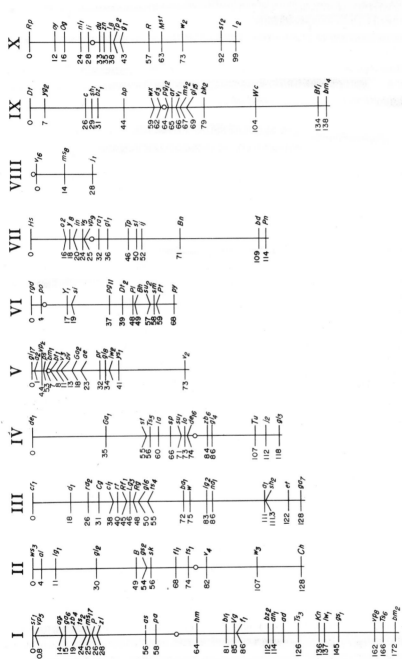

FIGURE 6-5. *Linkage map of the 10 maize (corn) chromosomes, showing locations of various known genes (symbolized by letters). Distances between genes are obtained by breeding experiments described in Chapter 6.* (After Neuffer, 1965. Courtesy De Kalb Agricultural Association, Inc.; De Kalb, Illinois. By permission.)

the probability of another crossover in an adjacent region is reduced. This phenomenon is called **interference.**

Interference appears to be unequal in different parts of a chromosome, as well as among the several chromosomes of a given complement. In general, interference appears to be greatest near the centromere and at the ends of a chromosome. Degrees of interference are commonly expressed as **coefficients of coincidence** or simply as **coincidence:**

$$\text{coincidence} = \frac{\text{actual frequency of double crossovers}}{\text{calculated frequency of double crossovers}}$$

In our *Drosophila* example coincidence is

$$\frac{0.002}{0.003} = 0.67$$

As interference decreases, coincidence increases. Coincidence values ordinarily vary between 0 and 1. Absence of interference gives a coincidence value of 1, whereas complete interference results in a coincidence of 0. Coincidence is generally quite small for short map distances. In *Drosophila*, coincidence is zero for distances of less than 10 to 15 map units, but gradually increases to 1 as distances exceed 15 map units. Furthermore, there seems to be no interference across the centromere from one arm of the chromosome to the other.

Similarly, interference is reported from a wide variety of organisms. For example, Hutchison (1922), who discovered the *c-sh* linkage in corn, reported map distances for three genes, *c* (colorless aleurone), *sh* (shrunken grains), and *wx* (waxy endosperm). His data indicated the following crossover frequencies:

"Region"	Genes	Percentage Crossovers	Map Distance (in map units)
I	c sh	3.4	3.4 + 0.1 = 3.5
II	sh wx	18.3	18.3 + 0.1 = 18.4
Double crossovers	c sh wx	0.1	

Again, if crossing-over in regions I and II were independent, we should predict 0.035 × 0.184 = 0.6 per cent double crossovers, whereas only 0.1 per cent was observed, giving a coefficient of coincidence of 0.167. On the other hand, especially in bacterial and bacteriophage genetics, double crossovers may be encountered in excess of random expectation, giving coincidence values > 1. This is referred to as *negative interference.*

Linkage Studies in Bacteria

Present evidence indicates that such bacteria as the common colon bacillus, *Escherichia coli*, contain one to several *nucleoids*, rich in deoxyribonucleic acid (Chapter 15), that appear in electron micrographs as areas of lesser density. Internal organization is not clearly discernible, but studies strongly suggest that each nucleoid contains long, continuous fibrils of DNA having no free ends. Such a closed DNA structure of the nucleoid is functionally comparable to the chromosome of higher organisms. It is, however, *not* a chromosome in the structural sense of eukaryotic organisms. As described in Appendix B, during conjugation between donor and recipient, this ring "chromosome" opens at a particular point and passes as a filament into the body of the recipient cell, the length of transferred segment depending upon the duration of the transfer process.

EVIDENCE FROM CONJUGATION

In such monoploid organisms, the usual technique of determining linkage distances by means of recombinational frequencies cannot ordinarily be employed. Instead, donor cells of known genotypes are mixed with a large number of recipients of a different genotype, then separated at predetermined times by agitating with a Waring blender. After separation, progeny of the recipient cells are tested by inoculating them on different deficiency media in order to detect various physiologically deficient strains (**auxotrophs**). Genes are found to be located sequentially on the chromosome and are transferred in order to the recipient cell. For example, if we use donor cells with four known genes:

pan	pantothenic acid (vitamin B_5) synthesis
arg D	synthesis of the enzyme ornithine transcarbamylase
lac Z	synthesis of the enzyme β galactosidase
gal A	synthesis of galactokinase

and interrupt conjugation at different times, we find genes transferred in this time sequence:

Duration of Conjugation (min.)	Genes Transferred
1.5	*pan*
5.0	*pan, arg D*
10.0	*pan, arg D, lac Z*
16.0	*pan, arg D, lac Z, gal A*

Therefore, the order of these genes is as given in the 16-minute sequence.

A particular experiment might involve, for instance, these strains:

(donor) *pan⁻ arg D⁻ lac Z⁺ gal A⁺* . . .
(recipient) *pan⁺ arg D⁺ lac Z⁻ gal A⁻* . . .

Here a minus sign indicates inability to synthesize the substance listed in the previous paragraph (i.e., auxotrophic for that substance) and a plus sign indicates ability to produce the given substance (**prototrophic**). The two strains are mixed in a tube of liquid (complete) medium, then, after conjugation, plated out in an agar plate containing a complete medium. Here a large number of progeny colonies, both auxotrophs and prototrophs, develop. After incubation, colonies are transferred to a series of deficiency media to detect recombinations. This is done by pressing the master plate onto a sheet of velvet whose fibers pick up individuals of each colony. The velvet is then pressed to a series of replica plates of a deficiency medium.

Many such experiments, usually utilizing triple auxotrophs (e.g., $- - - + + + \times + + + - - -$) to reduce to a very low value the probability of mutation as a factor, have resulted in a fairly complete genetic map of *Escherichia coli,* as shown in Fig. 6-6. Note that a total time of 89 minutes is indicated for transfer of the complete chromosome during conjugation. Reference to Appendix B will show that the circular chromosome opens out at the point where the F (fertility) factor is located. Although the fertility factor may be anywhere along the chromosome extent, opening of the chromosome is always on the same side of F. The F locus is the last to be transferred; hence the genes that enter the recipient first vary from strain to strain, although the *sequence* of genes is ordinarily identical.

Linkage Studies in Viruses

Bacteriophages, or phages, are viruses that infect bacteria. Their structure and "life cycle" are described in Chapter 15. That they, like all viruses, have genetic systems is evident from the fact that they show, for instance, host and symptom specificities. Moreover, they can be shown to change in such genetically controlled properties as virulence toward a particular host, or in the kind of coat proteins produced, in short, to *mutate.* Although early thought ran more toward vague concepts of adaptation and induction, Luria and Delbrück (1943) opened the way to an understanding of mutation and selection in bacteria and their phages. These workers, and a host of others in a flood of later papers, demonstrated that both bacteria and viruses have genetic material, just as do corn, mice, fruit flies, and men.[3] Adaptation and induction have given way to the same processes of mutation and selection that operate in higher organisms.

---

[3] Delbrück and Luria (along with Hershey) shared the 1969 Nobel Prize for physiology and medicine for this work.

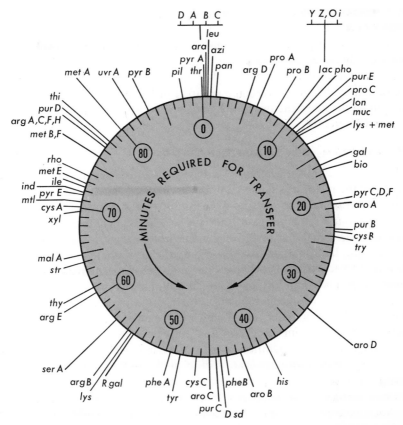

FIGURE 6-6. *Genetic map of the colon bacterium,* Escherichia coli, *based on recombination as determined by conjugation studies. The unit of map structure is the one-minute interval rather than the crossover frequency of higher organisms. A transfer time of 89 minutes for the entire "chromosome" is indicated. Symbols represent ability or inability to utilize or synthesize various substances.*

Virulent phages destroy, or lyse, their hosts. One mutant in T2 and T4 phages (the T series infects the common colon bacterium, *Escherichia coli*) accomplishes this host destruction more rapidly than does the "wild-type" phage. These rapid lysing strains are designated as T2r, T4r, and so on. If a culture of *E. coli* is infected by a mixture of T2 and T4r phages, *four* progeny types are recovered: the "parental" T2 and T4r, and also the recombinants T2r and T4!

Such recombination as this does not result from a sexual process, but rather by recombination and exchange of genetic material between the phages in conjunction with an involvement of the bacterial host's genetic material. We shall discuss the details of this process as soon as we know more about genetic material itself but, for our present purposes, it is important to note (1) that

transduction

viruses can be "crossed," and (2) as a result, a map of the virus genome can be constructed. Rather early in these investigations, Hershey proposed three different linkage groups for T2, but more work has shown that the phage map, like that of bacteria, is "circular" in that each marker is linked to another on either side. One whole complex of T4r mutants, known as rII (because they were originally assigned to linkage group II of Hershey), have thrown important light on the fine structure of the gene, and we shall examine them in more detail in Chapter 18. *p 427*

MECHANISM OF RECOMBINATION

The exact mechanism whereby the donated nucleoid segment becomes integrated into the recipient's nucleoid in bacterial conjugation is unknown, as are the precise mechanics of crossing-over in eukaryotes. Summarizing the problem in bacteria, Wollman, Jacob, and Hayes (1962) conclude that "in conjugation it is evident that a segment of donor chromosome . . . does not always participate in the formation of a recombinant chromosome and, when it does, that it is not necessarily incorporated as a whole. A mechanism analogous to crossing-over must therefore be assumed although its physical basis is likely to differ from one of simple breakage and reunion in view of the asymmetry of the parental components, one of which may be very small." Although a full statement of our present understanding must be deferred until we have considered the nature and operation of the genetic material at the molecular level (Chapter 15), it appears that some sort of break-and-exchange mechanism is operative. For our immediate purposes it will suffice to point out that the segment of donor nucleoid synapses with the homologous segment of the recipient's nucleoid, following which breaks in both donor and recipient may occur. A segment of the donor nucleoid then may replace a corresponding section of the recipient's nucleoid. If genes in the two segments differ (e.g., *gal A*$^+$ versus *gal A*$^-$), recombination will have occurred in the recipient cell.

In eukaryotes, the most suggestive evidence concerning the events of recombination is derived from certain plants that have a dominant monoploid phase —e.g., many algae, most true fungi, and all bryophytes (liverworts and mosses). In bryophytes, for example, meiosis results in a spherical tetrad of four *unordered* meiospores. But in such ascomycete fungi as *Neurospora*, the cells resulting from meiosis are *ordered;* that is, they are situated in line in the ascus (see life cycle, Appendix B). Such ordering reflects the pattern of chromosomal arrangement at each meiotic stage. Although meiosis in *Neurospora* produces the usual four meiospores, each of these divides once by mitosis to produce a total of eight ascospores, sequential pairs of which are genotypically identical. Each ascospore may be removed in order from the ascus and germinated to determine physiological phenotype, or examined visually for such morphological traits as color. Such an analysis is termed **tetrad analysis;** it clearly indicates that recombination via crossing-over must occur at the four-chromatid stage, and thus supports the concept of a break-and-exchange mechanism.

i.e. 2 homologous chromosomes each consisting of 2 chromatids

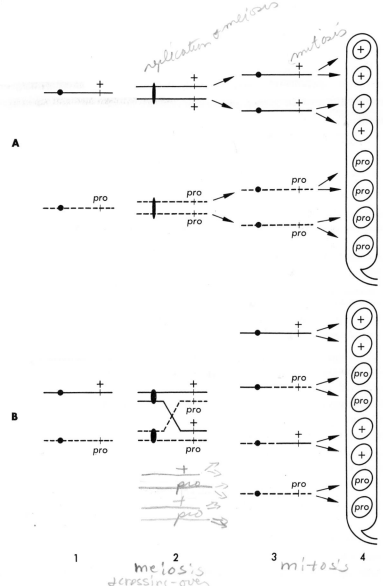

FIGURE 6-7. *Alternative arrangements of ascospores in ascus of* Neurospora. *In (A) crossing-over does not occur or does not involve genes + and pro; in (B) crossing-over occurs between the centromere and the + pro alleles (B 2). Only if crossing-over occurs in the four-strand stage, involving nonsister chromatids, can the sequence of ascospores shown at (B 4) be attained. In 1 of both (A) and (B), the chromosomes contributed to the zygote by each parental strain are shown; in 2, replication has taken place and it is at this stage that synapsis and crossing-over (if any) occurs; in 3, the chromosomes of each of the meiospores resulting from meiosis are depicted; in 4, a mature ascus and the genotypes of each of its ascospores are shown. Meiosis is taking place between 1 and 3; mitosis occurs between 3 and 4.*

From a cross between an auxotroph (e.g., *prolineless*) and the wild-type prototroph, both + and *pro* ascospores are found in equal numbers in each ascus. However, the sequential arrangement of these spores may be either + + + + *pro pro pro pro*, or + + *pro pro* + + *pro pro*. The latter arrangement is possible only if recombination by crossing-over occurs during the four-strand stage (Fig. 6-7).

If a double auxotroph for the linked traits *prolineless/serineless* (*pro/ser*) and the wild type (+/+) are crossed, the reciprocal recombinants (+/*ser* and *pro*/+) are produced with equal frequency, along with the parental types.

Even more can be learned from studies of segregation of blocks of three linked markers. Tetrad analysis clearly shows that each crossover can involve either of the two chromatids of each homologous chromosome so that three different basic types of double crossover tetrads are possible:

1. *Two-strand doubles*, in which the same two chromatids are involved in both crossovers (Fig. 6-8A).
2. *Three-strand doubles*, in which three chromatids are involved, one of them participating twice (Fig. 6-8B).
3. *Four-strand doubles*, in which each of the crossovers involves a different pair of chromatids (Fig. 6-8C).

At any given moment a visible chiasma does not necessarily indicate the point at which crossing-over occurs, because after their formation, chiasmata move toward the ends of the chromatids involved. There is, nevertheless, a close correspondence between their number and the frequency of crossovers in organisms whose cytology and genetics have been extensively investigated. Three- and four-strand doubles (Fig. 6-8B and C) can be explained only if breaks occur before chiasmata are formed and after replication of chromatids has occurred.

Summary. Thus, as we should expect, we see that

1. The number of genes exceeds the number of pairs of chromosomes as our knowledge of the organism's genetics develops.
2. Therefore certain blocks of genes are linked, and
3. The number of linkage groups is equal to the number of pairs of chromosomes (with appropriate exception for nonhomologous sex chromosomes), but
4. Linkage is not inviolable.
5. A reciprocal exchange of material between homologous chromosomes in heterozygotes is reflected in crossing-over;
6. The frequency of crossing-over appears to be closely related to physical distance between genes on a chromosome and serves as a tool in constructing genetic maps of chromosomes.
7. Crossing-over results basically from an exchange of genetic material between nonsister chromatids by break-and-exchange following replication.

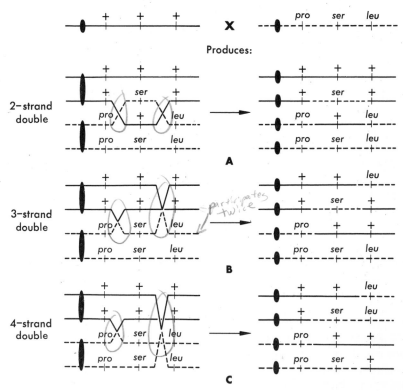

FIGURE 6-8. *Diagram showing possible types of double crossing-over involving (A) two chromatids only, (B) three chromatids. (C) all four chromatids, as inferred from tetrad analysis.*

REFERENCES

BATESON, W., and R. C. PUNNETT, 1905–1908. Experimental Studies in the Physiology of Heredity, Reports to the Evolution Committee of the Royal Society, 2, 3, and 4. Reprinted in J. A. Peters, ed., 1959. *Classic Papers in Genetics.* Englewood Cliff, N.J., Prentice-Hall.

BATESON, W., E. R. SAUNDERS, and R. G. PUNNETT, 1905. Experimental Studies in the Physiology of Heredity. *Rep. Evol. Comm. Roy. Soc.,* II, 1–55 and 80–99.

CREIGHTON, H. S., and B. McCLINTOCK, 1931. A Correlation of Cytological and Genetical Crossing-Over in *Zea mays. Proc. Nat. Acad. Sci. (U.S.),* **17:** 492–497. Reprinted in J. A. Peters, ed., 1959. *Classic Papers in Genetics.* Englewood Cliffs, N.J., Prentice-Hall.

GERMAN, J., and R. S. K. CHAGANTI, 1973. Mapping Human Autosomes: Assignment of the MN Locus to a Specific Segment in the Long Arm of Chromosome No. 2. *Science,* **182:** 1261–1262.

HALDANE, J. B. S., 1942. *New Paths in Genetics.* New York, Harper & Row.

HUTCHISON, C. B., 1922. The Linkage of Certain Aleurone and Endosperm Factors in Maize, and Their Relation to Other Linkage Groups. *Cornell Agr. Exp. Sta. Mem.,* **60.**

LURIA, S. E., and M. DELBRÜCK, 1943. Mutations of Bacteria from Virus Sensitivity to Virus Resistance. *Genetics,* **28:** 491–511. Reprinted in E. A. Adelberg, ed., 1966, 2nd ed. *Papers on Bacterial Genetics.* Boston, Little, Brown.

MARSH, W. L., R. S. K. CHAGANTI, F. H. GARDNER, K. MAYER, P. C. NOWELL, and J. GERMAN, 1974. Mapping Human Autosomes: Evidence Supporting Assignment of Rhesus to the Short Arm of Chromosome No. 1. *Science,* **183:** 966–968.

MCKUSICK, V. A., 1971. The Mapping of Human Chromosomes. *Sci. Amer.,* **224:** 104–113.

MCKUSICK, V. A., 1975. *Mendelian Inheritance in Man,* 4th ed. Baltimore, The Johns Hopkins Press.

MCMORRIS, F. A., T. R. CHEN, F. RICCIUTI, J. TISCHFIELD, R. CREAGAN, and F. RUDDLE, 1973. Chromosome Assignments in Man of the Genes for Two Hexosephosphate Isomerases. *Science,* **179:** 1129–1131.

MORGAN, T. H., 1910a. Sex-Limited Inheritance in *Drosophila. Science,* **32:** 120–122.

MORGAN, T. H., 1910b. The Method of Inheritance of Two Sex-Limited Characters in the Same Animal. *Proc. Soc. Exp. Biol. Med.,* **8:** 17.

MORGAN, T. H., 1911a. The Application of the Conception of Pure Lines to Sex-Limited Inheritance and to Sexual Dimorphism. *Amer. Nat.* **45:** 65.

MORGAN, T. H., 1911b. Random Segregation Versus Coupling in Mendelian Inheritance. *Science,* **34:** 384.

RENWICK, J. H., 1971. The Mapping of Human Chromosomes. In H. L. Roman, ed., *Annual Review of Genetics,* volume 5. Palo Alto, Calif., Annual Reviews, Inc.

RUDDLE, F., F. RICCIUTI, F. A. MCMORRIS, J. TISCHFIELD, R. CREAGAN, G. DARLINGTON, and T. CHEN, 1972. Somatic Cell Genetic Assignment of Peptidase C and the Rh Linkage Group to Chromosome A-1 in Man. *Science,* **176:** 1429–1431.

STERN, C., 1931. Zytologisch-genetische Untersuchungen als Beweise für die Morgansche Theorie des Faktorenaustauchs. *Biol. Zentralbl.,* **51:** 547–587.

SUTTON, W. S., 1903. The Chromosomes in Heredity. *Biol. Bull.,* **4:** 231–251. Reprinted in J. A. Peters, ed., 1959. *Classic Papers in Genetics.* Englewood Cliffs, N.J., Prentice-Hall.

WOLLMAN, E. L., F. JACOB, and W. HAYES, 1962. Conjugation and Genetic Recombination in *Escherichia coli* K-12. *Cold Spring Harbor Symp. Quant. Biol.,* **21:** 141–162. Reprinted in E. A. Adelberg, ed., 1966, 2nd ed. *Papers on Bacterial Genetics.* Boston, Little, Brown.

PROBLEMS

6-1. If, in sweet pea, the cross *R Ro/r ro* × *R Ro/r ro* is made, what would be the expected frequencies of (a) parental gametes of each of the possible genotypes, (b) *R Ro/R Ro* progeny, (c) purple long progeny?

6-2. Elliptocytosis (a rare but harmless condition in which the erythrocytes are ellipsoidal instead of the more common disk shape) and ability to produce rhesus antigen D are both due to dominant, linked genes. Persons producing rhesus antigen D are referred to as Rh positive. An Rh positive man exhibits elliptocytosis, as did his Rh negative mother. His father had normal erythrocytes and was Rh positive. (a) Using the symbol E for elliptocytosis and D for production of antigen D, give this man's genotype. (b) What kind of linkage configuration is this?

6-3. The following four pairs of genes are linked on chromosome 2 of tomato:

Aw, aw purple, green stems
Dil, dil normal green, light green leaves
O, o oval, spherical fruit
Wo, wo wooly, smooth leaves

Crossover frequencies in a series of two-pair testcrosses were found to be: wo-o, 14 per cent; wo-dil, 9 per cent; wo-aw, 20 per cent; dil-o, 6 per cent; dil-aw, 12 per cent; o-aw, 7 per cent. (a) What is the sequence of these genes on chromosome 2? (b) Why is not the wo-aw two-pair crossover frequency greater?

In the next four questions, the following facts will have to be used. In *Drosophila*, the following genes occur on chromosome III: e, ebony body; fl, fluted or creased wings; jvl ("javelin"), bristles cylindrical and crooked; obt ("obtuse"), wings short and blunt.

6-4. A series of dihybrid testcrosses shows the following crossover frequencies: jvl-fl, 3 per cent; jvl-e, 13 per cent; fl-e, 11 per cent. (a) What is the gene sequence? (b) How do you account for the fact that the sum of the fl-e and fl-jvl frequencies exceeds the jvl-e frequency?

6-5. Another cross discloses a crossover frequency of 19 per cent between jvl and obt. How well can you locate obt in the sequence established in problem 6-4?

6-6. If the e-obt crossover frequency is next found to be 7 per cent, where should gene obt be located in the sequence?

6-7. The fl-obt crossover frequency is determined by the cross + +/fl obt (♀) × fl obt/fl obt (♂) to be 17.5 per cent. (a) Does this confirm your answer to problem 6-6? (b) What should be the frequency of double crossovers in a trihybrid testcross involving genes fl, e, and obt if there is no interference?

6-8. As pointed out in the explanatory note preceding problem 6-4, genes e, fl, jvl, and obt are located on chromosome III, along with many other known genes, thus constituting part of one linkage group. How many linkage groups are there altogether in *Drosophila melanogaster* females?

6-9. Each individual of the Jimson weed (*Datura stramonium*) produces both sperms and eggs. After consulting Table 3-1, give the number of linkage groups in this plant.

6-10. How many linkage groups are there in the (a) female grasshopper, (b) male grasshopper, (c) human female, (d) human male?

6-11. Mendel studied seven pairs of contrasting characters in the garden pea. Why did he not discover the principle of linkage?

6-12. Referring back to problems 6-4 through 6-7, how many different gamete genotypes are produced by (a) the female in the cross + +/jvl fl (♀) × jvl fl/jvl fl (♂), (b) the male in the cross + +/jvl fl (♀) × + +/jvl fl (♂)?

6-13. How many different gamete genotypes are produced by (a) the female in the cross + + +/jvl fl e (♀) × jvl fl e/jvl fl e (♂), (b) the male in the cross + + +/jvl fl e (♀) × + + +/jvl fl e (♂)?

6-14. Two of the pairs of alleles known in tomato are

Cu, "curl" (leaves curled)
cu, normal leaves
Bk, "beakless" fruits
bk, "beaked" fruits, having sharp-pointed protuberance on blossom end of mature fruit

The cross of two doubly heterozygous "curl beakless" plants yields four phenotypic classes in the offspring, of which 23.04 per cent are "normal beaked." Are these two pairs of genes linked? How do you know?

6-15. From the data of problem 6-14 can you deduce (a) whether, in the parents, cu and bk were in the cis or trans configuration, (b) the distance between them in map units (assuming no interference)?

In the next four problems, use this information. Two of the many known pairs of genes in corn are

Pl   purple plant
pl   green plant
Py   tall plant (normal height)
py   pigmy (very dwarf)

These genes are 20 map units apart on chromosome 6. The cross Pl Py/pl py × Pl Py/pl py is made. Now answer the following four questions.

6-16. What is the gamete genotype ratio produced by each parent?

6-17. What percentage of the offspring has the genotype pl py/pl py?

6-18. What percentage of the progeny is purple pigmy?

6-19. What percentage of the offspring will be "true-breeding"?

6-20. Look again at the trihybrid testcross on page 126. If the cross + ss +/cu + sr × cu ss sr/cu ss sr had been made instead, what would be the percentage of (a) + ss + and cu + sr, (b) + + + and cu ss sr flies in the progeny, assuming the same crossover frequencies and the same interference?

6-21. In *Drosophila*, these genes occur on chromosome III:

+ wild   h   hairy (extra hairs on scutellars and head)
+ wild   fz   frizzled (thoracic hairs turn inward)
+ wild   eg   eagle (wings spread and raised)

The cross + + +/h fz eg × h fz eg/h fz eg yielded this F_1:

wild wild wild	393	wild wild eagle	28
hairy frizzled eagle	409	hairy frizzled wild	30
wild frizzled eagle	58	wild frizzled wild	1
hairy wild wild	80	hairy wild eagle	1

(a) Give the sequence of genes and the distances between them. (b) What is the coincidence?

6-22. The cross $+\ +\ +/abc$ ($♀$) $\times$ abc/abc ($♂$) in the fruit fly gives the following crossover results:

a-b single crossovers 5.75 per cent
b-c single crossovers 8.08 per cent
a-c double crossovers 0.25 per cent

What is the coincidence?

6-23. In tomato the following genes are located on chromosome 2:

+ tall plant   d dwarf plant
+ uniformly green leaves   m mottled green leaves
+ smooth fruit   p pubescent (hairy) fruit

Results of the cross $+\ +\ +/d\ m\ p \times d\ m\ p/d\ m\ p$ were

+	+	+	470	+	m	p	1
+	+	p	14	d	+	p	25
d	+	+	0	d	m	p	441
+	m	+	19	d	m	+	30

(a) Which groups in the progeny represent double crossovers? (b) What is the correct gene sequence? (c) What are the distances in map units between the first and second, and between the second and third genes? (d) Is there interference?

6-24. From the chromosome map for maize (corn), Figure 6-5, note that genes pg_{12}, gl_{15}, and bk_2 are all on chromosome 9. The testcross $+\ +\ +/pg_{12}\ gl_{15}\ bk_2 \times pg_{12}\ gl_{15}\ bk_2/pg_{12}\ gl_{15}\ bk_2$ is made. If there is complete interference, what is the frequency of (a) noncrossovers, (b) $pg_{12}\ gl_{15}$ single crossovers, (c) $gl_{15}\ bk_2$ single crossovers, (d) double crossovers in the progeny? (Assume all genotypes to be equally viable.)

6-25. With the cross of problem 6-24, what would be the frequencies of each of the eight progeny classes (a) if there is no interference, (b) if coincidence is 0.5? (Assume all genotypes to be equally viable, and crossover probabilities to be equal in all parts of the chromosome.)

6-26. As noted in this chapter, genes for elliptocytosis and the Rh blood antigen D are linked on human autosome 1. Some studies suggest a distance of three map units between these loci. If a doubly heterozygous man, known to have these two genes linked in the *cis* configuration, marries an Rh negative woman who has normal, disk-shaped red blood cells, and if coincidence is assumed to be 1, what is the probability of each of the following phenotypes among their children: (a) Rh+, with elliptocytosis, (b) Rh+, without elliptocytosis?

CHAPTER 7
Multiple Alleles and Blood Group Inheritance

Iɴ our discussion thus far, we have assumed that a particular chromosomal position, or **locus,** is occupied by either of two alleles. Many instances are now known, however, in which a given locus may bear any one of a series of several alleles, so that a diploid individual possesses any two genes of the series. When any of three or more genes may occupy the same locus in a given pair of homologous chromosomes they are said to constitute a series of **multiple alleles.**

The Concept of Multiple Alleles
COAT COLOR IN RABBITS

The coat of the ordinary (wild-type) rabbit is referred to as "agouti" or full color, in which individuals have banded hairs, the portion nearest the skin being gray, succeeded by a yellow band, and finally a black or brown tip (Fig. 7-1). Albino rabbits, totally lacking in pigmentation, have also long been known (Fig. 7-2). Crosses of homozygous agouti and albino individuals produce a uniform agouti F_1; interbreeding of the F_1 produces an F_2 ratio of

FIGURE 7-1. *Wild-type agouti rabbit* (Photo courtesy American Genetic Association.)

FIGURE 7-2. *Albino rabbit. Albino animals are totally lacking in pigment.* (Photo courtesy American Genetic Association.)

3 agouti : 1 albino. Two thirds of these latter agouti individuals can be shown by testcrosses to be heterozygous. Clearly then, this is a case of monohybrid inheritance, with agouti completely dominant to albino.

Other individuals, lacking yellow pigment in the coat, have a silvery-gray appearance because of the optical effect of their black and gray hairs. This phenotype is referred to as chinchilla (Fig. 7-3). Crosses between chinchilla and agouti produce all agouti individuals in the F_1 and a 3 agouti : 1 chinchilla ratio in the F_2. Thus genes determining chinchilla and agouti appear to be alleles, with agouti again dominant. If, however, the cross chinchilla × albino is made, the F_1 are all chinchilla, and the F_2 shows 3 chinchilla : 1 albino. Therefore genes for chinchilla and albino are also alleles, and agouti, chinchilla, albino are said to form a multiple allele series.

Still another phenotype is often encountered in pet shops. This is Himalayan (Fig. 7-4), in which the coat is white except for black extremities (nose, ears, feet, and tail). Eyes are pigmented, unlike albino. By appropriate crosses it can be shown that the gene for Himalayan is dominant to that for albino, but recessive to those determining agouti and chinchilla. Some of the possible crosses, with F_1 and F_2 progeny, are shown in Figure 7-5. Gene symbols often assigned are c^+ (agouti), c^{ch} (chinchilla), c^h (Himalayan), and c (albino). From these crosses, we see the following dominance interrelationships:

$$c^+ > c^{ch} > c^h > c$$

It is easy, of course, to predict F_1 and F_2 progeny for two crosses not shown in Figure 7-5, such as agouti × Himalayan or chinchilla × albino.

FIGURE 7-3. *Chinchilla rabbit.* (Photo courtesy American Genetic Association.)

FIGURE 7-4. *Himalayan rabbit.* (Photo courtesy American Genetic Association.)

Phenotypes and their associated genotypes, therefore, for this series in rabbit are as follows:

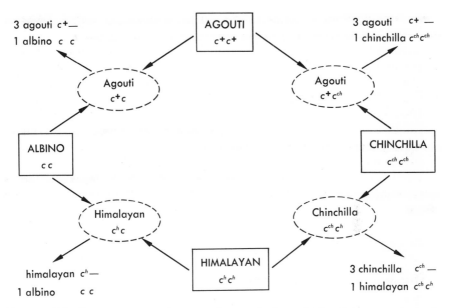

FIGURE 7-5. *Diagram representing several crosses in the multiple-allele coat series in the rabbit. Parental generations in solid rectangles. F_1 generations within dashed ovals. F_2 generations not enclosed.*

Phenotype	Genotype
Agouti	c^+c^+, c^+c^{ch}, c^+c^h, c^+c
Chinchilla	$c^{ch}c^{ch}$, $c^{ch}c^h$, $c^{ch}c$
Himalayan	c^hc^h, c^hc
Albino	cc

Ten different genotypes occur in this series. Earlier (Chapter 2), we saw that a single pair of alleles at a given locus produces three genotypes where dominance is complete. By the same token, a series of three multiple alleles produces six genotypes. Note that as the number of genes in a series of multiple alleles increases, the variety of genotypes rises still more rapidly:

Number of Alleles in Series	Number of Genotypes
2	3
3	6
4	10
5	15
n	$\frac{n}{2}(n + 1)$

With the number of possible genotypes increasing more rapidly than the number of alleles, a considerable increase in genetic variability ensues. Consider, for example, a hypothetical organism having only 100 loci, with a series of exactly four multiple alleles at each locus. Ten genotypes are possible at the first locus; these 10 can be combined with any of the 10 at the second locus, and so on. The total number of possible genotypes becomes 10^{100}, or 1 followed by 100 zeros!

Available evidence indicates that a given locus may mutate in several directions many times in the history of a species. The various members of the series in rabbit undoubtedly arose at different times and places as mutations of an ancestral gene, quite possibly c^+. Small wonder, then, that many apparent cases of one-pair differences ultimately turn out to involve series of multiple alleles! Other instances in animals and plants are referred to in the problems at the close of this chapter, though some interesting series in man are described in the following sections.

The Blood Groups in Man

THE ANTIGEN-ANTIBODY REACTION

Blood consists of two principal components: cells (red, white, and platelets) and liquid (plasma). Plasma, minus the clotting protein fibrinogen, is referred to as serum. In early attempts at transfusion, in fact as long ago as the eighteenth century, death of the recipient sometimes ensued for no determinable reason. But in 1901, Dr. Karl Landsteiner, working in a laboratory in Vienna, observed that red blood cells (erythrocytes) of certain individuals would clump together into macroscopically visible groups when mixed with the serum of some, but not all, other persons.

The basis for this clumping is the **antigen-antibody reaction,** an understanding of which is helpful at this point. Injection of a foreign substance (**antigen**) into the bloodstream of an animal brings about the production by some component of the blood of a characteristic **antibody** that reacts with the antigen. The antigen is ordinarily a protein, at least in part, and may be some plant or animal protein, a bacterial toxin, or even derived from pollen. The antibody is highly specific for a particular antigen (though cross reactions of varying degree may occur between one antibody and other closely similar antigen molecules). Such antibodies are termed *acquired,* because their production depends upon the entry of the foreign antigen; they are not otherwise produced. These form the basis of immunization practices as well as of allergic reactions. On the other hand, in a few cases, antibodies are produced naturally and normally by the blood, even in the absence of the appropriate antigen. These *natural antibodies* include several of those involved in human blood groups, particularly the important A-B-AB-O groups, which we shall discuss shortly.

Depending on the nature of the antigen and of its antibody, numerous sorts of each can be differentiated, each producing its own typical antigen-antibody reaction. If, for example, the antigen is a *toxin* (such as produced by typhoid, cholera, staphylococcus, whooping cough, and many other bacteria, or such substances as snake venom), neutralizing antibodies are called *antitoxins.* If the antigen is cellular in nature, the antibody may be a *lysin,* which lyses or disintegrates the invading cells, or an *agglutinin,* which causes clumping or agglutination of the cells. These are but a few of the recognized antibody types.

Following Landsteiner's discovery of agglutination of red blood cells and an understanding of the antigen-antibody reaction, further study by a number of investigators disclosed the occurrence of two natural antibodies in blood serum and two antigens on the surface of the erythrocytes. With regard to antigens, an individual may produce either, both, or neither; he may produce either, neither, or both antibodies. After some early and confusing multiplicity of nomenclature for these substances, the system in most general use today designates the antigens as A and B, and the corresponding antibodies as anti-A (or α) and anti-B (or β). Chemically, the A and B antigens are mucopolysaccharides, consisting of a protein and a sugar. The protein portion is identical in both antigens; it is the sugar that is the basis for the antigen-antibody specificity. An individual's blood group is denoted by the type of antigen he produces, as indicated in Table 7-1; blood tests for each of the four major groups are shown in Figure 7-6.

TABLE 7-1. Antigens and Antibodies of the Human
Blood Groups

Blood Group	Antigen on Erythrocytes	Antibody in Serum
A	A	anti-B
B	B	anti-A
AB	A and B	neither
O	neither	anti-A and anti-B

A rather uncommon subgroup of A was discovered in 1911 so that group A was subdivided into A_1 and A_2. More recently, a still rarer subgroup, A_3, has been found; an even less common subgroup, A_4, is now known. Three slightly different variants of B are also reported. Groups O and A are the most common in the United States population. Large samples show the approximate frequencies to be A, 0.42; B, 0.10; AB, 0.04; O, 0.44. Group A_1 is by far the most frequently encountered in A and AB persons; for example, of 1,210 genetics students at Ohio Wesleyan University tested over a nine-year period,

Anti-A   Anti-B   Blood Group

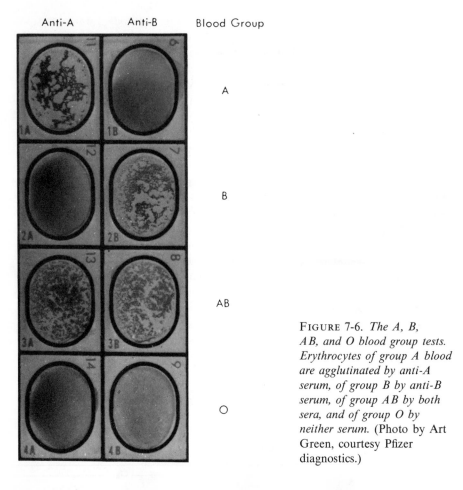

A

B

AB

O

FIGURE 7-6. *The A, B, AB, and O blood group tests. Erythrocytes of group A blood are agglutinated by anti-A serum, of group B by anti-B serum, of group AB by both sera, and of group O by neither serum.* (Photo by Art Green, courtesy Pfizer diagnostics.)

only one A_2 was encountered and that was an A_2B person. No A_3 or A_4 individuals were found, either as A or as AB. Although the A, B, AB, and O groups are important in transfusions, subgroups are relevant to certain legal problems. Therefore, we will direct our attention first to the genetics of the A, B, AB, and O blood groups.

INHERITANCE OF A, B, AB, AND O BLOOD GROUPS

Studies of large numbers of human pedigrees have shown that children produce the A antigen only if at least one parent also produces it. Similarly, the B antigen is found only in individuals where at least one parent has it. However, group O individuals may occur in the progeny of A and/or B parents, but O parents have only O children, suggesting recessiveness of the gene for group O. Yet marriages of A and B parents produce, in some cases, children having both A and B antigens, indicative of *codominance* of the genes for these latter antigens. Table 7-2 presents a summary of progeny phenotypes.

Pedigree analysis clearly shows that an individual possesses, in either the homozygous or heterozygous state, any two of a series of *multiple alleles*. Because the antigens involved are of the type known as isoagglutinogens (or isohemagglutinogens), these genes are often designated as I^A, I^B, and i. Neglecting for the moment the subgroups, dominance relationships of these three alleles may be represented thus: $(I^A = I^B) > i$. Additional studies that take into account the subgroups of the A antigen indicate that I^A may occur in at least four allelic forms, as we have noted; these are symbolized I^{A_1}, I^{A_2}, I^{A_3}, and I^{A_4}. I^{A_1} is dominant to all other I^A alleles, I^{A_2} is recessive to I^{A_1} but dominant to the other three, and so on. Considering four forms of I^A, one of I^B, and one of i, dominance within the series may be shown in this way:

$$[(I^{A_1} > I^{A_2} > I^{A_3} > I^{A_4}) = I^B] > i$$

Omitting the very rare I^{A_4}, this series of multiple alleles produces 15 genotypes and eight phenotypes:

Genotype	Phenotype	Genotype	Phenotype
$I^{A_1} I^{A_1}$		$I^{A_1} I^B$	A_1B
$I^{A_1} I^{A_2}$			
$I^{A_1} I^{A_3}$	A_1	$I^{A_2} I^B$	A_2B
$I^{A_1} i$			
		$I^{A_3} I^B$	A_3B
$I^{A_2} I^{A_2}$			
$I^{A_2} I^{A_3}$	A_2	$I^B I^B$	B
$I^{A_2} i$		$I^B i$	
$I^{A_3} I^{A_3}$			
$I^{A_3} i$	A_3	$i i$	O

Curiously enough, apparent changes in one's A-B-O phenotype are associated with certain pathologic conditions. Best known is a weakening of A or B antigenic reaction by a variable proportion of erythrocytes in persons suffering acute myeloblastic leukemia. This is especially well documented in the case of A persons; in a few instances the diminished response to anti-A serum is reported to be accompanied by a weak reaction to anti-B in leukemic persons of group A. During remissions normal antigenic response of the red blood cells rises somewhat, only to fall again during relapses. The basis of this change is not known, but is thought to be related to chromosome changes occurring in leukemia (Salmon et al., 1967). In addition to changes associated with leukemia, certain bacterial infections sometimes cause group A cells to acquire a

TABLE 7-2. Inheritance of Blood Group Phenotypes in Man*

Parental Groups →	A_1	A_2	A_3	B	O	A_1B	A_2B	A_3B
A_1	A_1 A_2 A_3 O	A_1 A_2 A_3 O	A_1 A_2 A_3 O	A_1 A_2 A_3 A_1B A_2B A_3B B O	A_1 A_2 A_3 O	A_1 A_1B A_2B A_3B B	A_1 A_2 A_1B A_2B A_3B B	A_1 A_2 A_3 A_1B A_2B A_3B B
A_2		A_2 A_3 O	A_2 A_3 O	A_2 A_3 A_2B A_3B B O	A_2 A_3 O	A_1 A_2B A_3B B	A_2 A_2B A_3B B	A_2 A_3 A_2B A_3B B
A_3			A_3 O	A_3 A_3B B O	A_3 O	A_1 A_3B B	A_2 A_3B B	A_3 A_3B B
B				B O	B O	A_1 A_1B B	A_2 A_2B B	A_3 A_3B B
O					O	A_1 B	A_2 B	A_3 B
A_1B						A_1 A_1B B	A_1 A_1B A_2B B	A_1 A_1B A_3B B
A_2B							A_2 A_2B B	A_2 A_2B A_3B B
A_3B								A_3 A_3B B

* (Group A is arbitrarily shown as consisting of three subtypes. Phenotypes of parents are listed across the top and at the left; phenotypes of possible children are shown in the body of the table.)

weak response to anti-B serum. This has been explained on the basis of adsorption of a bacterial polysaccharide, chemically similar to the B antigen, on the A red cells (Race and Sanger, 1968). Except for such unusual situations, one's blood antigen-antibody traits are constant and accurate reflections of his genotype.

[handwritten margin note: ?]

THE H ANTIGEN

[handwritten annotation: precursor ⟹ H- antigen $\xrightarrow{I^A \text{ and } I^B}$ antigen A or B / H ⟹]

Antigens A and B are synthesized from a <u>precursor</u> mucopolysaccharide in the presence of the dominant allele of another pair designated as *H* and *h*. With genotypes *HH* or *Hh*, the precursor is converted to an H antigen, which, in turn, in the presence of I^A and/or I^B, is *partly* converted to antigens A and/or B (Fig. 7-7). Gene *h* is termed an *amorph* in that it is responsible for no demonstrable product. So long as they are of genotype *H−*, group A individuals produce antigens A and H, group B persons test positively for antigens B and H, and group AB persons produce antigens A, B, and H. Individuals of group O produce only antigen H if their genotype is *iiH−*. On the other hand, blood of persons of genotype *− −hh* (Fig. 7-7) gives no reaction with anti-A, anti-B, or anti-H; this is the very rare *Bombay phenotype,* so named because it was first described in a family from that city. A similar case was described by Levine et al. (1955) in an American family of Italian descent, part of whose pedigree is shown in Figure 7-8. Gene *h* has an extremely low frequency; the Bombay phenotype has been described in the literature for fewer than 30 cases. Bhatia and Sanghvi (1962) estimate the frequency of the Bombay phenotype as only 1 in 13,000 in the Bombay area, and Race and Sanger (1968) report no cases have been found in England in more than 1 million persons tested.

Because the H antigen is partly converted to antigens A and/or B in the presence of suitable genotypes as shown in Figure 7-7, it should be expected that not all *− −H−* genotypes would exhibit an equally strong reaction to anti-H serum, and this is true. Blood of persons of genotype *iiH−* (group O) show the strongest anti-H reaction, those of AB the weakest. Within the subtypes of A, A_3 reacts more strongly than A_2, which, in turn, reacts more strongly than A_1. A_1 blood gives a stronger reaction to anti-H than does group B.

MEDICOLEGAL ASPECTS OF THE A-B-O SERIES

[handwritten annotation: since none of the H antigen was converted to either antigen A or B]

From Table 7-2 and a knowledge of the dominance relationships of the multiple allele series involved, from which the data of the table are derived, applications to cases of disputed parentage can readily be seen. Although mixups are rare today in hospitals, situations have occurred in the past where one or more sets of parents believed they were given someone else's child upon discharge of mother and new baby from the hospital. For example, a court case of some years ago involved just such a situation. Two sets of parents had taken babies home from a particular hospital at about the same time; in the process, the identification bracelet of the infant taken by family 1 had

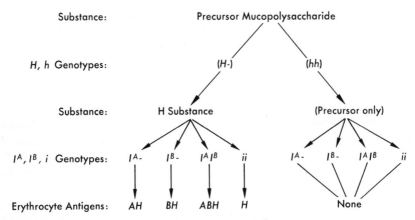

FIGURE 7-7. *Pathways leading to production of antigens on the red blood cells. The precursor mucopolysaccharide is converted into H substance in the presence of genotype H- and is, itself, partly converted into A and/or B antigens in the presence of genes* I^A *and/or* I^B, *with* I^{A_1} *more effective than* I^{A_2}, *and* I^{A_3} *least effective. The very rare gene* h *(for lack of H substance) is epistatic to the multiple alleles at the A-B-O locus. Cells of persons of --hh genotype give no reaction with anti-A or anti-B sera (even though they possess genes* I^A *or* I^B*; this is the rare Bombay phenotype).*

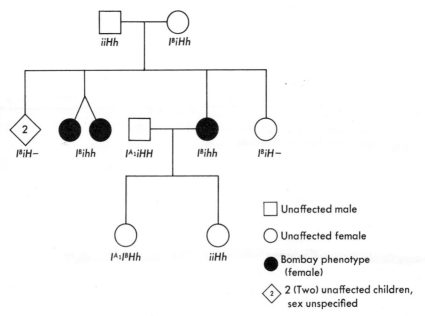

FIGURE 7-8. *Partial pedigree of a family with three children having the Bombay pheno-type. The fact that these three are all girls is merely coincidental.* (Based on work of Levine, Robinson, Celano, Briggs, and Falkinburg, 1955.)

become detached. Family 2 soon discovered family 1's name on the child they had received, but the latter family would not agree to an exchange. Fortunately, blood tests quickly demonstrated that neither child could have belonged to the family that had taken it home, but each *could* belong to the other parents:

	Parental Blood Groups	Blood Group of Child Taken Home
Family 1	A × AB	O
Family 2	O × O	B

An exchange thereupon satisfied both families. Obviously, the tests in this case did not *prove* that the child received by family 1 belonged to family 2, but only that it *could.* Suppose the two families had been A × B and O × B, and that the children given to each had been found to be O and B, respectively. With no additional information from other tests (some of which are described later in this chapter and in the next) it would be manifestly impossible to make a valid decision, because either family could have produced either child.

Quite clearly, too, blood tests are of considerable value in cases of illegitimacy. Again, tests cannot *prove* a man to be the father, but they can in some instances show that he could *not* be. Using only the ABO, MN, and Rh systems,[1] the probability of exclusion for a man wrongly accused is about 0.53. Adding the Kell, Lutheran, Duffy, and Kidd systems, plus the secretor trait (all described briefly later in this chapter), raises the probability to about 0.71. Making use of even more tests, and with sophisticated techniques, it is possible that the rate of exclusion could eventually reach 0.8 to 0.9.

Unfortunately, all courts do not accord this kind of evidence the same weight, although courts and legislatures have made considerable progress since 1950 in updating laws and court procedures in the light of scientific fact. In some states the results of blood tests, properly performed and presented by expert witnesses, are definitive, or conclusive, evidence if they establish non-paternity; in some, blood tests constitute admissible or introduced evidence, to be accorded no greater weight than any other. Some of the more realistic laws and practices concerning blood tests include those of

Alabama: admissible only if definite exclusion is established; courts are permitted to order blood tests (statute).
California: conclusive evidence (statute).
Colorado: defendant entitled to have blood tests received in evidence when exclusion is indicated (statute).
Connecticut: admissible only if definite exclusion is established (statute).

---

[1] The MN groups are described later in this chapter; the Rh in Chapter 8.

Kentucky: court authorized, at its discretion, to admit tests that show a possibility of paternity (statute).

Maine: admissible in combination with expert testimony when nonpaternity indicated (statute).

Massachusetts: admissible only if exclusion is established (statute).

Mississippi: conclusive evidence (statute).

New Hampshire: admissible if exclusion is established (statute).

New York: blood tests that demonstrate nonpaternity constitute conclusive evidence (statute).

North Carolina: admissible when presented by expert testimony (statute).

Ohio: admissible in combination with expert testimony when nonpaternity indicated (statute).

Pennyslvania: conclusive evidence if nonpaternity is established (statute).

Rhode Island: admissible if definite exclusion is established (statute).

Wisconsin: admissible if exclusion is established (statute).

Happily, the number of states in which blood tests may not even be introduced as evidence is decreasing.

It should be emphasized that, in states in which blood test evidence is not conclusive, considerable weight is given to the current or most recent decision. For example, in one opinion in Ohio, the judge wrote, "In accordance with the enlightened judicial acceptance of the high value of blood grouping tests properly conducted, I hold that, in the absence of any competent proof that blood grouping tests were not properly made, the results of such tests . . . should be given such great weight by the Court that the exclusion of the defendant as the father of the child follows irresistibly." In the juvenile court of one Ohio county, in a 14-year period, some 12,000 paternity suits were handled. Blood tests were made in 734 of these; in 104 exclusions were demonstrated. In 102 of these latter, the defendant won the decision or the case was dropped; of the remaining two, one won a second trial and the other (surprisingly) settled out of court.

Before legislatures and courts had "caught up" with scientific fact, many injustices had occurred. One of the most celebrated cases of this kind occurred late in 1944 in California. A widely known movie star was accused by a former starlet-protégé of being the father of her young daughter. The plaintiff's case rested on a remarkable memory for dates and details, a memory so precise that all other potential fathers were eliminated. Three physicians made blood tests of the alleged father, the mother, and the baby with these results:

	Blood Group
Alleged father	O ii
Mother	A $I^A I^A$ or $I^A i$
Daughter	B $I^B I^B$ or $I^B i$

A moment's reflection on the genetics of these phenotypes, or reference to Table 7-2, shows clearly that the defendant could not possibly have been the father (barring a highly improbable mutation). Rather, the real father must have belonged either to group B or AB. In spite of such scientific evidence, the jury in a second trial (the first ended in a "hung jury") found the defendant *guilty!* So far as can be determined, he was required to contribute for twenty-one years to the support of a child not his own.

The quantitative degree to which paternity cases will occupy the time and attention of the courts is probably already decreasing. The advent of oral contraceptives and other safe devices, and their increasing use, together with a decline in stigmatizing the "uniparental child" or its parent make it likely that, with the exception of forcible rape, only the ignorant or negligent woman will have an unwanted child. Nevertheless, rights of a wrongfully accused man, or a child whose future security is doubtful, should be safeguarded.

Blood tests involving the A-B-O series, as well as others to be described, may, of course, also be used to good advantage in cases of claimants to estates or in certain kinds of criminal proceedings, particularly since blood group may usually be determined from corpses. In fact, blood type can often be determined from mummies, and this has become an important anthropological tool in some investigations. Problems at the end of this chapter explore some hypothetical possibilities.

OTHER BLOOD PHENOTYPES

The Secretor Trait. As study of the A, B, AB, and O blood groups continued, it was noted that with some, but not all, persons whose erythrocytes bear A, B, and/or H antigens, these antigens could also be detected in such aqueous secretions as those from eyes, nose, and salivary glands. Persons with this trait are referred to as *secretors;* they produce water-soluble antigens. Several reports in the literature indicate that about 77 to 78 per cent of all persons tested are secretors. Individuals lacking this trait are termed *nonsecretors,* and their antigens are only alcohol soluble. Note that the secretor-nonsecretor phenotype can be determined only for *HH* or *Hh* genotypes. The blood group of *H—* secretors can be determined even from dried saliva.

Pedigree studies indicate a single pair of genes, *Se* and *se,* to be responsible. The secretor trait is completely dominant. This pair of genes markedly increases the number of blood phenotypes.

The M-N Series. In the course of their investigations of human blood antigens, Landsteiner and Levine in 1927 discovered two, M and N, which, when injected into rabbits or guinea pigs, stimulated antibody production in the serum of the experimental animal. Apparently, human beings do not produce their own antibodies for these antigens, so they are of no importance in transfusion. They are, however, of some interest in genetics, since inheritance of the trait depends upon a pair of codominant genes sometimes referred to as L^M and L^N (for Landsteiner), but more frequently now simply as *M* and

N, producing phenotypes as follows:

Phenotype	Antigen Produced
M	antigen M only
MN	antigens M *and* N
N	antigen N only

No allele for the absence of either antigen is known. Statistical analysis often reflects the 1:2:1 phenotypic ratio which may occur in instances of codominance. For example, one study of 6,129 individuals in this country included 29.2 per cent M, 49.6 per cent MN, and 21.2 per cent N type persons. It should be noted, however, that the *frequencies* of genes *M* and *N* must have been close to 0.5 each in the parents of these 6,129 individuals in order for this progeny ratio to be produced. Frequencies of the members of a pair of alleles, of course, need not necessarily be close to equality in a randomly mating group, and progeny ratios depend in large part on the frequencies with which the alleles occur. This topic is explored and extended in Chapter 14.

Another pair of antigens, S and s, with intimate genetic relation to the M-N series, was discovered in 1947. Unfortunately, the designation of antigens has developed rather randomly, so that we have here two sets of antigens, one denoted by two different capital letters (M and N), the other by the same letter in upper and lower case (S and s). Studies show that all human beings produce either antigen S, antigen s, or both; therefore another pair of codominant genes (now generally designated as *S* and *s*) is involved.

As indicated, there is a close genetic relationship between the *M-N* and *S-s* pairs of genes. For example, in families where the parents are phenotypically MNSs and NS, the children, with very rare exceptions, fall into either of two categories: (1) MNS and NSs, or (2) MNSs and NS. Clearly in the first of these two cases, children received genes for either MS or Ns from the heterozygous parent, whereas in the second they received either genes for Ms or NS. Earlier explanations favored a series of four multiple alleles at a single locus: M^S, M^s, N^S, and N^s and, indeed, such results as those just cited could rest on a multiple allele mechanism. However, Race and Sanger (1968), in their extensive studies on human blood groups, prefer an alternate explanation, i.e., two pairs of very closely linked codominant genes, *M-N*, and *S-s*. This conclusion is based on rare instances of recombination between the two suggested loci. On the basis of the latter assumption, the two cases described in this paragraph may be diagramed as follows:

$$\text{Case 1:}\quad \text{P}\quad MS/Ns \times NS/NS$$
$$\text{F}_1\quad \tfrac{1}{2}\,MS/NS + \tfrac{1}{2}\,Ns/NS$$

Case 2: P $\quad Ms/NS \times NS/NS$

$\qquad$ $F_1 \quad \frac{1}{2} Ms/NS + \frac{1}{2} NS/NS$

Upward of a dozen and a half other antigens, most of them quite uncommon, have been shown to be related to the M-N and S-s substances. The explanation probably lies in a sequential series of gene-mediated reactions on one or a few precursor substances.

Other Antigens. Many other blood antigens have been described in the literature, some quite rare. These are usually designated by the family name of the individual in whom the antigen or antibody was first demonstrated. Thus we have the Kidd factor (about 77 per cent of the tested United States population is reported to be Kidd-positive) and the Cellano, Duffy, Kell, Lervis, and Lutheran factors (all named for the family in which each was first discovered), to list but a few.

These and a large number of additional antigens and/or antibodies produce a great diversity of human blood groups. In one test of 475 persons in London for A_1-A_2-B antigens, the M-N-S-s series, Rh, Kell, Lutheran, and Lervis groups, 269 types were reported, of which 211 included only a single person each. Considering the presently known groups and the fact that additional ones continue to be reported, the already large number of blood phenotypes may someday rise to the point where an individual's blood group may identify him as certainly as do his fingerprints. Only identical twins, triplets, etc., who have identical genotypes, would then be indistinguishable by appropriate tests.

REFERENCES

BHATIA, H. M., and L. D. SANGHVI, 1962. Rare Blood Groups and Consanguinity: "Bombay" Phenotype. *Vox Sang.,* **7:** 245–248.

LANDSTEINER, K., 1901. Über Agglutinationserscheinungen Normalen Menschlichen Blutes. *Wien Klin. Wochenschr.,* **14:** 1132–1134. Reprinted (in English) in S. H. Boyer, ed., 1963. *Papers on Human Genetics.* Englewood Cliffs, N.J., Prentice-Hall.

LANDSTEINER, K., and P. LEVINE, 1927. A New Agglutinable Factor Differentiating Individual Human Bloods. *Proc. Soc. Exp. Biol. N.Y.,* **24:** 600–602.

LEVINE, P., E. ROBINSON, M. CELANO, O. BRIGGS, and L. FALKINBURG, 1955. Gene Interaction Resulting in Suppression of Blood Group Substance B. *Blood,* **10:** 1100–1108.

RACE, R. R., and R. SANGER, 1968. 5th ed. *Blood Groups in Man.* Philadelphia, F. A. Davis Company.

SALMON, C., A. JACQUET, C. KLING, and D. SALMON, 1967. Analogie d'Affinité entre un Antigène B, Modifié par la Leucémie chez un Sujet A_1B, et un Antigène B Partiel Induit par un Chromosome cis A_1B. *Nouv. Rev. r. Hemat.,* **7:** 755–764.

WALSH, R. J., and C. MONTGOMERY, 1947. A New Human Isoagglutinin Subdividing the MN Blood Group System. *Nature,* **160:** 504.

PROBLEMS

(handwritten margin note: cᶜʰxcᴴ, c̄c̄xc̄)

7-1. Is it possible to cross two agouti rabbits and produce both chinchilla and Himalayan progeny?

7-2. A series of rabbit matings, chinchilla × Himalayan, produced a progeny ratio
of 1 Himalayan:2 chinchilla:1 albino. What were the parental genotypes?

7-3. In the ornamental flowering plant nasturtium, flowers may be either single,
double, or superdouble. These differ in number of petals, superdouble having
the largest number. Crosses of superdouble × double sometimes yield 1 super-
double:1 double, and sometimes all superdouble. Superdouble × superdouble
produces all superdouble, or 3 superdouble:1 double, or 3 superdouble:1
single. Single × single produces only single. (a) How many multiple alleles occur
in this series? (b) Arrange the phenotypes in order of relative dominance.
(c) Another cross of superdouble × double produces progeny in the ratio of
1 double:2 superdouble:1 single. What do you know about the parental genotypes?

(handwritten margin notes: "must be multiple alleles"; "same one of progeny is normal?")

Use the following information in answering the next four problems. In the
Chinese primrose the flower has a center, or "eye," of a color different from
the remainder of the petals. Normally this eye is of medium size and yellow in
color. These variants also occur: very large yellow eye ("Primrose Queen"
variety), white eye ("Alexandra"), and blue eye ("Blue Moon"). Results of
certain crosses are the following:

P	F₁	F₂
Normal × Alexandra	Alexandra	3 Alexandra:1 Normal
Alexandra × Primrose Queen	Alexandra	(not reported)
Blue Moon × Normal	Normal	3 Normal:1 Blue Moon
Primrose Queen × Blue Moon	Blue Moon	(not reported)

7-4. Arrange these phenotypes in order of relative dominance.

7-5. How many genotypes can produce the "Alexandra" phenotype?

7-6. How many genotypes can produce the "Primrose Queen" phenotype?

7-7. How many different combinations of parental genotypes will produce a progeny
ratio of 3 Alexandra:1 normal?

Use the following information in the next four problems. A series of multiple
alleles for coat color is reported in the mouse. One series of breeding results was
as follows:

P	F₁
plain black × white-bellied	white-bellied
plain black × dark-bellied	dark-bellied
white-bellied × dark-bellied	white-bellied

7-8. What should be the F₁ of the cross plain black × plain black?

7-9. The cross of two heterozygous dark-bellied animals will produce what F₁
phenotypic ratio?

7-10. A cross between two white-bellied animals produced an F₁ ratio of 3 white-
bellied:1 dark-bellied. What were the parental genotypes?

7-11. Later, an additional allele in this series, producing black-and-tan coat, was discovered. Crosses of black-and-tan with white-bellied produce all white-bellied in the F_1; black-and-tan $\times$ dark-bellied produce all dark-bellied in the F_1. Crossing a different black-and-tan with plain black produced 3 black-and-tan:1 plain black. What is the order of relative dominance among these four phenotypes? *must be in F_2*

7-12. A hypothetical series of 20 multiple alleles is known for a certain locus. How many phenotypic classes are possible?

7-13. How many different genotypic classes are possible for the locus referred to in problem 7-12?

7-14. Considering human blood group A to include 3 subtypes, and groups B and O to include 1 each, how many phenotypes are included in the A-B-O series?

7-15. Considering three subtypes of group A, which of the blood groups making up your answer to the preceding question would you expect to give the weakest anti-H response? *A_1 weaker than A_2 or A_3*

7-16. A paternity case involves these facts: the woman is A_1, her child is O, and the alleged father is B. Could he be the father? Explain. *Yes*

7-17. In a paternity case it is determined that the woman is A_1, MS/NS, and a secretor. The man she charges is O, MS/NS, and a secretor. The child is A_2, MS/Ns, and a nonsecretor. Using only this information, can you eliminate the man as the father of this child? Explain.

$$I^{A_1} \frac{I^{A_1}}{i} \times I^{B} \frac{I^{B}}{i}$$

$$\downarrow$$

$$i\,i$$

$$\left.\begin{array}{c} w > b \\ d > b \\ w > d \end{array}\right\} \text{ white} > \text{dark} > \text{black}$$

7-8) $bb \times bb \rightarrow$ all black

9) $db \times db \rightarrow \frac{1}{4}dd \; \frac{1}{2}db \; \frac{1}{4}bb$
   $\underline{\frac{3}{4}dark} \; \frac{1}{4}black$

10) $wd \times wd \; \underline{\underline{or}} \; wd \times wb$

11) $w > bt$
   $d > bt$
   $bt\,bt \times bb \rightarrow bt\,b$ in F_1
   Then $bt\,b \times bt\,b \rightarrow \frac{3}{4}$ black + tan : $\frac{1}{4}$ black in F_2
   So $\;\; w > d > bt > b$

CHAPTER 8
Pseudoalleles and the Rh Blood Factor

MID-TWENTIETH-CENTURY research has uncovered evidence in a number of different organisms that many traits, originally thought to be governed by multiple-allele series, are due to separate but extremely closely linked loci. Some of the earliest and best-known of these cases are found in *Drosophila;* an important trait in man, the Rh blood factor, can be explained on the basis of at least three such closely linked loci. We shall examine both the Rh factor and some early illustrations from *Drosophila* in this chapter; from these examples a new concept of the nature of the gene emerges.

DROSOPHILA

Eye Color. In *Drosophila* the wild-type eye is red; this phenotype is produced by a completely dominant gene at about locus 1.5 on chromosome I (the X chromosome; see also Chapter 11). Many other eye colors occur, among them coral, cherry, apricot, eosin, ivory, and white. Early work suggested that genes for these and some other eye colors formed a multiple-allele series, all mapping at the same locus. Under this hypothesis $+$ (or w^+) designated the gene for wild-type red; w^a, apricot; and w, white. As data were accumulated for crosses where progeny were expected to be of only two phenotypic classes, apricot and white, an occasional wild-type red appeared. The frequency of these red-eyed flies was low, less than one per thousand, but too high to be explained by mutation. By using the "marker genes" y (yellow body) and *spl* (split thoracic bristles) at loci 0.0 and 3.0, respectively, in the heterozygous condition, it was found that genes for apricot and white were at different, although very close, loci and thus could not be part of a multiple-allele series. The symbol for apricot was therefore changed to *apr*. From this work it was soon determined that (1) *apr* $+$ / $+$ w flies had apricot eyes, (2) $+$ $+$/*apr* w flies had red eyes, and (3) linkage relationships of *apr* and w did not change unless y and *spl* did also. Thus $+$ $+$ $+$ $+$/y *apr* w *spl* flies were red-eyed, whereas $+$ $+$ w *spl*/y *apr* $+$ $+$ flies, for example, had apricot eyes, demonstrating separate loci for *apr* and w. Frequency of crossing-over between *apr* and w was found to be less than 0.001. Loci *apr* and w are thus *functionally allelic* in that they both affect the same general phenotypic character (eye color). They are, however, *structurally nonallelic,* that is, they can be shown by recombination to occupy different loci. In the *cis* configuration they produce one phenotype, red eyes, but the *trans* configuration results in a different phenotype, apricot. The two loci exhibit **complementation** in the *cis* position, but **noncomplementation** in the *trans* position. This is referred to as a *cis–trans* position effect. If the *cis* and *trans* configurations produce different phenotypes, the two forms are parts of the same gene; but should the *cis* and *trans* configurations produce identical phenotypes, the two are considered to be different genes (Fig. 8-1).

Genes that (1) govern different expressions of the same trait, (2) map at different but closely adjacent loci, (3) show a low frequency of recombination by crossing-over, and (4) exhibit the *cis-trans* position effect are called **pseudoalleles.** A cluster of pseudoalleles is referred to as a *complex locus.* Pseudoallelism suggests that the gene cannot be defined as a unit of *both* function and structure or, to put it another way, the structural gene may be smaller than the functional gene. Evidently some product of the functional gene must be produced *in toto* in the *cis* configuration, whereas the separate parts of this

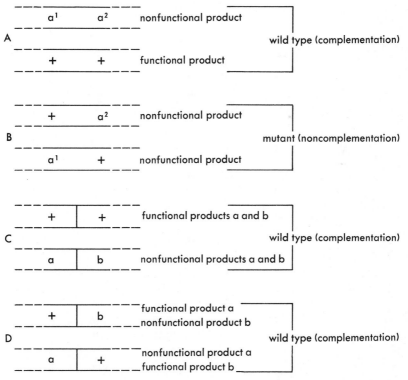

FIGURE 8-1. Cis-trans *effect. A functional product is produced only when the two loci are in the* cis *position* (A); *in the* trans *position no functional product is produced* (B). *Loci a¹ and a² exhibit complentation only in the* cis *configuration and are functionally allelic but structurally nonallelic. In* (C) *and* (D) *loci a and b exhibit complementation in both the* cis *and* trans *positions and are, therefore, considered to be different genes. See text for further details.*

product, produced in the *trans* configuration, cannot determine a normally functioning substance. The classical gene may therefore be subdivisible (Pontecorvo, 1968), consisting of many intragenic units of recombination and mutation. This concept and the light it sheds on the nature and operation of the gene will be more fully explored after we have examined the molecular nature of the genetic material.

The Lozenge Loci.   Study of complex loci began in 1940 with the work of C. P. Oliver on the so-called lozenge locus in *Drosophila*. In this organism the wild-type eye is broadly oval in shape (Fig. 8-2); in the recessive phenotype lozenge the eyes are narrowly ovoid with an irregular surface. The genes responsible are on chromosome I, mapping at about locus 27.7. Subsequent work, much of it by Green and Green (1949, 1961), showed four separate, closely linked loci (Fig. 8-3). These loci exhibit the *cis-trans* effect as in the preceding case. For example, the *cis* heterozygote $lz^s lz^k + +/+ + + +$ has the wild-type phenotype, but the *trans* heterozygote $lz^s + + +/+ lz^k + +$ is lozenge. Altogether more than a dozen alleles are known to occur at these four loci.

MAN: THE Rh FACTOR

History.   The now well-known Rh factor was discovered in 1940 by Landsteiner and Wiener, who reported that if a rabbit was injected with blood of the *Macaca rhesus* monkey, antibodies were formed by the rabbit. These antibodies would agglutinate the red blood cells of all rhesus monkeys. Thus monkey erythrocytes bear on their surfaces a specific antigen, designated as Rh. Tests of human beings show that most persons also produce this antigen; in fact some 85 per cent of white Americans and more than 91 per cent of black Americans do.[1] Such persons are designated as Rh-positive (Rh +); the much smaller percentage who do not produce the rhesus antigen are termed Rh-negative (Rh −).

Genetic Bases.   Evidence almost immediately indicated a genetic basis to the Rh phenotype, with Rh + the dominant trait. A single pair of genes, *R*, and *r*, was postulated, with Rh + persons having genotype *RR* or *Rr* and Rh − persons having genotype *rr*. Since the work of Landsteiner and Wiener more Rh antigens have been discovered; the number is now over 40, and the genetics is much more complex than originally believed. Race and Sanger (1968) and Giblett (1969) have presented detailed summaries of the complexity of the Rh blood groups. Two major explanatory theories have been developed, one by the American investigator Wiener, and a second, chiefly in England, by Fisher and others. Each has its own advantages and disadvantages.

The **Wiener** system postulates a series of at least 10 multiple alleles (Table 8-1), some common, others very rare, at a single locus. As noted in Chapter 6,

---

[1] Reasons for this kind of disparity are explored in Chapter 14. Actually, the rabbit and the human anti-Rh antibodies are not the same, a fact not important for our purposes.

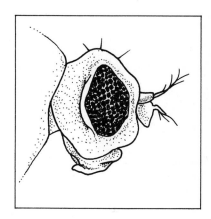

FIGURE 8-2. *Wild-type eye of* Drosophila melanogaster

this locus has since been identified as being on chromosome 1. Each, except the completely recessive gene *r*, is responsible for production of one or more of the Rh antigens. Gene *r*, for which no antigen has ever been conclusively demonstrated, is therefore termed an *amorph*.

The **Fisher** system, as elaborated by Race and Sanger (1968) and others, proposes a group of at least three very closely linked pseudoalleles *D, d, C, c, E,* and *e*. Evidence supports the sequence *D-C-E*. Instances of compound antigens (e.g., ce), a single reported case of possible crossing-over (Steinberg, 1965), and some apparent cases of position effect are more readily explained by the pseudoallele concept. For example, Race and Sanger point out that the *cis* heterozygote *Dce/DCE* produces compound antigen ce (among others), but in the *trans* configuration (e.g., *DCe/DcE*) this antigen is not produced.[2] Loci *c* and *e* thus illustrate the *cis-trans* position effect and are noncomplementary.

The case reported by Steinberg is interesting in that it can be explained either by recombination between closely linked pseudoalleles or by mutation within a single site, and Steinberg recognized this. In this instance the husband was determined to have genotype *DCe/dce* and the wife *dce/dce*. There were eight children, four *dce/dce*, three *DCe/dce*, and one *dCe/dce*. Steinberg was able to eliminate the possibility of illegitimacy of the *dCe/dce* child. Because evidence has become increasingly clear that the correct sequence of loci under the pseudoallele concept must be *D-C-E*, as first suggested by Fisher (1947), birth of this child must have been preceded by a crossover in the father.

Although difficulties are inherent in both the Wiener and Fisher systems, each also has its advantages, and both are in current use. A final judgment between the two cannot yet be made. A comparison of the two systems is shown in Table 8-1.

In the Wiener system each gene listed is responsible for formation of more than one antigen. Responsibility of a single gene for more than one phenotypic effect, often seemingly unrelated, is known as **pleiotropy.**

[2] Antigen ce is sometimes designated as antigen f.

TABLE 8-1. A Comparison of the Wiener and Fisher Rh Gene Systems and Their Antigens

Genes		Frequency*	Antigens									
Wiener	Fisher		Rh_0	rh′	rh″	hr′	hr″	rh^w	hr	rh			(Wiener)
			D	C	E	c	e	C^w	ce	Ce	CE	cE	(Fisher)
r	dce	0.385	−	−	−	+	+	−	+	−	−	−
r'	dCe	0.007	−	+	−	−	+	−	−	+	−	−
r''	dcE	0.007	−	−	+	+	−	−	−	−	−	+
r^y	dCE	rare	−	+	+	−	−	−	−	−	+	−
R^o	Dce	0.022	+	−	−	+	+	−	+	−	−	−
R^1	DCe	0.405	+	+	−	−	+	−	−	+	−	−
R^2	DcE	0.154	+	−	+	+	−	−	−	−	−	+
R^z	DCE	0.002	+	+	+	−	−	−	−	−	+	−
R^{1w}	DC^we	0.016	+	+	−	−	+	+	−	+	−	−
$R^{o''}$	D^uce	rare	weak	−	−	+	+	−	+	−	−	−

* Approximate, in white populations of western European affinities.

In this table a plus sign indicates production of a given antigen as well as agglutination of red cells by the corresponding antiserum. Some very rare genes are not included here.

Current practice designates as Rh+ any genotype that includes ability to produce antigen D (Rh_0); all others are designated Rh−.

Antigen C^w was first described by Callender and Race (1946).

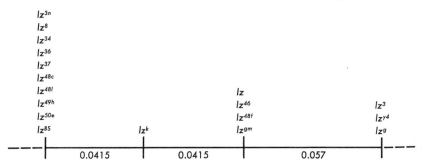

FIGURE 8-3. *The lozenge loci of* Drosophila melanogaster. *The horizontal line represents a portion of chromosome I, and the numbers below it show recombination frequencies. See text for details.*

Antigen D ($= Rh_o$) is the one that most commonly causes problems in transfusions and in certain pregnancies (see next section); therefore, under ordinary circumstances, persons are typed as Rh+ if their erythrocytes are agglutinated by anti-D (anti-Rh_o) antiserum. Thus an Rh− person (e.g., of genotype *dce/dce, dCE/dce,* and so on) does not produce antigen D (Rh_o), but does produce other Rh antigens, as shown in Table 8-1. Some interesting results of Rh typing can usually be observed when, say, a genetics class is tested. Erythrocytes of some persons are strongly agglutinated by anti-D antiserum (the one most commonly used, and alone, for this purpose), whereas blood of some other persons reacts much more weakly against anti-D. For example, red blood cells of persons of genotype *DCe/dce* (R^1/r) are strongly agglutinated by anti-D, but cells of persons having genotype *DCe/dCe* (R^1/r') react only weakly. The reason for this sort of difference is unknown; however, under the Fisher system, it can be classed as a position effect. The strong reaction given by *DCe/dce* could, then, be due to the fact that *d* and *c* are adjacent; the weaker reaction of *DCe/dCe* would result from *d* being next to *C*. It is not clear how such a position effect operates, however, although it would support the Fisher concept.

Some of the commoner genotypes for samples from the American population are listed in Table 8-2. Heiken and Rasmuson (1966) have published an extensive list of genotypic frequencies for a large sample of Swedish children (Table 8-3).

Erythroblastosis Fetalis. No cases are known of persons whose blood naturally contains anti-Rh antibodies, though Rh− individuals can and do develop them if exposed to the corresponding Rh antigen. Such exposure can occur by transfusion, and this, as noted earlier, is the reason the Rh type (ordinarily for production or nonproduction of the D, or Rh_o, antigen only) is now routinely determined for blood donors and recipients.

TABLE 8-2. Rh Phenotypes and Genotypes with Per Cent Frequencies
in the American Population

| Phenotype | Genotypes | | $n = 135$ | $n = 105$ | $n = 766$ |
	Wiener	Fisher	Black	Oklahoma Indian	White
Rh+	R^o/r	Dce/dce	45.9	2.9	2.2
Rh+	R^1/R^1	DCe/DCe	0.9	34.3	20.9
Rh+	R^1/r	DCe/dce	22.8	5.7	33.8
Rh+	R^2/R^2	DcE/DcE	16.3	17.1	14.9
Rh+	R^1/R^2	DCe/DcE	4.4	36.2	13.9
Rh+	R^1/R^z	DCe/DCE	0.0	2.9	0.1
Rh−	r/r	dce/dce	9.6	0.9	13.9

TABLE 8-3. Rh Phenotypes, Genotypes, and Per Cent Frequencies
in 8,297 Swedish Children

| Phenotype | Representative Probable Genotype | | Per Cent of Sample |
	Wiener	Fisher	
Rh+	R^o/r	Dce/dce	1.48
Rh+	R^2/R^2	DcE/DcE	3.08
Rh+	R^2/r	DcE/dce	12.50
Rh+	R^1/r	DCe/dce	32.72
Rh+	R^{1w}/r	DC^we/dce	1.45
Rh+	R^1/R^2	DCe/DcE	14.46
Rh+	R^z/R^2	DCE/DcE	0.05
Rh+	R^{1w}/R^2	DC^we/DcE	0.66
Rh+	R^1/R^1	DCe/DCe	16.16
Rh+	R^{1w}/R^z	DC^we/DCE	1.86
Rh+	R^z/R^1	DCE/DCe	0.05
		Total Rh+	84.47
Rh−	r/r	dce/dce	14.90
Rh−	r''/r	dcE/dce	0.22
Rh−	r'^w/r	dC^we/dce	0.02
Rh−	r'/r	dCe/dce	0.39
		Total Rh−	15.53

Based on work of Heiken and Rasmuson, 1966.

But anti-Rh antibody development can occur also in certain pregnancies,
often resulting in a fetal condition known as **erythroblastosis.** This is a hemolytic anemia in the fetus, often accompanied by jaundice as liver capillaries
become clogged with red blood cell remains and bile is absorbed by the blood.
The damaged erythrocytes are imperfect oxygen carriers and resemble the
immature cells of the marrow where they are normally formed. Death may
take place before birth or soon after unless appropriate corrective measures
are taken in time.

This disorder occurs only when a number of coincident conditions are met. The mother must be Rh−, the fetus Rh+; therefore only marriages of Rh− women and Rh+ men are involved. There must also be a placental defect whereby fetal blood, whose red cells carry the Rh antigen on their surfaces, passes from the embryo into the maternal circulation. This occurs largely just before or during birth. As a consequence, Rh antibody concentration is gradually built up in the mother; she will have been *sensitized*. In a second or subsequent pregnancy involving an Rh+ child these antibodies may return to the fetus where they destroy the antigen-carrying red cells. Buildup of antibodies in the mother is gradual; moreover, she is sensitized only at or just before birth of her first child (unless, of course, she had previously received an Rh+ transfusion).

Although fetal erythroblastosis is a severe and tragic condition, it is, fortunately, relatively infrequent. One study in Chicago showed only 91 affected children in 22,742 births in a seven-year period. This is a frequency of 0.004, a fairly average figure for erythroblastosis resulting from Rh incompatibility. The expected frequency of erythroblastotic births, in the absence of any complicating factors, can be calculated from the frequencies of genes *D* and *d,* following the methods introduced in Chapter 5. Disregarding loci *C* and *E*, Rh− persons are of genotype *dd,* and in the white American population occur with a frequency of about 0.15. This latter figure is the value of b^2 in the expansion of $(a + b)^2$. The frequency of the recessive gene *d* is then equal to $\sqrt{b^2}$ or $\sqrt{0.15} = 0.39$. Because we are dealing here with only one pair of alleles, $(a + b)^2 = 1$, and $a = 1 - b$, or 0.61, which is, then, the frequency of gene *D*. Only two types of marriages, $DD\male \times dd\female$ and $Dd\male \times dd\female$ can result in Rh+ children by Rh− women.[3] Frequencies of these marriages, based on the expectation of random mating, would be

$$DD\male \times dd\female: \quad a^2 \times b^2 = 0.61^2 \times 0.39^2 \qquad = 0.0566$$
$$Dd\male \times dd\female: \quad 2ab \times b^2 = 2(0.61 \times 0.39) \times 0.39^2 = 0.0723$$

In the $DD\male \times dd\female$ marriages, all the children are Rh+, whereas in the second the probability of an Rh+ child is 0.5; therefore the expected frequency of all erythroblastotic births (if all pregnancies come to term) is $0.0566 + 0.0361 = 0.0927$, or nearly 10 per cent of all pregnancies, neglecting the fact that in most cases the first child is unaffected. The 10 per cent figure is far greater than the observed frequencies of most studies.

The relative infrequency of erythroblastosis is in part due to other probably genetically based traits, such as the defective placenta. But another important factor is *ABO incompatibility.* ABO-compatible marriages are those in which

---

[3] The rare gene D^u is also implicated, but is not included in these calculations because of its very low frequency.

the husband is of suitable ABO group to donate blood to his wife; incompatible marriages are the reverse, namely, those in which the husband is not of an ABO group suitable for transfusion to his wife; that is,

	♀	♂
Compatible	A	A, O
	AB	A, B, AB, O
	B	B, O
	O	O
Incompatible	A	B, AB
	B	A, AB
	O	A, B, AB

Some studies show far fewer Rh-erythroblastotic children are born to ABO-incompatible marriages than to compatible ones. In ABO-incompatible marriages fetal erythrocytes, which bear an antigen for which the mother's blood contains the corresponding antibody and which cross the placenta, will be quickly destroyed before anti-Rh antibody formation can occur.

Discovery of two types of anti-D (anti-Rh$_0$) antibodies almost simultaneously in 1944 by Wiener and by Race has, fortunately, provided a simple preventive measure for Rh-erythroblastosis so that the condition need no longer be a cause of infant death or anemia. These two types of anti-D antibody are (1) *complete,* which agglutinates red cells carrying antigen D, and (2) *incomplete,* which does not. The incomplete antibody, however, attaches to receptor sites on Rh+ erythrocytes, "sensitizing" them and preventing them from acting antigenically. Intramuscular injection of human D (Rh$_0$) immune globulin, containing incomplete anti-D antibody, into Rh− women within 72 hours after giving birth to a $D-$ or D^u- child effectively blocks the child's red blood cells already in the mother's circulatory system from inducing antibody production on her part. The incomplete antibody is often obtained from Rh− males who have been injected with Rh+ blood.

ORIGIN OF PSEUDOALLELES

Should a chromosome suffer, for example, two breaks, with rejoining of the ends delayed until after chromosome replication, repair can result in a repeated, or duplicated, internal segment. Later these initially identical segments, perhaps consisting of a single gene each, might become functionally somewhat different through mutation and so constitute pseudoalleles. Other explanations are, of course, also possible. Mutation in one or both of two closely adjacent genes that produce different effects could result in two that perform similar functions. On the other hand, rearrangements of chromosomal segments following breaks might bring together in the same chromosome genes for similar products, genes originally located in different chromosomes.

REFERENCES

CALLENDER, S. T., and R. R. RACE, 1946. A Serological and Genetical Study of Multiple Antibodies Formed in Response to Blood Transfusion by a Patient with Lupus Erythematosus Diffusus. *Ann. Eugen.,* **13:** 103–117.

FISHER, R. A., 1947. The Rhesus Factor: A Study in Scientific Method. *Am. Scien.,* **35:** 95–103.

GIBLETT, E. R., 1969. *Genetic Markers in Human Blood.* Oxford, Blackwell Scientific Publications.

GREEN, M. M., 1961. Phenogenetics of the Lozenge Locus in *Drosophila melanogaster.* II. Genetics of Lozenge-Krivshenko (*lz^k*). *Genetics,* **46:** 1169–1176.

GREEN, M. M., 1963. Pseudoalleles and Recombination in *Drosophila.* In W. J. Burdette, ed. *Methodology in Basic Genetics.* San Francisco, Holden-Day.

GREEN, M. M., and K. C. GREEN, 1949. Crossing Over Between Alleles at the Lozenge Locus in *Drosophila melanogaster. Proc. Nat. Acad. Sci., (U.S.),* **35:** 586–591.

HEIKEN, A., and M. RASMUSON, 1966. Genetical Studies on the Rh Blood Group System. *Hereditas Lund,* **55:** 192–212.

LANDSTEINER, K., and A. S. WIENER, 1940. An Agglutinable Factor in Human Blood Recognized by Immune Sera from Rhesus Blood. *Proc. Soc. Exp. Biol. Med. N.Y.,* **43:** 223.

LEWIS, E. B., 1952. The Pseudoallelism of White and Apricot in *Drosophila melanogaster. Proc. Nat. Acad. Sci. (U.S.),* **38:** 953–961.

LEWIS, E. B., 1955. Some Aspects of Pseudoalleles. *Amer. Natur.,* **89:** 73.

PONTECORVO, G., 1958. *Trends in Genetic Analysis.* New York, Columbia University Press.

RACE, R. R., 1944. An 'Incomplete' Antibody in Human Serum. *Nature,* **153:** 771–772.

RACE, R. R., and R. SANGER, 5th ed., 1968. *Blood Groups in Man.* Philadelphia, F. A. Davis.

STEINBERG, A. G., 1965. Evidence for a Mutation or Crossing-over at the Rh Locus. *Vox Sang.,* **10:** 721.

WIENER, A. S., 1944. A New Test (Blocking Test) for Rh Sensitization. *Proc. Soc. Exp. Biol. Med. N.Y.,* **56:** 173–176.

PROBLEMS

8-1. What is the phenotype of each of the following genotypes in *Drosophila melanogaster:* (a) *lz^{BS}* + + +/+ *lz^k* + +; (b) *lz^{BS} lz^k* + +/+ + + +?

8-2. Consider the following genotypes and phenotypes in a hypothetical diploid organism:

Genotype	Phenotype
+ +/a b	wild
+ b/a +	wild
+ +/c d	wild
+ d/c +	mutant

Do any of these pairs of alleles appear to represent pseudoalleles? Explain.

8-3. Recall that in erythroblastosis fetalis the important genes are D and d and that, therefore, Rh+ persons can be represented as DD or Dd, and Rh− persons as dd. In the following pedigree the marriage shown is the first for both the man and the woman, and the woman has never had a blood transfusion. Solid symbols represent cases of erythroblastosis fetalis. After examining this pedigree give the Rh genotype of each member of the family.

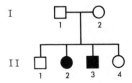

8-4. In the preceding problem why can you be sure of II-1's genotype and phenotype?

8-5. How should the pedigree of problem 8-3 be different had the mother received an injection of D-immune globulin within 72 hours of the birth of II-1?

8-6. In which of the following marriages is the risk of erythroblastosis fetalis greater: A − ♀ × O + ♂, or O − ♀ × A + ♂?

8-7. Considering human blood group A to include three subtypes, and groups B and O to include one each, how many phenotypes can be recognized if the A-B-O classes and the two Rh classes are all taken into account?

8-8. How many phenotypes are possible if A-B-O, Rh, and MN classes are all considered?

8-9. How many phenotypic classes are possible if the A-B-O, Rh, MN, Ss, and Hh phenotypes are considered together?

8-10. A woman of group A_2 charges a man of group A_2 as the father of her group A_3Rh+ child. Could he be?

8-11. Both the man and the woman of the preceding problem are Rh−. Does this change your judgment?

8-12. A couple believes they have brought the wrong baby home from the hospital. The wife is O+, her husband B+, and the child O−. Could the child be theirs?

8-13. An additional test discloses the husband and wife of the preceding problem to be of blood type M, whereas the child is MN. Does this added fact change your judgment?

8-14. An elderly couple is killed in an accident; no survivors are known. Their estate has been willed to charity. Later a man claims the estate on the grounds that he is their son who left home at an early age. It is known by friends that this couple had had a son, but he was thought to have died while quite young. Unfortunately, birth records for the period have been destroyed in a courthouse fire. Their child was born at home, the attending physician has been dead for some time, and his records are no longer available. From various hospital and medical records it is determined that the dead man was of blood type A, MS/Ns, Rh+; his wife was B, MS/NS, Rh−. The claimant's blood tests as O, Ms/Ns, Rh+. Does it appear that his claim is valid? Justify your conclusion.

8-15. In a court action a man claims that some of the six children purportedly belonging to him and his wife are not his children. Blood tests of the husband, wife, and six children gave the following information:

Husband: O,   *Dce/DcE, MS/Ms*
Wife: A_1,   *DcE/dce, MS/Ns*
Child 1: A_1,   *DcE/DcE, MS/MS*
Child 2:   O,   *Dce/dce, MS/Ns*
Child 3:   O,   *DcE/dce, Ms/Ns*
Child 4: A_1 *DcE/dce, NS/Ns*
Child 5:   O,   *dce/dce,   MS/NS*
Child 6: A_1B, *DCe/DcE, MS/NS*

Assuming all were, in fact, born to the wife, could all these children belong to the husband? Explain.

CHAPTER 9
Polygenic Inheritance

I F you think back over the heritable traits we have examined thus far in text and problems, you will note that phenotypic classes have always been distinct and easily separable from each other; that is, they have been sharply *discontinuous.* The traits themselves may be termed *qualitative* ones. Thus *Coleus* leaves have either regular or irregular venation; cattle have horns or they do not, and may be red, roan, or white; rabbits may be distinguished by coat color; people belong to one blood group or another, and so on. This has been the case whether we are dealing with form and structure, pigments, antigens and antibodies, and so on, and whether the genes involved show complete dominance, incomplete dominance, or codomiance.

Not all inherited traits are expressed in this discontinuous fashion, however. For example, in man height is a genetically determined trait. But if you were to attempt to classify a random sample of students on your campus according to height, you would quickly find that you were dealing with a trait showing essentially *continuous* phenotypic variation. Many other traits are expressed in a similar fashion, including intelligence, skin and eye color in man, color and food yield in various plants, size in many plants and animals, as well as degree of coat spotting in animals or of seed coat mottling in some plants. The essential difference between continuous and discontinuous inheritance is illustrated with generalized data from crosses of each type compared graphically in Figure 9-1. Clearly, our concepts of simple Mendelian inheritance, with the reappearance of the parental phenotypes as distinct and separate classes in the F_1 and F_2 generations, must be modified in order to explain continuous variation in *quantitative* characters, such as height, weight, intelligence, or color.

We are concerned not with just tall *versus* dwarf, but with *how* tall, that is, with *continuous characters of degree rather than discontinuous characters of kind.* Moreover, quantitative inheritance more often deals with a population in which all possible matings occur, and less often with individual matings.

Kernel Color in Wheat

Among the earliest investigations that gave a significant clue to the mechanism of quantitative inheritance was the work of Nilsson-Ehle (1909) with wheat. One of his crosses consisted of a red-kerneled variety × a white-kerneled strain. Grain from the F_1 was uniformly red, but of a shade intermediate between the red and white of the parental generation. This might suggest incomplete dominance, but by crossing members of the F_1 among

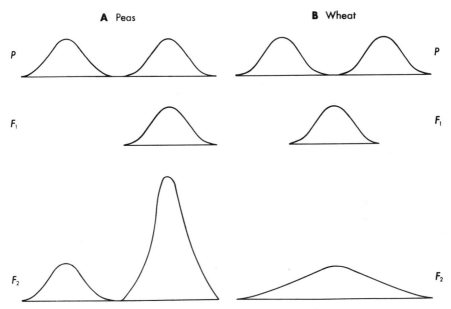

FIGURE 9-1. *Curves comparing results of crosses involving (A) height, a discontinuous trait in peas and (B) kernel color, a continuous trait in wheat, followed for three generations. Ordinates represent the number of individuals, abscissas the particular quantitative trait.*

themselves, Nilsson-Ehle produced an F_2 in which he was able to discern five phenotypic classes in a $1:4:6:4:1$ ratio. Noting that $\frac{1}{16}$ of the F_2 was as extreme in color as either of the parental plants (i.e., as red or as white as the P individuals), he theorized that two pairs of genes controlling production of red pigment were operating in this cross. If we symbolize the genes for red with the capital letters A and B and their alleles resulting in lack of pigment production by a and b, we can diagram this cross as follows:

$$P \quad AABB \quad \times \quad aabb$$
$$\text{dark red} \qquad \text{white}$$
$$F_1 \quad AaBb \quad (\times \, AaBb)$$
$$\text{intermediate red}$$
$$F_2 \quad \tfrac{1}{16}AABB + \tfrac{2}{16}AaBB + \tfrac{1}{16}aaBB$$
$$+ \tfrac{2}{16}AABb + \tfrac{4}{16}AaBb + \tfrac{2}{16}aaBb$$
$$+ \tfrac{1}{16}AAbb + \tfrac{2}{16}Aabb + \tfrac{1}{16}aabb$$

Assuming each "dose" of a gene for pigment production increases the depth of color, we can sort this F_2 out phenotypically according to the number of genes for red in the following way:

Genotype	Number of Genes for Red	Phenotype	Fraction of F_2
AABB	4	dark red	$\frac{1}{16}$
AABb, AaBB	3	medium red	$\frac{4}{16}$
AAbb, aaBB, AaBb	2	intermediate red	$\frac{6}{16}$
aaBb, Aabb	1	light red	$\frac{4}{16}$
aabb	0	white	$\frac{1}{16}$

Genes symbolized by capital letters, those *"contributing"* to red color in this case, are termed **contributing alleles.** Those that do not "contribute" to red color (here symbolized by lower case letters) may be designated **noncontributing alleles.** Some geneticists refer to these as "effective" and "noneffective" alleles, respectively. Here, then, we have a **polygene** series of as many as four contributing alleles. The term *polygene* was introduced by Mather (1954), who has summarized the modern interpretation of quantitative inheritance. This term has since found wide usage, and is supplanting the older term *multiple gene.*

In an effort to determine whether the mechanism of polygene inheritance is the same as that which we have seen operating in instances of qualitative characters, several simplifying assumptions must be made. We will *assume* that

1. Each contributing gene in the series produces an equal effect.
2. Effects of each contributing allele are cumulative or additive.
3. There is no dominance.
4. There is no epistasis among genes at different loci.
5. There is no linkage involved.
6. Environmental effects are absent or may be ignored.

Certainly as the number of pairs of genes increases, the probability of linkage rises, and the effect of environment can be ignored only in the most closely controlled experiments. Furthermore, some geneticists have disagreed with the first two of these assumptions. But many polygene effects do appear to operate in a manner consistent with the first four points, and our task is eased considerably if we can make all six as *simplifying assumptions.*

Calculating the Number of Polygenes

In another cross in wheat reported by Nilsson-Ehle, a different red variety was used with the result that $\frac{1}{64}$ of the F_2 was as extreme as either parent with seven classes in a $1:6:15:20:15:6:1$ ratio. A little reflection will serve to suggest the operation here of *three* pairs of genes. From earlier chapters on monohybrid inheritance, we can see that, if only one pair of genes were involved, one fourth of the F_2 should be as extreme as either parent. If information for one, two, and three pairs of polygenes is tabulated, we can see a pattern emerging:

Number of Pairs of Polygenes in Which Two Parents Differ	Fraction of F_2 Like Either Parent	Number of Genotypic Classes in F_2	Number of Phenotypic Classes in F_2
1	$\frac{1}{4}$	3	3
2	$\frac{1}{16}$	9	5
3	$\frac{1}{64}$	27	7
n	$(\frac{1}{4})^n$	3^n	$2n + 1$

Thus, with four pairs of polygenes, $\frac{1}{256}$ of the F_2 is as extreme as either parent, with 5 only $\frac{1}{1,024}$, with 10 the fraction drops to $\frac{1}{1,048,576}$, and with 20 pairs only 1 in 1,099,511,627,776 of the F_2 will have measurements like one parent or the other! The number of genotypic classes increases, of course, with startling rapidity as the number of pairs of polygenes becomes larger: for four pairs of genes there are 81 F_2 classes, five pairs of genes produce 243 F_2 genotypes, 10 pairs 59,049, and 20 pairs 3,486,784,401! Thus as the number of polygenes governing a particular trait goes up, the progeny very quickly form a continuum of variation in which class distinctions become virtually impossible to make.

If we wished to calculate the number of contributing alleles instead of the number of pairs of polygenes, we would use the expression $(\frac{1}{2})^n$. Dividing the total quantitative difference by the number of contributing alleles, of course, indicates the amount contributed by each effective allele. For example, in the following hypothetical case in pumpkin, how many contributing alleles are operating and how much does each contribute?

P   5 lb fruits $\times$ 21 lb fruits
F_1 13 lb fruits
F_2 $\frac{3}{750}$ 5 lb fruits . . . $\frac{3}{750}$ 21 lb fruits

Note that $\frac{3}{750}$, the fraction of the F_2 that has fruits as light or as heavy as the P generation, simplifies to $\frac{1}{250}$. Considering the formula $(\frac{1}{4})^n$, we see that $(\frac{1}{4})^n = \frac{1}{256}$ if $n = 4$ (*pairs* of genes), and $\frac{1}{250}$ is close only to this fraction $(\frac{1}{256})$ in the series. Therefore, we must be dealing with a case of four pairs of genes, where the plants producing the heaviest fruits have all *8* contributing alleles. Since the total weight difference is 16 pounds (21 − 5), $\frac{16}{8} = 2$ pounds contributed by each effective allele. Alternatively, we could solve by using $(\frac{1}{2})^n$ to determine directly the number of contributing alleles (instead of the number of *pairs*). In this case $(\frac{1}{2})^n = \frac{1}{256}$ only if $n = 8$. In our illustration with weight of pumpkin fruits, a weight of 5 lb is termed the *base weight*, suggesting that of all the polygenes which may be involved in fruit weight in this species, the two parental strains were homozygous and alike for all but the four pairs our calculations showed.

Calculating Phenotypic Classes and Ratios

It is also apparent that the number of F_2 phenotypic classes follows a pattern, but one which produces a less dramatic increase as the number of pairs of polygenes becomes larger. Thus with one pair, the F_2 of the cross $AA \times aa$ includes three phenotypic classes, which corresponds to the number of genotypic classes (*AA, Aa,* and *aa*). The two examples from Nilsson-Ehle's work on wheat indicate that the number of F_2 *phenotypic classes* is one more than twice the number of pairs of polygenes, or $2n + 1$.

Note that the phenotypic *ratio* also follows a pattern. One pair gives rise to a $1:2:1$ F_2 ratio, two pairs to $1:4:6:4:1$, and three to $1:6:15:20:15:6:1$. You will recognize that these ratios are the same as the sequence of *coefficients* in binomial expansions of a power equal to twice the number of pairs of multiple genes. Thus expansion of $(a + b)^2$ gives a coefficient sequence of $1:2:1$, which is the same as the F_2 phenotypic ratio for one pair of polygenes; expanding $(a + b)^4$ produces the coefficient series $1:4:6:4:1$; and so on. So we can determine an F_2 phenotypic ratio for any number of pairs of polygenes by thinking of the sequence of coefficients given by expanding a binomial raised to the power $2n$ where, again, n represents the number of pairs of polygenes.

HUMAN EYE COLOR

People clearly differ in eye color, that is, in the amount of melanin pigment in this iris. Except for albinos no one is without some eye pigmentation. Those with the least pigment have eyes that appear blue, those with the greatest have eyes that appear brown. Blue eyes owe their color to the scattering of white light by the nearly colorless superficial cells of the iris. This effect is greatest in the shorter (blue) wavelengths of the visible spectrum, giving the iris its blue appearance. Close inspection of the apparently nonbrown iris of some persons discloses small to very small flecks of brown or somewhat orange coloration, because the small amount of pigment is most abundant in discrete groups of cells. In other persons the pigment is more evenly distributed, and the eyes appear uniformly blue.

Clearly, though, there is a gradation in eye color, ranging from the lightest blue to the darkest brown ("black"). Human beings simply do not fall into "either-or," blue or brown, categories, but, rather, large samples form a continuum of variation, strongly indicative of polygenic inheritance. The number of phenotypic classes recognized is therefore arbitrary and depends in part on the observational techniques and equipment used, and in part on the observer.

Although the inheritance of eye color is complex and only incompletely understood, at least nine phenotypic classes[1] may be recognized as a matter of convenience. In order of increasing amounts of melanin pigmentation, these can be designated as light blue, medium blue, dark blue, gray, green, hazel, light brown, medium brown, and dark brown. Recalling that the number of phenotypic classes is one more than twice the number of pairs of polygenes (page 177), nine classes would result from the action of four pairs of genes. Under this hypothesis, a simplified basis for human eye color would be as follows:

Number of Contributing Alleles in Genotype	Eye Color
8	Dark brown
7	Medium brown
6	Light brown
5	Hazel
4	Green
3	Gray
2	Dark blue
1	Medium blue
0	Light blue

Regardless of the number of pairs of polygenes postulated (and it must be emphasized that studies to date do not permit a definitive determination of the number of pairs), it is clear that eye color is due to polygenes, some of which may interact in poorly understood ways. No conclusive data are yet available on linkage of genes affecting this trait, although Brues (1946) and others have suggested X-linkage for some.

OTHER HUMAN TRAITS

Skin color in man also depends upon relative amounts of melanin. Studies by Davenport and Davenport (1910) and by later investigators, which attempt to relate sample frequencies of various degrees of pigmentation to models based on different numbers of pairs of polygenes, show best agreement between observation and theoretical expectation for four, five, or six pairs rather than for fewer or for more. As might be expected, some Caucasians in the samples were darker than some blacks and vice versa. Although polygene inheritance is clearly suggested in man for many quantitative traits, such as height, intelligence, and hair color (except for red versus nonred), no complete

---

[1] However, Davenport (1913) designated five phenotypic classes, and Hughes (1944) recognized seven.

hypothesis setting forth the exact number of pairs of genes and their individual and collective effect has yet been developed for any of these traits.

Transgressive Variation

Some progeny may be more extreme than either parent or grandparent. You are probably familiar with examples among your own acquaintances where, say, some children are shorter or taller than either parent or any of their more remote ancestors. The same phenomenon sometimes occurs, too, with respect to intelligence, skin color, and eye color. Such examples illustrate **transgressive variation.**

One of the earliest instances of transgressive variation to be reported in the literature was one described in 1914 and 1923 by Punnett and Bailey. They crossed the large Golden Hamburg chicken with the smaller Sebright Bantam. The F_1 was intermediate in weight between the parents and fairly uniform, but a few of the F_2 birds were heavier or lighter than either of the parental individuals. Their results suggested to Punnett and Bailey four pairs of genes, with the Golden Hamburg being of, say, genotype *AABBCCdd* and the Sebright Bantam being of genotype *aabbccDD*. Likewise, children may have darker or lighter eyes than either of their parents. Some of the problems at the end of this chapter deal with some interesting illustrations of transgressive variation.

Other Organisms

Instances of polygene inheritance are known from many other plant and animal species. One of the more suggestive was discovered in tomato by Lindstrom (1924 and 1926). Crosses between the larger-fruited Golden Beauty (average fruit weight 166.5 g) and the smaller-fruited Red Cherry (average fruit weight 7.3 g) varieties produced an F_1 having fruits intermediate in size between the parental types, but distinctly closer to the smaller-fruited variety (average weight 23.9). Such results could be explained by assuming dominance or unequal effect among at least some of the noncontributing alleles.

In cattle both solid color and spotting occur. Solid color is due to a dominant gene, *S,* spotting to its recessive allele, *s.* Studies indicate that the degree of spotting in *ss* individuals depends on a rather large series of polygenes (Fig. 9-2). Those cattle that are $S-$ will, of course, not be spotted, regardless of the remainder of their genotype.

Chai (1970) has shown that the difference between high and low leukocyte count in the mouse is a polygenic trait. He suggests dominance, however, for low count and goes on to make the interesting point that "the evidence is accumulating . . . that a quantitative trait is an aggregate of effects from different biological systems, each of which contributes specific effects that differ in magnitude and biological effect under the control of individual genes.

FIGURE 9-2. *Variation in degree of spotting in a herd of cattle. Amount of spotting depends upon a series of polygenes, but these are hypostatic to* S (*solid color*). (Photo courtesy Ayrshire Breeders' Association.)

The present results, although not considered as definitive, further indicate this to be the case."

CONCLUDING STATEMENT

In summary, the basic mechanisms operative in quantitative inheritance appear to be the same as those for qualitative characters. Such studies as are reported here also further emphasize the fact that many traits are the result of *interaction* of one kind or another among several pairs of genes.

It should be emphasized in your thinking about such *quantitative characters* as we have dealt with here that the possible effect of environment must be considered and carefully regulated in any controlled experiment. For example, height in many plants (e.g., corn, tomato, pea, marigold) is a genetically controlled character, but we also know that such environmental factors as soil fertility, texture, and water, the temperature, the duration and wavelength of incident light, the occurrence of parasites, to name just a few, also affect height. Or, with identical twins (who have identical genotypes) growing up in different kinds of environments, many classical studies indicate intelligence quotients sufficiently different that environment surely played a significant role. In summary, *genotype* determines the range an individual will occupy with regard to a given quantitative character; environment determines the *point* within the genetically determined range at which an individual's measurement will fall.

Study and evaluation of polygene inheritance requires certain statistical treatments, particularly those which describe populations. These we shall examine in the next chapter.

REFERENCES

BRUES, A. M., 1946. A Genetic Analysis of Human Eye Color. *American Journal of Physical Anthropology* (New Series) **4:** 1–36.

CHAI, C. K., 1970. Genetic Basis of Leukocyte Production in Mice. *Jour. Hered.,* **61:** 61–71.

DAVENPORT, C. B., 1913. Heredity of Skin Color in Negro-White Crosses. *Carnegie Institution of Washington,* Publication **188:** 1–106.

DAVENPORT, G. C., and C. B. DAVENPORT, 1910. Heredity of Skin Pigmentation in Man. *Amer. Natur.,* **44:** 641–672.

HUGHES, B. O., 1944. The Inheritance of Eye Color—Brown and Nonbrown. *Contributions from the Laboratory of Vertebrate Biology, University of Michigan,* No. **27:** 1–10.

LINDSTROM, E. W., 1924. A Genetic Linkage Between Size and Color Factors in the Tomato. *Science,* **60:** 182–183.

LINDSTROM, E. W., 1926. Hereditary Correlation of Size and Color Characters in Tomatoes. *Iowa Agr. Exper. Sta. Research Bull.,* **93.**

LINDSTROM, E. W., 1929. Linkage of Qualitative and Quantitative Genes in Maize. *Amer. Natur.,* **63:** 317–327.

MATHER, K., 1954. The Genetical Units of Continuous Variation. *Proc. IX International Cong. Genet.,* **Part I:** 106–123.

NILSSON-EHLE, H., 1909. *Lunds Univ. Arsskrift N.F. Avd.,* **2:** Bd. 5.

PUNNETT, R. C., 1923. *Heredity in Poultry.* New York, Macmillan (pp. 44 ff.).

PROBLEMS

9-1. Which of the following human phenotypes would appear to be based on polygene inheritance: intelligence, absence of incisors, height, phenylketonuria (inability to metabolize the amino acid phenylalanine), ability to taste phenylthiocarbamide, skin color, cryptophthalmos (failure of eyelids to separate in embryonic development), eye color?

9-2. Show, by means of appropriate genotypes, how parents may have children taller than themselves.

9-3. Suppose another race of wheat is discovered in which kernel color is determined to depend on the action of six pairs of polygenes. From the cross *AABBCCDD-EEFF* × *aabbccddeeff,* (a) what fraction of the F_2 would be expected to be like either parent? (b) How many F_2 phenotypic classes result? (c) What fraction of the F_2 will possess any six contributing alleles?

9-4. Two races of corn, averaging 48 and 72 in. in height, respectively, are crossed. The F_1 is quite uniform, averaging 60 in. tall. Of 500 F_2 plants, two are as short as 48 in. and two as tall as 72 in. What is the number of polygenes involved, and how much does each contribute to height?

9-5. Two other varieties of corn, averaging 48 and 72 in. in height, respectively, are crossed. The F_1 is again quite uniform at an average height of 60 in. Out of 3,100 F_2 plants, four are as short as 36 in., and two as tall as 84 in. How many polygenes are involved, and how many inches of height are contributed by each effective allele?

9-6. In problems 6-15 to 6-18 it was pointed out that Pl — corn plants are purple, whereas $pl\ pl$ individuals are green. Assume now that the parents in problem 9-5 are also $Pl\ pl$. A breeder wishes to recover a pure-breeding green, 84-in. variety from the cross given in problem 9-5. What fraction of the F_1 will satisfy his requirement?

9-7. Two 30-in. individuals of a hypothetical species of plant are crossed, producing progeny in the following ratio: one 22-in., eight 24-in., twenty-eight 26-in., fifty-six 28-in., seventy 30-in., fifty-six 32-in., twenty-eight 34-in., eight 36-in., and one 38-in. What are the genotypes of the parents? (Start with the first letter of the alphabet and use as many more letters as needed.)

9-8. Two different 30-in. plants of the same species are crossed and produce all 30-in. progeny. What parental genotypes are possible?

9-9. Mr. A has dark brown eyes; his wife's eyes are light blue. Based on the hypothesis that four pairs of polygenes ($A, a; B, b; C, c; D, d$) are responsible for human eye color, give the genotype of (a) Mr. A and (b) his wife. (c) Give the phenotype(s) of children they could have.

9-10. A daughter of Mr. and Mrs. A marries a man (Mr. B) having the same genotype as she. What is the probability that they could have a (a) dark-brown-eyed child, (b) dark-blue-eyed child, (c) hazel-eyed child?

9-11. (a) What eye color is most likely to occur in any child of Mr. and Mrs. B? (b) What is the probability of a child having this eye color?

9-12. The children referred to in problem 9-10 illustrate what phenomenon?

9-13. As noted in the text, spotting in certain breeds of cattle is dependent on interaction of $S-$ (solid color) or ss (spotted) and a number of polygenes for degree of spotting. Assume four pairs of the latter (which is almost certainly too low), designated as $A, a; B, b; C, c; D, d$. If data are accumulated from enough $SsAaBbCcDd \times SsAaBbCcDd$ crosses to give a total of, say, 1,024 calves from such matings, how many of these should be unspotted?

9-14. Among ss animals how many different degrees of spotting could be found, using information from problem 9-13?

9-15. Assume height in a particular plant to be determined by two pairs of unlinked polygenes, each effective allele contributing 5 cm to a base height of 5 cm. The cross $AABB \times aabb$ is made. (a) What are the heights of each parent? (b) What height is to be expected in the F_1 if there are no environmental effects? (c) What is the expected phenotypic ratio in the F_2?

9-16. If each pair of alleles in problem 9-15 exhibited complete dominance instead of an additive effect, (a) what are the heights of each parent? (b) What height would be expected in the F_1? (c) What is the expected phenotypic ratio in the F_2?

CHAPTER 10

Statistical Concepts and Tools

I N the preceding chapter we saw that polygenes govern continuously variable, quantitative traits. Analysis of this kind of inheritance requires application of certain techniques from the branch of mathematics called statistics. Statistics are useful in two fundamental problems commonly encountered in scientific research: (1) what can be learned about a population from measurements of a sample of it and (2) how much confidence can be placed in judgments about that population.

There is a clear difference between **population** and **sample.** A population consists of an infinite group of individuals, measured for some variable quantitative character. In this context the important aspect of a population is a series of numbers that represent such a variable, quantitative trait. Thus figures showing, for example, gains in weight of adult laboratory rats that have been fed a specific diet for a certain number of days represent a biological population, as do height measurements of college males or females, or weights of ripe pumpkin fruits. The statistician's concern is not the rats, the people, or the pumpkins themselves, but *figures* representing a particular quantitative character they possess. A population is usually infinite in size, and often so large as to be theoretical. For example, height measurements of people or pumpkin fruit weights include all individuals who have been or will be alive. The rat population really consists of an infinite number of measurements obtained from an infinite number of experimental rats. Such populations, of course, never actually exist. Even a more discrete problem dealing with a population of an endemic species on an isolated small island includes *all* the individuals of the species on that island. For obvious reasons, it is either impractical or impossible to accumulate measurements for such a group; rarely is the population sufficiently finite for *all* individuals to be measured.

So descriptions of the population must generally be formulated from *samples* of it. To be useful in making estimates of a population, the sample must have been drawn as randomly as possible. In dealing with heights or weights, for example, sample measurements must not be selected more from the taller, or shorter, or heavier, or lighter individuals, but should reflect the same kind and degree of variability as does the population. This it will do if it is large and chosen at random.

Actual values for populations are constants called **parameters;** estimates of populations based on samples are **statistics.** The statistics are, of course, subject to some degree of chance error resulting from sampling practices, but once a statistic is determined, it is possible to state the range of the corresponding parameter with a particular degree of confidence. The geneticist needs to be able not only to estimate the parameters with which he is concerned, but also to determine how likely it is that he is dealing with individuals from either the same or different populations.

In the genetic context, then, the uses of statistics listed in the first paragraph of this chapter may be rephrased to state that statistics provide

1. A concise description of the quantitative characteristics of the sample,
2. An estimate of
   a. The quantitative characteristics of the population from which the sample was drawn,
   b. How well the sample represents that population,
3. An expression of the probability that two samples differ significantly (or, conversely, only within limits set by chance alone) in terms of a particular theory as to the reason for observed differences.

Basic Statistics

Five principal statistics will provide the estimates and descriptions just referred to.

1. THE MEAN

A very elementary statistic, and one with which you are undoubtedly familiar, is the average or **mean.** Calculation of the mean, symbolized by $\bar{x}$ ("x-bar"), may be represented by the formula

$$\bar{x} = \frac{\Sigma x}{n} \tag{1}$$

where Σ, the upper case Greek letter sigma, directs us to sum all following terms, x the individual measurements, and n the number of individuals in the sample. Often, when n is quite large, it becomes convenient to *group* data by *classes.* Thus in computing the average grade on a quiz in a large class it might be more practical to tally the number or *frequency* of individuals scoring between 96 and 100 in one class or group, those between 91 and 95 in another, and so on. A "class value" midway between the extremes of each class range is also entered. The tabulated data would then be arranged as shown in the table on page 186.

If the data are grouped in this way, the mean will, of course, be given by

$$\bar{x} = \frac{\Sigma fx}{n} \tag{2}$$

Formula (2) loses a little in accuracy over formula (1), but the much simpler arithmetic of its method more than offsets this.

Class Range	Class Value x	Frequency f	fx
96–100	98	1	98
91–95	93	4	372
86–90	88	8	704
81–85	83	12	996
76–80	78	18	1,404
71–75	73	25	1,825
66–70	68	17	1,156
61–65	63	10	630
etc.	etc.	etc.	etc.

Although the mean is a necessary statistic, it is a rather uninformative one in that a comparison of means of different samples reflects nothing of their spread, or *variability*. Consider as an example three students, the first having grades of 75, 75, 75; the second, 65, 75, 85; and the third, 50, 75, and 100. Obviously the mean for each is 75, but the distributions reflect quite different spreads. There is no variability in the first student's record and quite a bit in the last. Furthermore, the mean is greatly affected by a few extreme values.

2. VARIANCE

Variability of the population is measured by the **variance**, σ^2:

$$\sigma^2 = \frac{\Sigma(x - \mu)^2}{N} \tag{3}$$

where x represents each individual measurement in the population, μ the population mean, and N the number of individuals comprising the population. Of course, the population mean and the total number of individuals generally are not determinable because it is impractical or impossible to measure every individual in a population of infinite size. These difficulties are avoided by substituting sample values for those of the population:

$$s^2 = \frac{\Sigma(x - \bar{x})^2}{n} \tag{4}$$

But formula (4) is biased in the direction of underestimating σ^2 *because, in using the* sample *mean, the number of* independent *measurements is* $n - 1$. For example, the series of six values $4 + 3 + 6 + 2 + 1 + 2$ has a mean of 3, which is calculated in such a way that one value is fixed by the sum of the others. That is, given

$$\frac{4 + 3 + 6 + 2 + 1 + x}{6} = 3$$

the value of x can only be 2. To obviate the bias in formula (4), it is multiplied by the correction factor

$$\frac{n}{n-1},$$

giving an unbiased estimate of the sample variance:

$$s^2 = \frac{\Sigma(x - \bar{x})^2}{n-1} \tag{5}$$

If data are grouped by classes, as in our calculation of the mean (pages 185–186), formula (5) becomes

$$s^2 = \frac{\Sigma f(x - \bar{x})^2}{n-1} \tag{6}$$

The sample variance, s^2, provides an unbiased estimate of the population variance (σ^2). But variance is in *squared* units, and we do not wish to express height in square feet, or weight variability in square pounds. This difficulty is resolved by the next statistic to be described.

3. STANDARD DEVIATION

To avoid expressing variability in squared units of measurement, we simply extract the square root of the variance. This statistic is the **standard deviation,** s, which is given by the formula

$$s = \sqrt{\frac{\Sigma f(x - \bar{x})^2}{n-1}} \tag{7}$$

In effect, the standard deviation reflects the extent to which the mean represents the entire sample. If all individuals had exactly the same value, there would be no variability and the mean would represent the sample perfectly. Examination of formula (7) indicates that the standard deviation would then be *zero*. As the sample becomes more variable, the mean serves progressively less well as an index of the entire sample, and as departures from the mean, class by class, increase, so does the standard deviation. However, extracting the square root of the variance reintroduces bias. Nevertheless, the standard deviation has the necessary advantage of expression in the units of measurement, as well as usefulness in determining other statistics.

To see how this statistic may be calculated and what it discloses regarding the sample, consider some length measurements of 200 hypothetical F_1 plants resulting from a particular cross. Calculation will be facilitated if data are grouped by classes and tabulated as follows:

1	2	3	4	5	6
Class Value (cm) x	Fre-quency f	fx	Deviation from Mean $(x - \bar{x})$	Squared Deviation $(x - \bar{x})^2$	$f(x - \bar{x})^2$
48	8	384	-4.75	22.56	180.50
50	32	1,600	-2.75	7.56	242.00
52	75	3,900	-0.75	0.56	42.19
54	52	2,808	$+1.25$	1.56	81.25
56	28	1,568	$+3.25$	10.56	295.75
58	5	290	$+5.25$	27.56	137.81
	$n = 200$	$\Sigma fx = 10,550$			$\Sigma f(x - \bar{x})^2 = 979.50$

$$\bar{x} = \frac{\Sigma fx}{n} = \frac{10,550}{200} = 52.75 \text{ cm}$$

$$s = \sqrt{\frac{\Sigma f(x - \bar{x})^2}{n - 1}} = \sqrt{\frac{979.5}{199}} = \sqrt{4.92} = 2.218 \cong 2.22$$

This calculation provides a mean, plus or minus a standard deviation; that is, $\bar{x} = 52.75 \pm 2.22$. To understand the meaning of this expression and the information it conveys, "curves of distribution" must be examined.

As data for large samples are plotted with a quantitative measurement, such as length, along the abscissa and numbers of individuals (frequency) along the ordinate, the resulting curve is frequently bell-shaped; the variation is rather symmetrical about the largest class (or mode), as indicated in Figure 10-1. Such a curve is a **normal curve** or a curve of normal distribution. If data are carefully plotted and a perpendicular erected from the abscissa at a value equal to the mean, it will intersect such a curve at the latter's highest point and will divide the area under the curve into two equal parts (Fig. 10-1) and, therefore, the sample into two groups of equal size. Now if perpendiculars to the abscissa are erected on it at points having values equal to $\bar{x} + s$ and $\bar{x} - s$, the area under the curve between $\bar{x} - s$ and $\bar{x} + s$ is 68.26 per cent of the area under the curve (Fig. 10-2).

Similarly, the area under the curve between $\bar{x} - 2s$ and $\bar{x} + 2s$ is 95.44 per cent of the total area; for $\bar{x} \pm 3s$, the area included is 99.74 per cent of the total (Fig. 10-3). This means that, *in a normal distribution,* about 68 per cent (or roughly two thirds) of the individuals will have values between $\bar{x} - s$ and $\bar{x} + s$, about 95 per cent between $\bar{x} - 2s$ and $\bar{x} + 2s$, and so on. Therefore if an individual is chosen *at random* from a normally distributed population, the probability is about 0.68 that it will belong to that part of the population lying in the range $\bar{x} \pm s$. Similarly, there is a 0.95 probability that the individual selected will lie within the limits $\bar{x} \pm 2s$ or, to phrase it another way, only a 0.05 probability that the randomly chosen individual will lie outside

FIGURE 10-1. *The curve of normal distribution. A perpendicular erected from the abscissa at a value equal to the mean intersects the curve at its highest point and divides the area under the curve into areas of equal size.*

FIGURE 10-2. *The curve of normal distribution with perpendiculars to the abscissa erected at points showing values of $\bar{x} + s$ and $\bar{x} - s$. Areas a and b each comprise 34.13 per cent of the area under the curve.*

FIGURE 10-3. *Curve of normal distribution with perpendiculars to the abscissa erected at values $\bar{x} \pm 1s$, $\bar{x} \pm 2s$, and $\bar{x} \pm 3s$. Areas under the curve are as follows: a + b = 68.26 per cent; (a + b) + (c + d) = 95.44 per cent; (a + b) + (c + d) + (e + f) = 99.74 per cent.*

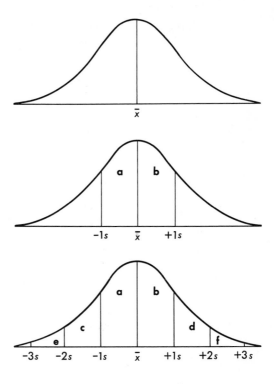

those limits. The percentages of the sample determined by the mean plus or minus different multiples of the standard deviation are shown more fully in Table 10-1. Thus the standard deviation is a useful description of the variability of the sample and, if the sample is large and randomly chosen, a good indicator of the variability of the population. As variability of the sample increases, so does its standard deviation.

4. THE STANDARD ERROR OF THE SAMPLE MEAN

If a series of samples were to be drawn from the same population, their means and standard deviations would probably not be the same. Page 188 carried measurements for a sample of hypothetical plants having a mean of 52.75 cm. Another group of 200 F_2 plants from the same population would be expected, by chance, to have a somewhat different mean. Yet if a series of samples of 200 individuals each were to be drawn at random, it is fair to assume from our knowledge of the laws of probability that only a few samples would have relatively low means and a few relatively high, but that most would be intermediate. In fact, if a large number of these successive sample means were plotted, they would be found to form a normal curve, giving a fairly clear

TABLE 10-1. Percentages of the Sample Falling Within Given Multiples of the Standard Deviation from the Mean

Mean ± Values of s	Per Cent of Sample Included	Mean ± Values of s	Per Cent of Sample Included
0.1	7.96	2.0	95.44
0.2	15.86	2.1	96.42
0.3	23.58	2.2	97.22
0.4	31.08	2.3	97.86
0.5	38.30	2.4	98.36
0.6	45.14	2.5	98.76
0.675	50.00	2.58	99.00
0.7	51.60	2.6	99.06
0.8	57.62	2.7	99.30
0.9	63.18	2.8	99.48
1.0	68.26	2.9	99.62
1.1	72.86	3.0	99.74
1.2	76.98	3.1	99.80
1.3	80.64	3.2	99.86
1.4	83.84	3.3	99.90
1.5	86.64	3.4	99.94
1.6	89.04	3.5	99.96
1.645	90.00	3.6	99.96
1.7	91.08	3.7	99.98
1.8	92.82	3.8	99.98
1.9	94.26	3.9	99.99
1.96	95.00	4.0	99.99

picture of the population distribution. The population mean and standard deviation could then be calculated. Practical limitations preclude doing exactly this, but the standard deviation of means, or the **standard error of the sample mean,** can be calculated from one good-sized sample. This will give a good picture of the population.

The standard error of the sample mean, $s_{\bar{x}}$, represents an estimate of the standard deviation of the means of many samples which might be taken, and is a measure of the closeness with which the sample mean, $\bar{x}$, represents the population mean, μ. The size of the sample used and the variability of the population affect the reliability of $\bar{x}$ as an estimate of μ. The greater the variation in the population, the larger the sample needed to provide an adequate representation of the population. Both these factors are taken into account in the formula for calculating the standard error of the sample mean:

$$s_{\bar{x}} = \frac{s}{\sqrt{n}} \tag{8}$$

where s is the standard deviation of the sample and n, of course, the number of individuals composing the sample.

In the hypothetical sample of 200 F_1 plants (page 188), where $\bar{x} = 52.75$ cm and $s = 2.22$, the standard error of the sample mean becomes

$$s_{\bar{x}} = \frac{2.22}{\sqrt{200}} = \frac{2.22}{14.14} = 0.157, \text{ or about } 0.16$$

Because $s_{\bar{x}}$ represents the standard deviation of a *series* of sample means, and recalling the areal relationships of a normal curve (page 189) where $\bar{x} \pm s$ includes 68.26 per cent of the sample, $\bar{x} \pm 2s$ includes 95.44 per cent of the sample, etc., the value $s_{\bar{x}} = 0.16$ indicates that there is about a 0.68 probability that μ lies in the range 52.75 cm $\pm$ 0.16, i.e., between 52.59 cm and 52.91 cm. Likewise, the probability that μ is in the range $\bar{x} \pm 2s_{\bar{x}}$, or between 52.43 and 53.07, is about 0.95. Thus $\bar{x} = \mu \pm s_{\bar{x}}$ 68.26 per cent of the time by *chance alone*, $\bar{x} = \mu \pm 2s_{\bar{x}}$ in 95.44 per cent of the cases, and $\bar{x} = \mu \pm 3s_{\bar{x}}$ 99.74 per cent of the time. If the sample is large (>100), other confidence levels may be determined from Table 10-1. For example, there is a 0.9 probability that μ lies in the range 52.75 $\pm$ 1.645 $s_{\bar{x}}$, or 52.75 $\pm$ 0.263 cm; a 0.5 probability that it is within the range 52.75 $\pm$ 0.675$s_{\bar{x}}$; and so on. Obviously, the smaller the standard error, the more reliable the estimate of the population means. As sample size, *n,* increases, the magnitude of the standard error decreases. Therefore it is desirable to use samples as large as possible in order to determine characteristics of the population.

In cases involving polygene inheritance, the standard error of the sample mean will give a measure of the population mean. It will also help to fix, for example, a parental mean to which a given fraction of the F_2 may be compared. In other words, in the example of pumpkin fruits (page 177), the question could be raised as to what value between 4 and 6 or 19 and 23 is acceptable as representative of the parental strains so that certain F_2 individuals can be designated as being as extreme as either parent. If the standard error of the mean is calculated for each parental sample, a range is arrived at within which there is 0.68, 0.95, or 0.99 confidence that the population mean lies and, therefore a valid representative of the parental populations against which we can compare the F_2.

5. STANDARD ERROR OF THE DIFFERENCE IN MEANS

It is often necessary to determine whether the difference in the means of two samples is statistically significant—that is, to determine the likelihood that two sample means represent genetically different populations rather than chance differences in two samples from the same population. A judgmental answer to this problem is provided by a statistic known as the **standard error of the difference in means** (S_d):

$$S_d = \sqrt{(s_{\bar{x}_1})^2 + (s_{\bar{x}_2})^2} \tag{9}$$

For example, consider the statistics developed for the hypothetical group of 200 plants ("sample 1") as compared with like statistics for a second hypothetical sample:

Sample 1	*Sample 2*
$n_1 = 200$	$n_2 = 200$
$\bar{x}_1 = 52.75$	$\bar{x}_2 = 55.87$
$s_1 = 2.218$	$s_2 = 3.150$
$s_{\bar{x}_1} = 0.16$	$s_{\bar{x}_2} = 0.22$

Substituting in formula (9) to determine S_d for these two samples gives

$$\begin{aligned} S_d &= \sqrt{(0.16)^2 + (0.22)^2} \\ &= \sqrt{0.026 + 0.048} \\ &= \sqrt{0.074} \\ &= 0.272 \end{aligned}$$

The meaning of a value of $S_d = 0.272$ can be seen by comparing the difference in sample means, here $\bar{x}_2 - \bar{x}_1$, with the standard error of the difference in means, S_d:

$$\frac{\bar{x}_2 - \bar{x}_1}{S_d} \tag{10}$$

Substituting values, formula (10) becomes

$$\frac{3.12}{0.27} = 11.55$$

(The smaller sample mean is always subtracted from the larger so as to give 'a positive numerator.)

What, then, does the value of S_d and its relation to the difference in sample means signify? Remember that, in a normal curve, the area under the curve equal to $\bar{x} \pm 2s$ comprises about 95 per cent of the total area, which is to say that 95 per cent of the individuals will have a quantitative value between $\bar{x} - 2s$ and $\bar{x} + 2s$. Now a standard error represents the standard deviation of a series of means, so in a standard error of the difference in sample means, we are still dealing with this same relationship between $\pm 1s$, $\pm 2s$, and so on, assuming, of course, a normal curve. So *if the difference in sample means is greater than twice the standard error of the difference of the sample means, then the difference in sample means is considered significant.* Significance begins whenever $\bar{x}_1 - \bar{x}_2 > 2S_d$. Here *significance* means two different populations.

Would a difference in means of, say, *exactly* twice the value of S_d mean that two different populations are not represented by the two samples? No! But the values of $\bar{x}_1 - \bar{x}_2 = 2S_d$ would give us 95 *per cent confidence* that two populations are involved. Or, to state this another way, there would then be only a 5 per cent probability that the two samples happened to have been drawn from the same population. *When the probability that two samples have come from the same population falls BELOW 5 per cent ($P < 0.05$), the difference in means is considered significant.* So to be significant, $\bar{x}_1 - \bar{x}_2$ must *exceed* $2S_d$. In the example here, $\bar{x}_1 - \bar{x}_2 = 11.55 \ S_d$, hence $\bar{x}_1 - \bar{x}_2$ is considered highly significant. The probability that only one population is represented is so very small that we reject it.

Note, incidentally, that a significant difference between two sample means is inherently neither "good" nor "bad." In some situations we may be pleased to find significance; in others we may be just as happy to find a lack of significance.

These five statistics are summarized in Appendix D.

Applications of Statistics to Genetic Problems

Two illustrations of the application of statistics to genetic problems will serve to point up the usefulness of such analyses.

In the first instance, assume a commercial producer of hybrid seed corn wishes to market grains that will produce plants having ears of very uniform length. He has two varieties, A and B, both of which produce ears of just under 8 in., which is a satisfactory length for his marketing purposes. Although variety A averages closer to 8 in. than does B, it appears to be more variable. He grows several acres of each variety in as uniform an environment as possible, then analyzes 100 ears from each. Variety A has the following statistics:

$$\bar{x} = 7.95 \text{ in.}$$
$$s = 0.52$$
$$s_{\bar{x}} = 0.05$$

Although the sample mean is very close to the desired length, the standard deviation indicates that two thirds of the ears of this variety may be expected to vary up to 0.52 in. from the mean of 7.95 in. Of course, one third will deviate more than this.

This is more variability than the grower would prefer, so, for comparison, 100 ears of variety B are similarly analyzed. This variety has these statistics:

$$\bar{x} = 7.88 \text{ in.}$$
$$s = 0.23$$
$$s_{\bar{x}} = 0.02$$

Although the ears average slightly shorter than those of variety A, B has a much narrower range of variation. Two thirds of the ears of the latter may be expected to fall within the range 7.65 to 8.11 in., as compared with 7.43 to 8.47 for A. Therefore the breeder elects to use B.

The standard errors of the two samples are useful in indicating to the grower just how much the mean length of his samples might be expected to vary from the mean lengths of *all* plants of the two varieties. His samples are sufficiently large, and the standard errors are quite small, showing that the samples do reflect reliably the magnitude of variability in the two varieties.

Another example will show an application of statistics to a more theoretical type of problem (Table 10-2). Data were accumulated on days to maturity for two varieties of tomato (Burpeeana Early Hybrid, P_1, and Burpee Big Boy, P_2) and their hybrids (F_1 and F_2). Maturation time is dependent on both heredity and environment, so environmental differences must be minimized. This is often done in randomized plots, whereby different varieties are distributed randomly in the field. The time elapsing between setting the plants in the

TABLE 10-2. Data for Two Parental and Two Progeny Strains of Tomato Based on Days Required to Reach Maturity

Days	P_1	P_2	F_1	F_2
55	1			
56	6			
57	9			1
58	40			1
59	28			2
60	14			3
61	2		3	4
62			8	9
63			20	10
64			31	12
65			19	14
66			10	20
67			8	7
68			1	4
69				3
70				3
71				1
72				2
73				1
74				1
75		4		1
76		12		1
77		20		
78		35		
79		15		
80		10		
81		3		
82		1		
$\bar{x}$	58.38	77.92	64.22	65.14
s	1.14	1.43	1.49	3.43
$s_{\bar{x}}$	0.114	0.143	0.149	0.343
S_d		0.183		0.374

field and the ripening of the first fruit was recorded as "days to maturation" and data recorded for samples of 100 plants of each of the four varieties, all grown in the same season in randomized plots.

Inspection of the data in Table 10-2 clearly suggests that P_1 and P_2 represent different populations, and this is amply confirmed by the standard deviation and standard error of each, as well as by the standard error of the difference in sample means. In fact, the difference in sample means is 19.54 days, which is about 108 times the standard error of the difference in means! Remembering that a difference in sample means of more than $2S_d$ is considered significant, the difference here is highly significant.

The same statistics for the F_1 and F_2 generations also show that although differences in maturation times are considerably less than for the parental strains, they are significantly different. One therefore has more than 95 per cent confidence that they represent two genetically different populations. The difference in means for these two samples is only 0.92 day, but this is about 2.5 times the standard error of the difference in the sample means.

These data clearly suggest polygenes. The F_1 is intermediate between the parents, as is the F_2, but the F_2 has a wider range than the F_1. If P_1 and P_2 are assumed to be completely homozygous, then the variability each shows must be wholly environmental. The F_1 would then be completely and uniformly heterozygous, and its variability again environmental. The F_2 would be expected to segregate so that genetic variation is superimposed on environmental.

The F_2 data may be used to furnish a very rough estimate of the number of contributing alleles. Eleven of the 100 F_2 were as extreme as P_1, and 2 as extreme as P_2. Now $\frac{11}{100}$ simplifies to about $\frac{1}{9}$, and $\frac{2}{100}$ to $\frac{1}{50}$. Recalling the formula for computing numbers of effective alleles from the F_2 data (page 177), it is seen that $\frac{1}{16}$, which indicates four contributing alleles, is between the two extremes of $\frac{1}{9}$ and $\frac{1}{50}$. This approach is really too simple and probably gives too low an estimate, for it would take many hundreds or thousands of F_2 individuals to provide a reasonable chance of recovering the maximum extremes possible.

Sewall Wright has shown that the number of *pairs* of polygenes (n) can be calculated from the formula

$$n = \frac{R^2}{8(s^2_{F_2} - s^2_{F_1})} \tag{11}$$

where R is the range (greatest difference) between the mean values of extreme phenotypes (whether found in the parental generations or elsewhere), $s^2_{F_2}$ is the *variance* of the F_2, and $s^2_{F_1}$ the variance of the F_1. Applying this formula to the data of Table 10-2 and substituting, we obtain

$$n = \frac{(77.92 - 58.38)^2}{8(11.76 - 2.22)} = \frac{(19.54)^2}{8(9.54)} = \frac{381.81}{76.32} = 5.003$$

or about five pairs of polygenes.

In using this formula we have tacitly assumed (1) no environmental effects, (2) no dominance, (3) no epistasis, (4) equal, additive contributions by all loci, (5) no linkage, and (6) complete homozygosity in each parent, with complete heterozygosity in the F_1. With these assumptions, the F_1 mean would be expected to fall midway between the parental means. That it does not suggests that not all of our assumptions may be completely valid in this case and/or that the sample is too small. Moreover, the fact that a crude estimate of the number of pairs of genes from the fraction of the F_2 as extreme as either parent is lower than the value given by the Wright formula also points up the need for a considerably larger sample.

These are practical contributions of statistical analyses to genetic problems. But application of statistics provides additional, more subtle benefits. In many cases it has been possible to separate genetic mechanisms from sampling errors, and in others to differentiate between genetic and environmental variation. Above all, critical attitudes toward design of experiments and treatment of data have been sharpened.

PROBLEMS

The following data were obtained on the weight in pounds of a given sample of pumpkin fruits:

30	26	22	16	24
24	24	30	22	14
28	22	16	26	22
24	20	28	14	28
22	24	22	22	26

10-1. What is the mean to the nearest tenth of a pound?

10-2. What is the variance?

10-3. (a) What is the standard deviation to the nearest tenth of a pound? (b) What does this value tell you about the sample?

10-4. If an individual is selected at random from this sample, what should be the probability that it will weigh more than 18 but less than 28 lb, assuming a normal curve?

10-5. (a) What is the standard error of the sample mean? (b) What information does this give you?

10-6. What is the probability that the population mean lies between about 21 and 25 lb?

Data on a second sample of 25 individuals were collected and the following statistics arrived at:

mean of sample 2	= 21.8 lb
standard deviation of sample 2	= 2.0
standard error of the sample mean	= 0.8

10-7. What is the standard error of the difference in sample means?

10-8. What is the approximate probability that samples 1 and 2 represent two different populations?

10-9. In view of your answer to the preceding question, is the difference in sample means significant? Why?

10-10. Two parental strains having mean heights of 64.29 and 135 cm, respectively, were crossed. The F_1 and F_2 that resulted had the following statistics:

	$\bar{x}$	s
F_1	99.64	10.000
F_2	102.11	13.346

If we neglect any possible environmental effects, dominance, linkage, or epistasis, and assume each effective allele to make the same additive contribution, how many pairs of polygenes are involved? (Note: the difference in parental means represents the range between extreme phenotypes here.)

CHAPTER 11
Sex Determination

TWO types of sexually reproducing animals and plants may be recognized: (1) **monoecious** (from the Greek *monos,* "only," and *oikos,* "house") in which each individual produces two kinds of gametes, sperm and egg, and (2) **dioecious** (Greek prefix *di-,* "two," and *oikos*) in which a given individual produces only sperms or eggs. In dioecious organisms, the **primary sex difference** concerns the kind of gamete and the primary sex organs by which these are produced. Each sex also exhibits many **secondary sex characters.** In humans these include voice, distribution of body fat and hair, and details of musculature and skeletal structure; in *Drosophila* these include the number of abdominal segments, presence (♂) or absence (♀) of sex combs, and so on (Fig. 11-1). Our immediate problem is that of examining mechanisms whereby the sex of an individual is determined. We shall discuss two basic types, chromosomal and genic, though the distinction is not always a sharp one.

Sex Chromosomes

DIPLOID ORGANISMS

XX-XO System. Chromosomal differences between the sexes of several dioecious species were found early in the course of cytological investigations. Henking, a German biologist, in 1891 noted that half the sperms of certain insects contained an extra nuclear structure, which he called the "X body." The significance of this structure was not immediately understood, but in 1902 an American, McClung, reported the somatic cells of the female grasshopper contained 24 chromosomes, whereas those of the male had only 23. Three years later, Wilson and Stevens succeeded in following both oogenesis and spermatogenesis in several insects, and it was realized that the X body was a chromosome, so the X body became known as an X chromosome. Thus in many insects there is a chromosomal difference between the sexes, females being referred to as XX (having two X chromosomes) and males as XO ("X-oh," having one X chromosome). As a result of meiosis, all the eggs of such species carry an X chromosome, whereas only half the sperms have one, the other half having none.

XX-XY System. In the same year, 1905, Wilson and Stevens found a different arrangement in other insects. In these cases females were again XX, but males had, in addition to one X chromosome, an odd one of a different size, which was called the Y chromosome; males thus were XY. Half the sperms carry an X, and half a Y. The so-called XY type occurs in a wide variety of

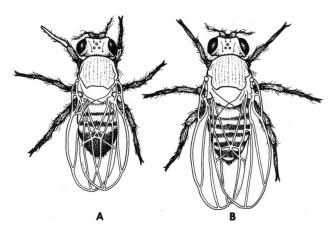

FIGURE 11-1. Drosophila melanogaster, (*A*) *male*; (*B*) *female*.

$$XX \; \partial \; XY$$

animals, including *Drosophila* and mammals, as well as in at least some plants (e.g., the angiosperm genus *Lychnis*) (Fig. 11-2).

A distinction can thus be made between the X and Y chromosomes associated with sex and those that are alike in both sexes. The X and Y chromosomes are called **sex chromosomes;** the remaining ones of a given complement, which are the same in both sexes, are **autosomes.** In both the XX-XO and XX-XY types described thus far, all the eggs have one X chromosome, whereas the sperms are of two kinds, X and O, or X and Y. In each case the male is the **heterogametic** sex (producing two kinds of sperms), whereas the female is the **homogametic** sex (producing but one kind of egg).

ZZ-ZW System. A final major type of chromosomal difference between the sexes is that in which the female is heterogametic and the male homogametic. The sex chromosomes in this case are often designated as Z and W to avoid confusion with instances in which the female is homogametic. Females are thus ZW and males ZZ. Birds (including the domestic fowl), butterflies, and some fishes belong to this group.

MONOPLOID ORGANISMS

Liverworts. Like all sexually reproducing plants, liverworts (division *Bryophyta*) are characterized by a well-marked alternation of generations (see Appendix B), in which a monoploid, sexually reproducing phase or generation (the gametophyte) alternates in the life history with a diploid, asexually reproducing individual (the sporophyte). Allen (1919) reported the chromosome complement of the sporophyte of the liverwort *Sphaerocarpos* to consist of seven matching pairs, plus an eighth pair in which one of the two chromosomes was much larger than the other. The larger member of this eighth pair has been designated the X chromosome, its smaller partner the Y chromosome.

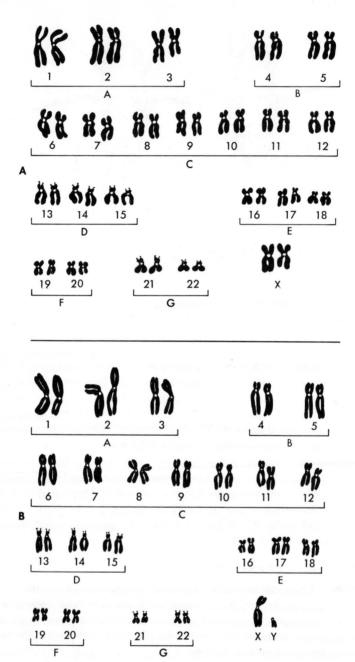

FIGURE 11-2. *Karyotypes of human beings showing sex differences. (A) female; (B) male. Chromosome pairs 1–22 are autosomes, customarily designated by letter groups. The X and Y are sex chromosomes; structurally, X is similar to group C, and Y to group G.*

At meiosis, which terminates the diploid sporophyte generation, X and Y chromosomes are segregated, so that of the four meiospores produced from each meiocyte, two receive an X chromosome and two a Y chromosome. Meiospores containing an X chromosome develop into female gametophytes, those with a Y into males. Thus females are X, males Y, and asexual sporophytes XY.

SUMMARY OF SEX CHROMOSOME TYPES

The various types of chromosomal differences between the sexes may be summarized as follows:

♀	♂	Examples
XX	XY	*Drosophila*, man and other mammals, some dioecious angiosperm plants
XX	XO	Grasshopper; many Orthoptera and Hemiptera
ZW	ZZ	Birds, butterflies, and moths
X	Y	Liverworts

Determination of Sex Under the Chromosomal System

Such chromosomal differences as these raise certain fundamental questions. For example, is a *Drosophila* individual a male because he has a Y or because he has only one X chromosome? Is an individual female because of the absence of the Y or because of the presence of two X chromosomes? Do the autosomes have anything to do with sex determination? Is the system identical in *Drosophila* and man, both of which have the XX-XY sex difference? What do genes have to do with the situation? What causes sex reversal, in which an individual of one sex becomes one of the other sex? Or, what operates to produce individuals that are part male and part female in species where the sexes are ordinarily separate and distinct? At least partial answers are available for all these questions.

DROSOPHILA

Primary Nondisjunction of X Chromosomes.   Work on the genetics of *Drosophila melanogaster* showed that sex determination, at least in that animal (where $2n = 8$), was far less simple than the mere XX-XY difference would suggest. From some unusual breeding results in Bridges' laboratory, he was able, in a masterpiece of inductive reasoning, to lay the groundwork for a complete understanding of sex determination in *Drosophila* (Bridges, 1916a, 1916b, and 1925).

The gene for wild type red eyes (+) is carried on the X chromosome; a recessive allele (*v*) produces vermilion eyes in homozygous females and in all males (which, of course, have only one X chromosome). Ordinarily, vermilion-eyed females mated to red-eyed males produce only red-eyed daughters and vermilion-eyed sons:

$$
\begin{array}{lccc}
\text{P} & vv & \times & + \ Y \\
\text{P gametes:} & \female \quad v & & v \\
& \male \quad \tfrac{1}{2}+ & + & \tfrac{1}{2}\ Y \\
\hline
\text{F}_1 & \tfrac{1}{2}+v & + & \tfrac{1}{2}\ vY \\
& (\text{red } \female) & & (\text{vermilion } \male)
\end{array}
$$

However, in rare instances, crosses of this type produced unexpected vermilion-eyed daughters and red-eyed sons with a frequency of one per 2,000 to 3,000 offspring. Bridges surmised that these unusual progeny were due to a failure of the X chromosomes in an XX female to disjoin during oogenesis. Such *primary nondisjunction* (Bridges, 1916a), he reasoned, would produce three kinds of eggs, the majority containing the normal single X chromosome and a small number bearing either two X chromosomes or no X at all. Symbolizing each X chromosome as either X^+ (carrying the dominant gene for red eyes) or X^v (bearing the recessive gene for vermilion eyes), and each *set of three* autosomes as A, Bridge's cross may be represented in this way:

$$
\begin{array}{ll}
\text{P} & AAX^vX^v \times AAX^+\ Y \\
& (\text{vermilion } \female)\ (\text{red } \male) \\
\text{P gametes: } \female & AX^v\ (\text{numerous}) + AX^vX^v\ (\text{rare}) + AO\ (\text{rare}) \\
\quad\quad\quad\ \ \male & \quad\quad\quad\quad AX^+ + AY \\
\hline
\end{array}
$$

$$
\begin{array}{ll}
\text{F}_1 \quad AAX^+\ X^v & \text{red } \female \text{ (numerous; normal)} \\
\quad\quad\ AAX^+\ X^vX^v & \text{``metafemale'' (rare; die)} \\
\quad\quad\ AAX^+\ O & \text{sterile red } \male \text{ (rare)} \\
\quad\quad\ AAX^vY & \text{vermilion } \male \text{ (numerous; normal)} \\
\quad\quad\ AAX^vX^vY & \text{vermilion } \female \text{ (rare)} \\
\quad\quad\ AAOY & \text{die (rare)}
\end{array}
$$

The metafemales ($AAXXX$) are weak, seldom living beyond the pupal stage; $AAOY$ individuals die in the egg stage. Note that the presence of a Y chromosome does not determine maleness itself, though males lacking it are sterile.

   Secondary Nondisjunction.   Bridges next mated the exceptional vermilion-eyed females (AAX^vX^vY) that arose as a result of primary nondisjunction to normal red-eyed males ($AAX^+\ Y$), obtaining progeny in these frequencies:

$$
\begin{array}{ll}
0.46 & \text{red } \female \\
0.02 & \text{vermilion } \female \\
0.02 & \text{red } \male \\
0.46 & \text{vermilion } \male \\
0.02 & \text{metafemales} \\
0.02 & OYY
\end{array} \Big\} \text{die}
$$

Occurrence of the vermilion-eyed females and red-eyed males is due to *secondary nondisjunction*. Meiosis in XXY females would be expected to be somewhat irregular because of pairing problems and, indeed, Bridges' results indicate that it is. In oogenesis, synapsis may involve either the two X chromosomes (XX type) with the Y chromosome remaining unsynapsed, or one X and the Y (XY type) with the other X remaining free. Bridges found XY synapsis to occur in about 16 per cent of the cases, and XX synapsis in about 84 per cent. After XY synapsis, disjunction segregates the X and Y synaptic partners to opposite poles. The unsynapsed X may go to *either* pole so that XY synapsis produces four kinds of eggs with a frequency of 0.04 each: XX, Y, X, and XY (Fig. 11-3). On the other hand, XX synapsis is followed by disjunction of the two previously synapsed X chromosomes and their movement to opposite poles. The free Y may, of course, go to either pole, but the result is only two kinds of eggs, X and XY, with a frequency of 0.42 each. The overall

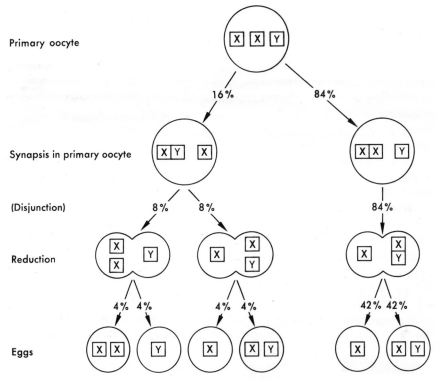

FIGURE 11-3. *Diagram of secondary nondisjunction in XXY female* Drosophila, *resulting in 46 per cent X, 46 per cent XY, 4 per cent XX, and 4 per cent Y eggs. If the egg is, for example, XX, then the polar bodies are XX (1) and Y (2); if the egg is Y, then polar bodies are Y (1) and XX (2), and so forth. For full explanation see text.* (Based on work of Bridges, 1916a.)

result of secondary nondisjunction is four kinds of eggs, in these frequencies:

$$0.46 \ X^vY + 0.46 \ X^v + 0.04 \ X^vX^v + 0.04 \ Y$$

Fertilization by sperm from a cytologically normal, red-eyed male ($X^+ \ Y$) produces eight types of zygotes, which may be grouped in six classes:

Frequency	Zygote	Phenotype
0.23	$X^+ \ X^vY$ ⎫	
0.23	$X^+ \ X^v$ ⎬	red ♀
0.02	$X^+ \ X^vX^v$	metafemale (die)
0.02	$X^+ \ Y$	red ♂
0.23	X^vYY ⎫	
0.23	X^vY ⎬	vermilion ♂
0.02	X^vX^vY	vermilion ♀
0.02	YY	die

The chromosomal constitution of each of the six viable genotypes was verified by Bridges.

As Bridges used the terms, *primary nondisjunction* may occur in either XX females or XY males. In the former it leads to the production of XX and O eggs. If it occurs in the first meiotic division of males, XY and O sperm are produced. Should it take place during the second division, XX, YY, and O sperms result. *Secondary nondisjunction,* on the other hand, occurs in XXY females where it gives rise to XX, XY, X, and Y eggs. As the term *nondisjunction* implies, production of these aberrant gametes results only from a failure of the sex chromosomes to disjoin after synapsis; they are not physically attached.

Attached-X Flies. Another strain of flies in which nondisjunction of the X chromosomes occurred in *all* females was discovered by L. V. Morgan (1922). When such females were mated to cytologically normal males carrying a recessive gene on the X chromosome, all viable male progeny were phenotypically like the father (but sterile), whereas all viable female offspring were like the mother. In addition, one fourth of the total progeny were metafemales and another fourth ($AAOY$) died in the egg stage. Morgan reasoned that in these attached-X females ($\widehat{XX}$) the two X chromosomes were physically attached so that only two kinds of eggs, $A\widehat{XX}$ and $AO,$ were produced. This explanation was soon confirmed cytologically.

Polyploid Flies. Experimentally produced triploid (three whole sets of chromosomes, or $3n$) and tetraploid ($4n$) flies were next incorporated into Bridges' work, so that many kinds of flies with respect to chromosome complements were ultimately produced. As this work was carried forward, it became increasingly clear that, in *Drosophila* at least, the important key to the sex of the individual was provided by the *ratio of X chromosomes to sets of autosomes.*

The Y, then, has nothing to do with sex determination but does govern male fertility. These results are summarized in Table 11-1.

normally A = 3 [handwritten]

TABLE 11-1. Summary of Chromosomal Sex Determination
in Drosophila (After Bridges)

Number of X Chromosomes	Number of Sets of Autosomes	X/A Ratio	Sex Designation
3	2	1.50	Metafemale *(super)* [handwritten]
4	3	1.33	Triploid metafemale
4	4	1.00	Tetraploid female
2	2	1.00	Female
3	4	0.75	Tetraploid intersex
2	3	0.67	Triploid intersex
1	2	0.50	Male
1	3	0.33	Triploid metamale
1	4	0.25	Tetraploid metamale

Metamales (or supermales) are to the male sex as metafemales (or super-females) are to the female sex. That is, they are weak, sterile, underdeveloped, and die early. Intersexes are sterile individuals displaying secondary sex characters between those of the male and female (Fig. 11-4).

From all these results the mechanism of sex determination in *Drosophila* may be summarized:

1. Sex is governed by the ratio of the number of X chromosomes to sets of autosomes. Thus, from Table 11-1, females have an X/A ratio = 1.0, males = 0.5. This relationship applies even to polyploid flies so long as the appropriate X/A ratio is maintained. *sex characteristics* [handwritten]
2. Genes for maleness per se are apparently carried on the autosomes, those for femaleness on the X chromosome.
3. The Y chromosome governs male *fertility,* rather than sex itself, because AAXY and AAXO flies are both male as to secondary sex characters but only the former produce sperm; it has no effect in AAXXY flies, which have an X/A ratio of 1.0 and are female.
4. An X/A ratio > 1.0 or < 0.5 results in certain characteristic malformations (*metafemales* and *metamales*).
5. An X/A ratio < 1.0 but > 0.5 produces individuals intermediate between females and males (*intersexes*). The degree of femaleness is greater where the X/A ratio is closer to 1.0, and the degree of maleness is greater where that ratio is closer to 0.5.

Other workers have confirmed and extended these observations on intersexes in *Drosophila*. By using X-rays to fragment chromosomes, it is possible to develop lines of flies having extra fragments of the X chromosome. Thus

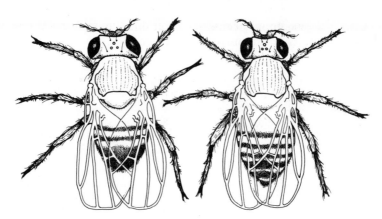

FIGURE 11-4. *Triploid intersexes (X/A ratio 0.67) in* Drosophila.

an individual with, say, about two-and-two-thirds X chromosomes and three sets of autosomes (X/A ratio = 2.67/3 = 0.89), although an intersex, displayed secondary sex characters more female in nature than did one with two-and-a-third X and three sets of autosomes (X/A ratio = 2.33/3 = 0.78), and so on.

The same system where the X/A ratio is critical is reported for the angiosperm weed, dock (*Rumex acetosa*), by Warmke (1946).

Temperature and Intersexes.   As we have noted, intersexes vary from a strong resemblance (in secondary sex characters) to males through degrees of intermediacy to a condition closely approximating femaleness. Yet in a given series of crosses where this difference occurred, the chromosomal situation could be shown to be identical in all pedigrees. That is, in some crosses AAAXX intersex flies closely approximated males, in others females, but often they were somewhat intermediate. Temperature was found to be the critical factor with higher temperatures shifting the balance toward femaleness and lower ones shifting it toward maleness.

The Transformer Gene.   One additional complicating factor in sex determination in *Drosophila* is worth examining briefly. A recessive gene, *tra* on the third chromosome (an autosome), when homozygous, "transforms" normal diploid females (AAXX) into sterile males. The XX *tra tra* flies have many sex characters of males (external genitalia, sex combs, and male-type abdomen) but, as noted, are sterile.

Gynandromorphs.   Concepts of sex determination as developed for *Drosophila* are verified by the occasional occurrence of **gynandromorphs** (or gynanders). These are individuals in which part of the body expresses male characters whereas other parts express female characters. A bilateral gynandromorph, for example, is male on one side (right or left) and female on the other. The male portions of such flies would be expected to have a male chromosomal composition. By ingenious experiments using known "marker" genes on the X chromosome, it has been shown that this is indeed so. In *Drosophila* right and left body halves are determined at the first cleavage of the zygote. Lagging of the X chromosome at this first mitosis can result in two daughter cells of the chromosomal complement AAXX and AAXO when the laggard X fails to be incorporated in a daughter nucleus. The portion of the body developing from the former cell will be normal female, that from the latter (sterile) male. Gynandromorphs represent one kind of *mosaic*, or organisms part of which are composed of cells genetically different from the remaining part.

Sex As a Continuum.   So we see that sex in *Drosophila* is far from the simple, either-or, male or female, XX or XY condition it was initially thought to be. Instead, sex may be viewed as a continuum, ranging from supermaleness through maleness, intersexes of varying degree, to femaleness and on to superfemaleness. Where an individual places in such a continuum is seen to be related to the ratio of individual X chromosomes to sets of autosomes. Of course, it is not the chromosomes as gross structures that are the deciding factors, but, rather, *genes* on the chromosomes. Thus genes for maleness are associated with the autosomes, those for femaleness with the X chromosomes. Yet this entire sex-determining arrangement may, in some cases, be upset by a single pair of recessive autosomal genes (*tra*)!

Man

In normal human beings males are XY and females are XX, just as in *Drosophila*. But is sex here also determined by the X/A ratio? With regard to sex, does the Y chromosome bear genes for male fertility as in *Drosophila* or for male sex per se? To answer these questions it will be helpful to examine (1) the sex chromosomes, (2) the concept of sex differentiation, and (3) human sex anomalies and their chromosomal makeup.

THE SEX CHROMOSOMES

X Chromosome.   The human X chromosome is a medium-length, submetacentric one, intermediate in length between autosomes 7 and 8 (Table 11-2). Its centromere is more nearly central than any of the larger chromosomes of the C group. With the development of fluorescent banding techniques (Chapter 3), visual differentiation of the X chromosomes and morphologically similar autosomes became exact (Fig. 11-5). In mitotic metaphase spreads the X chromosome measures approximately 4.5 to 5.5 μm, depending upon the preparation.

Y Chromosome.   The Y chromosome in most human males averages roughly 1.8 μm in length, very slightly longer than the members of the shortest or G group of autosomes (Table 11-2). It is, however, more variable in length among different men; in some it is regularly as long as or longer than members of the F group (autosomes 19 and 20), whereas in others it may be less than

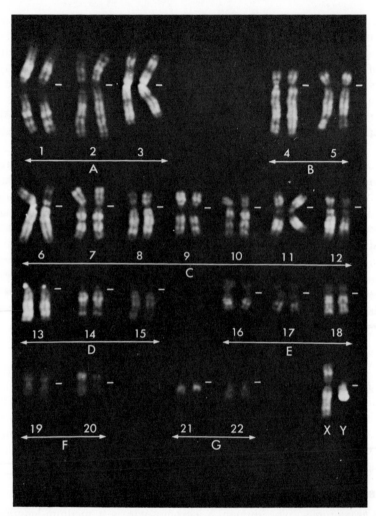

FIGURE 11-5. *Karyotype of normal human male (46, XY), Q-banding. Banding techniques permit much greater accuracy in matching autosomes and identifying the sex chromosomes.* (Courtesy Dr. C. C. Lin.)

half the length of the members of the G group. No particular phenotype or syndrome has been consistently associated with either a long or a short Y chromosome. The Y chromosome is acrocentric, even more so than the G autosomes, its longer arms generally lying close together. Unlike autosomes 21 and 22, the Y has no satellites. With fluorescent banding techniques the longer arm fluoresces brilliantly in good-quality preparations (Fig. 11-5).

X- and Y-Chromatin in Interphase Nuclei.   A clue to the sex chromosomal complement of an individual may be obtained quite simply by examining

TABLE 11-2. Measurements of Relative Lengths of Human
Chromosomes in Percentage of Total
Monoploid Autosome Length

Chromosome Number	A	B	C
1	9.08	9.11 ± 0.53	8.44 ± 0.433
2	8.45	8.61 ± 0.41	8.02 ± 0.397
3	7.06	6.97 ± 0.36	6.83 ± 0.315
4	6.55	6.49 ± 0.32	6.30 ± 0.284
5	6.13	6.21 ± 0.50	6.08 ± 0.305
6	5.84	6.07 ± 0.44	5.90 ± 0.264
7	5.28	5.43 ± 0.47	5.36 ± 0.271
X	5.80	5.16 ± 0.24	5.12 ± 0.261
8	4.96	4.94 ± 0.28	4.93 ± 0.261
9	4.83	4.78 ± 0.39	4.80 ± 0.244
10	4.68	4.80 ± 0.58	4.59 ± 0.221
11	4.63	4.82 ± 0.30	4.61 ± 0.227
12	4.46	4.50 ± 0.26	4.66 ± 0.212
13	3.64	3.87 ± 0.26	3.74 ± 0.236
14	3.55	3.74 ± 0.23	3.56 ± 0.229
15	3.36	3.30 ± 0.25	3.46 ± 0.214
16	3.23	3.14 ± 0.55	3.36 ± 0.183
17	3.15	2.97 ± 0.30	3.25 ± 0.189
18	2.76	2.78 ± 0.18	2.93 ± 0.164
19	2.52	2.46 ± 0.31	2.67 ± 0.174
20	2.33	2.25 ± 0.24	2.56 ± 0.165
21	1.83	1.70 ± 0.32	1.90 ± 0.170
22	1.68	1.80 ± 0.26	2.04 ± 0.182
Y	1.96	2.21 ± 0.30	2.15 ± 0.137

Column A: Previous Denver-London Conference data.
Column B: Data from 10 cells by Drs. T. Caspersson, M. Hultén, J. Lindsten, and L. Zech. Cells stained with orcein.
Column C: Data from 95 cells provided by Drs. H. Lubs, T. Hostetter, and L. Ewing from 11 normal subjects (6 to 10 cells per person). Average total length of chromosomes (diploid set) per cell: 176 μm. Cells stained with orcein or Giemsa 9 technique.
Percentages for autosomes do not add to 100 because they are averages for a given chromosome in several metaphase spreads.
From *Paris Conference (1971): Standardization in Human Cytogenetics.* In *Birth Defects: Orig. Art. Ser.,* ed. D. Bergsma. Published by The National Foundation–March of Dimes, White Plains, N.Y., Vol. VIII (7), 1972. Used by permission.

squamous epithelial cells from scrapings of the lining of the cheek. When stained, most somatic cells of normal females show a characteristic structure, the *Barr body,* so named after its discoverer, Murray Barr, who first described it in 1949 (Fig. 11-6). Females are therefore termed *sex chromatin positive;* normal males, whose cells do not contain a Barr body, are *sex chromatin negative.* Because of various cytological factors, as well as some of the techniques employed, frequency of sex chromatin positive cells in scrapings of oral mucosa from normal females have been reported to range from 36 to 80 per cent of cells examined. The Barr body is a small structure (about 1 μm in greatest

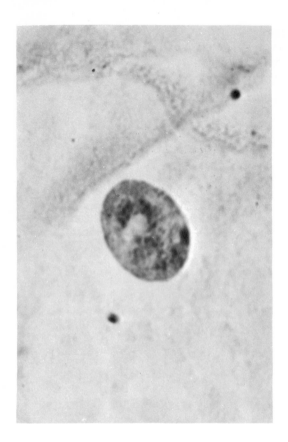

FIGURE 11-6. *Nucleus of normal human female squamous epithelial cell showing prominent, dark Barr body against the nuclear membrane.* (Courtesy Carolina Biological Supply Co.)

dimension), hemispherical, disk shaped, rod shaped, or triangular in outline. It ordinarily lies appressed to the inner surface of the nuclear membrane or, in nerve cells, may be associated with the nucleolus. In diploid cells of females the number of Barr bodies is one less than the number of X chromosomes.

In somatic cells having two X chromosomes, one replicates later than the other. Mary Lyon (1961, 1962) and others have proposed that the late-replicating X chromosome becomes inactivated early in embryonic life and forms the Barr body. Whether the maternal or paternal X chromosome is inactivated in any given cell depends on chance, but once this has occurred in an embryo cell, it appears always to be the same one that becomes the Barr body in all cells derived therefrom. Thus, incidentally, females heterozygous for genes located on the X chromosome are *mosaics;* some patches of tissue express the dominant phenotype, others the recessive.

Staining of XY (or XYY) cells with such fluorescent dyes as quinacrine hydrochloride discloses the brightly fluorescing Y chromosome(s), not only in mitotic metaphase, but also in interphase (Fig. 11-7). The same technique

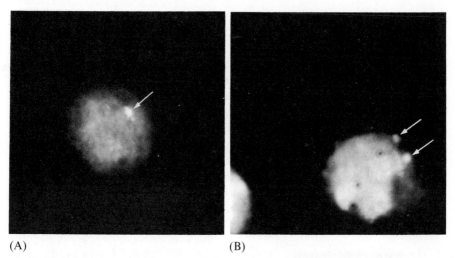

(A)   (B)

FIGURE 11-7. *The brightly fluorescing Y chromosome* (arrows) *in quinacrine-stained interphase.* (*A*) *46, XY male.* (*B*) *47, XYY male.* (Courtesy Dr. C. C. Lin.)

can be used successfully on sperm (Sumner, Robinson, and Evans, 1971) and on cells in the amniotic fluid (Rook, Hsu, Gertner, and Hirschhorn, 1971).

SEX DIFFERENTIATION

Genetic Sex.   Normal females ordinarily have two X chromosomes, normal males one X and one Y. As noted earlier, it is, of course, genes on these chromosomes that determine femaleness or maleness. Thus one can speak of females as having the *genetic sex* designation XX and males as having the genetic sex designation XY, although exceptional cases, to be described, occur.

Gonadal Sex.   Chemical substances (inductors) produced by embryonic XX cells act on the *cortical* region of undifferentiated gonads to bring about development of ovarian tissue. In XY embryos, however, inductors stimulate production of testes from the *medulla* of the undifferentiated gonads. Hence the XX genetic sex is ordinarily associated with ovarian *gonadal sex,* and XY with testicular gonadal sex.

Genital Sex.   The embryonic gonads produce hormones that, in turn, determine the morphology of the external genitalia and the genital ducts. XX embryos normally develop ovaries, female external genitalia, and Müllerian ducts. XY embryos, on the other hand, ordinarily develop testes, male external genitalia, and Wolffian ducts. In XX embryos Wolffian ducts are suppressed; in XY embryos the Müllerian ducts remain undeveloped. Thus there is a distinction between male and female *genital sex.*

Somatic Sex.   Production of gonadal hormones continues to increase until, at puberty, *secondary sex characters* appear. These include amount and distribution of hair (e.g., facial, body, axillary, pubic); pelvic dimensions; general body proportions; subcutaneous fat over hips and thighs, and breast development in the female, as well as increased larynx size and deepening of the voice in the male.

Sociopsychological Sex.   In most individuals genetic sex, gonadal sex, genital sex, and somatic sex are consistent; XX persons, for example, develop ovaries, female genitalia, and female secondary sex characters. Ordinarily these persons are raised as females and adopt the feminine gender role under whatever cultural pattern has been established in the society of which they are members. A similar consistency from genetic sex to sociopsychological sex is seen for XY individuals. On the other hand, some individuals display an inconsistency of some kind or degree among these levels of sexuality. Discordance involving the genetic and anatomical result in intersexuality, which we shall examine briefly later in this chapter.

HUMAN SEX ANOMALIES

The Klinefelter Syndrome.   One in about 500 "male" births produces an individual with a particular set of abnormalities known collectively as the *Klinefelter syndrome* (Fig. 11-8). These persons have a general male phenotype; external genitalia are essentially normal in gross morphology. Although there is some variability in other characteristics, testes are typically small, sperms are usually not produced, and most such men are mentally retarded. Arms are longer than average, some degree of breast development is common, and the voice tends to be higher pitched than in normal males. Klinefelters are sex chromatin positive; the karyotype shows 47 chromosomes, i.e., 47,XXY.

The XXY individual may arise through fertilization of an XX egg by a Y sperm, or through fertilization of an X egg by an XY sperm. Although the majority of Klinefelters are born to mothers under the age of 30, this is in large part a reflection of the age group in which most births of all types occur (Fig. 11-9). After a drop in Klinefelter births from ages 27 to 32, there is a small increase again after age 32, whereas total births decrease rapidly in this age group. This fact suggests nondisjunction of the X chromosomes in aging oocytes as a somewhat more important factor than XY nondisjunction during spermatogenesis.

Less frequently, Klinefelters have more than two X chromosomes (Terheggen et al., 1973), even more than one Y (Table 11-3). Generally the greater the number of X chromosomes, the greater the degree of mental retardation.

As a clue to the nature of the sex-determining mechanism in human beings, the important point here is the fact that persons with at least one Y chromosome have the general phenotype of a male, even in the presence of any number of X chromosomes. Thus it begins to appear that "maleness," however one defines it, is dictated by the presence of at least one Y chromosome.

The Turner Syndrome.   A second major sex anomaly is represented by the *Turner syndrome,* in which the individual is phenotypically female, but with poorly developed ovaries, and sterile. Characteristically such persons exhibit

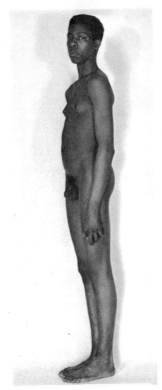

FIGURE 11-8. *The Klinefelter syndrome in man. Such persons are AAXXY, with 47 chromosomes as shown in the karyotype. External genitalia are male-type, but there is usually some female-like breast development as in this case.* (Photo courtesy Dr. Victor A. Mc-Kusick; karyotype redrawn from V. A. McKusick, Medical Genetics, *Journal of Chronic Diseases,* **12**: 1–202, 1960 by permission of The C. V. Mosby Co.)

TABLE 11-3. Sex Chromosome Anomalies and Their Estimated Frequencies

Designation	Chromosome Constitution	Sex Chromatin	Somatic Chromosome Number	Sex Phenotype	Usual Fertility	Estimated Frequency per 1,000	Estimated Number in the U.S.*
Klinefelter	AAXXY	+	47	♂	−	2.0	420,000
Turner	AAXO	−	45	♀	−	0.2	42,000
Triplo-X	AAXXX	++	47	♀	+	0.75	157,500
Tetra-X	AAXXXX	+++	48	♀	Unknown	Very low	Very low
Triplo-X,Y	AAXXXY	+++	48	♂	−	Very low	Very low
Tetra-X,Y	AAXXXXY	+++	49	♂	−	Very low	Very low
Penta-X	AAXXXXX	++++	49	♀	−	Very low	Very low
XYY	AAXYY	−	47	♂	±	4.0	850,000
Klinefelter XXYY	AAXXYY	+	48	♂	Not reported	Very low	Very low
Klinefelter XXXYY	AAXXXYY	++	49	♂	Not reported	Very low	Very low

* Based on a population of 210 million.

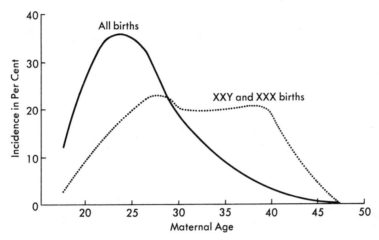

FIGURE 11-9. *The relationship between maternal age and XXY and XXX births.*

"webbing" of the neck, low-set ears, broad chest, underdeveloped breasts, but often only slightly below average intelligence (Fig. 11-10). Although Turners are usually sterile, one normal birth has been reported to an apparent Turner female. Turners are sex chromatin negative, which suggests a single X chromosome instead of the expected pair. This is confirmed by cytological studies that show the somatic chromosome number to be 45, i.e., 45,X (AAXO). No maternal age effect has been demonstrated; in fact, Ferguson-Smith (in McKusick and Claiborne, eds., 1973) reports that in 72 per cent of XO births, the single X was contributed by the mother. Thus paternal nondisjunction of the X and Y chromosomes was the cause in nearly three out of four live XO births. The frequency of live Turner births has been estimated at 2 to 3 per 10,000 female births, but Battin and Serville (1973), in a study of 139 cases of gonadal dysgenesis, place the incidence at only 1 in 10,000. This low incidence is associated with a high rate of intrauterine mortality (90 per cent or more) of XO fetuses. Some 20 to 30 per cent of all spontaneously aborted fetuses are reported to be XO. The conclusion to be drawn from the Turner syndrome is that, in the absence of a Y chromosome, the general sex phenotype is female.

Poly-X Females. In 1959 Jacobs and others reported the first known case of a *triplo-X* individual, i.e., 47,XXX. This was clearly a female in general sex phenotype, but at age 22, she had infantile external genitalia and marked underdevelopment of internal genitalia and breasts. She was somewhat retarded mentally. Since that time many more XXX females have been described, and it is estimated that between 1 in 1,000 and 1 in 2,000 live female births is triplo-X. Some XXX females are essentially normal, but others are retarded and/or show abnormalities of primary and secondary sex characters. Apparently all are fertile, but according to Stern (1973), among more than 30

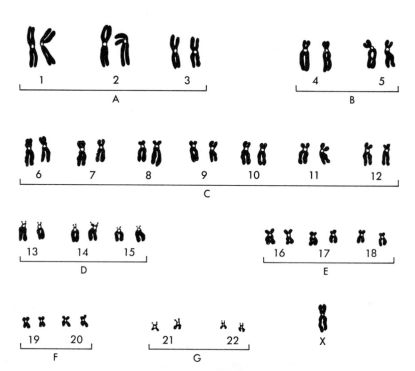

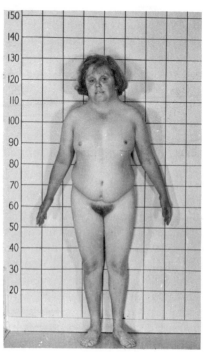

FIGURE 11-10. *The Turner syndrome. Such persons are AAXO, with only 45 chromosomes as shown in the karyotype. Note external female genitalia, webbed neck, broad chest, underdeveloped breasts, and short stature. Turner individuals have a small uterus and vestigal ovaries.* (Photo courtesy Dr. Victor A. Mc-Kusick; photo and karyotype reproduced from V. A. McKusick, Medical Genetics, *Journal of Chronic Diseases,* **12:**1–202, 1960, by permission of The C. V. Mosby Co.)

children of triplo-X mothers, all were XX or XY except for one Klinefelter. A very few tetra-X (48,XXXX) and penta-X (49,XXXXX) persons have been described; manifestations are similar to those of triplo-X individuals, but more marked. In general, as the number of X chromosomes increases, intelligence is progressively reduced, as with Klinefelters. For our immediate concerns, it is important to note that multiple-X individuals are female in general phenotype. As in the case of Klinefelters, there is a maternal age effect (Fig. 11-9).

The XYY Male. Particular interest has recently focused on behavior patterns in XYY individuals. These are all males, above average in height ($>$ 72 in.), often below normal in intelligence (reported range from I.Q. 80 to about 118), and with a history of severe facial acne during adolescence. Abnormalities of internal and external genitalia have been noted in some, but no consistent major anomalies occur. This kind of individual first came to notice when Jacobs et al. (1965) reported a high incidence of XYY males (9 of 315 persons) housed in the maximum security section of a Scottish criminal institution. Many similar reports followed during the latter half of the 1960s,[1] leading to an initial notion that XYY males are more aggressive and more likely than the XY male to commit crimes of violence. This conclusion is now viewed as having been premature, for after additional studies of the general population, it has become clear that many XYY males do make satisfactory social adjustment; some are indeed of very retiring personality. Lubs and Ruddle (1970), in a study of 4,366 consecutive births at Yale-New Haven Hospital, found XYY births to occur with a frequency of 0.69 per thousand.

A number of widely publicized criminal cases have served to direct attention to legal aspects of chromosomal aberrations, particularly the XYY male (Table 11-4).

TABLE 11-4. Some Criminal Cases Involving XYY Males

Accused	Location	Charge	Plea	Disposition
Daniel Hugon	Paris	Murder	Not guilty by reason of XYY finding	Seven years' imprisonment
Richard Speck	Chicago	Multiple murder	Not guilty by reason of insanity	Sentenced to death; appealed
Lawrence Hannell	Melbourne	Murder	Not guilty by reason of insanity	Not guilty by reason of insanity
Robert Tait	Melbourne	Murder	Not guilty	Sentenced to hang; commuted to life imprisonment

[1]Borgaonkar (1969) has compiled an extensive bibliography on the XYY syndrome.

Aside from the legal and social questions raised by the XYY individual, it is important to note here that the general sex phenotype is male.

Other Sex Chromosome Anomalies.   A number of other sex chromosome anomalies have been reported, most of them quite rare (Table 11-3). In every case, though, the presence of even one Y chromosome serves to produce the male sex phenotype, regardless of the number of Xs that may be present. No clear relationship between sex chromosome anomalies in children and any particular parental characteristics, except for some suggestion of a higher risk in mothers with thyroid disorder (hyperthyroid and hypothyroid).

Hermaphroditism.   No account of sex chromsome anomalies would be complete without mention of hermaphroditism and pseudohermaphroditism. *True hermaphrodites,* by generally accepted definition, are individuals possessing both ovarian and testicular tissue (Fig. 11-11). The external genitalia are ambiguous, but often more or less masculinized; secondary sex characters vary from more or less male to more or less female. All are sterile. Some are reared as males, some as females.

Polani (1970) summarized 108 cases of true hermaphroditism, 59 of which were XX in chromosomal constitution, 21 XY, and 28 mosaics. All but two of the latter were found to possess some Y chromosome cell lines. Accounting for development of testicular tissue in the absence of a Y chromosome presents some difficulties, of course. XX true hermaphrodites could be accounted for without conflicting with the apparent requirement of a Y chromosome for testicular development by assuming loss of the Y chromosome from an XXY fetus at an appropriate stage of embryo development. The most informative evidence may well come from the XX/XY mosaics, a condition that suggests that most, or possibly all, true hermaphrodites did originally have both XX and XY cell lines.

Pseudohermaphrodites have either testicular or ovarian tissue, usually very rudimentary, but not both. On the basis of chromosomal constitution, two major classes, *male* and *female,* can be distinguished. The former are XY, with testicular tissue, but almost always sterile. External genitalia are ambiguous, tending toward the male in the *masculinizing* variety of male pseudohermaphroditism (Fig. 11-12), but more female in morphology in the *feminizing* variety (Fig. 11-13). Feminizing male pseudohermaphrodites often have large breasts and a blind vagina; some lead an essentially normal female sex life (though they are XY!).

Female pseudohermaphrodites (Fig. 11-14) are XX, with ambiguous external genitalia and immature ovaries. The general appearance is ordinarily masculine.

Mechanism of Sex Determination in Man.   Table 11-3 clearly indicates that individuals having at least one Y chromosome are male, at least as to external genitalia, though they may be sterile. Contrariwise, persons with one or more X chromosomes are phenotypically female as long as no Y is present, though, again, they are sometimes infertile. So it would seem that genes for maleness

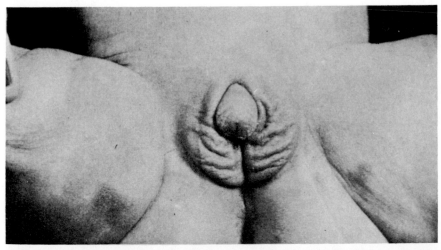

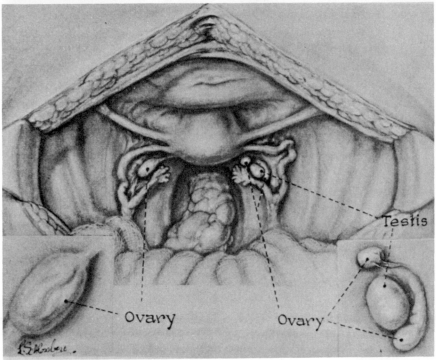

FIGURE 11-11. *True hermaphroditism in a five-month-old child. (A) Ambiguous external genitalia. The clitoris is enlarged, resembling a penis; the labio-scrotal folds are fused. (B) Drawing of operative findings, showing an ovary (left) and an ovo-testis (right).* (From M. Bartalos and T. A. Baramki, 1967. *Medical Cytogenetics.* Baltimore, The Williams & Wilkins Company. Used by permission of authors and publisher.)

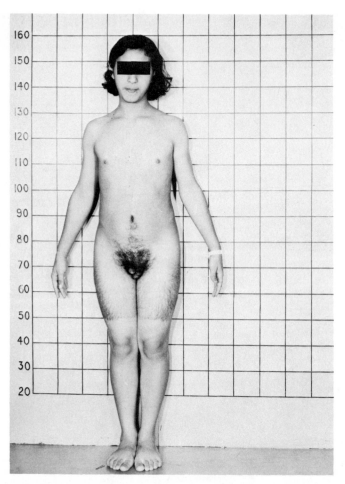

FIGURE 11-12. *A 16-year-old male pseudohermaphrodite of the masculinizing variety (46, XY), reared as a female. Note body hair, lack of breast development, and ambiguous external genitalia.* (Courtesy Dr. T. A. Baramki.)

are carried on the Y chromosome in man, those for femaleness on the X. One Y chromosome offsets several Xs, so that even XXXXY persons are male, though sterile. Thus sex in man appears to be determined by presence of X and Y chromosomes.

The question of whether or not autosomes play any part is seen in individuals with exceptional numbers of autosomes. Although these will be dealt with more fully under chromosomal aberrations (Chapter 13), suffice it to say here that most variations in number of autosomes involve persons all of whose so-

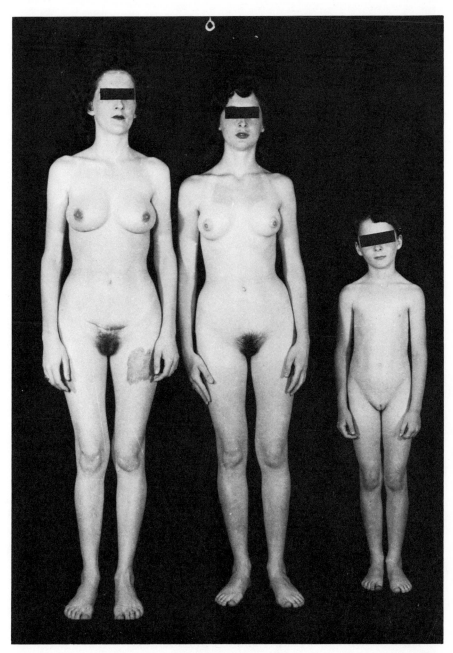

FIGURE 11-13. *Feminizing male pseudohermaphroditism (46, XY) in three siblings.* (From M. Bartalos and T. A. Baramki, 1967. *Medical Cytogenetics.* Baltimore, The Williams & Wilkins Company. Used by permission of authors and publisher.)

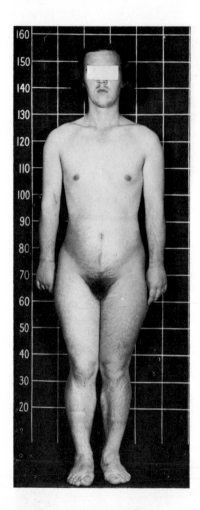

FIGURE 11-14. *A 16-year-old female pseudo-hermaphrodite (46, XX), showing short, stocky build, body hair, and lack of breast development. External genitalia are ambiguous.* (From M. Bartalos and T. A. Baramki, 1967. *Medical Cytogenetics.* Baltimore, The Williams & Wilkins Company. Used by permission of authors and publisher.)

matic cells contain one extra of a particular autosome, not additional whole sets. No complete living polyploids are known in man, though an interesting **mosaic** was reported some years ago in Sweden. This was a young male, at first reported to be triploid (69 chromosomes). It was later determined that he was a mosaic for $2n$ and $3n$ tissues. Phenotypically, this individual was male. In both autosomal trisomics (individuals with 47 chromosomes, in which the forty-seventh is a specific autosome) and in the Swedish mosaic, sex appears to be determined only by the X and Y chromosomes.

On the other hand, several cases are reported in the literature of spontaneously aborted, completely triploid fetuses. With regard to the sex chromosomes, some of these were XXX, some XXY, and a few XYY. A very small number of completely tetraploid, aborted fetuses is also known. From available evidence, it would appear that the genic imbalance produced by extra whole genomes in man is lethal.

We can then summarize the situation in man in this way:

1. Autosomes play no part in determining sex.
2. The Y chromosome determines maleness, with even a single one outweighing any number of Xs.
3. The X chromosome determines femaleness *in the absence of any Ys.*

Sex determination in man, then, although related to the X-Y chromosome makeup, operates differently than in *Drosophila.* Mammals generally appear to follow the mechanism outlined for human beings.

PLANTS

One flowering plant, the wild campion (*Lychnis dioica,* formerly *Melandrium*) of the pink family (Caryophyllaceae), has been extensively investigated. It is worth reporting here because it illustrates still another variation of the XX-XY system. In angiosperms, the plant we recognize by name is the asexual, diploid sporophyte generation. The sexual phase is microscopic and contained largely within the tissues of the sporophyte, on which it is parasitic. Flowers contain either or both of two essential organs, stamens and/or pistils. The former produce microspores that develop into male-gamete-bearing plants (at one stage these are the well-known pollen grains). Pistils produce and contain the egg-bearing sexual plant. Many plants have so-called perfect flowers, which contain both stamens and one or more pistils. *Lychnis* has imperfect flowers, which bear either stamens *or* pistils. The species is, moreover, dioecious, so that there are staminate (male-producing) individuals and pistillate (female-producing) ones. Though not accurate, staminate plants are often referred to as male plants and pistillate as female.

In *Lychnis,* staminate plants are XY and pistillate plants are XX. Warmke (1946) found the X/A ratio to bear no relation to "sex" but, through studies of polyploid strains, determined the X/Y ratio to be critical, X/Y ratios of 0.5, 1.0, and 1.5 were found in plants having only staminate flowers; in plants whose X/Y ratio was 2.0 or 3.0, occasional perfect flowers occurred among otherwise staminate flowers. In plants having four sets of autosomes, four X chromosomes and a Y, flowers were perfect but with an occasional staminate one. These results are summarized in Table 11-5, and illustrated in Fig. 11-15.

Westergaard (1948) extensively investigated the cytology of *Lychnis* and, from studies of plants from which portions of the X or Y chromosome were deleted, arrived at a comparison of the X and Y chromosomes, as shown in Fig. 11-16. The sex chromosomes are quite dissimilar in size, the Y being larger than the X, and each larger than any autosome. Only a small portion of the X is homologous with a similar small bit of the Y, as suggested by their synaptic

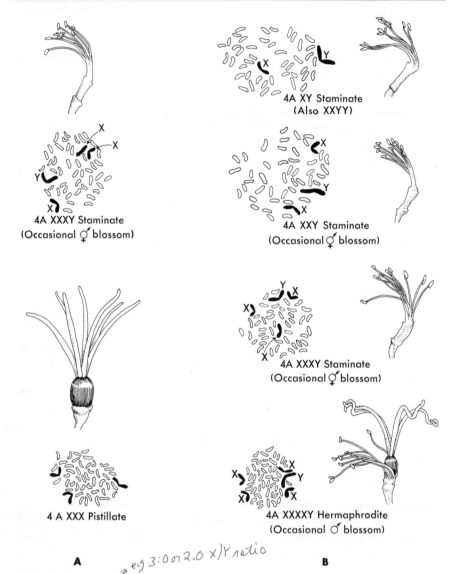

A $eg\ 3:0\ or\ 2.0\ X/Y\ ratio$ **B**

FIGURE 11-15. (A) Reproductive parts and camera lucida drawings of the somatic chromosomes from 4A XXXY and 4A XXX plants of Lychnis (Melandrium). Note that a single Y chromosome produces a staminate plant having occasional perfect flowers; in the absence of the Y chromosomes, plants are pistillate. (B) Reproductive parts and camera lucida drawings of the somatic chromosomes of Lychnis in a series showing pistillate-determining tendency of the X chromosome. All plants shown are tetraploid with respect to autosomes, but differ in the number of X chromosomes present. Tendency to produce pistillate flowers increases as the number of X chromosomes rises. (Redrawn from H. W. Warmke, 1946, Sex Determination and Sex Balance in Melandrium. American Journal of Botany, **33**: 648–660, by permission of the author and the Botanical Society of America, Inc., Washington.)

TABLE 11-5. "Sex" and X/Y Ratios in *Lychnis*

Chromosome Constitution	X/Y ratio	"Sex"
2 AXYY	0.5	♂
2 AXY		
3 AXY	1.0	♂
4 AXY		
4 AXXXYY	1.5	♂
2 AXXY		
3 AXXY		
4 AXXY	2.0	♂ with occasional ⚥ flower
4 AXXXXYY		
3 AXXXY		
4 AXXXY	3.0	♂ with occasional ⚥ flower
4 AXXXXY	4.0	⚥ with occasional ♂ flower

♂ = staminate; ♀ = pistillate; ⚥ = perfect.
Each A = one set of 11 autosomes, each X = an X chromosome, each Y = a Y chromosome.

figures in meiosis. When region I is deleted, perfect-flowered plants are produced, whereas loss of region II produces pistillate plants (that is, whereas AAXY plants are staminate, AAXY^{-II} plants are pistillate), and deletion of region III produces sterile staminate plants with aborted stamens.

Genic Determination of Sex

Sex does not appear to be controlled in all dioecious organisms by numerous genes on two or more chromosomes, however. Several examples will serve to illustrate some variations on the so-called chromosomal method.

Asparagus. Rick and Hanna (1943) presented evidence to show that sex in asparagus is determined by a single pair of genes, with "maleness" (i.e., formation of staminate flowers) the dominant character. Asparagus is normally dioecious, some plants producing only staminate flowers, others only pistillate. Rudimentary pistils occur in staminate flowers, and abortive stamens in pistillate blossoms. These rudimentary organs are ordinarily nonfunctional. However, the pistils in staminate flowers do very rarely function to produce viable seeds. Flower structure in this species is such that these seeds most likely result from self-pollination. In effect, this is a "male × male" cross from the genetic standpoint. The basis for the occasional functional pistils in staminate plants is not known, but both genetic and environmental factors have been suggested.

Rick and Hanna germinated 198 seeds from these uncommon functional pistils produced on staminate plants, finding a progeny ratio of 155 staminate to 43 pistillate. This is a close approximation to a 3:1 ratio, as a chi-square test will indicate. One third of the staminate progeny, when crossed with normal

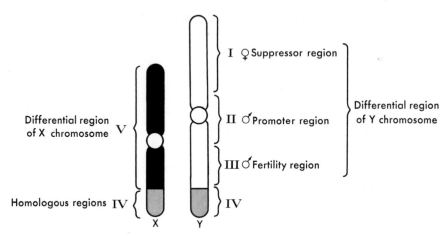

FIGURE 11-16. *Comparison of X and Y chromosomes of the plant,* Lychnis. *Regions I, II, and III bear holandric genes, and region V sex-linked genes. Genes in region IV are termed incompletely sex linked.* Lychnis *is unusual in that its Y chromosome is the larger of the two sex chromosomes.*

pistillate plants, yielded only staminate offspring. The rest proved to be heterozygous, producing staminate and pistillate offspring in a ratio close to 1:1.

If "maleness" is represented as $A-$ and "femaleness" as aa, the original "male × male" cross would be

$$P \quad Aa \times Aa$$
$$F_1 \quad \tfrac{3}{4}A- + \tfrac{1}{4}aa$$

On the basis of monohybrid inheritance examined in Chapter 2, one third of the $A-$ (staminate) plants should be AA and two thirds Aa. Crosses between these genotypes and normal pistillate (aa) individuals would produce the results reported by Rick and Hanna.

Honeybee. Worker and queen bees are diploid females, with 32 chromosomes. Drones, on the other hand, are males having but 16 chromosomes and develop from unfertilized eggs. Whiting (1945), through extensive studies on the parasite wasp *Habrobracon,* postulated that femaleness in such hymenopterans as these is determined by heterozygosity of a number of different chromosome segments. The situation is analogous to a series of multiple alleles but is based on chromosomal segments containing several genes each. Monoploid individuals, hatching from unfertilized eggs, cannot be heterozygous; hence they are male. As you might suspect, Whiting was able to produce diploid males by developing individuals homozygous in a sufficient number of chromosomal segments.

Corn. Unlike the organisms so far described, corn (*Zea mays*) is monoecious. The "tassel" consists of staminate flowers, and the ear of pistillate flowers. Among several controls over "sex" in this plant are two interesting pairs of genes. The genotype *bs bs* ("barren stalk") results in plants having no ears at all, though a normal tassel is present (Fig. 11-17). Such individuals are staminate ("male"). Gene *ts* ("tassel seed"), when homozygous, converts the tassel to pistillate flowers, so that ears develop at the top of the plant (Fig. 11-18). Therefore *ts ts* individuals are pistillate ("female"). If we let ♂ represent staminate flowers, and ♀ pistillate ones, the various genotypes and phenotypes are as follows:

Genotype	Phenotype	
Bs– Ts–	Normal monoecious	
bs bs Ts–	Staminate	
Bs– ts ts	Pistillate; ears terminal *and* lateral	
bs bs ts ts	Pistillate; ears terminal only	

Plants of genotype *bs bs Ts–* plus either *Bs– ts ts* or *bs bs ts ts* comprise a dioecious race of corn. Such a race may easily be produced by making the cross *bs bs ts ts* × *bs bs Ts ts* which will segregate pistillate and staminate in a 1 : 1 ratio. Perhaps some sort of development like this has occurred in the evolution of dioecism. The practical advantage of the dioecious state to the breeder is obvious.

Bacteria. The discovery by Lederberg and Tatum (1946) of conjugation in the colon bacillus (*Escherichia coli*) opened the way to a whole new area of understanding in genetics. We shall consider some of its implications later, but as a variant in "sex" determination, it should be noted here.

Conjugation in bacteria involves a process in which two cells become joined by a bridge (Fig. 11-19) through which one cell "donates" to the other part or all of its single nucleoid (Appendix B). Conjugation results in a temporary diploidy of the recipient cell. This diploidy is complete for all genes only if the entire nucleoid of the donor passes into the recipient; otherwise it is partial, involving only certain genes. The usual monoploid condition is restored at the next cell division (Appendix B).

FIGURE 11-17. *Barren stalk corn,* bsbs. *Note the absence of ears.* (From *The Ten Chromosomes of Maize,* DeKalb Agricultural Association, DeKalb Illinois. Reproduced by permission.)

The donor cell may be thought of as analogous to a male, the recipient cell to a female. "Maleness," in this sense, is determined by possession of an **episome** or fertility factor, F, "femaleness" by its absence. The F factor may be located in the cytoplasm, physically separate from the nucleoid, or it may be integrated into the bacterial "chromosome." If the F factor is transferred during conjugation, the recipient cell is converted from "female" to "male." Much has been learned about the nature of the genetic material itself from studies of recombination resulting from conjugation in bacteria, and we shall explore this matter in Chapter 15.

Chlamydomonas.   The common microscopic unicellular green alga *Chamydomonas* (Fig. 11-20) reproduces sexually by isogametes. These are gametes in which no morphological differences can be observed. Vegetative cells and gametes are monoploid. But syngamy is selective in that the gametes of all individuals are not equally capable of fusing. Mating strains, designated as + and −, rather than male and female, are recognized. Plus strains conjugate only with minus. Mating type appears to be related to physiological differences within the cell and is, in turn, known to be genetically determined. Thus zy-

FIGURE 11-18. *Tassel seed corn,* ts ts. *Note silks of developing ear at top of stem in place of tassel.* (From *The Ten Chromosomes of Maize,* DeKalb Agricultural Association, DeKalb, Illinois. Reproduced by permission.)

gotes produce four cells by meiosis; two become plus adults and two become minus. Ebersold (1967) was able to develop diploid adults from zygotes by mitosis. All of these unusual 2n individuals were of the minus mating strain, indicating dominance of this trait. Progeny ratios resulting from crosses involving diploids confirm this. So, in this simple alga where evolution of sex has not progressed very far, sex appears to be controlled by a single pair of genes responsible for physiological rather than morphological differences.

CONCLUSION

Sex represents something of a continuum in many organisms, but is basically gene-determined. Species differ with respect to (1) the number of genes that appear to play a part and (2) the locations of those genes (X chromosome, Y chromosome, autosomes, cytoplasmic). The genes an individual receives at the moment of syngamy determine which kind of gonads, and therefore of gametes, will be formed. Later processes of sex differentiation are quite distinct from the initial event of sex determination. In insects the critical factors in sex differentiation appear to be intracellular. In man and other mammals, however, sex differentiation is hormonal. Sex hormones, produced by the gonads, interact with endocrine glands elsewhere in the body to affect the

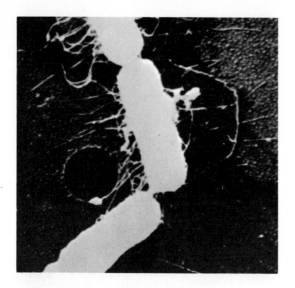

FIGURE 11-19. *Electron micrograph of conjugating* E. coli *cells. The bridge connecting the two conjugants may be seen at the angle between upper and lower partners. It happens that the conjugating cells are both in late stages of fission as well, and one of them is coincidentally being attacked by bacteriophages.* (Photo courtesy Dr. Thomas F. Anderson, Institute for Cancer Research, Philadelphia.)

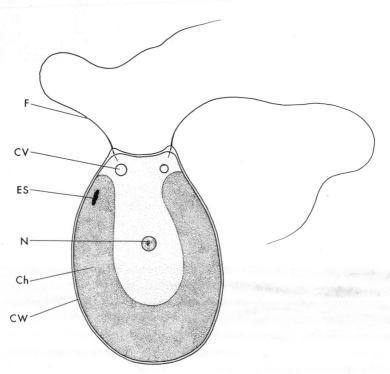

FIGURE 11-20. *Diagram of vegetative cell of the green alga,* Chlamydomonas: *F, flagellum; CW, cell wall; CV, contractile vacuole; ES, eye spot; Ch, chloroplast; N, nucleus.*

multiplicity of secondary sex characters. These latter may be markedly affected by hormone injections, suggesting that each individual has some potentialities for either sex.

REFERENCES

ALLEN, C. E., 1919. The Basis of Sex Determination in *Sphaerocarpos*. *Proc. Amer. Phil. Soc.*, **58**.

ALLEN, C. E., 1935. The Genetics of Bryophytes. *Bot. Rev.*, **1**: 269–291.

BARR, M. L., 1966. The Sex Chromosomes in Evolution and Medicine. *Can. Med. Assoc. Jour.*, **95**: 1137–1148.

BARR, M. L., F. R. SERGOVICH, D. H. CARR, and E. L. SHAVER, 1969. The Triplo-X Female: An Appraisal Based on a Study of 12 Cases and a Review of the Literature. *Can. Med. Assoc. Jour.*, **101**: 247–258.

BATTIN, M. J., and F. SERVILLE, 1973. Étude d'une Série de 139 Cas de Dysgénésies Gonadiques de Phénotype Féminin Recueillis en Milieu Pédiatrique. *Bord. Med.*, **6**(14): 2045–2061.

BORGAONKAR, D. S., 1969. 47, XYY Bibliography. *Ann. Genet.*, **12**: 67–70.

BORGAONKAR, D. S., H. M. HERR, and J. NISSIM, 1970. DNA Replication Pattern of the Y Chromosome in XYY and XXYY Males. *Jour. Hered.*, **61**: 35–36.

BRIDGES, C. B., 1916a. Nondisjunction as Proof of the Chromosome Theory of Heredity. *Genetics*, **1**: 1–52.

BRIDGES, C. B., 1916b. Nondisjunction as Proof of the Chromosome Theory of Heredity (concluded). *Genetics*, **1**: 107–163.

BRIDGES, C. B., 1925. Sex in Relation to Chromosomes and Genes. *Amer. Natur.*, **59**: 127–137.

CHILDS, B., 1965. Genetic Origin of Some Sex Differences Among Human Beings. *Pediatrics*, **35**: 798–812.

COHEN, M. M., and M. W. SHAW, 1965. Two XY Siblings with Gonadal Dysgenesis and a Female Phenotype. *New England Jour. Med.*, **272**: 1083–1088.

EBERSOLD, W. T., 1967. *Chlamydomomas reinhardii:* Heterozygous Diploid Strains. *Science*, **157**: 447–449.

FERGUSON-SMITH, M. A., 1970. Chromosomal Abnormalities II: Sex Chromosome Defects. In V. A. McKusick and R. Claiborne, eds. *Medical Genetics.* New York, HP Publishing Co.

GERMAN, J., 1967. Autoradiographic Studies of Human Chromosomes. In J. F. Crow and J. V. Neel, eds. *Proc. Third Int. Cong. Human Genetics.* Baltimore, The Johns Hopkins Press.

GERMAN, J., 1970. Studying Human Chromosomes Today. *Amer. Scientist*, **58**: 182–201.

JACOBS, P. A., A. G. BAIKIE, W. M. COURT BROWN, T. N. MACGREGOR, N. MACLEAN, and D. G. HARNDEN, 1959. Evidence for the Existence of the Human "Superfemale." *Lancet*, **2**: 423.

JACOBS, P. A., M. BRUNTON, M. M. MELVILLE, R. P. BRITTAIN, and W. F. MCCLERMONT, 1965. Aggressive Behavior, Mental Subnormality and the XYY Male. *Nature*, **208**: 1351–1352.

LEDERBERG, J., and E. L. TATUM, 1946. Gene Recombination in *Escherichia coli. Nature,* **158:** 558.

LUBS, H. A., and F. H. RUDDLE, 1970. Chromosomal Abnormalities in the Human Population: Estimation of Rates Based on New Haven Newborn Study. *Science,* **169:** 495–497.

LYON, M. F., 1961. Gene Action in the X-Chromosome of the Mouse (*Mus musculus* L.). *Nature,* **190:** 372–373.

LYON, M. F., 1962. Sex Chromatin and Gene Action in Mammalian X-Chromosomes. *Amer. Jour. Human Genetics,* **14:** 135–148.

MCKUSICK, V. A., and R. CLAIBORNE, eds., 1973. *Medical Genetics.* New York, HP Publishing Co.

MITTWOCH, U., 1964. Sex Chromatin. *Jour. Med. Genetics,* **1:** 50–76.

MORGAN, L. V., 1922. Non-Criss-Cross Inheritance in *Drosophila melanogaster. Biol. Bull.,* **42:** 267–274.

POLANI, P. E., 1970. Hormonal and Clinical Aspects of Hermaphroditism and the Testicular Feminizing Syndrome in Man. *Phil. Trans. Roy. Soc. London,* **B259:** 187–204.

PRICE, W. H., J. A. STRONG, P. B. WHATMORE, and W. F. MCCLERMONT, 1966. Criminal Patients with XYY Sex-Chromosome Complement. *Lancet,* **1:** 565–566.

RICK, C. M., and G. C. HANNA, 1943. Determination of Sex in *Asparagus officinalis* L. *Amer. Jour. Bot.,* **33:** 711–714.

ROOK, A., L. Y. HSU, M. GERTNER, and K. HIRSCHHORN, 1971. Identification of Y and X Chromosomes in Amniotic Fluid Cells. *Nature,* **230:** 53.

STERN, C., 1973. *Principles of Human Genetics,* 3rd ed. San Francisco, W. H. Freeman.

STERNBERG, W. H., and D. L. BARCLAY, 1967. Women With Male Sex Chromosomes: The Syndromes of Testicular Feminization and XY Gonadal Dysgenesis. *Jour. Amer. Med. Women's Assn.,* **22.**

SUMNER, A. T., J. A. ROBINSON, and H. J. EVANS, 1971. Distinguishing Between X, Y, and YY-Bearing Spermatozoa by Fluorescence and DNA Content. *Nature New Biol.,* **229:** 231–233.

SUMNER, A. T., H. J. EVANS, and R. A. BUCKLAND, 1971. New Techniques for Distinguishing Between Human Chromosomes. *Nature New Biol.,* **232:** 31–32.

TEPLITZ, R. L., and E. BEUTLER, 1966. Mosaicism, Chimerism, and Sex-Chromosome Inactivation. *Blood,* **27:** 258–271.

TERHEGGEN, H. G., R. A. PFEIFFER, H. HAUG, M. HERTL, A. DIGGINS, and W. SCHÜNKE, 1973. Das XXXXY-Syndrome. Bericht über 7 Neue Fälle und Literaturübersicht. *Z. Kinderheilkd.,* **115**(3): 209–233.

TURNER, C. D., 1964. Special Mechanisms in Anomalies of Sex Differentiation. *Amer. Jour. Obstet. and Gyn.,* **90**(No. 7, pt. 2): 1208–1226.

WARMKE, H. E., 1946. Sex Determination and Sex Balance in *Melandrium. Amer. Jour. Bot.,* **33:** 648–660.

WESTERGARD, M., 1948. The Relation Between Chromosome Constitution and Sex in the Offspring of Triploid *Melandrium. Hereditas,* **34:** 257–279.

WHITING, P. W., 1954. The Evolution of Male Haploidy. *Quart. Rev. Biol.,* **20:** 231–260.

ZEUTHEN, E., J. NIELSEN, and H. YDE, 1973. XYY Males Found in a General Male Population: Cytogenetic and Physical Examination. *Hereditas,* **74**(2): 283–290.

PROBLEMS

11-1. What is the sex designation for each of the following fruit flies (each A = one set of autosomes, each X = one X chromosome): (a) AAXXXX, (b) AAAAAXX, (c) AAXXXXXX, (d) AAAAAXXX, (e) AAAAXXXX, (f) AAAXY?

11-2. What is the sex designation for the following human beings: (a) AAXXX, (b) AAXXXYYY, (c) AAXO, (d) AAXXXXXY, (e) AAXYYY?

11-3. What phenotypic ratio results in corn from selfing *Bs bs Ts ts* plants?

11-4. In *Drosophila*, what fraction of the progeny of the cross *Tra tra* XX × *tra tra* XY are "transformed"?

11-5. What is the sex ratio in the progeny of the cross given in problem 11-4?

11-6. In poultry, the dominant gene, *B*, for barred feather pattern is located on the Z chromosome. Its recessive allele, *b*, produces nonbarred feathers. What is the genotype of a (a) nonbarred female, (b) barred male, (c) barred female, (d) nonbarred male?

11-7. Give the phenotypic and sex ratios in the progeny of the following crosses in poultry: (a) nonbarred ♀ × heterozygous barred ♂, (b) barred ♀ × heterozygous barred ♂.

11-8. In poultry, removal of the ovary results in the development of the testes. Thus a female can become converted to a male, producing sperm and developing male secondary sex characters. If such a "male" is mated to a normal female, what sex ratio occurs in the progeny?

11-9. An XXY *Drosophila* female is mated to a normal male. The progeny include 5 per cent metafemales. If secondary nondisjunction occurred, what was the frequency of XX eggs?

11-10. From your answer to 11-9, and further assuming Y eggs to occur with a frequency of 0.1, what percentage of the progeny of the cross of problem 11-9 will be phenotypically normal females?

11-11. An attached-X female *Drosophila* is mated to a normal male. (a) What fraction of the zygotes become viable females? (b) What fraction of the zygotes are metafemales in chromosomal constitution? (c) What fraction of the viable, mature progeny are sterile males?

11-12. If the female parent of problem 11-11 is AAX$^+$X^w and the male AAX^{w}Y, what will be the eye color of (a) viable, mature male progeny and (b) viable, mature female progeny?

11-13. How can the human Y chromosome be distinguished in suitably stained preparations?

11-14. If you assume parents to be AAXX and AAXY, how could you account in man for children of each of the following types: (a) AAXYY, (b) AAXXY, (c) AAXO?

11-15. In the literature numerous cases of human mosaics are reported. How could you account cytologically for an AAXX-AAXO mosaic?

11-16. No individuals, either live births or spontaneously aborted fetuses (abortuses), completely lacking any X chromosomes (e.g., AAOY) have ever been discovered. Why is this to be expected?

11-17. From the standpoint of survival of the species, which system of sex determination, that of *Asparagus* or of man, seems to offer the greater advantage? Why?

11-18. For man give the genetic sex of (a) most true hermaphrodites, (b) masculinizing male pseudohermaphrodites, (c) feminizing male pseudohermaphrodites, (d) female pseudohermaphrodites.

CHAPTER 12
Inheritance Related to Sex

THE fact that males in *Drosophila* and in man have an X and a Y chromosome, whereas females have two Xs and no Y, raises some interesting genetic possibilities. This is especially so when it is realized that the sex chromosomes do not bear genes affecting the same traits throughout their entire lengths and do carry some genes other than sex determiners. Therefore we should expect to find inheritance patterns related to the sex of the individual and somewhat different from those examined previously. For example, genes occurring only on the X chromosome will be represented twice in females, once in males; recessives of this type might be expected to show up phenotypically more often in males. Genes located exclusively on the X chromosome are called **sex-linked,** or **X-linked, genes.** On the other hand, genes occurring only on the Y chromosome can produce their effects only in males; these are **holandric genes** (Greek, *holos,* "whole," and *andros,* "man"). Still other mechanisms are known whereby a given trait is limited to one sex (**sex-limited genes**), or even in which dominance of a given allele depends on the sex of the bearer (**sex-influenced genes**). Genes occurring on homologous portions of the X and Y chromosomes are called **incompletely sex-linked.** We shall examine the first four of these types of sex-related inheritance in this chapter.

Sex Linkage

since unable to be overshadowed by a dominant normal allele

Drosophila   From 1904 to 1928 T. H. Morgan taught at Columbia University, attracting many students who were later to become brilliant geneticists. Most of these men worked in what became known fondly as "the fly-room." This must have been a remarkable and stimulating group; as one of them (Sturtevant, 1965) describes it, "There was an atmosphere of excitement in the laboratory, and a great deal of discussion and argument about each new result as the work rapidly developed."

For example, in a long line of wild-type, red-eyed flies, an exceptional white-eyed male was discovered. To the Columbia group this was a new character, apparently arising through mutation or change of the gene for red eyes (Fig. 12-1). Morgan and his students crossed this new male to his wild-type sisters; all the offspring had red eyes, indicating that white was recessive. An F_2 of 4,252 individuals was obtained: 3,470 were red-eyed and 782 white-eyed. This is not a good representation of the expected 3:1 ratio, but it is now known that white-eyed flies do not survive as well as their wild-type sibs and are therefore less likely to be counted. However, the important point here is that all 782 white-eyed F_2 flies were males! About an equal number of males had red eyes.

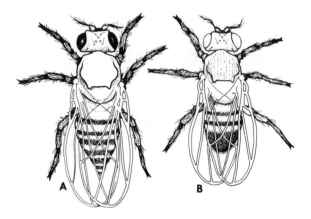

FIGURE 12-1. *Eye color mutation in* Drosophila melanogaster. *(A) Red-eyed female; (B) white-eyed male.*

From Chapter 11 you will recall that these genes for eye color are on the X chromosome. Again using Y to represent the Y chromosome (which does not carry a gene for eye color), $+$ for red eyes, and w for white eyes (alleles on the X chromosome), Morgan's crosses may be represented thus:

$$P \quad + + \quad \times \quad w\,Y$$
$$\text{red } ♀ \qquad \text{white } ♂$$
$$F_1 \quad \tfrac{1}{2} + w \ \text{and} \ \tfrac{1}{2} + Y$$
$$\text{red } ♀ \qquad\quad \text{red } ♂$$
$$F_2 \quad \tfrac{1}{4} + +, \tfrac{1}{4} + w, \tfrac{1}{4} + Y, \tfrac{1}{4} w\,Y$$
$$\text{red } ♀ \qquad \text{red } ♂ \ \text{white } ♂$$

Morgan correctly predicted that white-eyed females would be produced by the cross $+ w \times w\,Y$. A perpetual stock of white-eyed flies of both sexes was then established by mating white-eyed males and females.

An important characteristic of sex-linked inheritance emerges from an examination of the original Morgan cross. Note that the F_2 white-eyed males have received their recessive gene from their F_1 mothers, and these, in turn, have received gene w from their own white-eyed fathers. This "crisscross" pattern from father to heterozygous daughter (often termed a *carrier*) to son is typical for a recessive sex-linked gene. Moreover, in this particular pedigree, about half the sons in the F_2 show the trait, whereas none of the daughters do. Of course, in the cross $+ w \times w\,Y$ half the daughters as well as half the sons have the recessive phenotype. Notice that normal females (AAXX) carry two of these genes and thus may be either homozygous ($+ +$ or ww) or heterozygous ($+ w$), but normal males (AAXY) can only be **hemizygous** ($+ Y$ or $w\,Y$).

Sex Linkage in Man. McKusick, in *Mendelian Inheritance in Man* (4th edition, 1975), a catalog of human autosomal and X-linked phenotypes, lists 93 confirmed and 78 probable X-linked loci, for a total of 171. Most of these are due to recessive genes. Red-green color blindness was the first such to be described, and also appears to be the most commonly encountered sex-linked trait in human beings. It is reported that about 2 out of every 25 white males are red-green color blind, but only about 1 in 150 women is so affected.

Hemophilia, a well-known disorder in which blood clotting is deficient because of a lack of the necessary substrate thromboplastin, is likewise a sex-linked recessive condition. Two types of sex-linked hemophilia are recognized:

1. *Hemophilia A,* characterized by lack of antihemophilic globulin (Factor VIII). About four fifths of the cases of hemophilia are of this type.
2. *Hemophilia B,* or "Christmas disease," after the family in which it was first described in detail, resulting from a defect in plasma thromboplastic component (PTC, or Factor IX). This is a milder form of the condition.

From an unusual family in which both types of hemophilia were segregating, Woodliff and Jackson (1966) concluded that the two loci involved were far apart on the X chromosome. Incomplete evidence suggests a distance of more than 40 map units. Hemophilia is well known in the royal families of Europe, where it is traceable to Queen Victoria, who must have been heterozygous (Fig. 12-2). No hemophilia is known in her ancestry, hence it is surmised that she arose from a mutant gamete.

Male hemophiles occur with a frequency of about 1 in 10,000 male births, and heterozygous females may be expected in about twice that frequency. The method of calculating this is explored on pages 311 and 312. Under a system of random mating, hemophilic females would be expected to occur once in $10,000^2$, or 100 million, births. But this probability is reduced by the likelihood of male hemophiles dying before reaching reproductive age. Moreover, a hemophilic girl would almost certainly die by adolescence. Consequently, few cases of female hemophiles are known, though they have been reported in some pedigrees of first-cousin marriages. Because clotting time in different hemophiles varies somewhat, it has been suggested that the condition is affected by a number of modifying genes. Heterozygous women can be detected by a small increase in clotting time as well as lower levels of Factor VIII. Whissell et al. (1965) find that Factor VIII levels in heterozygous women are distributed on a normal curve whose mean is lower than that for homozygous normal females.

Deficiency of the enzyme glucose-6-phosphate dehydrogenase (G6PD) is another, and important, trait resulting from action of a recessive X-linked gene, estimated to affect up to 100 million persons. The enzyme is directly involved in a minor glycolysis pathway in red blood cells. G6PD deficiency is common in blacks (about 10 per cent of American black males) and in persons in Mediterranean areas (10 to 40 per cent of Sardinian males in malarial areas, but less than 1 per cent in Sardinian males of nonmalarial areas). The disorder is rare in whites where malaria is not indigenous. Aside from the apparent advantage of G6PD deficiency in affording some protection against malaria, the condition is noteworthy for the destruction of erythrocytes, with

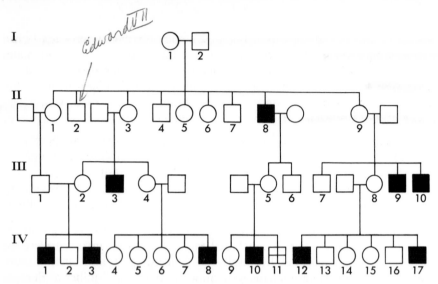

FIGURE 12-2. *Pedigree of some of the descendants of Queen Victoria showing incidence of hemophilia (shaded symbols). (I-1) Queen Victoria; (II-8) Leopold, Duke of Albany; (III-3) Frederick of Hesse; (III-8) Victoria Eugenie, who married Alfonso XIII of Spain; (III-9) Lord Leopold of Battenberg; (III-10) Prince Maurice of Battenberg; (IV-1) Waldemar of Prussia; (IV-3) Henry of Prussia; (IV-8) Tsarevitch Alexis of Russia; (IV-10) Rupert, Viscount Trematon; (IV-12) Alfonso of Spain; (IV-17) Gonzolo of Spain. (IV-11) died in childhood; (II-2) represents Edward VII of England, great-grandfather of Elizabeth II.*

consequent severe hemolytic anemia, when certain drugs are administered. These drugs include para-amino salicylic acid, the sulfonamides, napthalene, phenacetin, and primaquine (primaquine is an antimalarial agent). Inhalation of the pollen of the broad bean (*Vicia faba*) or ingestion of its seeds produces the same result. Anemia caused by the broad bean is known as favism. In the absence of these incitants, neither hemizygous recessive males nor heterozygous and homozygous recessive females suffer ill effects.

Many variants of this recessive gene are known, each affecting a different portion of the polypeptide chain of the enzyme (*polymorphism*). These variants range in effect from severe through mild to no enzyme deficiency; some even produce increased enzyme activity. The basis for polymorphism is elaborated in Chapter 18.

Some of the other sex-linked traits in man include two forms of diabetes insipidus, one form of anhidrotic ectodermal dysplasia (absence of sweat glands and teeth), absence of central incisors, certain forms of deafness, spastic paraplegia, uncontrollable rolling of the eyeballs (nystagmus), a form of cataract, night blindness, optic atrophy, juvenile glaucoma, juvenile muscular dystrophy, and white forelock (a patch of light frontal hair on the head). Most of these are fairly clearly due to recessive genes. On the other hand, hereditary enamel hypoplasia (hypoplastic amelogenesis imperfecta), in which tooth enamel is abnormally thin so that teeth appear small and wear rapidly down to the gums, is due to a dominant sex-linked gene. Women, with their two X chromosomes, have twice as great a probability of receiving this gene as men in the same family.

Sex Linkage in Other Organisms. Sex linkage in XX-XO species is, of course, just as it is in *Drosophila* and man, because sex-linked genes are, by definition, those on the X chromosome. Here again only females can be heterozygous.

In birds, where the female is the heterogametic sex, the situation is reversed, although the mechanism is unchanged. Sex-linked genes follow the "crisscross" pattern, but from mother through heterozygous sons to granddaughters.

Although the barred feather pattern, resulting from action of a dominant sex-linked gene in Plymouth Rock chickens (Fig. 12-3), is a frequently cited illustration, a pair of alleles governing speed of feather growth is a more interesting one. Both traits have been used for identifying chick sex, but feather color patterns are sometimes modified by other genes. Gene k, for slow feather growth, eliminates such complications and can be detected within hours after hatching. For example, remembering males to be homogametic, the cross $+ W(♀) \times kk(♂)$ results in two kinds of progeny, $+ k$ (males with normal feather growth) and kW (females with slower feather growth). Gene k has no effect on other characters of commercial value.

Sex-Linked Lethals. It might occur to you that the gene for hemophilia is actually a recessive, sex-linked *lethal*, for it may often cause death. Slight scratches or accidental injuries, which would not be serious in normal persons, often result in fatal bleeding for the hemophile. Sex-linked lethals, then, may alter the sex ratio in a progeny as soon as they bring about death.

Duchenne (or progressive pseudohypertrophic) muscular dystrophy is a disease in which the affected individual, though apparently normal in early childhood, exhibits progressive wasting away of the muscles, resulting in confinement to a wheel chair by about age 12, and death in the teen years (Fig. 12-4). Like hemophilia, it is due to a recessive sex-linked gene. At present, no means of arresting or preventing the disease is known; a given genotype dooms the bearer at conception to death in adolescence. The gene responsible is, then, to be considered a lethal, and will change the sex ratio in a given group of offspring over time. Letting $+$ represent the normal (dominant) gene and d the gene for muscular dystrophy, consider children of many marriages between heterozygous women and normal men, $+d \times +Y$. The offspring would be expected in a ratio of $\frac{1}{4} + +$, $\frac{1}{4} + d$, $\frac{1}{4} + Y$, and $\frac{1}{4} dY$ at birth, but individuals of the last genotype die before age 20. So, an initial approximately 1:1 female-to-male ratio will later become 2:1 female-to-male. If a recessive sex-linked lethal kills before birth, the ratio of female to male live births is changed from nearly 1:1 to 2:1. A 2:1 female-to-male ratio is always a strong indication of a sex-linked recessive lethal.

FIGURE 12-3. *Barred feather pattern, caused by a dominant sex-linked gene, in Plymouth Rock female (left) and male (right).*

HOLANDRIC GENES

since men have only Y a / chromosome

Other than the incompletely understood male-determining genes on the rather small Y chromosome in human beings, few are clearly established as having loci on that part of the Y chromosome that is not homologous with the X chromosome. Traits that are due to such genes can occur only in men and, moreover, must appear in all the sons of an affected father. It should, therefore, be relatively easy to detect such **holandric** genes in man, but partly because of the relative genetic inertness of Y and partly because of the difficulty of distinguishing them from sex-limited genes (see next section), no completely unequivocal case has been established in man. The best candidate for the holandric pattern in man is one causing excessive hair development on the ears (hypertrichosis). This trait is seen rather frequently in India (Fig. 12-5), but also occurs in Caucasians, Australian aborigines, and Japanese. Pedigrees have been published showing hypertrichosis in every man in the male line of descent from a common affected ancestor. Some complication is injected, however, by the fact that expression of the trait is somewhat variable, ranging from complete absence through a few stiff hairs to heavy development. Nevertheless, Stern (1973) finds no compelling evidence against the holandric nature of the gene for hairy ears.

Other traits have been suggested as having Y chromosome loci, but Stern (1973) argues convincingly against their holandric nature. The best known of these is the so-called porcupine man, born in England in 1716. In this male, the skin turned yellow at the age of a few weeks; it gradually blackened and became covered with rough scales having bristly outgrowths. This abnormal skin, which occurred over the whole body except for face, palms, and soles, is reported in some accounts to have been sloughed at intervals, only to have the condition recur. Original reports indicate that the trait appeared in all

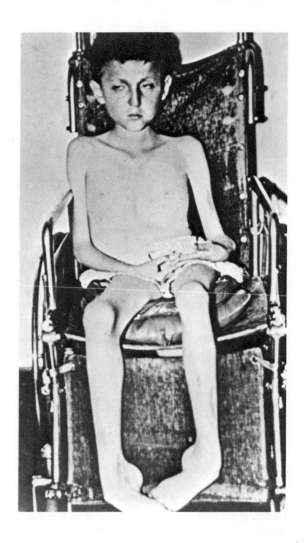

FIGURE 12-4. *Duchenne pseudohypertrophic) muscular dystrophy, caused by a recessive sex-linked gene. This case will terminate in death.* (Courtesy Muscular Dystrophy Association of America, Inc., New York.)

six sons of this man, as well as in all their male descendants, but in no females. This pattern of inheritance is certainly explainable on the basis of a holandric gene. However, careful investigation by Stern of the pedigree showed some affected females and some unaffected males, leading him to suggest an autosomal dominant.

chromosome not associated with the sex of the individual

Sex-Limited Genes

Sex-limited genes are those whose phenotypic expression is determined by the presence or absence of one of the sex hormones. Their phenotypic effect is thus *limited* to one sex or the other.

(i.e, whether shows at all or not)

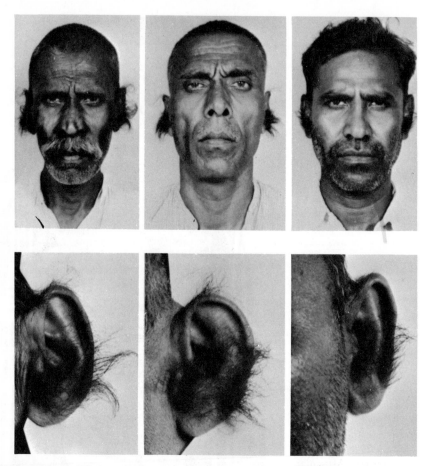

FIGURE 12-5. *"Hairy ears," a trait common in certain parts of the world. It may be caused by a holandric gene.* (From C. Stern, W. R. Centerwall, and S. S. Sarker, New Data on the Problem of Y-Linkage of Hairy Pinnae, *American Journal of Human Genetics,* **16:** 455–471, 1964. By permission Grune and Stratton, Inc. Photo courtesy Dr. Curt Stern.)

Perhaps the most familiar example occurs in the domestic fowl where, as in many species of birds, males and females may exhibit pronounced differences in plumage. In the Leghorn breed, males have long, pointed, curved, fringed feathers on tail and neck, but feathers of females are shorter, rounded, straighter, and without the fringe (Fig. 12-6). Thus males are cock-feathered and females hen-feathered. In such breeds as the Sebright bantam, birds of both sexes are hen-feathered. However, in others (Hamburg or Wyandotte) both hen- and cock-feathered males occur, but all females are hen-feathered. Leghorn females are hen-feathered, males all cock-feathered.

FIGURE 12-6. *Hen-feathering (left) and cock-feathering (right) in domestic fowl. Cock-feathering is characterized by long, pointed, curving neck and tail feathers.*

It has been shown that hen-feathering results from a single gene, *H,* and cock-feathering from its allele, *h:*

Genotype	♀	♂
HH	hen-feathered	hen-feathered
Hh	hen-feathered	hen-feathered
hh	hen-feathered	cock-feathered

Thus Sebright bantams are all *HH,* Hamburgs and Wyandottes may be *H —* or *hh,* but Leghorns are all *hh.* Cock-feathering, where it occurs, is *limited* to the male sex.

The basis for the differential behavior of gene *h* is seen in chickens from which the ovaries or testes have been removed. These gonadectomized birds all become cock-feathered at the next molt following surgery, regardless of genotype! The particular kind of plumage thus appears to depend upon a specific combination of genotype and sex hormone. The action of the two alleles may be summarized thus:

H produces hen-feathering in the presence of either sex hormone, and cock-feathering in the absence of any hormone; dominant to

h produces cock-feathering if female hormone is absent, hen-feathering if it is present.

Sex-limited inheritance patterns are quite different from those of sex-linked genes. The latter may be expressed in either sex, though with differential frequency. Sex-limited genes express their effects in only one sex or the other, and their action is clearly related to sex hormones. They are principally responsible for secondary sex characters. Beard development in human beings is such a sex-limited character, men normally having beards, women normally not. Yet studies indicate no significant difference between the sexes in number of hairs per unit area of skin surface, only in their development. This appears to depend on sex hormone production, changes in which may result in a genuine bearded lady.

Sex-Influenced Genes

In contrast to sex-limited genes where one expression of a trait is limited to one sex, sex-influenced genes are those whose dominance is *influenced* by the sex of the bearer.

Although baldness may arise through any of several causes (e.g., disease, radiation, thyroid defects), "pattern" baldness exhibits a definite genetic pattern. In this condition, hair gradually thins on top, leaving ultimately a fringe of hair low on the head (Fig. 12-7). Pattern baldness is more prevalent in males than in females, where it is rare and usually involves marked thinning rather than total loss of hair on the top of the head. Numerous pedigrees clearly show that pattern baldness is due to a pair of autosomal genes operating in this fashion:

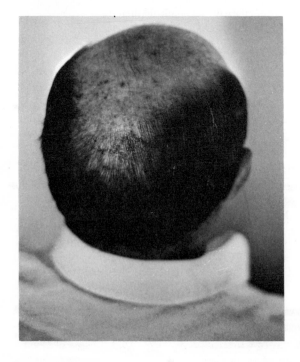

FIGURE 12-7. *Pattern baldness in a man, a sex-influenced trait dominant in males and recessive in females.*

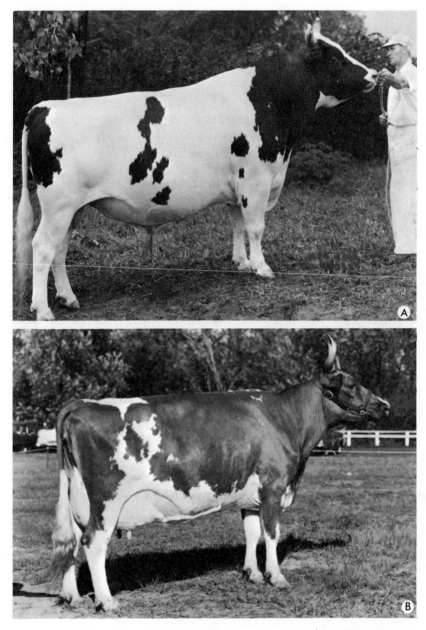

FIGURE 12-8. *A sex-influenced trait in cattle. (A) Mahogany and white, dominant in males, recessive in females; (B) red and white, dominant in females, recessive in males.* (Courtesy Ayrshire Breeders' Association, Brandon, Vermont.)

Genotype	♂	♀
BB	Bald	Bald *since homozygous*
Bb	Bald	Not bald
bb	Not bald	Not bald

recessive

i.e., to be dominant

Gene *B* behaves as a dominant in males and as a recessive in females, appearing to exert its effect in the heterozygous state only in the presence of male hormone. A number of reports in the literature dealing with abnormalities leading to hormone imbalance or with administration of hormone support this view (Hamilton, 1942, 1951). Some authors prefer to classify pattern baldness as sex-limited because the trait is generally less completely manifested in females, but it seems best to restrict the concept of sex-limited genes to those cases where one of the phenotypic expressions is *limited entirely* to one sex or the other because of anatomy.

A few well-known cases of sex-influenced genes occur in lower animals—e.g., horns in sheep (dominant in males) and spotting in cattle (mahogany and white dominant in males, red and white dominant in females, Fig. 12-8). Some hypothetical crosses involving these characters are explored in problems at the end of this chapter.

REFERENCES

HAMILTON, J. B., 1942. Male Hormone Stimulation Is Prerequisite and an Incitant in Common Baldness. *Amer. Jour. Anat.,* **71**: 451–480.

HAMILTON, J. B., 1951. Patterned Loss of Hair in Man: Types and Incidence. *Ann. N.Y. Acad. Sci.,* **53**: 708–728.

MC KUSICK, V. A., 1975. *Mendelian Inheritance in Man,* 4th ed. Baltimore, The Johns Hopkins University Press.

MORGAN, T. H., 1910. Sex Limited Inheritance in *Drosophila. Science,* **32**: 120–122. Reprinted in J. A. Peters, ed., 1959. *Classic Papers in Genetics.* Englewood Cliffs, N.J., Prentice-Hall.

STERN, C., 1973. *Principles of Human Genetics,* 3rd ed. San Francisco, W. H. Freeman.

STURTEVANT, A. H., 1965. *A History of Genetics.* New York, Harper & Row.

WHISSELL, D. Y., M. S. HOAG, P. M. AGGELER, M. KROPATKIN, and E. GARNER, 1965. Hemophilia in a Woman. *Amer. Jour. Med.,* **38**(1): 119–129.

WOODLIFF, H. J., and J. M. JACKSON, 1966. Combined Haemophilia and Christmas Disease. A Genetic Study of a Patient and His Relatives. *Med. Jour. Aust.,* **53**: 658–661.

PROBLEMS

12-1. "Bent," a dominant sex-linked gene, *B*, in the mouse, results in a short, crooked tail; its recessive allele, *b*, produces normal tails. If a normal-tailed female is mated to a bent-tailed male, what phenotypic ratio should occur in the F_1?

12-2. Nystagmus is a condition in man characterized by involuntary rolling of the eyeballs. The gene for this condition is incompletely dominant and sex-linked. Three phenotypes are possible: normal, slight rolling, severe rolling. A woman who exhibits slight nystagmus and a normal man are considering marriage and ask a geneticist what the chance is that their children will be affected. What will he tell them?

12-3. "Deranged" is a phenotype in *Drosophila* in which the thoracic bristles are disarranged and the wings vertically upheld, producing a characteristic appearance. Crosses between deranged females and normal males results in a 1:1 ratio of normal females to deranged males in the progeny. What is the mode of inheritance and how would you describe the dominance of this gene?

12-4. In poultry, sex-linked gene *B*, producing barred feather pattern, is completely dominant to its allele, *b*, for nonbarred pattern. Autosomal gene *R* produces rose-comb; its recessive allele, *r*, produces single comb in the homozygous state. A barred female, homozygous for rose-comb is mated to a nonbarred, single-comb male. What is the F_1 phenotypic ratio?

12-5. Members of the F_1 from problem 12-4 are then crossed with each other. What fraction of the F_2 is barred rose, and are these male or female?

12-6. In what ratio does (a) barred, nonbarred and (b) rose, single segregate in the F_2 of problem 12-5?

12-7. A family has five children, three girls and two boys. One of the latter died of muscular dystrophy at age 15. The others have graduated from college and are concerned over the probability that their children may develop the disease. What would you tell them?

12-8. "Jimpy" is a trait in the mouse characterized by muscular incoordination which results in death at an age of three to four weeks. Crosses between heterozygous females and normal males produce litters in which half the males are jimpy. What type of gene is jimpy?

12-9. At locus 0.3 on the X chromosome of *Drosophila* there occurs a recessive gene, *l*, which is lethal in the larval stage. A heterozygous female is crossed to a normal male; what F_1 adult sex phenotypic ratio results?

12-10. A woman with defective tooth enamel and normal red-green color vision, who had a red-green blind father with normal tooth enamel and a mother who had defective tooth enamel and normal red-green vision, marries a red-green blind first cousin with normal tooth enamel. What is the probability of their having a child with normal tooth enamel and red-green color blindness if crossing-over does not occur?

12-11. What is the probability, if they have three children, that these will be two red-green blind girls with normal teeth and one boy with defective teeth but normal red-green color vision?

12-12. The bald phenotype can sometimes be distinguished at a relatively early age. A nonbald, red-green blind man marries a nonbald, normal-visioned woman whose mother was bald and whose father was red-green blind. What is the probability of their having each of the following children: (a) bald girls, (b) bald normal-visioned boys, (c) nonbald boys, (d) red-green blind children of either sex?

12-13. In addition to the allelic pair determining pattern baldness (*B,b*) described in this chapter, consider early baldness to be due to another autosomal gene (*E*) on a different pair of chromosomes, and also dominant in males. The phenotype for *ee* may be either late baldness or nonbaldness, depending on sex and the genotype for the *B,b* alleles. Two doubly heterozygous persons (*BbEe*) marry. (a) What is or will be the phenotype of the male parent? (b) What is or will be the phenotype of the female parent? (c) What is the phenotypic ratio among *male* children of couples such as this one? (d) What is the phenotypic ratio among *female* children of couples such as this one?

12-14. A mahogany-and-white cow has a red-and-white calf. What is the sex of the calf?

12-15. The cross of a horned female ewe and a hornless ram produces a horned offspring. What is the sex of the F₁ individual?

12-16. In the clover butterfly, males are always yellow, but females may be either yellow or white. What kind of inheritance is operating?

12-17. In the clover butterfly, white is dominant in females. Crossing of two heterozygotes produces what F₁ phenotypic ratio?

12-18. Milk yield in cattle is governed by several pairs of autosomal genes. Of the types of sex-related inheritance described in this chapter, which fits milk yield?

12-19. Length of index finger relative to that of the fourth (ring) finger in human beings has been ascribed to a pair of sex-influenced genes, with short index finger dominant in males. Give the expected phenotypic ratios to be expected in children of the following marriages: (a) homozygous short ♂ × short ♀, (b) homozygous short ♂ × homozygous long ♀, (c) heterozygous short ♂ × heterozygous long ♀, (d) heterozygous short ♂ × short ♀.

12-20. How could you differentiate between a sex-linked recessive trait and a holandric one?

12-21. Why is nothing known regarding dominance of holandric genes?

12-22. How could you differentiate between a sex-linked dominant gene and a holandric one?

12-23. If hairy ears (hypertrichosis) is a holandric trait, what kind of children can be produced by a hairy-eared man?

12-24. In poultry, a particular cross produced an F₁ ratio of 3 hen feathering : 1 cock feathering. One third of the hen-feathered progeny were males. What were the parental genotypes?

12-25. Which of the individuals in Figure 12-2 of the text was heterozygous for hemophilia?

12-26. What is the probability that any woman shown in generation IV of Figure 12-2 was heterozygous?

12-27. King Edward VII (Fig. 12-2) married Princess Alexandra of Denmark; their son was George V, who married Princess Mary of Teck. George VI, son of George V, married Lady Elizabeth Bowes-Lyon; one of their children is the present Queen Elizabeth II, whose husband is Prince Philip. None of these men were hemophilic, and the disease did not occur in the families of their wives.

[handwritten top margin:] $\frac{F}{f}\frac{A}{a} \times \frac{F}{f}\frac{A}{Y}$ } Treat like $(F_b\times F_b)$ $(Aa \times AY)$ $(\frac{1}{4}FF\frac{1}{2}Ff\frac{1}{4}ff)(\frac{1}{4}AA\frac{1}{4}Aa\frac{1}{4}AY\frac{1}{4}aY)$

In men, Ff & FF are short

in women, only Ff is short

a) $(\frac{1}{4}FF)(\frac{1}{4}aY)+(\frac{1}{2}Ff)(\frac{1}{4}aY)=\frac{3}{16}$

b) $(\frac{1}{4}ff)(\frac{1}{4}aY)=\frac{1}{16}$ c) all males

Would it have been possible for Queen Elizabeth II to have had any hemophilic children? *sex-influenced*   *sex-linked*

12-28. A woman, heterozygous for long index finger and for hemophilia A, marries a nonhemophilic man who is heterozygous for short index finger. What is the probability of a child of theirs being a hemophile with (a) short index finger or (b) long index finger? (c) Of what sex(es) will such children be? *good — do again*

12-29. Recall from Chapter 11 the XX-XY chromosomal situation in the flowering plant *Lychnis*. A dominant X-linked gene, *N*, produces normal broad leaves; its recessive allele, *n*, produces narrow, almost grasslike leaves. A heterozygous broad-leaved pistillate plant is pollinated by a broad-leaved staminate plant. (a) Give the genotypes of the two parents. (b) What fraction of the progeny should have narrow leaves? (c) Will these narrow-leaved plants be pistillate, staminate, or some of each?

12-30. A woman, heterozygous for hemophilia A but homozygous normal at the B locus, marries a first cousin who suffers from hemophilia B only. (a) What type(s) of hemophilia could occur in their children? (b) Would this be in daughters or in sons? (c) Would any of the daughters be carriers for either type of hemophilia? Explain. (d) What is the probability of their having a normal child who will not transmit either disorder to future generations?

[handwritten right:] $\frac{AB}{aB} \times \frac{Y}{Ab}$

$\frac{AB}{Y}$ $\frac{AB}{Ab}$

$\frac{aB}{Y}$ $\frac{aB}{Ab}$

12-31. In the two hypothetical pedigrees given below, shaded symbols represent persons exhibiting G6PD deficiency. In each pedigree III-2 is a Turner. Determine for each pedigree whether it is the maternal or the paternal X chromosome that is lacking in the Turner child.

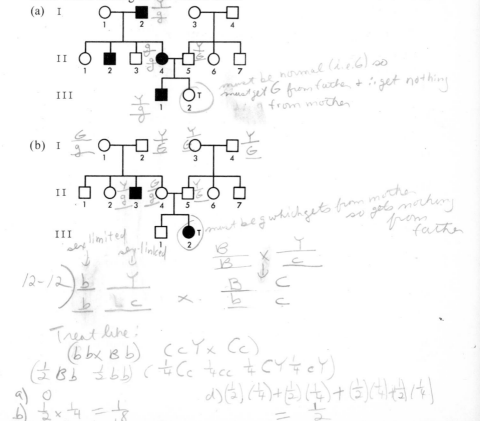

[handwritten near (a) III-2:] must be normal (i.e. G) so must get G from father & ∴ get nothing / G from mother

[handwritten near (b) III-2:] must be g which gets from mother / so gets nothing from father

[handwritten (b) III labels:] III   sex-limited   sex-linked

[handwritten bottom:] $\frac{B}{B} \times \frac{Y}{c}$

$\frac{B}{b} \quad \frac{C}{c}$

12-12) $\frac{b}{b} \quad \frac{Y}{c}$ × $\frac{B}{b} \quad \frac{C}{c}$

Treat like: $(bb \times Bb)$ $(cY \times Cc)$

$(\frac{1}{2}Bb \frac{1}{2}bb)(\frac{1}{4}Cc \frac{1}{4}cc \frac{1}{4}CY \frac{1}{4}cY)$

a) 0

b) $\frac{1}{2} \times \frac{1}{4} = \frac{1}{8}$

c) $\frac{1}{2}\times\frac{1}{4}+\frac{1}{2}\times\frac{1}{4}=\frac{2}{8}$

d) $(\frac{1}{2})(\frac{1}{4})+(\frac{1}{2})(\frac{1}{4})+(\frac{1}{2})(\frac{1}{4})+(\frac{1}{2})(\frac{1}{4}) = \frac{1}{2}$

CHAPTER 13
Chromosomal Aberrations

In earlier chapters it was pointed out that each species of plant and animal is characterized by a particular chromosome complement or set, represented once in monoploid cells (e.g., gametes and spores) and twice in diploid cells. Possession of such sets of chromosomes, or **genomes,** gives to each species a specific chromosome number (see Table 3-1). But irregularities sometimes occur in nuclear division, or "accidents" (as from radiation) may befall interphase chromosomes so that cells or entire organisms with aberrant genomes may be formed. Such chromosomal aberrations may include whole genomes, entire single chromosomes, or just parts of chromosomes. Thus cytologists recognize (1) **changes in number of whole chromosomes (heteroploidy)** and (2) **structural modifications.** Heteroploidy may involve either entire extra sets of chromosomes (**euploidy**), or loss or addition of single chromosomes (**aneuploidy**). Each of these may produce phenotypic changes, modifications of phenotypic ratios, or alteration of linkage groups. Many are of some evolutionary significance.

Changes in Chromosome Number

EUPLOIDY

Monoploids. Euploids are characterized by possession of entire sets of chromosomes, monoploids carrying one genome (n); diploids, two ($2n$); and so on. Monoploidy is rare in animals (the male honeybee is an outstanding exception), but common in plants. In most sexually reproducing algae and fungi and in all bryophytes (liverworts and mosses), the monoploid phase represents the dominant part of the life cycle and is the plant we recognize by species name. In vascular plants this stage is short-lived and microscopic, though occasionally adult monoploid vascular plants may be recognized in natural populations. These are ordinarily weak, small, and highly sterile. Blakeslee and Belling appear to have reported the first such case in 1924, in the Jimson weed; there is good evidence that these developed from unfertilized eggs. Sterility in monoploids is due to extreme irregularity of meiosis because of the impossibility of chromosomal pairing and the very low probability of their distribution in complete sets to daughter nuclei. Thus viable gametes are rarely formed; their occurrence depends on the chance movement of the whole monoploid set to one pole at meiosis. This is a highly improbable event.

Polyploidy. Euploids with three or more complete sets of chromosomes are called **polyploids.** This condition is rather common in the plant kingdom but rare in animals. Whereas one $4n$ plant, for example, would produce $2n$

250

gametes and could, in many species, be self-fertilized to produce more $4n$ progeny, the probability of *two* such *animals* (one of each sex) occurring and mating is extremely low. Furthermore, the imbalance of sex-determining mechanisms that would result from polyploidy would be expected to result in sterility because of aberrant meiosis, or to produce individuals that, for many morphological reasons, might be at a considerable disadvantage in mating.

By contrast, many plant genera consist of species whose chromosome numbers constitute a euploid series. The rose genus, *Rosa,* includes species with the somatic numbers 14, 21, 28, 35, 42, and 56. Notice that each of these numbers is a multiple of 7. Thus this is a euploid series of the basic monoploid number 7, giving diploid, triploid, tetraploid, pentaploid, hexaploid, and octoploid species. All but the diploids may be collectively referred to as polyploids. One authority has estimated that at least two thirds of all grass species are polyploids.

In many instances in plants, morphological differences between diploids and their related polyploids are not great enough for taxonomists to give the latter forms species rank, although the polyploids can usually be distinguished visually. Thus Burns (1942) found that a large amount of morphological variation within a single species of swamp saxifrage (*Saxifraga pensylvanica*) was associated with polyploidy and that diploids, triploids, and tetraploids differed consistently in several respects. The greater chromosome number of polyploids was found to be reflected in larger cell size (Fig. 13-1). Measurement of lower leaf epidermal cell size, for example, of randomly selected diploids and tetraploids showed mean cell area of diploids to be 1,608 μm^2 and, for tetraploids, 2,739 μm^2. Burns found the greater cell size of the tetraploids to be associated with larger size of plant and plant parts, and a lower length-to-width ratio of leaves, giving $2n$ and $4n$ plants distinctly different appearances in the field. Leaves of tetraploids are noticeably wider for their length than those of the $2n$ plants (Fig. 13-2). Such differences between diploid and tetraploid were found to be significant; statistical tests provided a very high degree of confidence in the premise that two different populations exist, as seen in the following comparison of length-to-width ratios of leaves:

	Diploids	Tetraploids
$\bar{x} =$	4.67	3.44
$s =$	1.109	0.837
$s_{\bar{x}} =$	0.158	0.119
$S_d =$	0.194	

The difference in sample means is $4.67 - 3.44 = 1.23$, which is 6.34 times greater than the standard error of the difference in sample means. Chromosomes of the various members of the series are illustrated in Figure 13-3.

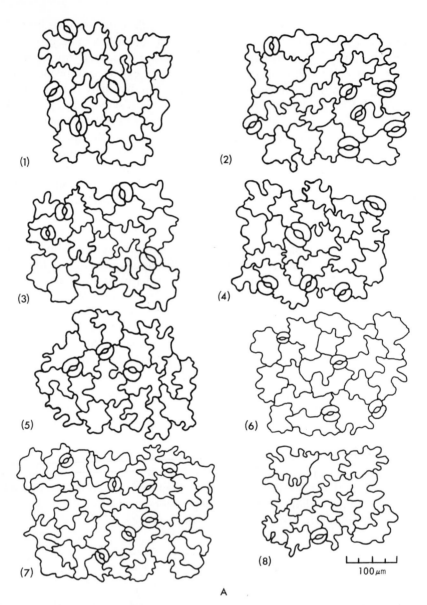

FIGURE 13-1. *Leaf epidermal cells of saxifrage* (Saxifraga pensylvanica). (*A*) *Diploids; mean cell area ranges from 1,147 μm² to 1,897 μm².*

In general, tetraploids are often hardier, more vigorous in growth, able to occupy less favorable habitats, and/or have larger flowers and fruits. Burns (1942) reported tetraploid swamp saxifrage to extend a little farther west into drier habitats in Minnesota, and Dean (1959) found the tetraploids of the

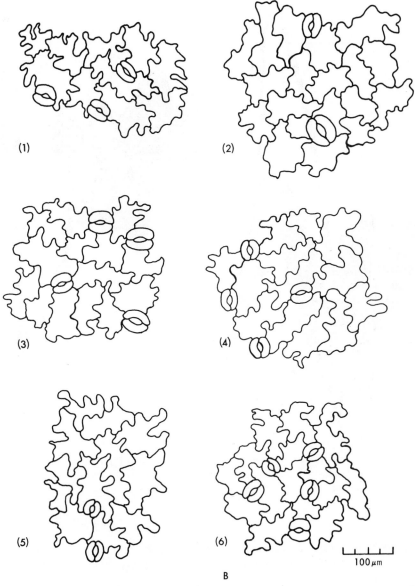

(1)

(2)

(3)

(4)

(5)

(6)

100 μm

B

FIGURE 13-1 (continued). (*B*) *Tetraploids; mean cell area ranges from 2,378 μm² to 3,408 μm².*

spiderwort plant (*Tradescantia ohioensis*) to occupy drier and more disturbed areas than their diploid relatives.

In many cultivated plants tetraploid varieties are commercially more desirable than their diploid counterparts and are commonly available from seed

FIGURE 13-2. *Herbarium specimens of* Saxifraga pensylvanica. (*A*) *Diploid;* (*B*) *tetraploid. Note differences in leaf length-width ratio. Other characteristic dissimilarities occur in shapes and sizes of flower parts, fruits, and seeds.*

or plant suppliers (Fig. 13-4). Many of our most important crop plants are polyploids or include polyploid (usually tetraploid) forms: alfalfa, apple, banana, barley, coffee, cotton, peanut, potato, strawberry, sugar cane, tobacco, and wheat are a few examples. As might be expected, polyploid varieties with an even number of genomes (e.g., tetraploids) are often fully fertile, whereas those with an odd number (e.g., triploids) are highly sterile. This latter fact is made use of in marketing of seeds containing triploid embryos. Triploid watermelons, for example, are nearly seedless and are listed in a number of seed catalogs (Fig. 13-5). Triploids are commonly created by crossing the normal diploid, whose gametes are n, with a tetraploid, gametes of which are $2n$.

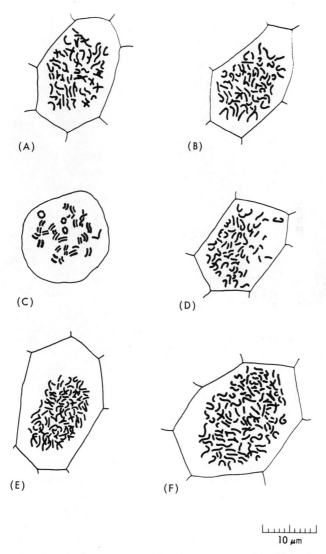

FIGURE 13-3. *Camera lucida drawings of the chromosomes of* Saxifraga pensylvanica. *(A), (B), and (D) diploids, 2n = 56; (C) microsporocyte at synapsis, n = 28; (E) triploid, 3n = 84; (F) tetraploid, 4n = 112. All but (C) show polar views of mitotic metaphase chromosomes in root tip cells.*

Production of Polyploids. Polyploids may arise naturally or be artificially induced. In plants it appears that diploidy is more primitive, and that polyploids have evolved from diploid ancestors. In natural populations this may arise as the result of interference with cytokinesis once chromosome replication has occurred, and may occur either (1) in somatic tissue, giving tetraploid

FIGURE 13-4. (A) Diploid and (B) tetraploid snapdragons. Note larger and more numerous flowers in the tetraploid (the somewhat open character of the diploid flowers is not related to chromosome number). (Courtesy Burpee Seeds.)

branches, or (2) during meiosis, producing unreduced gametes. It has been found that chilling may accomplish this in natural populations (Belling, 1925).

Application of the alkaloid colchicine, derived from the autumn crocus (*Colchicum autumnale*), either as a liquid or in lanolin paste, induces polyploidy. Although chromosome replication is not interfered with, normal spindle formation is prevented and the double number of chromosomes becomes incorporated within a common nuclear membrane. Subsequent nuclear divisions are normal, so that the polyploid cell line, once initiated, is maintained. Polyploidy may also be induced by other chemicals (acenaphthene and veratrine) or by exposure to heat or cold.

Autopolyploidy and Allopolyploidy. If a tetraploid is developed by the colchicine treatment, for example, its cells contain four genomes, all of the same species. Such a polyploid is an **autotetraploid.** Autopolyploids may, of course, exist with any number of genomes. The same situation results if an individual is formed from the fusion of two diploid gametes; the four sets of its chromosomes all belong to the same species.

FIGURE 13-5. (*A*) *Diploid watermelon with numerous seeds;* (*B*) *Triploid variety with few and imperfect seeds.* (Courtesy Burpee Seeds.)

Synapsis in autotetraploids usually involves groups of four homologs, though these ordinarily associate only in pairs for given segments. The result is a characteristic quadrivalent configuration in prophase-I (Fig. 13-6). Subsequent disjunction and passage of chromosomes to spindle poles may be highly irregular, so that very few functional gametes are formed. Such sterility is not universal in autotetraploids, however; synapsis apparently occurs normally in some so that only bivalent chromosomes are seen. Fertility then is unimpaired.

On the other hand, polyploids may develop (and be developed) from hybrids between different species. These are **allopolyploids;** the most commonly encountered type is the allotetraploid, having two genomes from each of the two ancestral species. The Russian cytologist Karpechenko (1928) synthesized a new genus from crosses between vegetables belonging to different genera, the radish (*Raphanus*) and the cabbage (*Brassica*). These plants are fairly closely related and belong to the mustard family (*Cruciferae*). Each has a somatic chromosome number of 18, but those of radish have many genes not occurring in cabbage chromosomes, and vice versa. Karpechenko's hybrid had in each of its cells 18 chromosomes, 9 from radish and 9 from cabbage. Members of the very unlike genomes failed to pair, and the hybrid was largely sterile. A few 18-chromosome gametes were formed, however, and a few allotetraploids were thereby produced as an F_2. These were completely fertile, because two sets each of radish and cabbage chromosomes were present and pairing between homologs occurred normally. The allotetraploid, or **amphidiploid,** was named *Raphanobrassica*. Unfortunately it has the root of cabbage and leaves of radish and is of no direct economic importance. The method does, however, offer a means of producing fertile interspecific or intergeneric hybrids.

Polyploidy in Man. Complete polyploid human beings are, as might be expected, quite rare and the few cases known are either spontaneously aborted fetuses (**abortuses**) or stillborn. A few live for a matter of hours. Gross and multiple malformations, reflecting the extreme genic imbalance of these individuals, occur in all cases. Table 13-1 presents a sampling of human polyploids reported in the literature. In a 1963 paper Carr described the chromosome complement of 227 abortuses. Some chromosome anomaly was found in 50 of these; two were triploids and one was tetraploid. None of the latter were mosaics so far as could be determined. Stern (1973) estimates that about 15 per cent of all spontaneously aborted fetuses are either triploids or tetraploids. A larger number of abortuses and live-born children that die very early are mosaics for diploid-polyploid cell lines.

Origin of polyploid human embryos is difficult to explain satisfactorily. Although fusion of a normal monoploid gamete with a diploid one would, of course, produce a triploid zygote that might even progress to some stage of embryo development, little concrete evidence exists for occurrence of unreduced gametes in mammals. These could be produced in theory, however,

FIGURE 13-6. *Synapsis in an autotetraploid. Pairing is in twos, but all four chromosomes are included in each synaptic figure. An actual case would be far more complicated in appearance because of the intertwining of chromatids, omitted here for simplicity.*

TABLE 13-1. Polyploidy in Human Abortuses and Live Births

Reference	69,XXY	69,XXX	69,XYY	92,XXXX	92,XXYY
Penrose and Delhanty (1961)	2				
Makino et al. (1964)	3				
Szulman (1965)	4	1			
Carr (1965)	6	2	1	1	1
Patau et al. (1963)		2			
Schindler and Mikamo (1970)	1*				
Butler et al. (1969)		1*			

*Born alive; died within hours.

by incorporation of the chromosomes of the first or second polar body within the egg nucleus, or by an analogous abnormality during spermatogenesis, or by meoisis of an exceptional tetraploid oocyte or spermatocyte. Fertilization of a normal monoploid egg by more than one sperm has also been suggested, because polyspermy has been found to occur occasionally in rabbits and rats, but there is no clear evidence for this in man. In summary, polyploidy in man, whether complete or as a mosaic, leads to gross abnormalities and death.

Evolution Through Polyploidy. Interspecific hybridization combined with polyploidy offers a mechanism whereby new species may arise suddenly in natural populations. Cytogenetic investigations of such instances of speciation in many cases have involved real "detective work" and even culminated in the artificial production of the new species.

An excellent example is furnished by the work of Huskins (1930) and of Marchant (1963) on the saltmarsh grass *Spartina*. A European species, *S. maritima*, occurs along the Atlantic coast of Europe and adjacent Africa. *S.*

A

B

C

FIGURE 13-7. Spartina alterniflora. (*A*) *Habitat at edge of salt marsh. The 50-foot-wide border directly adjacent to the water is a pure growth of* S. alterniflora. (*B*) *Closer view of pure stand of* S. alterniflora *in bloom bordering the salt water.* (*C*) *Single flowering spike of* S. alterniflora. (All photos by the author in Cape Cod National Seashore Park, Eastham, Mass.)

alterniflora, an eastern North American species (Fig. 13-7), was introduced accidentally into Great Britain in the eighteenth century. The two species are morphologically quite different from each other and have different chromosome numbers, $2n = 60$ for *S. maritima* and 62 for *S. alterniflora.* The American species gradually spread after introduction in western Europe, and the two species grew intermixed. In the 1870s a sterile putative F_1 hybrid (now called *S. x Townsendii*) was collected. This latter form has a somatic chromosome number of 62, rather than the expected 61 ($= \frac{60}{2} + \frac{62}{2}$), apparently because of meiotic abnormality in one of its parents. Chromosomes in this genus are small and not easily distinguished from each other, hence it is difficult or impossible to determine the exact origin of the chromosome complement of *S. x Townsendii* with reference to *S. alterniflora* and *S. maritima.* In any event, the hybrid is sterile because of the lack of meiotic pairing in its mixture of *alterniflora* and *maritima* chromosomes. In the 1890s a new, vigorous, fertile *Spartina,* which quickly spread over the British coasts and into France, was discovered. This new form, named *S. anglica,* has a somatic chromosome number of 124 in most cases, although some with 122 and 120, respectively, have also been reported. From its morphology and cytology, *S. anglica* appeared to be an amphidiploid of the cross *S. alterniflora* × *S. maritima.* This did prove to be the case, for *S. anglica* has been created (or re-created) in experimental plots, a triumph of cytological research (Fig. 13-8).

Similarly, other workers investigated the origin of New World cotton (*Gossypium*). Briefly, Old World cotton has 26 rather large chromosomes, whereas a Central and South American species has 26 much smaller ones. The cultivated cotton has 52, of which 26 are large and 26 smaller, and was suspected of being an allotetraploid of a cross between Old and New World species. Beaseley was able to reconstitute the cultivated species by crossing the two putative parents and using colchicine to double its chromosome number.

But in considering polyploidy as a mechanism of evolution, a note of caution must be inserted. Basically, polyploidy adds no new genes to a gene pool but, rather, results in new combinations, especially in allopolyploids. Phenotypic effects of autopolyploidy generally represent merely exaggerations of existing characters of the species. Possession of multiple genomes reduces the likelihood that a recessive mutation will express itself until and unless its frequency becomes quite high in the population. Polyploidy, therefore, has the potential for actually decreasing genetic variation. So although entities that we recognize as new species do arise through allopolyploidy, and although vigor and often geographic range may be increased by autopolyploidy, it is nevertheless wise to view polyploidy in its proper perspective as both a positive and negative force in evolution.

ANEUPLOIDY

Trisomy. The Jimson weed (*Datura stramonium*) shows a considerable amount of morphological variation in many traits but particularly in fruit

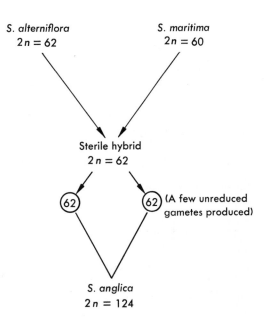

FIGURE 13-8. *Development of Spartina anglica, an allotetraploid (amphidiploid). The cross S. alterniflora × S. maritima produced a sterile hybrid (named S. × townsendii by Marchant, 1963) having 62 chromosomes instead of the expected 61. Production of infrequent gametes having all 62 chromosomes, followed by their fusion, results in the fertile hybrid S. anglica. Some amphidiploid plants with 120 and 122 chromosomes have also been reported.*

characters. The normal chromosome number for this plant is $2n = 24$, but in a now classical study, Blakeslee and Belling (1924) showed that each of several specific morphological variants had 25 chromosomes. One of the 12 kinds of chromosomes was found to be present in triplicate; that is, the somatic cells were $2n + 1$. Such a **trisomic** plant has three of each of the genes of the extra chromosome. Because the Jimson weed has 12 pairs of chromosomes, 12 recognizable trisomics should be possible, and Blakeslee and his colleagues succeeded in producing all of them (Fig. 13-9). Trisomics usually arise through ~~maternal~~ nondisjunction so that some gametes contain two of a given chromosome.

When trisomics are crossed, ordinary Mendelian ratios do not result. For example, the trisomic "poinsettia" (Fig. 13-9) has an extra ninth chromosome (i.e., it is triplo-9), which carries gene P (purple flowers) or its allele p (white flowers). One possible genotype, therefore, for a purple-flowering "poinsettia" is PPp. Crossing two such plants produces a $17:1$ F_1 phenotypic ratio. Pollen (and hence sperms) carrying either more or less than 12 chromosomes is nonfunctional. Megaspores (and eggs that develop from them), however, are not so affected. Meiosis in trisomics ordinarily results in two of the three homologs going to one pole, and one to the other, giving rise to some gametes carrying various combinations of two homologs and others carrying but one. With this in mind we can represent the cross $PPp \times PPp$ as follows:

P PPp ♀   × PPp ♂
        "purple poinsettia"   "purple poinsettia"

P gametes:

♀ $\frac{1}{6}$ ea.: $P + P + Pp + Pp + PP + p$

♂ $\frac{1}{3}$ ea.: $P + P + p$

F_1 $\frac{4}{18} PP$ homozygous purple diploid

$\frac{4}{18} Pp$ heterozygous purple diploid

$\frac{5}{18} PPp$ heterozygous purple trisomic ("poinsettia")

$\frac{2}{18} Ppp$ heterozygous purple trisomic ("poinsettia")

$\frac{2}{18} PPP$ homozygous purple trisomic ("poinsettia")

$\frac{1}{18} pp$ homozygous white diploid

Purple and white segregate in a 17:1 ratio. Other ratios are, of course, possible with different parental genotypes; some of these are taken up in problems at the end of this chapter.

Other Aneuploids. Aneuploids other than trisomics are reported in the literature, but the best-known ones occur in the Jimson weed because of the extensive work of Blakeslee and his associates and because most of them are lethal in many organisms. They are summarized in Table 13-2, along with the other modifications of chromosome number considered in this chapter.

TABLE 13-2. Summary of Variations in Chromosome Number (Heteroploidy)

Type	Designation	Chromosome Complement (where one set consists of four chromosomes, numbered 1, 2, 3, and 4)
Euploids		
Monoploid	n	1-2-3-4
Diploid	$2n$	1-2-3-4 1-2-3-4
Triploid	$3n$	1-2-3-4 1-2-3-4 1-2-3-4
Autotetraploid	$4n$	1-2-3-4 1-2-3-4 1-2-3-4 1-2-3-4
Allotetraploid	$4n$	1-2-3-4 1-2-3-4 1'-2'-3'-4' 1'-2'-3'-4'
etc.		
Aneuploids		
Trisomic	$2n + 1$	1-2-3-4 1-2-3-4 1
Triploid tetrasome	$3n + 1$	1-2-3-4 1-2-3-4 1-2-3-4 1
Tetrasomic	$2n + 2$	1-2-3-4 1-2-3-4 1-1
Double trisomic	$2n + 1 + 1$	1-2-3-4 1-2-3-4 1-2
Monosomic	$2n - 1$	1-2-3-4 2-3-4
Nullisomic	$2n - 2$	2-3-4 2-3-4

TRISOMY IN MAN

Down's Syndrome. An important and tragic instance of trisomy in man involves Down's syndrome,[1] or mongoloid idiocy (Fig. 13-10). The incidence

[1]Named for the nineteenth-century British physician J. Langdon Down, who first described the syndrome in 1866.

Normal
DIPLOID

FIGURE 13-9. *Fruits of normal diploid Jimson weed (top) and its 12 possible trisomics below. Each of the latter was produced experimentally by Blakeslee and Belling.* (Redrawn from A. B. Blakeslee and J. Belling, Chromosomal Mutations in the Jimson Weed, *Datura stramonium. Journal of Heredity,* **15:** 195–206, 1924, by permission of the American Genetic Association, Washington, D.C.)

in the general population is roughly 1 in 600 live births. Manifestations of Down's syndrome include (1) mental retardation (I.Q. in the range 25 to 75, with most in the lower 40s), (2) below-average height, (3) a specific peculiarity of the upper eyelid (epicanthal fold) that suggests Oriental eyes, (4) somewhat sloping forehead, (5) low or flattened nose bridge, (6) low-set ears, (7) short, broad hands with characteristic abnormalities of the palm prints, and often (8) cardiac malformations. The mouth is usually open, with the tongue protruding. Life expectancy is reduced. Sexual maturity is generally not attained;

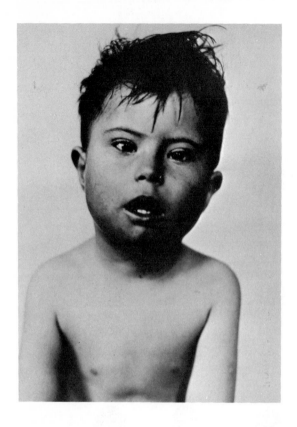

FIGURE 13-10. *Down's syndrome* (*Mongoloid idiocy*) *in a child.* (Courtesy National Foundation, New York.)

males appear to be sterile, but a few females have been reported to have borne children.

Mongoloids are trisomic for one of the G group of autosomes (Fig. 13-11), originally identified as number 21, the second smallest of the autosomes. Recent work with fluorescence microscopy, however, has shown that it is the smallest (formerly designated 22) that is present in triplicate. Because the notation *triplo-21* has become so entrenched in the literature, it has been generally agreed to renumber the two members of the G group, with 22 becoming 21 and vice versa. Although this is contrary to the otherwise standard system of assigning successively higher numbers to successively shorter autosomes, it obviates confusion between older and newer literature on Down's syndrome. The important fact is the presence of an extra one of the shortest autosomes, regardless of the numerical designation.[2]

From the cytological viewpoint, two types of Down's syndrome may be recognized: (1) *triplo-21*, in which the affected individual is trisomic for autosome 21, with a total somatic complement of 47 chromosomes, and (2) *trans-*

[2]Williams et al. (1975) have localized the determinants of the Down phenotype to trisomy of the segment of the long arm of chromosome 21 distal to the proximal part of band q21 (see p. 66).

FIGURE 13-11. *Karyotype of trisomic mongoloid idiot having 47 chromosomes (triplo-21).* (Courtesy National Foundation, New York.)

location, in which the extra twenty-first chromosome has become attached to another autosome, most frequently one of the D group, now generally agreed to be number 14. More rarely the translocation involves 15 and 21, or even one of the G group, probably another 21. A karyotype of a translocation Down's syndrome, therefore, shows 46 chromosomes, one of them 14-21 (Fig. 13-12). Translocations will be discussed more fully in the next section, but suffice it to point out here that both types involve an extra number 21; in one case it is a separate entity, in the other it has become attached to another chromosome. Phenotypes of both triplo-21 and translocation mongoloids are identical.

Incidence of Down's syndrome is closely related to maternal age (Fig. 13-13). No consistent correlation with paternal age has been discerned. Triplo-21s arise through nondisjunction, chiefly in oogenesis. A female is born with all

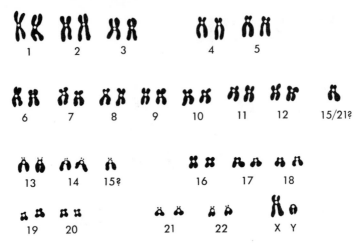

FIGURE 13-12. *Karyotype of translocation mongoloid. This individual has 46 chromosomes, but only one normal 15 and a large one formed by union of a 21 and the other 15. Two normal 21s are also present, giving the individual three "doses" of chromosome 21.*

the oocytes she will ever produce. These remain in an arrested early prophase-I of meiosis from before birth until ovulation, generally at the rate of one per menstrual cycle (about 28 days) after puberty. A given developing ovum may remain in this state of suspended development for some 12 to 45 years or so. Spermatocytes, on the other hand, go through the meiotic cycle in 48 hours or less.

Nondisjunction during oogenesis is thus a function of senescence of the oocytes whereby separation of homologs is interfered with in some way. This may be through the breakdown of chromosomal fibers, or through deterioration of the centromere. Presence of a virus, as well as radiational damage (whose cumulative effect would be greatest in older individuals) have been suggested as causative factors. Cells are especially susceptible to damage by viruses and by radiation during division. Of course, the greater the age of the woman, the longer these agents will be able to act on her oocytes. Pantelakis et al. (1970) report a higher incidence of maternal history of infectious hepatitis, a viral disease, in 22 trisomy-21 cases among 10,412 live births over a two-year period. Doxiadis et al. (1970) report the risk of having a child with Down's syndrome to be about three times greater, age group for age group, in mothers who have had clinically infectious hepatitis prior to pregnancy.

Some pedigrees suggest a specific gene that interferes with disjunction; such genes are known in some organisms. In certain cases it appears that autoimmune diseases may play a part because evidence shows about a threefold increase in risk of a Down's syndrome child at any given maternal age

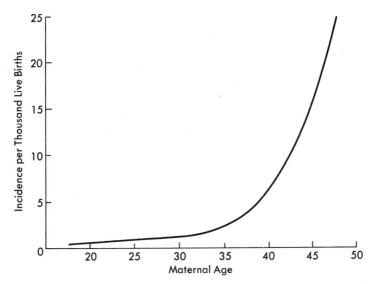

FIGURE 13-13. *Relationship between maternal age and birth of children with Down's syndrome.*

where high thyroid autoantibody levels are present. Fialkow et al. (1971) report some form of clinical thyroid disorder in 30 of 177 mothers of mongoloids, as compared with only 11 of 177 control mothers (i.e., those without Down's syndrome children).

Another explanation may lie in the fact that an egg will degenerate in the fallopian tube within about one day if it is not fertilized. If fertilization should occur toward the close of the period of viability, degenerative changes in the egg do lead to chromosomal abnormalities in the zygote in some animal species, probably including man. The correlation of trisomy-21 with increased maternal age under this explanation probably is a reflection of less-frequent intercourse in older couples. That this is not a *principal* factor is seen in the shape of the curve in Figure 13-13. Delayed fertilization, acting alone, would be expected to produce a linear curve, but as shown in Figure 13-13, the increased incidence of Down's syndrome in the later maternal years is exponential. In any event, though the causes are not fully clear, the maternal age effect is well established.

Mosaics with Down's syndrome, who exhibit the classical symptoms in varying degree, probably arise through failure of the 21's to segregate at mitosis during embryogeny, producing a fetus with cells of three types: (1) normal 21, 21, (2) triplo-21, and (3) monosomic-21. The latter appear unable to survive, leading to a disomic-trisomic mosaic. Should mitotic segregation of the 21s fail at the first division of the zygote, however, two daughter cells, one trisomic and the other monosomic, would result. Death of the latter cell would then lead to production of a wholly trisomic embryo.

Examination of parents of translocation mongoloids almost always discloses one of them to have only 45 chromosomes, including one 21, one 14, and a fused 14-21. Such "carriers" are phenotypically normal inasmuch as their genetic material is present in proper amount. Interestingly, it is far more often the *mother* of a translocation Down's syndrome child who is the carrier. The reason why carrier fathers rarely produce translocation children is unknown. Eicher (1973) reports a greater chance of translocation trisomic offspring also in mice where the maternal parent is the carrier. If the centromere of the maternal translocation chromosome is that of number 14, four kinds of eggs are produced, leading apparently to three kinds of children:

Egg	Sperm	Zygote	Progeny
14-21, 21	14, 21	14, 14-21, 21, 21	Translocation Down's syndrome
14-21	14, 21	14, 14-21, 21	Translocation carrier
14, 21	14, 21	14, 14, 21, 21	Normal
14	14, 21	14, 14, 21	Presumably lethal

Edwards' Syndrome.   First described in 1960 by Edwards and his colleagues, trisomy-18 (Fig. 13-14) is now well known in more than 50 published cases.[3] Incidence is about 0.3 per 1,000 births. It is characterized by multiple malformations, chief among which are malformed, low-set ears; small, receding lower jaw; flexed, clenched fingers; cardiac malformations; and various deformities of skull, face, and feet. Harelip and cleft palate often occur. Death takes place generally around three to four months of age, but may be delayed to nearly two years. One mentally retarded female, age 15, was reported in 1965 by Hook et al. Triplo-18 children give evidence of severe mental retardation. The defect is about three times more frequent in females; the reasons for this are not clear.

As in Down's syndrome, there is a pronounced maternal age effect, although plotting frequency against maternal age produces a bimodal curve (Fig 13-15). The secondary peak in the early 20s reflects the normal maternal age group of maximum births, whereas the pronounced peak from 35 to 45 years is clearly related to greater age of the mother.

Trisomy-13 (Patau's Syndrome).   In 1960 Patau and associates described a case of multiple malformations in a newborn child, established as a trisomic for one of the D-group of autosomes, now generally agreed upon as number 13. Individuals appear to be markedly mentally retarded, and very frequently have sloping forehead, harelip, and cleft palate; the last two malformations

---

[3]Edwards et al. first described this condition as trisomy-17.

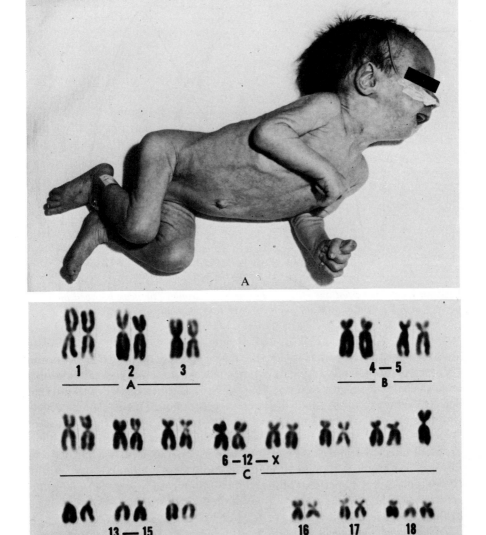

FIGURE 13-14. *Edwards' syndrome (trisomy-18). (A) Young triplo-18 child, showing low-set ears, small, receding lower jaw, and flexed, clenched fingers. (B) Karyotype of same child; note the seven E-group autosomes.* (Photos courtesy Dr. Richard C. Juberg.)

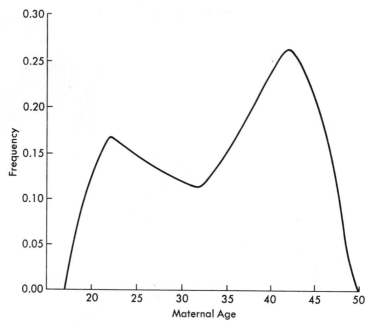

FIGURE 13-15. *Maternal age effect in Edwards' syndrome. The peak between maternal ages 35 and 45 is the significant one here. See text for details.*

are so strongly developed as to give the face a severely deformed appearance. Polydactyly (both hands and feet) is almost always present; the hands and feet are also characteristically deformed (Fig. 13-16). Cardiac and various internal defects (kidneys, colon, small intestine) are common. Death usually occurs within hours or days, but some live for a few months. In an extreme case of longevity, a living triplo-13, aged 5 years, has been reported. There is a slight excess of affected females. Age of the mother is not as clearly a factor as in Edwards' syndrome, although tabulated cases do form a bimodal curve. Smith (1964) reports that 40 per cent of the triplo-13s in his study were born to mothers over 35, whereas this age group accounted for less than 12 per cent of all births. Incidence is of the order of 0.2 per 1,000 births.

Trisomy and Spontaneous Abortion. It appears that trisomy for autosomes other than 13, 18, and 21 may well be lethal. Boué and Boué (1973) examined 1,457 abortuses, of which 892 (61 per cent) had some chromosomal abnormality, including XO, trisomy D, trisomy E, and triploidy. McCreanor et al. (1973) reported spontaneous abortion of an empty embryo sac, wall cells of which were trisomic for autosome 7, as revealed by Giemsa banding. Kajii and associates (1973) report that of the 22 possible human autosomal trisomics, they have found all but those for numbers 1, 5, 12, 17, and 19 in abortuses. Either these five are so rare as to have escaped detection or they may well be

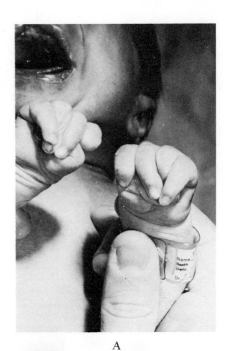

A

FIGURE 13-16. *Patau's syndrome (tri-somy-13). (A) Clenched fists. (B) Deformities of feet. See text for details.* (Photos courtesy Dr. Richard C. Juberg.)

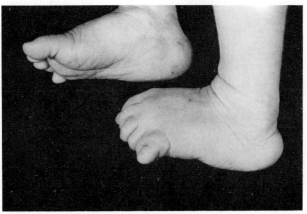

B

lethal so early that spontaneous abortion of the very immature fetus passes unnoticed. A C-trisomy is shown in Figure 13-17.

NONHUMAN TRISOMICS

That Down's syndrome may also occur in other animals is clear from the work of McClure et al. (1969), who have described a trisomic young female chimpanzee. This animal displayed clinical and behavioral features similar to those of human mongolism, scoring below normal in a variety of behavioral

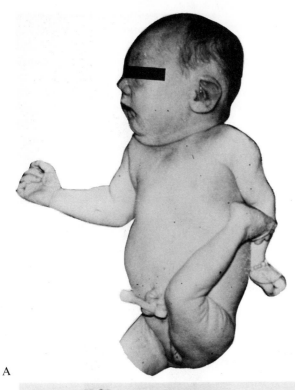

FIGURE 13-17. *C-trisomy in a female. (A) External features; note ears, jaw, hands, and leg position. (B) Karyotype of same child showing 17 chromosomes in the C, X group instead of the normal 16, for a total of 47.* (From R. C. Juberg, E. F. Gilbert, and R. S. Salisbury, 1970. Trisomy-C in an Infant with Polycystic Kidneys and Other Malformations. *Jour. Pediatr.*, **76:** 598–603. Used by permission of Dr. Juberg, who supplied the photos, and The C. V. Mosby Co.)

A

1 2 3   4 — 5
———— A ————   —— B ——

6 — 12 and X
———— C ————

13 —— 15   16   17   18
———— D ————   ——— E ———

19 — 20   21 — 22
——— F ———   ——— G ———

B

and postural tests. Chimpanzees have a diploid chromosome number of 48 (Table 3-1) but in the animal in question most of the blood cells examined had 49, with an extra small acrocentric chromosome that matched pair 22 (the second smallest of the autosomal pairs). This same chromosome was identifiable in triplicate even in the few cells where only a total of 46, 47, or 48 chromosomes could be counted. Both parents were cytologically and behaviorally normal. McClure and his colleagues note that "a comparable condition has not been reported in nonhuman primates. The occurrence of this condition in a lower primate again emphasizes the close phylogenetic relation between man and the great apes and may provide a model for studying this relatively frequent human syndrome."

Structural Changes in Chromosomes

TYPES OF STRUCTURAL CHANGES

For all their complexities of structural organization, chromosomes are far from indestructible. Through such agents as radiation and chemicals they can suffer breakage, which may result in genetic damage to subsequent generations. Suitable precautions, of course, must be taken in X-ray diagnoses, and nuclear weaponry constitutes a definite hazard. Lisco and Conard (1967), for example, report chromosomal aberrations associated with breaks in 23 of 43 Marshall Islanders 10 years after accidental exposure to radioactive fallout following testing of a high-yield nuclear device at Bikini in the Pacific. A wide variety of chemical substances, some formerly or presently in common use by man, have been implicated or suspected in inducing chromosomal damage. We shall examine the possible effects of some of these later in this chapter. Breaks may also occur "naturally," for no assignable cause. They may produce any of several cytological and genetic consequences; they may be detected in almost any dividing cell, though certain kinds offer special advantage. Particularly useful in this regard are the salivary gland chromosomes of dipteran insects such as *Drosophila* (see next section). Genetic effects of chromosomal aberrations are observable in unexpected phenotypes, altered linkage relationships, reduced fertility, and an increase in spontaneous abortion.

Salivary Gland Chromosomes.   Structural changes in chromosomes resulting from breakage are most profitably studied in (1) salivary gland chromosomes of dipteran insects (such as *Drosophila*) and (2) meiocytes during the process of meiosis. Because of their large size and banded structure, salivary gland chromosomes offer particularly favorable material for such studies. So, before discussing the several kinds of structural modifications, let us examine these unusual chromosomes more closely.

Nuclei of the salivary gland chromosomes of the larvae of dipterans like *Drosophila* have unusually long and wide chromosomes, 100 or 200 times the size of the chromosomes in meiosis or mitosis of the same species (Fig. 13-18). This is particularly surprising, because the salivary gland cells do not divide

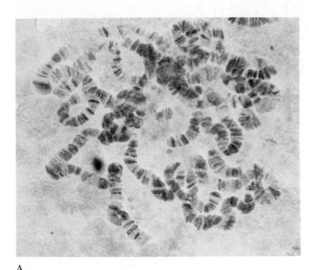

A

FIGURE 13-18. *Salivary gland chromosomes of* Drosophila melanogaster *showing the banding that permits rather accurate cytological mapping. (A) Entire complement. (B) Partial complement, enlarged.* [(A) Courtesy Miss Chris Arn. (B) Courtesy Mr. John Derr.]

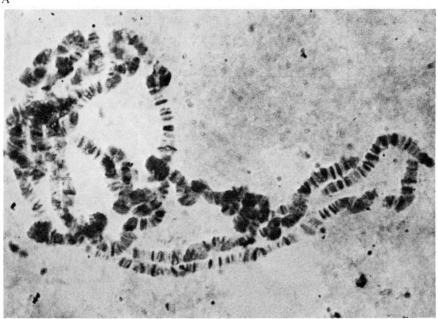

B

after the glands are formed, yet their chromosomes replicate several times and become unusually long as well. Moreover, these chromosomes are marked by numerous cross bands that are apparent even in unstained nuclei. These bands are widely assumed to represent chromomeres, side by side, of the replicated chromatids; they are quite constant in size and spacing for a given normal chromosome. These multiple chromosomes appear to be in a perpetual prophase and are synapsed. Thus any difference in banding between homologs can be easily compared.

TYPES OF STRUCTURAL CHANGES

A chromosomal break may or may not be followed by a "repair." If segments are rejoined as they were, the break ordinarily passes undetected. Should such repair not be effected, one or more of the aberrations listed in Table 13-3 occur. Note that a deletion in one chromosome may be accompanied by a translocation or a duplication in another. Reciprocal translocations, in which two nonhomologs *exchange* segments, often unequal in length, may also occur.

TABLE 13-3. Types of Aberrations Produced by Chromosomal Breaks

Type	Description	Gene Changes
Normal	(*ABCDEFGH*)	None
Deletion	No rejoining, chromosomal segment lost	*ABFGH, CDEFGH*, etc.
Inversion	Broken segment reattached to original chromosome in reverse order	*ABFEDCGH*, etc.
Duplication	Broken segment becomes attached to homolog that has experienced a break; homolog then bears one block of genes in duplicate	*ABCDEFGEFGH*
Translocation	Broken segment becomes attached to a non-homolog resulting in new linkage relations	*LMNOPQRCDEFGH*, etc.

DELETIONS

Deletions and Cytological Mapping. Perhaps the simplest result of breakage is the loss of a part of a chromosome. Portions of chromosomes without a centromere (*acentric fragments*) lag in anaphasic movement and are lost from reorganizing nuclei. Such loss of a portion of a chromosome is called a **deletion** or a deficiency.

Deletions often make possible *cytological mapping* of chromosomes. For example, if large numbers of fruit flies of autosomal genotype *ABCDEF . . . /ABCDEF . . .* (i.e., homozygous dominant for several traits) are subjected to X-irradiation, a few of the individuals may suffer a break and deletion in one of the chromosomes bearing these genes. If a deletion for genes C and D occurs in one primary spermatocyte, then two of the four sperms produced from it would have an *intercalary deletion* in this particular autosome:

Primary spermatocyte   *AB——EF...*
  (irradiated)   *ABCDEF...*

Secondary sperm-
  atocytes   *ABEF..., ABCDEF...*
Spermatids and
  sperms   *ABEF..., ABEF..., ABCDEF..., ABCDEF...*

Mating such males to recessive females, *abcdef.../abcdef...*, wil produce off-spring, some of which will have the genotype abcdef.../ABEF. Therefore these flies will express the recessive phenotype for genes *c* and *d*, whereas those receiving an *ABCDEF...* sperm from the male parent will express the dominant phenotype for all six genes. The expression of genes *c* and *d* where they would have been obscured by genes *C* and *D* had these been present is called **pseudodominance.** This is really not a good term, but it is a useful one and is firmly established in the literature.

Next, flies showing pseudodominance are mated with normal *abcdef.../abcdef...* individuals. Half the offspring will now exhibit pseudodominance for the same genes. Examination of the salivary gland chromosomes of larvae of this cross will quickly disclose the *abcdef.../ABEF* individuals. Remember, these chromosomes are banded and pair exactly, band for band. The segment that includes genes *C* and *D* will also include some of the bands, and the giant chromosomes of *abcdef.../ABEF...* would appear somewhat as shown in Figure 13-19, the normal, unaltered one showing a characteristic **deletion loop.**

Another strain of flies with an *overlapping deficiency* such as *ABC——F...* /*ABCDEF...* might be developed and mated as in the preceding case. Giant chromosomes of *abcdef.../ABCF...* individuals will again display the characteristic deficiency loop. Comparison with the loops produced by the *abcdef...* /*ABEF...* individuals permits easy detection of the precise chromosomal region which bears gene *D;* its locus can be seen to be associated with a particular band (Fig. 13-19).

Detailed *cytological maps* of *Drosophila* have been prepared in just this way. The geneticist now has visible bands within which genes appear to be located, and he can measure their distances in ordinary units. Comparison of cytological and genetic maps confirms the linear sequence given by the latter. Cytological maps, however, show known genes to be more uniformly distributed over the chromosome than is suggested by the genetic map. This results from the greater degree of interference near the centromere and termini of the chromosome. For example, in chromosome II of *Drosophila* the centromere is at locus 55.0. Painter's map (Fig. 6-4) shows gene *pr* (purple eyes) to be at locus 54.5, just 0.5 unit to the left of the centromere. This is about 1 per cent of the 55 map units between the centromere and the end of the left arm. Yet on the salivary gland map gene *pr* is about 12 genetic map units to the left of the centromere, or nearly 22 per cent of the total distance involved. Figure 13-20 compares genetic and cytological map locations for several genes of chromosome II.

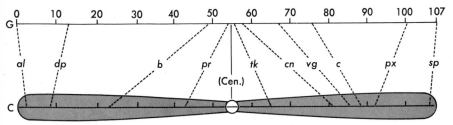

FIGURE 13-20. *Diagrammatic comparison of genetic map (G) and cytological map (C) of chromosome II of Drosophila melanogaster. Note particularly the much closer spacing of genes near the centromere on the genetic map, as opposed to the distances between them on the cytological map. This reflects the interference effect of the centromere.* (Cen = centromere.)

minus sign, added segments by a superscript plus sign. Thus a karyotype carrying the Philadelphia chromosome is designated as Gq$^-$. Ph^1 is acquired during life, perhaps as a result of radiation exposure or of viral infection; it is not transmitted from parent to offspring. *since doesn't affect gonads, only somatic cells?*

Juberg and Jones (1970) reported an 18-month-old mongoloid male who died of acute myeloblastic leukemia at the age of 21 months. Cytological investigation disclosed two cell lines in the bone marrow and in the peripheral blood, one of 47 chromosomes, the other of 51 (Fig. 13-23A, B). The 47-chromosome karyotype shows three G-group chromosomes (designated number 22 in Fig. 13-23A), two of them with deletions of all of the short arm (Gp$^-$). In the 51-chromosome cell line, the G group consisted of seven autosomes, three of them with no detectable short arms (Fig. 13-23B). In addition, this cell line showed two extra chromosomes in other groups, one in the C group and another in the F group. Cytological examination of the parents disclosed the mother to be a carrier of the Gp$^-$ chromosome (Fig. 13-23C); she was phenotypically normal. The father's karyotype was normal. This abnormal G chromosome was first reported in 1962 in two siblings suffering from chronic lymphocytic leukemia, and was designated the Christchurch chromosome (Ch^1) from the New Zealand city where the case was found (Gunz et al., 1962).

A number of phenotypic abnormalities were present in the case reported by Juberg and Jones, but abnormalities, when present, appear not to be constant from one case to another. In 45 persons described to 1970, only 17 had some abnormality, according to Juberg and Jones, and only 2 of 30 carriers displayed any phenotypic abnormality. Further study is necessary before the nature of the inconstant effect of the Ch^1 chromosome can be clarified.

Deletion of part of the short arm of one of the X chromosomes produces a typical Turner syndrome. Deletion of part of the long arm of an X, on the other hand, results in an atypical, Turnerlike syndrome. The latter individuals are of normal height, but have vestigial gonads and are sterile. No other anomalies are consistently associated with this cytological condition. By contrast,

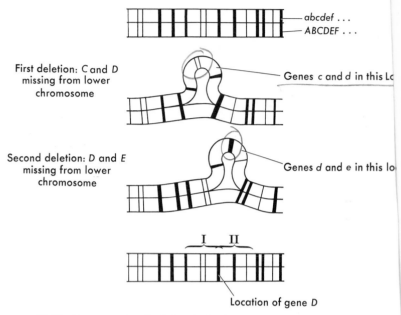

First deletion: C and D
missing from lower
chromosome

Genes c and d in this Lc

Second deletion: D and E
missing from lower
chromosome

Genes d and e in this lo

I   II

Location of gene D

FIGURE 13-19. *Determining which band of salivary gland chromosomes bears a*
gene is done by a series of overlapping deficiencies. See text for details.

Deletions in Man. Probably the best-known disorder to be associated d
nitely with a deletion in man is the *cri du chat*, or cat-cry, syndrome descril
by Lejeune (1963). Symptoms include severe mental retardation, a rou
"moon face," and a characteristic, plaintive cry, similar to that of a cat,
infancy. Well over twenty cases have been studied; in each a large portion
the shorter arm of autosome 5 is deleted (Fig. 13-21). Several other deletic
are known and new instances are being reported in the literature from tii
to time. German (1970) described a male infant having a deletion of the sho
arm of a number 4 autosome (Fig. 13-22). This child was "unusually sma
(with) severe psychomotor retardation, convulsions, a wide, flat nasal bridg
a prominent forehead, cleft palate, and congenital heart disease," accordii
to German. Some mentally defective patients have been found to have oth
chromosomal deletions; de Grouchy and others, for example, reported or
such child in 1963 who lacked the short arm of chromosome 18.

Patients with chronic granulocytic (myelogenous) leukemia carry a sho
chromosome, named the Philadelphia or Ph[1] chromosome, in leukocytes an
bone marrow cells (only). This has been identified as a G-group chromosome
apparently number 21, which has sustained loss of most of the distal part o
its longer arm. The standard designation for the longer arm of a chromosom
is q (that for the shorter arm is p); deletions are indicated by a superscrip

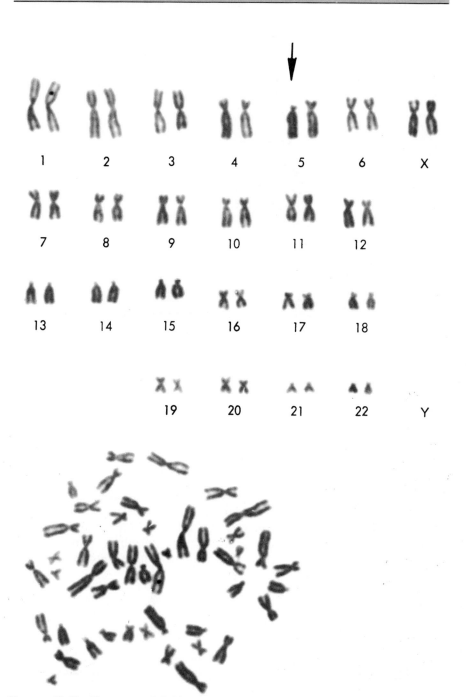

FIGURE 13-21. *Karyotype of child with* cri du chat *syndrome. Note deletion of the short arm of one of the number 5 autosomes.* (Photo courtesy Dr. James German, New York Blood Center.)

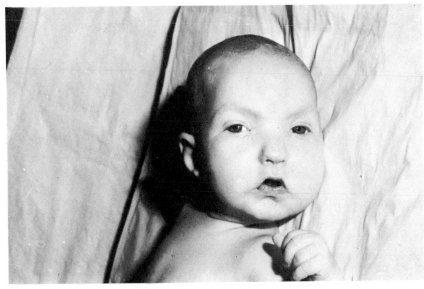

A

FIGURE 13-22. (A) *Infant with a deletion of the short arm of one of the number 4 autosomes. The syndrome is described in the text.* (B) *Karyotype of the child in Figure 12-16(A). Note missing segment of the short arm of one of the number 4 autosomes* (*arrow*). (Photos courtesy Dr. James German, New York Blood Center. Copyright 1970 by The Society of the Sigma Xi. Used by permission.)

apparent deletions in the Y chromosome are not known to produce any serious consequences, probably because of the relative genetic inertness of the Y.

Increasing use is being made of cultured amniotic fluid cells obtained by amniocentesis[4] to determine fetal karyotypes. Amniotic cells appear to show a rather high frequency of tetraploidy, unaccompanied by any cytological abnormality of the fetus, but structural aberrations, trisomy, and the like, are fairly easy to detect from these cells around the sixteenth week of gestation. This information can, of course, be extremely useful to parents and physician in considering a possible termination of the pregnancy.

INVERSIONS

Numerous **inversions** have been described in *Drosophila*. These probably arise when breaks occur at a place where a chromosome forms a tight loop during synapsis. An inversion heterozygote, in which only one of two homologs has a particular inversion, forms a characteristic **"inversion loop"** in prophase-I

---

[4]Amniocentesis involves the withdrawal of fluid, containing sloughed fetal epithelial cells, from the amniotic cavity. Chromosomal analysis as well as enzyme studies can then be made from subcultured cells.

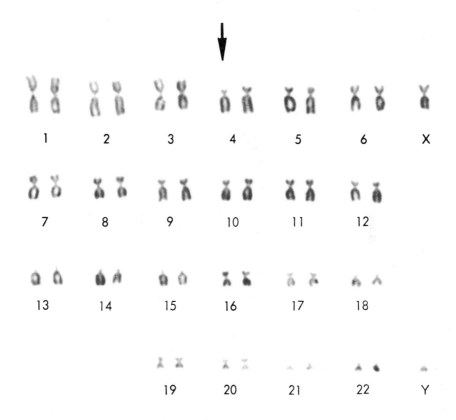

| 1 | 2 | 3 | 4 | 5 | 6 | X |

| 7 | 8 | 9 | 10 | 11 | 12 |

| 13 | 14 | 15 | 16 | 17 | 18 |

| 19 | 20 | 21 | 22 | Y |

B

of meiosis (Fig. 13-24). With a little care these can be distinguished easily from deficiency loops.

An inversion requires two breaks in a chromosome, followed by reinsertion of the segment in the opposite direction. A particular block of genes thus occurs

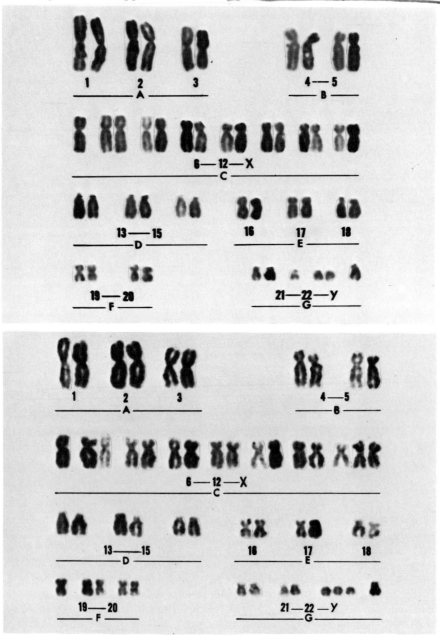

A

B

in reverse sequence (Table 13-3). Inversions may either include the centromere (*pericentric inversions*) or not include the centromere (*paracentric inversions*). Homologous chromosomes, with identical inversions in each member, pair and undergo normal distribution in meiosis. On the other hand, should only one member of the chromosome pair experience a given inversion, synapsis produces a characteristic inversion loop (Fig. 13-25). If a chiasma forms within a paracentric inversion, for example, a dicentric bridge is produced at anaphase-I, resulting in the loss of an acentric fragment (Fig. 13-25). The bridge itself breaks as anaphase-I progresses, resulting in additional aberrations such as deletions or duplications. Although inversions have been referred to as "crossover suppressors," it is the *products* of crossing-over that are eliminated. Crossing-over itself is not actually suppressed.

TRANSLOCATIONS

As indicated on page 277, **translocations** involve the shift of a part of one chromosome to another, nonhomologous chromosome. We can recognize

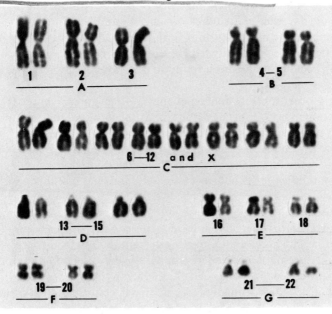

C

FIGURE 13-23. *Karyotype showing abnormal G chromosome (Ch¹, or Christchurch) from a case of acute myeloblastic leukemia. (A) From peripheral blood cell of patient; 47 chromosomes, with 5 members of the G group, two of which have short-arm deletions (Gp⁻). (B) From bone marrow cell of patient, showing 51 chromosomes. (C) From peripheral blood cell of patient's mother, showing a Gp⁻ chromosome (#22).* [From R. C. Juberg and B. Jones, 1970. The Christchurch Chromosome (Gp⁻): Mongolism, Erythroleukemia, and an Inherited Gp⁻ Chromosome (Christchurch). *New Eng. Jour. Med.,* **282:** 292–297. Photos courtesy Dr. Richard C. Juberg. Used by permission of authors and publisher.]

FIGURE 13-24. *Diagrammatic representation of synapsis of a pair of homolous chromosomes, the lower of which has sustained an inversion.*

two principal types of translocation:

1. *Simple*, in which, following breaks, a segment of one chromosome is transferred to another, nonhomologous chromosome where it occupies an intercalary location.
2. *Reciprocal* (interchange), in which segments, which need not be of the same size, are exchanged between nonhomologous chromosomes.

Reciprocal translocations are well known in animals and plants that have been studied extensively. They have been produced frequently by radiation in *Drosophila* (as well as in other organisms) and have occurred widely in natural populations of the evening primrose. (*Oenothera*), the plant whose resulting phenotypic variability led De Vries to formulate his mutation theory. Simple translocations have been found in man where they often lead to gross phenotypic deviations.

Two reciprocal translocation types may be recognized: homozygotes and heterozygotes (Fig. 13-26). The former may have normal meiosis and, in fact, be difficult to detect cytologically unless morphologically distinctive chromosome segments are involved. Genetically they are marked by altered linkage groups and by the fact that a gene with "new neighbors" may produce a somewhat different effect in its new location. We shall examine this **position effect** shortly.

Translocation heterozygotes, however, are marked by a considerable degree of meiotic irregularity. Peculiar and characteristic formations occur at synapsis because of the difficulty of attaining a pairing of homologous parts. Typically, a cross-shaped formation is seen in prophase-I; this often opens out into a ring as chiasmata terminalize (Figs. 13-27, 13-28, 13-29).

In addition to altered linkage groups, translocation heterozygotes are frequently partially sterile because between half and two thirds of gametes (in animals) or meiospores (in plants) fail to receive the full complement of genes required for normal development. Semisterility resulting from reciprocal translocations is easily observed in such plants as corn. Ears lack about half the kernels, and these are arranged irregularly (Fig. 13-30). Abortive pollen is seen to be reduced in size.

Translocations in Man. Translocations in addition to the 14–21 associated with Down's syndrome have been reported in a few instances for human beings. Those most likely to receive attention involve chromosomal segments

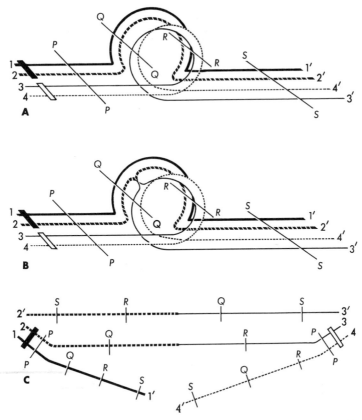

FIGURE 13-25. *Crossing-over within the inversion loop decreases the probability of functional, chromosomally normal gametes. (A) Synapsis of homologous chromosomes; the lower chromosome has sustained the inversion. (B) Crossing-over within the inversion loop between chromatids 2-2' and 3-3'. (C) The resulting production of one acentric chromosome (2'-3') and a dicentric complex. The former is quickly lost from reorganizing nuclei, the latter either breaks (losing some genes from the reorganizing nuclei) or behaves in some other irregular fashion, disturbing the normal gene balance in the nuclei that result from meiosis.*

present in triplicate. Among others, D/D, G/G, B/D, and C/E have been detected. When it is not easy to differentiate members of the several groups of autosomes, translocations in man are often designated by two letters separated by a slash, the first letter indicating the chromosome group supplying the translocated segment and the second the group to which it has been transferred.

In most cases there are numerous phenotypic abnormalities, and spontaneous abortion, or death within a few months after birth, usually ensues. Those who do live longer frequently exhibit mental retardation. Bray and

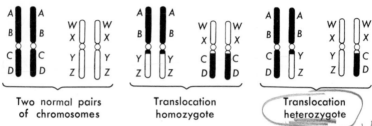

FIGURE 13-26. *Translocation homozygotes and heterozygotes compared.* problems pairing

Josephine (1964) reported a probable unbalanced translocation in which a large part of one of the D-group chromosomes was present in triplicate and a small part of one of the B chromosomes was lacking. There were numerous anatomical abnormalities and death of the subject occurred at seven months. Forty-six chromosomes were present, including an unusually long one believed to be a B + D. Both parents were cytologically normal.

A C/E translocation with effects resembling mongolism was apparently first detected in 1969 at a Boston hospital. The child had one extra long chromosome in its complement of 46, interpreted as an E with added material from a C-group member, which material was thus present in triplicate. This diagnosis was made on the basis of the mother's karyotype, which showed one shortened member of group C (believed to be number 8) and a longer than normal E chromosome (probably number 18). The mother was phenotypically normal, as would be expected because she had the normal amount of chromosomal material. mongoloid child: ♀ ↓ ↓ so 3 c's ♂ E ↓c

DUPLICATIONS AND THE POSITION EFFECT

Duplications occur when a portion of a chromosome is represented more than twice in a normally diploid cell. The extra segment may be attached either

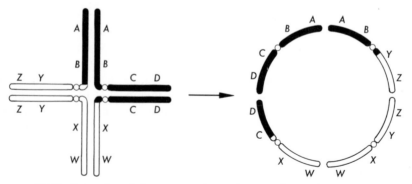

FIGURE 13-27. *Cross-shaped figure occurring in late prophase-I because of translocation heterozygosity. Such a figure often opens out into a ring.*

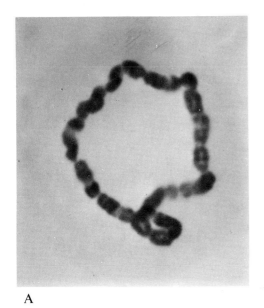

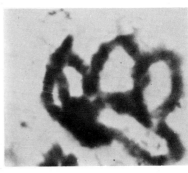

A   B

FIGURE 13-28. *Ring chromosomes in the tropical plant* Rhoeo discolor. *(A) Ring of 12 chromosomes. (B) Ring of several chromosomes.* [(A) Courtesy Dr. L. F. LaCour, John Innes Institute; (B) Courtesy Mr. Gary Bock.]

to the chromosome whose loci are repeated or to a different linkage group, or it may even be present as a separate fragment. Duplications are useful in studying the quantitative effects of genes normally present only in pairs in diploid cells.

The first duplication to receive critical study was the *bar eye* variant in *Drosophila*. The wild-type eye is essentially oval in shape; the bar eye phenotype is characterized by a narrower, oblong, bar-shaped eye with fewer facets. The now classical studies of Bridges (1936) showed this trait to be associated with the duplication of a segment of the X chromosome called section 16A, as observed in salivary gland chromosomes. Each added section 16A intensifies the bar phenotype. However, the narrowing effect is greater if the duplicated segments are on the same chromosome. Letting *A* represent one section 16A in a given X chromosome, we can recognize the genotypes and phenotypes listed in Table 13-4. Other arrangements are of course also possible, but these show clearly that the bar effect of a given number of duplicated 16A sections is intensified if the duplications occur in one X chromosome rather than being divided between the two of the female. Compare heterozygous ultrabar and homozygous bar eyes, for example.

Such a change in the effect of the gene or genes in a chromosomal segment is known as the **position effect.** In bar eye, each added segment narrows the eye still farther, and this effect is enhanced as more duplications occur in one chromosome. Other duplications are known that produce the opposite effect, counteracting the effect of mutant genes. Moreover, duplications need not always be immediately adjacent to exert this position effect.

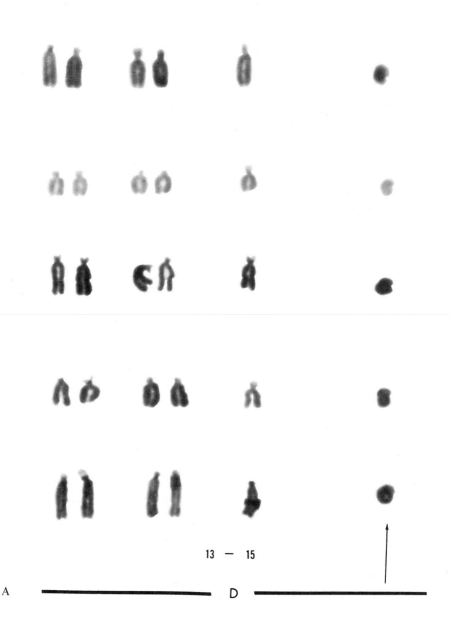

13 — 15

A   D

POSSIBLE CAUSATIVE CHEMICAL AGENTS
OF CHROMOSOMAL ABERRATIONS IN MAN

As noted earlier in this chapter, many chemical substances have been suspected as causal agents of chromosomal damage in man. We shall review the often contradictory evidence relating to LSD, marijuana, nicotine, cyclamates, DDT, and caffeine.

LSD and Chromosome Damage. Because of the therapeutic use of LSD (lysergic acid diethylamide) in treatment of some mental disorders and its frequent illicit use, many investigators have recently turned their attention to the question of whether this drug produces chromosomal and genetic damage. Studies have been carried on in vivo and in vitro with man, other animals, and plants. Conclusions are contradictory, as summed up by Maugh (1973) in these words: "Many studies have shown, for example, that LSD can cause chromosome damage in cultured human cells, that it can cause birth defects

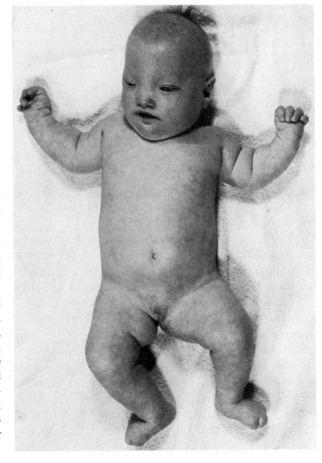

FIGURE 13-29. *Ring-D chromosome in a human female. (A) Partial karyotype of the D group showing ring chromosomes at the right. (B) Phenotypic abnormalities associated with the ring-D chromosome.* (Courtesy Dr. Richard C. Juberg. From R. C. Juberg, M. S. Adams, W. J. Venema, and M. G. Hart, 1969. Multiple Anomalies Associated with a Ring-D Chromosome. *Jour. Med. Gen.,* **6**:314–321. Used by permission of senior author and publisher.)

B

FIGURE 13-30. *Partial sterility results from a reduction in functional gametes caused by multiple translocations.* (Courtesy DeKalb Agricultural Association, Inc., DeKalb, Illinois.)

in laboratory animals and humans, and that it is carcinogenic. An equally large number of studies have shown that there is no evidence for such effects, particularly at LSD concentrations that might be encountered in drug abuse." Recent studies certainly indicate that in at least some individuals of some species an increased frequency of chromosome breaks is found.

Cohen et al. (1967) reported that addition of LSD to cultured human leukocytes resulted in a considerable increase in chromosomal abnormalities, principally a significantly larger number of breaks, most of them involving number 1, the largest of the autosomes. This team also observed a similar increase in a paranoid schizophrenic who had been extensively treated for four years with the drug. Subsequently, numerous studies on plants and animals, including man, have appeared; about half provide evidence for chromosome damage, the remainder either do not or are inconclusive.

Irwin and Egozcue (1967) found a significant increase in frequency of breaks in a large number of illicit users of LSD and other drugs. They conclude that

TABLE 13-4. Comparison of Genotypes and Phenotypes for Bar Eye in *Drosophila* Females

X Chromosome	Phenotype		Mean Number of Facets
A/A	Normal		779
AA/A	Heterozygous bar eye		358
AA/AA	Homozygous bar eye		68
AAA/A	Heterozygous Ultrabar		45
AAA/AAA	Homozygous Ultrabar		25

compare the numowing, where the 4 A's are divided so more are on 1 of the chromosoms eyes are narrower

A = One section 16A of the X chromosome.

LSD, used by all the subjects, was the probable principal cause. Singh and his colleagues (1970) report a high incidence of breaks in root tip cells of barley, about half of them in the vicinity of the centromere. Sadasivaiah et al. (1973) also found an increased frequency of chromosome breaks in mitotic cells of the root tips of onion, rye, and barley. Tetraploid rye, they report, was not appreciably more sensitive than diploid strains. Breaks, fragments, and univalents were also observed by this team in meiotic material of rye. Similarly, meiotic chromosomes of mice that had previously been injected with heavy doses of LSD showed a considerable increase in breaks and fragments over control animals (Skakkebaek et al., 1968). But Amarose and others (1973) found no increase in chromosomal breaks in rabbits treated with LSD, nor did Nicholson and his colleagues (1973) in hamsters, or Emerit et al. (1972) in either mothers or embryos in rats.

Bender and Sankar (1968), commenting on the great concern of psychiatrists with the question of chromosomal and genetic damage by LSD because of its medical use in treatment of such disorders as schizophrenia, object to the conclusions of Irwin and Egozcue, in part because the users could not be examined for possible breaks prior to their use of the drug. In studies on schizophrenic children under their care, they report no difference in break frequencies between treated children and other subjects who had not received the drug. However, LSD treatment had been discontinued in their treated group 20 to 48 months before leukocyte screening was done. But Irwin and Egozcue (1968), in response to Bender and Sankar, point out a lack of data to *eliminate* LSD as a causative agent in chromosome damage. On the other hand, Fernandez et al. (1973) found a statistically nonsignificant difference in proportions of chromosomal aberrations in a group of 32 psychiatric patients who had been treated with known amounts of pure LSD and a control group of 32 other psychiatric patients who had received psychopharmacological medication other than LSD. They further report a *lack* of cytogenetic evidence in their study that might suggest that pure LSD in therapeutic dosage increases chromosomal damage.

Long (1972), in a review of the literature involving a total of 161 children of parents who took LSD, concluded that LSD does not cause chromosome breakage in man. Earlier, Loughman and others (1967) had also found no significant difference in incidence of chromosome aberrations in leukocytes of admitted users who reported their dosages from memory. But they do point out that "other tissues of the body must be examined before ruling out the possibility of chromosome damage to cells actively dividing in vivo." Sparkes and his colleagues (1968) were unable to detect significant amounts of chromosomal damage in small samples of users, but suggest that some of the disagreement among various studies may arise because of the possibility that some persons may be more susceptible than others to chromosomal damage by LSD. They note that data before and after LSD use in the same individuals would be far more informative than comparisons of different individuals with different physiologies, some of whom had used or been treated with the drug for short or long periods and others of whom had never been exposed to it.

Many of the studies in the literature are not, and cannot be, based on standardized doses because they involve illicit users of probably quite variable drug concentration and purity. Furthermore, some of the subjects admit to use of other drugs as well. Much remains to be learned from in vivo studies of various kinds of cells, especially gametes, under rigidly controlled experimental conditions, and more data are needed on children born to persons using or receiving LSD. In such a controlled experiment, Hungerford et al. (1968) first established aberration frequencies in cultured leukocytes from a control group, and also from an experimental group whose members were later to receive LSD therapy. Aberrations scored included acentric fragments, dicentrics, deletions, breaks, and gaps. These researchers report that administration of three intravenous doses (usually of 200 μg each) was followed by some increase in aberration frequency, as well as the appearance of new abnormalities in the experimental group. However, they found a return to the lower control levels ensued within one to six months after the last dose.

Birth of a child with several congenital abnormalities to parents who had both used LSD prior to conception, but not during pregnancy, has been reported by Hsu and others (1970). Although the infant was found to have 46 chromosomes, she was trisomic for autosome 13 and exhibited a translocation involving members of the D group. Hsu and his colleagues suspect drug-induced damage to maternal germ cells; the child, they theorize, resulted from fertilization of an egg having an unbalanced chromosome complement.

At the present time, then, risk of chromosomal damage in users and of damage to their descendants cannot be eliminated from consideration, although the period of susceptibility may be limited. But Dishotsky and colleagues (1971), after surveying the literature, conclude that, in man, "pure LSD in moderate doses does not damage chromosomes in vivo (and) does not cause detectable genetic damage." Certainly man would not necessarily react to the same amount of LSD in the same way as barley, for example, and different persons very likely react differently to the same dosage. Perhaps Titus (1973), reviewing contradictory results of a number of studies on the effect of LSD on chromosomes, best sums up the present status of our knowledge when he states that a definite conclusion cannot be drawn yet.

Marijuana. The active ingredient in the smoke of leaves of marijuana (*Cannabis sativa*) is Δ^9-tetrahydrocannabinol (Δ^9-THC or THC). Considerable interest attaches to the question of possible cytological and genetic effects of marijuana because of its frequent illicit use. Experimental evidence to answer this question is less than that for LSD, but, again, it is somewhat contradictory. In the work of Nicholson et al. (1973), previously referred to in connection with LSD, hamsters were given daily subcutaneous injections of marijuana extract distillate containing 17.1 per cent Δ^9-THC for 10 consecutive days in amounts of 1, 10, and 100 mg per kilogram of body weight. Another group of hamsters was given the same dosages of Δ^9-THC in combination with 1,000 μg LSD per kilogram of body weight. They report no significant elevation of leukocyte chromosome breaks above control levels. Stenchever and colleagues (1972) exposed cultured leukocytes from four healthy human donors to 0.1, 10, and 100 μg Δ^9-THC per milliliter of culture solution and found no increase in incidence of chromosome breaks over control levels. They reported, however, that the 100 μg culture cells did not grow and showed no mitoses at all.

Leuchtenberger at al. (1973a, 1973b) exposed human adult lung fibroblast cells to four puffs of 25 ml each per day of smoke from cigarettes containing 1.8 g of marijuana, and another set of cultured cells to two puffs per day of smoke from tobacco cigarettes. They found various cytotoxic effects, including death and a considerable decrease in mitoses, in both types of cultures, but these were more marked in cells exposed to tobacco smoke than to marijuana smoke! Cells of both cultures showed various mitotic abnormalities up to 45 or more days after exposure. Leuchtenberger and his associates conclude that smoke from both marijuana and tobacco cigarettes does produce disturbances in the mitotic process and in the chromosomal complement and, therefore, in the genetic equilibrium.

Other Substances.   A wide variety of other substances in wide use by man have been implicated or suspected in the induction of chromosomal damage. In addition to the work of the Leuchtenberger group on nicotine, referred to in the preceding paragraph, Bishun et al. (1972) report a variety of cytotoxic effects of nicotine on human leukocytes in vitro, but no chromosomal aberrations. Gross abnormalities were observed, however, in mouse cells.

The once widely used artificial sweetener sodium cyclamate is reported to produce chromosomal breaks and gaps in cultured human leukocytes by Stoltz et al. (1970). Majumdar and his associates (1971) report calcium cyclamate produces both aneuploidy and chromosome breaks in bone marrow cells of the gerbil. Leonard et al. (1972) found no chromosome anomalies, however, in spermatocytes of mice that had received sodium cyclamate in drinking water. They conclude that the effect on reproductive cells may not be the same as that on somatic cells.

The insecticide DDT [1,1,1-trichloro-2,2-bis(p-chlorophenyl)ethane] was found by Kelly-Garvert and colleagues (1973) not to produce a significant increase of chromosome breaks in hamster cells in vitro. On the other hand, Johnson and others (1973) report a significant elevation in incidence of chromosome aberrations in mice. This, they point out, is one of the few records of DDT-induced chromosome damage.

Caffeine has been suspected of causing chromosome damage, but Weinstein et al. (1972) find no significant increase in damage levels in cultured leukocytes from human volunteers who had ingested caffeine in four 200-mg lots per day for a month. However, Loprieno et al. (1973) reports a reduction in crossover frequency in yeasts treated with 0.1 per cent caffeine in the nutrient medium, and Kaul et al. (1973) find an increase in chromosome breaks, especially in the region of the centromere, in cells of the broad bean (*Vicia faba*).

CHROMOSOMAL ABERRATIONS AND EVOLUTION

Speciation, the evolutionary divergence of segments of a population to the point where they are no longer able to combine their genes through sexual reproduction, is a complex process with no single causative mechanism. The diversification on which evolution is built does, in every case, however, require alterations in the genetic material itself (mutation, Chapter 18), and/or changes in its arrangement (chromosomal aberrations), both leading to reproductive incompatibility. The effectiveness of both of these factors is increased if the population is broken into two or more isolated groups. In this way fre-

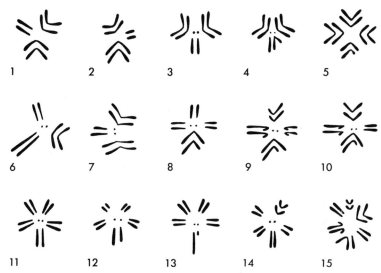

FIGURE 13-31. *Male karyotypes in several species of* Drosophila. *The X and Y chromosomes are at the bottom of each drawing.* (*1*) D. willistoni; (*2*) D. prosaltans; (*3*) D. putrida; (*4*) D. melanogaster; (*5*) D. ananassae; (*6*) D. spinofemora; (*7*) D. americana; (*8*) D. pseudoobscura; (*9*) D. azteca; (*10*) D. affinis; (*11*) D. virilis; (*12*) D. funebris; (*13*) D. repleta; (*14*) D. montana; (*15*) D. colorata.

quencies of mutant genes may often be spread more rapidly, probability of *different* mutations arising in separated groups is increased, and differing groups are often prevented from crossing by various isolating mechanisms, which are discussed more fully in Chapter 14.

Each of the chromosomal aberrations described here is, if not lethal or disabling, related to the development of new taxonomic entities, though each rarely acts alone over time. Thus trisomy in the Jimson weed leads to morphological differences that are recognizable and forms the basis for an aneuploid series. Certainly aneuploid series, which may arise in a variety of ways, do exist in many plant families and genera and are associated with interspecific morphological differences. The role of polyploidy, especially allopolyploidy, in what one writer has called "cataclysmic evolution," has already been described.

But structural changes too are important in a gradual, step-by-step change. Many inversions are known to differentiate the several species of *Drosophila*. Thus *D. pseudoobscura* and *D. persimilis*, morphologically very similar, produce sterile male but fertile female hybrids and differ in four major inversions. On the other hand, *D. pseudoobscura* and *D. miranda* produce completely sterile offspring when crossed. The very complex pairing arrangements assumed by giant chromosomes in hybrid larvae of this interspecific cross indicate repeated and extensive inversions. Translocations in the evening primrose, *Oenothera*, were largely responsible for the differences on which DeVries based his muta-tion theory. A series of many reciprocal translocations in isolated populations may lead to complete reproductive incompatibility, as well as definitive mor-phological differences, so that speciation may be said to have occurred. Com-parison of the chromosome complements of several species of *Drosophila* (Fig. 13-31) suggests all manner of chromosomal aberrations, including trans-locations (even unions of whole chromosomes) and deficiencies, at least. Note the aneuploid series 3, 4, 5, 6.

On the other hand, the disabling and lethal effects of many types of chro-mosomal aberrations, especially in higher animals such as man, serve to remind us that aneuploidy, polyploidy, and structural changes in chromosomes may carry a high negative selection value.

REFERENCES

AMAROSE, A. P., C. R. SHUSTER, and T. P. MULLER, 1973. An Animal Model for the Evaluation of Drug-Induced Chromosome Damage. *Oncology, 27:* 550–562.

BELLING, J., 1925. Production of Triploid and Tetraploid Plants. *Jour. Hered., 16:* 463–464.

BENDER, L., and D. V. SIVA SANKAR, 1968. Chromosome Damage Not Found in Leukocytes of Children Treated with LSD-25. *Science, 159:* 749.

BISHUN, N. P., N. LLOYD, R. W. RAVEN, and D. C. WILLIAMS, 1972. The in Vitro and in Vivo Cytogenetic Effects of Nicotine. *Acta Biol. Acad. Sci. Hung., 23:* 175–180.

BLAKESLEE, A. F., and J. BELLING, 1924. Chromosomal Mutations in the Jimson Weed, *Datura stramonium. Jour. Hered., 15:* 195–206.

BOUÉ, A., and R. BOUÉ, 1973. Évaluation des Erreurs Chromosomiques au Moment de la Conception. *Biomedicine, 18:* 372–374.

BRAY, P. E., and SISTER A. JOSEPHINE, 1964. Partial Autosomal Trisomy and Trans-location. Report of an Infant with Multiple Congenital Abnormalities. *Jour. Amer. Med. Assn., 187:* 566.

BRIDGES, C. B., 1936. The Bar "Gene," a Duplication. *Science, 83:* 210–211. Reprinted in J. A. Peters, ed., 1959. *Classic Papers in Genetics.* Englewood Cliffs, N.J., Prentice-Hall.

BURNS, G. W., 1942. The Taxonomy and Cytology of *Saxifraga pensylvanica L.* and Related Forms. *Amer. Midl. Nat., 28:* 127–160.

BUTLER, L. J., C. CHANTLER, N. E. FRANCE, and C. G. KEITH, 1969. A Liveborn Infant with Complete Triploidy (69, XXX). *Jour. Med. Genet., 6:* 413–421.

CARR, D. H., 1963. Chromosome Studies in Abortuses and Stillborn Infants. *Lancet, 2:* 603–606.

CARR, D. H., 1965. Chromosome Studies in Spontaneous Abortions. *Obstet. Gynecol., 26:* 308–326.

CARTER, C. O., and K. A. EVANS, 1961. Risk of Parents Who Have Had One Child with Down's Syndrome (Mongolism) Having Another Child Similarly Affected. *Lancet, 2:* 785–787.

COHEN, M. M., M. J. MARINELLO, and N. BACK, 1967. Chromosomal Damage in Human Leukocytes Induced by Lysergic Acid Diethylamide. *Science,* **155:** 1417–1419.

CONEN, P. E., and B. ERKMAN-BALIS, 1966. Frequency and Occurrence of Chromosomal Syndromes: 1. D-Trisomy, 2. E-Trisomy. *Amer. Jour. Hum. Genet.,* **18:** 374–398.

DAY, R. W., 1966. The Epidemiology of Chromosome Aberrations. *Amer. Jour. Hum. Genet.,* **18:** 70–80.

DEAN, D. S., 1959. Distribution of Tetraploid and Diploid *Tradescantia ohioensis* in Michigan and Adjoining Areas. *Amer. Midl. Nat.,* **61:** 204–209.

DE GROUCHY, J., M. LAMY, S. THIEFFRY, M. ARTHUIS, and C. SALMON, 1963. Dysmorphie Complexe avec Oligophrénie: Délétion des Bras Courts d'un Chromosome 17–18. *Compt. Rendus Acad. Sci. Paris,* **256:** 1028–1029.

DISHOTSKY, N. I., W. D. LOUGHMAN, R. E. MOGAR, and W. R. LIPSCOMB, 1971. LSD and Genetic Damage. *Science* **172:** 431–440.

DOXIADIS, S., S. PANTELAKIS, and T. VALAES, 1970. Down's Syndrome and Infectious Hepatitis. *Lancet,* **1:** 897.

EDWARDS, J. H., D. G. HARNDEN, A. H. CAMERON, V. M. GROSSE, and O. H. WOLFF, 1960. A New Trisomic Syndrome. *Lancet,* **1:** 787–789.

EICHER, E. M., 1973. Translocation Trisomic Mice: Production by Female but Not Male Translocation Carriers. *Science,* **180:** 81.

EMERIT, I., C. ROUX, and J. FEINGOLD, 1972. LSD: No Chromosomal Breakage in Mother and Embryos During Rat Pregnancy. *Teratology,* **6:** 71–74.

FERNANDEZ, J., I. W. BROWNE, J. CULLEN, T. BRENNAN, H. MATHEU, and I. FISCHER, 1973. LSD . . . An In Vivo Retrospective Chromosome Study. *Ann. Hum. Genet.,* **37:** 81–91.

FIALKOW, P. J., H. C. THULINE, F. HECHT, and J. BRYANT, 1971. Familial Predisposition to Thyroid Disease in Down's Syndrome: Controlled Immunologic Studies. *Amer. Jour. Hum. Genet.,* **23:** 67–86.

GERMAN, J., 1970. Studying Human Chromosomes Today. *Amer. Scientist,* **58:** 182–201.

GUNZ, F. W., P. H. FITZGERALD, and A. ADAMS, 1962. An Abnormal Chromosome in Chronic Lymphocytic Leukemia. *Brit. Med. Jour.,* **2:** 1097–1099.

HOOK, E. B., R. LEHRKE, A. ROESNER, and J. J. YUNIS, 1965. Trisomy-18 in a 15-Year-Old Female. *Lancet,* **2:** 910–911.

HSU, L. Y., L. STRAUSS, and K. HIRSCHORN, 1970. Chromosome Abnormality in Offspring of LSD User. *Jour. Amer. Med. Assn.,* **211:** 987–990.

HUNGERFORD, D. A., K. M. TAYLOR, C. SHAGASS, G. U. LaBADIE, G. B. BALABAN, and G. R. PATON, Cytogenetic Effects of LSD 25 Therapy in Man. *Jour. Amer. Med. Assn.,* **206:** 2287–2291.

HUSKINS, C. L., 1930. The Origin of *Spartina townsendii. Genetica,* **12:** 531–538.

IRWIN, S., and J. EGOZCUE, 1967. Chromosomal Abnormalities in Leukocytes from LSD-25 Users. *Science,* **157:** 313–314.

IRWIN, S., and J. EGOZCUE, 1968. (Untitled reply to Bender and Sankar, 1968.) *Science,* **159:** 749.

JOHNSON, G. A., and S. M. JALAL, 1973. DDT-Induced Chromosome Damage in Mice. *Jour. Hered.,* **64:** 7–8.

JUBERG, R. C., M. S. ADAMS, W. J. VENEMA, and M. G. HART, 1969. Multiple Congenital Anomalies Associated with a Ring-D Chromosome. *Jour. Med. Genet.,* **6:** 314–321.

JUBERG, R. C., E. F. GILBERT, and R. S. SALISBURY, 1970. Trisomy C in an Infant with Polycystic Kidneys and Other Malformations. *Jour. Pediatrics,* **76:** 598–603.

JUBERG, R. C., and B. JONES, 1970. The Christchurch Chromosome (Gp⁻). *New Eng. Jour. Med.,* **282:** 292–297.

KAJII, T., N. NIIKAWA, A. FERRIER, and H. TAKAHARA, 1973. Trisomy in Abortion Material. *Lancet,* **2:** 1214.

KARPECHENKO, G. D., 1928. Polyploid Hybrids of *Raphanus sativus L.* X *Brassica oleracea L. Ztschr. ind. Abst. Vererb.,* **48:** 1–83.

KAUL, B. L., and U. ZUTSCHI, 1973. On the Production of Chromosome Breakage in *Vicia faba* by Caffeine. *Cytobios,* **7:** 261–264.

KELLY-GARVERT, F., and M. S. LEGATOR, 1973. Cytogenetic and Mutagenic Effects of DDT and DDE in a Chinese Hamster Line. *Mutation Res.,* **17:** 223–229.

LEJEUNE, J., 1963. Trois Cas de Délétion du Bras Court d'un Chromosome 5. *Compt. Rendus Acad. Sci. Paris,* **257:** 3098–3102.

LEONARD, A., G. LINDEN, and C. R. SEANCES, 1972. Observations sur les Propriétés Mutagéniques des Cyclamates chez les Mammifères. *Biol. Fil.,* **166:** 468–470.

LEUCHTENBERGER, C., R. LEUCHTENBERGER, U. RITTER, and N. INNUI, 1973a. Effects of Marijuana and Tobacco Smoke on DNA and Chromosomal Complement in Human Lung Explants. *Nature,* **242:** 403–404.

LEUCHTENBERGER, C., R. LEUCHTENBERGER, and A. SCHNEIDER, 1973b. Effects of Marijuana and Tobacco Smoke on Human Lung Physiology. *Nature,* **241:** 137–139.

LISCO, H., and R. A. CONARD, 1967. Chromosome Studies on Marshall Islanders Exposed to Fallout Radiation. *Science,* **157:** 445–447.

LONG, S. Y., 1972. Does LSD Induce Chromosomal Damage and Malformations? A Review of the Literature. *Teratology,* **6:** 75–90.

LOPRIENO, N., R. BARALE, and S. BARONCELLI, 1973. Genetic Effects of Caffeine. *Mutation Res.,* **21:** 275.

LOUGHMAN, W. D., T. W. SARGENT, and D. M. ISRAELSTAM, 1967. Leukocytes of Humans Exposed to Lysergic Acid Diethylamide: Lack of Chromosomal Damage. *Science,* **158:** 508–510.

MAJUMDAR, S. K., and M. SOLOMAN, 1971. Chromosome Changes in Mongolian Gerbil Following Calcium Cyclamate Administration. *The Nucleus,* **14:** 168–170.

MAKINO, S., M. S. SASAKI, and T. FUKUSCHIMA, 1964. Triploid Chromosome Constitution in Human Chorionic Lesions. *Lancet,* **1:** 1273.

MARCHANT, C. J., 1963. Corrected Chromosome Numbers for *Spartina x townsendii* and Its Parent Species. *Nature,* **199:** 929.

MAUGH, T. H., 1973. LSD and the Drug Culture: New Evidence of Hazard. *Science,* **179:** 1221–1222.

MCCLURE, H. M., K. H. BELDEN, W. A. PIEPER, and C. B. JACOBSON, 1969. Autosomal Trisomy in a Chimpanzee: Resemblance to Down's Syndrome. *Science,* **165:** 1010–1011.

MCCREANOR, H. R., F. M. O'MALLEY, and R. A. REID, 1973. Trisomy in Abortion Material. *Lancet,* **2:** 972–973.

MILLER, J. R., and F. J. DILL, 1965. The Cytogenetics of Mongolism. *International Psychiatry Clinics,* **2:** 127–152.

NICHOLSON, M. T., H. B. PACE, and W. M. DAVIS, 1973. Effects of Marihuana and Lysergic Acid Diethylamide on Leukocyte Chromosomes of the Golden Hamster. *Res. Commun. in Chem. Pathol. and Pharmacol.,* **6:** 427–434.

PANTELAKIS, S. N., O. M. CHRYSSOSTOMIDOU, D. ALEXIOU, T. VALAES, and S. A. DOXIADIS, 1970. Sex Chromatin and Chromosome Abnormalities Among 10,412 Liveborn Babies. *Arch. Dis. Childhood,* **45:** 87–92.

PATAU, K., S. L. INHORN, and E. THERMAN, 1963. Lethal Chromosome Constitutions in Man. In S. J. Geerts, ed. *Genetics Today: Proceedings of the XI International Congress on Genetics,* volume 1. New York, Macmillan.

PATAU, K., D. W. SMITH, E. THERMAN, S. L. INHORN, and H. P. WAGNER, 1960. Multiple Congenital Anomaly Caused by an Extra Autosome. *Lancet,* **1:** 790–793.

PENROSE, L. S., and J. D. A. DELHANTY, 1961. Triploid Cell Cultures from a Macerated Foetus. *Lancet,* **1:** 1261–1262.

REGAN, J. D., and J. B. SMITH, 1965. Triploidy in a Human Cell Line. *Science,* **149:** 1516–1517.

SADASIVAIAH, R. S., G. B. COLLINS, and D. L. DAVIS, 1973. Effect of LSD on Mitotic and Meiotic Plant Chromosomes. *Chromosoma,* **44:** 309–318.

SCHINDLER, A. -M., and K. MIKAMO, 1970. Triploidy in Man: Report of a Case and a Discussion on Etiology. *Cytogenetics,* **9:** 116–130.

SINGH, M. P., C. S. KALIA, and H. K. JAIN, 1970. Chromosomal Aberrations Induced in Barley by LSD. *Science,* **169:** 491–492.

SKAKKEBAEK, N. E., J. PHILIP, and O. J. RAFAELSON, 1968. LSD in Mice: Abnormalities in Meiotic Chromosomes. *Science,* **168:** 1246–1248.

SMITH, D. W., 1964. Autosomal Abnormalities. *Amer. Jour. Obstet. Gynec.,* **90:** 1055.

SPARKES, R. S., J MELNYK, and L. P. BOZZETTI, 1968. Chromosomal Effect in vivo of Exposure to Lysergic Acid Diethylamide. *Science,* **160:** 1343–1344.

STENCHEVER, M. A., and M. ALLEN, 1972. The Effect of Delta-9-Tetrahydrocannabinol on the Chromosomes of Human Lymphocytes in Vitro. *Amer. Jour. Obstet. Gynecol.,* **114:** 819–821.

STERN, C., 1973. *Principles of Human Genetics,* 3rd ed. San Francisco, W. H. Freeman.

STOLTZ, D. R., K. S. KHERA, R. BENDALL, and S. W. GUNNER, 1970. Cytogenetic Studies with Cyclamate and Related Compounds. *Science,* **167:** 1501–1502.

SZULMAN, A. E., 1965. Chromosomal Aberrations in Spontaneous Human Abortions. *New Eng. Jour. Med.,* **272:** 811–818.

TITUS, R. J., 1973. Lysergic Acid Diethylamide: Its Effect on Human Chromosomes and the Human Organism In Utero. *Intnl. Jour. Addictions,* **7:** 701–714.

TURPIN, R., and J. LEJEUNE, 1969. *Human Afflictions and Chromosomal Aberrations.* New York, Pergamon Press.

WARKANY, J., E. PASSARGE, and L. B. SMITH, 1966. Congenital Malformations in Autosomal Trisomy Syndromes. *Amer. Jour. Dis. Children,* **6:** 502–517.

WEINSTEIN, D., I. MAUER, and H. M. SOLOMON, 1972. The Effect of Caffeine on Chromosomes of Human Lymphocytes. In Vivo and in Vitro Studies. *Mutation Res.,* **16:** 391–399.

WILLIAMS, J. D., R. L. SUMMITT, P. R. MARTENS, and R. A. KIMBRELL, 1975. Familial Down Syndrome due to t(10; 21) Translocation: Evidence That the Down

Phenotype Is Related to Trisomy of a Specific Segment of Chromosome 21. *Amer. Jour. Hum. Genet.*, 27: 478–485.

tetraploids don't produce gametes like (DDD) & (D). since they, unlike trisomies, can divide evenly into (DD) & (DD)

PROBLEMS

13-1. Application of colchicine to a vegetative bud of a homozygous tall diploid tomato plant (*DD*) causes development of a tetraploid branch. What is the genotype of the somatic cells of this branch?

13-2. Flowers are produced on the tetraploid branch of the plant in problem 13-1. What is the genotype of the gametes?

13-3. Pollinating one of the flowers of problem 13-2 with pollen from a diploid dwarf plant produces embryos of what genotype?

13-4. If the plant in problem 13-1 had been heterozygous tall (*Dd*), what would be the genotype of the somatic cells of the tetraploid branch?

difficult **13-5.** Give the gamete genotypes produced on the tetraploid branch of problem 13-4. In what ratio are they produced? *not like regular cross Dd × Dd where D can only combine with the other d from other parent*

13-6. (a) Self-pollinating flowers on one of the tetraploid branches referred to in problem 13-4 would produce embryos with what specific kind of ploidy? (b) What is the probability of a dwarf plant in the progeny? *can't combine with its own d*

13-7. If an autotetraploid plant heterozygous for two pairs of genes, e.g., of genotype *AAaaBBbb*, is self-pollinated, what is the probability of an *aaaabbbb* plant in the progeny?

13-8. If an autotetraploid plant of genotype *AAaaBBbbCCcc* is self-pollinated, what is the probability of an *aaaabbbbcccc* progeny individual?

13-9. If an autotetraploid plant, heterozygous for *n* pairs of genes (i.e., *AAaaBBbbCCcc* . . . and so on for *n* pairs of alleles) is self-pollinated, what mathematical expression would you solve to determine the probability of a completely recessive progeny individual? $\left(\frac{1}{6}\right)^{2n}$

good question **13-10.** Considering only pairs of alleles that show complete dominance, what is the effect of tetraploidy on phenotypic variability? *reduces it since aaaa less than aa will occur oftener*

13-11. A number of species of the birch tree have a somatic chromosome number of 28. The paper birch (*Betula papyrifera*) is reported as occurring with several different chromosome numbers, individuals with the somatic numbers 56, 70, and 84 being known. With regard to chromosome number, how should the 28, 56, 70, and 84 chromosome individuals be designated? *all multiples of 14. 28-diploid*

13-12. The sugar maple (*Acer saccharum*) and the box elder (*Acer negundo*) each have diploid chromosome numbers of 26. Note that they are different species of the same genus. However, hybrids between the two are sterile. What explanation can you offer? *normal pairing impossible*

13-13. Should it be possible to secure a fertile hybrid of the cross sugar maple × box elder? How? *by producing an allotetraploid containing 2 sets each of maple & elder chromosomes*

13-14. Different species of rhododendron have somatic chromosome numbers of 26, 39, 52, 78, 104, and 156. By what means does evolution appear to be taking place in this genus? *polyploidy*

13-15. What appears to be the basic monoploid chromosome number in rhododendrons?

13-16. How many sets are represented in the species having 156 chromosomes?

13-17. Referring back to Table 13-1, what was the sex phenotype of the triploid child reported by (a) Schindler and Mikamo, (b) Butler et al.?

13-18. If polyspermy were demonstrated to occur in human beings, what would be the chromosomal complement of an embryo resulting from polyspermy involving an X and a Y sperm? *don't forget the* **autosomes**

13-19. Suppose the first mitosis in a normal 46,XY zygote were abnormal so that it became a tetraploid cell. (a) What would be the chromosomal complement of that tetraploid cell? (b) Do you find any evidence in this chapter or its references to suggest that a viable child would result?

13-20. In the Jimson weed what gamete ratios are produced by (a) Ppp ♀, (b) Ppp ♂, (c) PPP ♀, (d) PP ♂, (e) PPp ♂? *in trisomies, generally all gametes will have 2 & some 1*

13-21. What is the F_1 phenotypic ratio produced in Jimson weed by crossing (a) purple Ppp ♀ × purple PPp ♂, (b) PPp ♀ × Pp ♂, (c) Ppp ♀ × Ppp ♂?

13-22. "Eyeless" (eyes small or absent) is a recessive character whose gene is located on the small chromosome IV of *Drosophila melanogaster*. Both triplo-IV eggs and triplo-IV sperms are functional. The dominant gene for normal eyes is designated +, the recessive for eyeless *ey*. What is the F_1 phenotypic ratio produced by each of the following crosses: (a) + *ey ey* × + *ey ey*, (b) + + *ey* × + *ey ey*, (c) + + *ey* × + + *ey*? *trisomy*

13-23. Suggest a meiotic configuration at synapsis for a situation where six chromosomes have undergone reciprocal translocations of about the same length.

13-24. In relation to the information in Table 13-4, where would AAA/AA and $AAAA/AAA$ be properly placed? Note that progressively narrower eyes are shown from the top to the bottom of Table 13-4. *Is answer deletion because x is recessive?*

13-25. In *Drosophila*, e is a gene at locus 70.7 on chromosome III; *ee* flies have ebony bodies, much darker than wild-type flies of genotype + + or + e. If the cross + + × *ee* yields a small percentage of ebony flies, but greater than could be accounted for by the known mutation frequency of this gene, (a) what would you suspect as the cause, (b) what would you look for as confirmation?

13-26. Evaluate the probable cytological and genetic effect on human beings of (a) LSD, (b) marijuana.

13-27. What explanation could you offer for the fact that although human trisomics are known for a number of different autosomes, no autosomal monosomics are reported among live births? *missing genes lethal*

13-28. The Jimson weed (*Datura*) has twelve pairs of chromosomes, and twelve different trisomics are known. How would you explain the fact that the only trisomics known in living human beings are those for chromosomes 13, 18, 21, X, and Y? *die before birth* *good question*

13-29. In a hypothetical organism, gene *a* maps genetically three map units from the centromere. Would a cytological map be expected to show this gene farther from, closer to, or the same distance from the centromere? Why?

interference decreases # of actual crossovers thru recombination, so genetic map distance is less

$\frac{1}{2}P \ \frac{1}{6}p \ \frac{1}{6}p \ \frac{1}{6}Pp \ \frac{1}{6}pp \ \frac{1}{6}P$

$\frac{1}{3}P \ \frac{1}{3}P \ \frac{1}{3}p$

$(\frac{2}{3}P : \frac{1}{3}p)(\frac{1}{6}P : \frac{2}{6}p : \frac{2}{6}P : \frac{1}{6}p)$

$\frac{1}{18}PPP \quad \frac{2}{18}pp$

CHAPTER 14

Population Genetics

UP to this point we have been largely concerned with the results of experimental breeding programs from which we have seen certain now familiar genotypic and phenotypic patterns emerge. In producing $1:2:1$ and $3:1$ F_2 ratios, however, note that we began with two homozygous parental strains, such as AA and aa; that is, we introduced the alleles A and a *in equal frequency*. Similarly, in all the other experiments with which we have dealt, gene frequency was intentionally included among the controlled factors. But in natural populations, frequencies of alleles may vary considerably. We have noted, for example, that the gene for polydactyly is dominant, yet the polydactylous phenotype is fairly infrequent among newborn infants even though it is not known to play any part in survival. Apparently the *frequency* of the dominant gene here is lower than that of its recessive allele; they evidently do not exist in the population in the $1:1$ ratio so commonly encountered in the laboratory. But the frequencies of these and other genes might well be expected not to be equal in different populations, and even to differ in the same population at different times. Population genetics deals with the analysis of changes in gene frequencies in a population over time.

Just as gene frequency is controlled in the genetics laboratory, so is the mating pattern, generation after generation. But outside the laboratory, mating is often largely a chance or random affair. In natural populations, as a result, an ultimate **equilibrium frequency** is attained by alleles, governed by such factors as

1. **Breakdown of isolating mechanisms** whereby additional numbers of one allele or the other are introduced from outside the group.
2. **Frequency of mutation** (change in the nature of the gene).
3. **Selection** (environmental and reproductive).
4. **Random genetic drift.**

The ultimate equilibrium frequency, which approximates the binomial distribution, is then maintained, subject only to random genetic drift, until and unless admixture from outside the population occurs, mutation frequency changes, and/or selection forces change.

In 1908, G. H. Hardy, a British mathematician, and the German physician W. Weinberg independently developed a relatively simple mathematical concept, now referred to as the Hardy-Weinberg theorem, to describe this genetic equilibrium. This principle is the foundation of population genetics. In essence it states that *in the absence of migration, mutation, and selection,* gene frequencies and genotypic frequencies remain constant, generation after

304

generation, in a large, randomly mating population. The Hardy-Weinberg theorem may be used to determine the frequency of each allele of a pair or of a series, as well as frequencies of homozygotes and heterozygotes in the population. In this chapter we shall examine applications of this theorem and the forces that alter gene frequencies.

Calculating Gene Frequency

Codominance. The M-N blood type furnishes a useful example of a series of phenotypes due to a pair of codominant genes. None of the three possible phenotypes, M, MN, and N, appears to have any selection value; in fact, most persons do not know their M-N type. We shall calculate frequencies of the two alleles involved for samples from two different groups.

On page 158 a study of M-N types for 6,129 white Americans living in New York City, Boston, and Columbus, Ohio, was cited. Because the closely related alleles S and s (see page 158) were not included in the data, we can group these individuals according to genotypes on the M-N system alone:

Genotype	Number
MM	1,787
MN	3,039
NN	1,303
	6,129

To calculate frequencies of the two alleles M and N, remember that these 6,129 persons possess a total of $6,129 \times 2 = 12,258$ alleles. The number of M alleles, for example, is $1,787 + 1,787 + 3,039$. Thus calculation of the frequencies for M and N may be worked out in this way:

$$(1) \quad M = \frac{1,787 + 1,787 + 3,039}{12,258} = \frac{6,613}{12,258} = 0.5395$$

$$(2) \quad N = \frac{1,303 + 1,303 + 3,039}{12,258} = \frac{5,645}{12,258} = 0.4605$$

So frequencies of the two alleles in this sample are almost equal, and this is reflected in the close approximation to a $1:2:1$ ratio we so often see in laboratory results.

Gene frequencies expressed as decimals may be used directly to state probabilities. If we can assume this sample to be representative of the population, then there is a probability of 0.5395 that of the chromosomes bearing this pair of alleles, any one selected randomly will bear gene M, and 0.4605 that it will bear N. Thus the frequencies of the three phenotypes to be expected in the population are as follows:

Genotype	Phenotype	Phenotypic Frequency
MM	M	$0.5395 \times 0.5395 = 0.2911$
$\left.\begin{array}{l} MN \\ NM \end{array}\right\}$	MN	$2(0.5395 \times 0.4605) = 0.4968$
NN	N	$0.4605 \times 0.4605 = 0.2121$
		1.0000

Notice that these genotypic frequencies follow a binomial distribution. If we let p represent the frequency of M, and q that of N, then the total array of probabilities of the three genotypes (MM, MN, and NN) is

$$p^2 + 2pq + q^2$$

which is the expansion of $(p + q)^2$. Note also that $p + q = 1$ and, of course, so does $(p + q)^2$. Thus the probability of a type-M individual, for example, under a system of random mating is given by the expression p^2, where we have already calculated the value of p to be 0.5395. Similarly, probability of MN persons (the heterozygotes) is given by $2pq$, and of N persons by q^2.

Alternatively, we could calculate the frequencies of M and N by another approach. Recall that an individual of genotype NN, for example, represents the simultaneous occurrence of two events of equal probability, namely the fusion of two gametes, each with the genotype N. Thus q^2 has the value $1{,}303/6{,}129 = 0.2126$, and $q = \sqrt{0.2126}$, or 0.46. Because $p + q = 1$, $p = 1 - q$, or $1 - 0.46 = 0.54$.

For comparison let us look briefly at a sample of 361 Navaho Indians from New Mexico:

Phenotype	Number
M	305
MN	52
N	4
	361

This sample is far from a laboratory-type $1:2:1$ ratio. Does it mean that the Navahos do not conform to the same genetic laws as the previous sample? The answer is "no" as soon as it is recalled that the $1:2:1$ proportion is based on an equal frequency of the alleles in the population. The raw data clearly suggest that M is considerably more frequent in Navaho Indians than is its allele. Let us, then, calculate gene frequencies by the same method as was used for the earlier sample:

$$\text{let } p = \text{frequency of } M = \frac{305 + 305 + 52}{722} = 0.9169$$

$$\text{let } q = \text{frequency of } N = \frac{52 + 4 + 4}{722} = \frac{0.0831}{1.0000}$$

Applying these gene frequencies to the sample, we have:

Genotype	Gene Frequencies	Genotype Probability	Genotypes in Sample	
			Expected	Observed
MM	$p^2 = (0.9169)^2$	0.8407	303.5	305
MN	$2pq = 2(0.9169 \times 0.0831)$	0.1524	55.0	52
NN	$q^2 = (0.0831)^2$	0.0069	2.5	4
		1.0000	361.0	361

A chi-square test of the data for the Navaho sample shows $\chi^2 = 1.071$; for the earlier sample of 6,129 persons, $\chi^2 = 0.0237$. So deviation from expectancies based on the calculated gene frequencies is well below the level of significance in each case. Remember that although there are three phenotypic classes there is but *one* degree of freedom, because only two alleles, *M* and *N*, are involved. Therefore only one phenotypic class can be set at random. For example, in a sample of 400 persons having just 200 *M* alleles and, therefore, 600 *N* alleles, any number of individuals *up to* 100 may be of genotype *MM*. If there are, say, 60 persons of type M, the other two classes are thereby automatically determined:

Phenotype	Number of Persons	Number of Alleles	
		M	*N*
M	60	120	—
MN	80	80	80
N	260	—	520
Totals	400	200	600

Complete Dominance. An interesting phenotypic trait having no known selection value is the ability or inability to taste the chemical phenylthiocarbamide ("PTC," $C_7H_8N_2S$), also called phenylthiourea. This was reported by Fox in 1932, who found a similar situation for several other thiocarbamides. The test is a simple one that can easily be performed by any genetics class.

The usual procedure is to impregnate filter paper with a dilute aqueous solution of PTC (about 0.5 to 1 g per liter), allow it to dry, then place a bit of the treated paper on the tip of the tongue. About 70 per cent of the white American population can taste this substance, generally as very bitter, rarely as sweetish. Although the physiological basis is unknown, tasting ability does depend on a completely dominant gene, which we will designate as T. Thus tasters are $T-$ (i.e., TT or Tt); nontasters are tt.[1]

Of 280 genetics students in the author's classes in one year, 198 were tasters and 82 were nontasters. From such data the frequencies of genes T and t in the sample may be readily calculated. The 82 nontasters (29.29 per cent of the sample) are persons of genotype tt, and in the Hardy-Weinberg theorem may be represented by q^2. Therefore

$$q^2 = 0.2929 \text{ and}$$
$$q = \sqrt{0.2929} = 0.5412$$

which is the frequency of gene t. Because only a pair of alleles are involved, frequencies of the two genes must total 1, i.e., $p + q = 1$, and $p = 1 - q$. Therefore in this example p, the frequency of gene T, equals $1 - 0.5412$, or 0.4588. Frequencies of homozygous and heterozygous tasters may now be computed. Using the binomial expansion $p^2 + 2pq + q^2$, we obtain

$$
\begin{aligned}
p^2 = TT &= (0.4588)^2 &&= 0.2105 \\
2pq = Tt &= 2(0.4588 \times 0.5412) &&= 0.4966 \\
q^2 = tt &= (0.5412)^2 &&= \underline{0.2929} \\
& && 1.0000
\end{aligned}
$$

By testing representative samples of different populations, the frequencies of T and t in those groups may similarly be calculated. We shall examine the significance of different frequencies of the same alleles in different populations in the concluding sections of this chapter.

Multiple Alleles. The binomial $(p + q)^2 = 1$ can be used only when two alleles occur at a particular locus. For cases of multiple alleles we simply add more terms to the expression. Recall that the A–B–O blood groups are determined by a series of three multiple alleles, I^A, I^B, and i, if we neglect the various subtypes. Hence in a gene-frequency analysis, we can here let

$$
\begin{aligned}
p &= \text{frequency of } I^A \\
q &= \text{frequency of } I^B \\
r &= \text{frequency of } i \\
\text{and } p &+ q + r = 1
\end{aligned}
$$

---

[1] Blakeslee and Salmon (1935) report the lowest concentration detectable by tasters to vary from 1:500,000 to 1:5,000, with a mean around 1:80,000. Sensitivity among tasters decreases with age, and females are slightly more sensitive than males.

Thus genotypes in a population under random mating will be given by $(p + q + r)^2$.

To see how this trinomial is applied, consider the following sample of 23,787 persons from Rochester, New York:

Phenotype	Number	Frequency
A	9,943	0.418
B	2,379	0.100
AB	904	0.038
O	10,561	0.444
	23,787	1.000

The frequency of each allele may now be calculated from these data, remembering that we have let p, q, and r represent the frequencies of genes I^A, I^B, and i, respectively. The value of r, that is, the frequency of gene i, is immediately evident from the figures given:

$$r^2 = 0.444, \text{ hence}$$
$$r = \sqrt{0.444} = 0.6663 \ (= \text{frequency of } i)$$

The sum of A and O phenotypes is given by $(p + r)^2 = 0.418 + 0.444 = 0.862$; therefore

$$p + r = \sqrt{0.862} = 0.9284$$

so $p = (p + r) - r = 0.9284 - 0.6663 = 0.2621 \ (= \text{frequency of } I^A)$. Because $p + q + r = 1$, $q = 1 - (p + r) = 1 - 0.9284 = 0.0716 \ (= \text{frequency of } I^B)$; we can now calculate genotypic frequencies as shown in Table 14-1. The probability figures arrived at for this large sample check quite closely with those arrived at for other samples of the general United States population.

TABLE 14-1. Calculation of Genotypic Frequencies for a Sample of 23,787 Persons Living in Rochester, New York

Phenotypes	Genotypes	Genotypic Frequencies	Population Probability Based on Sample	
O	ii	r^2		0.4440
A	$I^A I^A$	p^2	0.0687 ⎫	0.4180
	$I^A i$	$2pr$	0.3493 ⎭	
B	$I^B I^B$	q^2	0.0051 ⎫	0.1005
	$I^B i$	$2qr$	0.0954 ⎭	
AB	$I^A I^B$	$2pq$		0.0375
				1.0000

Samples taken from other races and/or nationalities, however, may show quite different frequencies for these alleles. In a sample of Navaho Indians, the following gene frequencies were obtained in one study:

$$I^A = 0.1448$$
$$I^B = 0.0020$$
$$i = 0.8532$$

If this sample is representative of the Navaho population, the percentage of individuals of each genotype and phenotype may be calculated from these frequencies as shown in Table 14-2. Thus we may either calculate gene frequencies from numbers of each phenotype or compute percentages of the population having each phenotype if we know gene frequencies.

TABLE 14-2. Probabilities of Genotypes and Phenotypes for a Sample of Navaho Indians

Phenotypes	Genotypes	Genotypic Frequencies	Population Probability Based on Sample	
O	ii	r^2	$(0.8532)^2 =$	0.7280
A	$I^A I^A$	p^2	$(0.1448)^2 = 0.020967$	
	$I^A i$	$2pr$	$2(0.1448 \times 0.8532) = 0.247087$	0.2680
B	$I^B I^B$	q^2	$(0.002)^2 = 0.000004$	
	$I^B i$	$2qr$	$2(0.002 \times 0.8532) = 0.003413$	0.0034
AB	$I^A I^B$	$2pq$	$2(0.1448 \times 0.002) =$	0.0006
				1.0000

The two preceding examples clearly suggest that different frequencies of genes I^A, I^B, and i may occur in different populations. That this is true, especially for I^A and I^B, is shown in Figures 14-1 and 14-2.

Sex Linkage. Thus far, in considering gene frequencies, we have dealt exclusively with autosomal genes. The same techniques, with one small modification, may be used in treating sex-linked genes, however. Because human males have only one X chromosome, they cannot reflect a binomial distribution for random combination of pairs of sex-linked genes as do females. Equilibrium distribution of genotypes for a sex-linked trait, where $p + q = 1$, is given by

$$(\male)p + q$$
$$(\female)p^2 + 2pq + q^2$$

Consider, for example, red-green color blindness. This trait is due to a sex-linked recessive, which we may designate r. About 8 per cent of all males are red-green color blind. This tells us at once that q, the frequency of gene r, is 0.08 and p, the frequency of its normal allele, R, 0.92. Thus the frequency of color blind females is expected to be $q^2 = 0.0064$. This is about what is found.

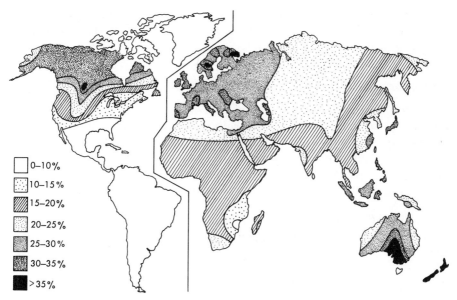

FIGURE 14-1. *Generalized world distribution of* I^A *for native populations.*

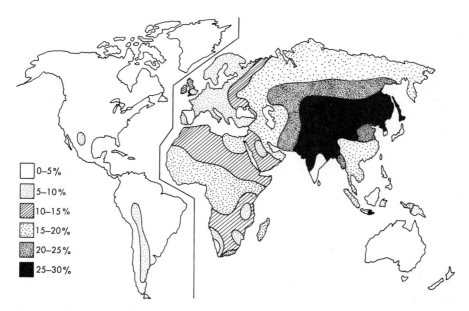

FIGURE 14-2. *Generalized world distribution of* I^B *for native populations.*

Sex-linked dominants may be handled in the same way; in the case of normal color vision, with value of $p = 0.92$, the incidence of normal women is $p^2 + 2pq = 0.9936$.

Factors Affecting Gene Frequency

Isolating Mechanisms. Any mechanism that prevents gene exchange is termed an *isolating mechanism.* Broadly considered, these may be either (1) geographic or physical, such as great distances or mountain or oceanic barriers that keep populations apart, or (2) other mechanisms that effectively prevent gene exchange between populations in the same area. Because geographic separation keeps populations physically isolated, there is a question of whether in all cases such groups would remain reproductively isolated if brought together. In fact, it is clear in many cases that physically isolated populations do exchange genes when the isolation is ended, as may be deduced from Table 14-3.

TABLE 14-3. Percentage of PTC Tasters in Various Human Populations

Population	Place	Sample Size	Per Cent Tasters	Gene Frequencies	
				T	t
Welsh	Five towns	237	58.7	0.36	0.64
Eskimo (unmixed)	Labrador and Baffin	130	59.2	0.36	0.64
Arab	Syria	400	63.5	0.40	0.60
American white	Montana	291	64.6	0.41	0.59
Eskimo (mixed)	Labrador and Baffin	49	69.4	0.45	0.55
American white	Columbus, Ohio	3,643	70.2	0.45	0.55
American students	Delaware, Ohio	280	70.7	0.46	0.54
American black	Alabama	533	76.5	0.52	0.48
Flathead Indians (mixed)	Montana	442	82.6	0.58	0.42
Flathead Indians (unmixed)	Montana	30	90.0	0.68	0.32
American black	Ohio	3,156	90.8	0.70	0.30
African black	Kenya	110	91.9	0.72	0.28
African black	Sudan	805	95.8	0.80	0.20
Navaho Indians	New Mexico	269	98.2	0.87	0.13

Comparison of frequencies for mixed and unmixed Eskimo and Flathead Indian samples shows that gene frequencies in a population may be changed by admixture of genes from other populations. Presumably the "mixing" in these cases is from the western European-American complex where percentage of tasters averages roughly 62–72 per cent with corresponding frequencies of T of 0.384 to 0.471. Note how the frequency of this gene has been increased by admixture in the Eskimo. The same sort of change, though in the opposite direction, has occurred by outbreeding of Flathead Indians where the frequency of T is high (0.683) in the unmixed group and somewhat lower (0.583) in the mixed population. The lower frequency of T in the American black as compared to African blacks has undoubtedly come about in similar fashion. Data for the multiple allele series I^A, I^B, and i, for the Rh alleles D and d, and for the M–N pair show similar population differences (Table 14-4).

TABLE 14-4. Gene Frequencies for Three Blood Group Loci for Selected Populations

Population	I^A	I^B	i	D	d	M	N
U.S. white	0.28	0.08	0.64	0.61	0.39	0.54	0.46
U.S. students*	0.26	0.07	0.67	0.65	0.35	—	—
U.S. black	0.17	0.14	0.69	0.71	0.29	0.48	0.52
West African	0.18	0.16	0.66	0.74	0.26	0.51	0.49
American Indian	0.10	0.00	0.90	1.00	0.00	0.76	0.24

* 1,098 Ohio Wesleyan University students.

Because of the actual or possible exchange of genes between previously separated populations once they are permitted to intermingle, it is often preferred to confine the concept of isolating mechanisms to those that prevent gene exchange between populations occupying the same area. These include one or more of the following: restriction to quite different habitats (especially in plants), reproductive maturation at different times, mating behavioral differences and/or physical incompatibility of genitalia (animals), destruction of sperm (principally in animals) or lack of development of pollen tubes (plants), and/or death of the zygote or of the embryo. In addition, some populations produce sterile hybrids, as in the mule (donkey × horse) and in many horticultural varieties of plants, effectively preventing establishment of new self-perpetuating genetic lines.

In summary, different populations are often characterized by particular gene frequencies, producing phenotypic frequencies that may be expected to fluctuate narrowly around a mean until something occurs to alter gene frequencies. Figure 14-3 indicates how frequencies of homozygotes and heterozygotes are changed by shift in gene frequencies.

On page 304 we noted that in large populations changes in gene frequency may come about not only through alteration of isolating mechanisms but also through mutation and selection. But in populations of finite size, an additional factor comes into play. This is random fluctuation in gene frequency, or **genetic drift.** We should now briefly examine each of these latter three forces as mechanisms of change in gene frequency.

Mutation. Basically, a mutation is a sudden, random alteration in the genotype of an individual. Strictly speaking, it is a change in the genetic mate-

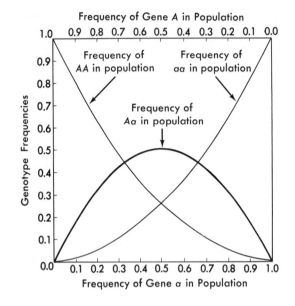

Frequency of Gene A in Population

1.0   0.9   0.8   0.7   0.6   0.5   0.4   0.3   0.2   0.1   0.0

Frequency of AA in population

Frequency of aa in population

Frequency of Aa in population

Genotype Frequencies

Frequency of Gene a in Population

FIGURE 14-3. *The effect of changes in gene frequencies on frequencies of genotypes in a population.*

rial itself, but the term is often loosely extended to include chromosomal aberrations of the sorts already considered (Chapter 13). Its importance in the genetics of populations is to provide new material on which selection can operate as well as to alter gene frequencies.

If, for example, gene T mutates to t, the relative frequencies of the two alleles are changed. If the mutation $T \rightarrow t$ recurs consistently, T could disappear from the population. But not only is mutation recurrent, it is also reversible, with a known frequency in many cases. These *back mutations* will at least slow the otherwise inexorable shift of T to t and perhaps prevent the total disappearance of T. But notice that the mutation $T \rightarrow t$ inevitably would shift gene frequencies over time (1) unless the rate of back mutation ($T \leftarrow t$) equals the rate of forward mutation ($T \rightarrow t$), and (2) only if possession of gene t either gives its bearer an advantage or confers no disadvantage. Thus mutation as a force in altering gene frequencies can scarcely be considered apart from the factor of selection, and we shall consider their *combined* effect later in this chapter.

Equilibrium resulting from mutation is easily derived algebraically. Because the rate of mutation $T \rightarrow t$ generally does not equal the rate of change $T \leftarrow t$, we can show this relationship as

$$T \underset{v}{\overset{u}{\rightleftarrows}} t$$

where u is the rate of mutation $T \rightarrow t$, and the back mutation $T \leftarrow t$ occurs at rate v, with $u \neq v$. Generally $u > v$; the reasons for this will be clear when we have examined the molecular structure of the genetic material and its operation.

trihybrid cross p88

pheno:

$\left(\frac{3}{4}\text{ deep} : \frac{1}{4}\text{ shal}\right)\left(\frac{3}{4}\text{ in} : \frac{1}{4}\text{ reg}\right)\left(\frac{3}{4}\text{ nonwhite} : \frac{1}{4}\text{ white}\right)$

$\left(\frac{9}{16}\,\frac{\text{deep}}{\text{in}} : \frac{3}{16}\,\frac{\text{shal}}{\text{in}} : \frac{3}{16}\,\frac{\text{deep}}{\text{reg}} : \frac{1}{16}\,\frac{\text{shal}}{\text{reg}}\right)\left(\frac{3}{4}\text{ nonwhite} : \frac{1}{4}\text{ white}\right)$

↓

8 phenotypes

(3:1)(3:1)(3:1) } $2^3 = 8$

(9:3:3:1)(3:1)

27:9:9:3:9:3:3:1

8 phenotypes

geno

(1:2:1)(1:2:1)(1:2:1) } $3^3 = 27$

(1:2:1:2:4:2:1:2:1)(1:2:1)

(1:2:1:2:4:2:1:2:1:2:4:2:4:8:4:
2:4:2:1:2:1:2:4:2:1:2:1)

27 genotypes

Letting q represent the frequency of mutating gene t in any one generation and p the frequency[2] of its mutating allele T, the change in frequency of t resulting from mutation will be governed by

1. **The addition of t, as determined by**
   a. the rate of forward mutation (u), and
   b. the frequency of T (p); that is,
   c. to the limit set by up.
2. **The loss of t, as determined by**
   a. the rate of back mutation (v), and
   b. the frequency of t (q); that is,
   c. to the limit set by vq.

Frequencies of T and t will be in equilibrium under mutation when additions and losses balance, i.e., when $up = vq$. The *rate* of change in q under mutation, or Δq_m, can be stated as

$$\Delta q_m = up - vq \tag{1}$$

Thus $\hat{q}$, the equilibrium frequency of t under mutation, will equal the rate $T \rightarrow t$, divided by the sum of the rates $T \rightarrow t$ and $T \leftarrow t$:

$$\hat{q} = \frac{u}{u + v} \tag{2}$$

If, for example, T mutates to t three times as often as t mutates to T, then

$$u = 3v$$

and

$$\hat{q} = \frac{3v}{3v + v} = \tfrac{3}{4}$$

So the population will reach equilibrium with regard to the frequencies of the mutating alleles T and t when $\hat{q}$, the frequency of t, is 0.75, and p, the frequency of T, is $1 - 0.75$, or 0.25.

Complete Selection. The common colon bacillus (*Escherichia coli*) is usually killed by the antibiotic streptomycin; the normal, wild-type phenotype is thus "streptomycin-sensitive." The alleles for this trait are designated str^+ (streptomycin-sensitive) and str (streptomycin-resistant). Because vegetative cells of *E. coli* are monoploid, any given cell has the genotype str^+ or str. In a freely growing culture of this organism, about one in 10 million cells can be shown by appropriate culture techniques to be streptomycin-resistant. Gene str^+ mutates to str with a frequency of about 10^{-7}. Now in its normal streptomycin-free environment, neither str^+ nor str confers any advantage or disadvantage. Hence the frequency of str would reach equilibrium after a

---

[2] Some authors prefer to designate the frequency of the dominant allele as $1 - q$.

number of generations. However, let streptomycin be introduced into the environment, and possession of *str* becomes a distinct advantage; without it an individual does not survive. So with a specific environmental change, *str* individuals survive, *str*+ cells do not. One might say an ordinarily inconsequential gene (*str*) has thus suddenly assumed a very high positive selection value and, at the same time, *str*+ has acquired a strong negative selection value.

Sickle-cell anemia affords an interesting and simple case of selection as a force in maintaining, in certain human populations, a surprisingly high frequency of an otherwise deleterious gene, Hb^S. Briefly, this gene is most frequent in almost precisely those regions where falciparum malaria is prevalent. These areas include principally the Mediterranean basin, central Africa from roughly 10° north latitude to 25° south latitude, the valley of the Nile, Madagascar, and portions of the Arabian peninsula. There is evidence that S-hemoglobin produces conditions unfavorable to the growth of the malarial parasite in the blood. Hence an environment that includes this disease confers some selective advantage on individuals having S-hemoglobin. Homozygous "normal" persons, $Hb^A Hb^A$, do not suffer sickle-cell anemia, of course, but are more susceptible to malaria; $Hb^S Hb^S$ individuals are resistant to malaria but develop the anemia and die before reaching reproductive maturity. Heterozygotes, $Hb^A Hb^S$, have some advantage in respect to both malaria and sickle-cell anemia.

The impact of selection may also be seen in an extreme example from tomato. In this plant several linkage groups contain dominant genes which confer resistance to different strains of the leaf mold fungus (*Cladosporium*). The disease is not ordinarily a problem in outdoor culture, but frequently becomes severe in commercial greenhouses where temperature and humidity may be consistently high enough to cause severe infestation of susceptible plants. In such cases, homozygous recessives, which are susceptible, are quickly killed and may not live to reproductive age. For simplicity, let us consider this situation as resulting from a single pair of alleles, *C* and *c*, with *cc* plants being susceptible to the fungus. As noted, the outdoor environment is generally not conducive to fatal infections in northern parts of the country. We shall assume a population of plants in which the frequencies of *C* and *c* are each 0.5 at the outset; we shall further imagine continuous greenhouse culture in which seed for new plantings is obtained from the crop being grown.

Initially, then, we can represent the parental generation as follows:

Frequency of $C = p = 0.5$; frequency of $c = q = 0.5$
P Genotypes: *CC*   *Cc*   *cc*
Genotypic frequencies: $p^2 + 2pq + q^2$
                            $= 0.25 + 0.50 + 0.25$

If *cc* individuals are unable to reproduce in this environment, the breeding population is reduced to *CC* and *Cc* plants now occurring in the ratio of 1:2:

$$CC: \quad \frac{0.25}{0.25 + 0.50} = 0.33 \text{ (new genotypic frequency)}$$

$$Cc: \quad \frac{0.50}{0.25 + 0.50} = 0.67 \text{ (new genotypic frequency)}$$

This leaves only $C- \times C-$ crosses possible, with consequent changes in phenotypic and genotypic frequencies in the next generation (Table 14-5).

TABLE 14-5

Crosses	Frequencies		F_1 Genotypic Frequencies		
			CC	Cc	cc
$CC \times CC$	$(0.33)^2$	$= 0.11$	0.11		
$CC \times Cc$	$2(0.33 \times 0.67)$	$= 0.44$	0.22	0.22	
$Cc \times Cc$	$(0.67)^2$	$= 0.45$	0.11	0.23	0.11
			0.44	0.45	0.11

The frequency of c has declined in a single generation from 0.5 to 0.33 $(= \sqrt{0.11})$, and C's frequency has risen from 0.5 to 0.67.

In effect, we have been considering *complete selection* against a recessive lethal. We can represent the frequency of such a gene after any given number of generations as

$$q_n = \frac{q_0}{1 + nq_0} \tag{3}$$

where q_n = the frequency of the recessive lethal after n additional generations, and q_0 = the initial frequency of that gene. Thus in our tomato example, where $q_0 = 0.5$ and $n = 1$,

$$q_1 = \frac{0.5}{1 + (1 \times 0.5)} = \frac{0.5}{1.5} = 0.33$$

In a closed population, with no admixture from other groups, only mutation can prevent the frequency of a gene so radically selected against from dropping quickly toward zero. The curve for complete selection against a recessive gene, however, is hyperbolic, with the final slope being determined less by mutation than by the fact that the deleterious gene persists in relatively rare heterozygotes that more frequently mate with homozygous dominant individuals. The number of additional generations (n) required to reduce q from a value of q_0 to some desired value, q_n, is given by the equation

$$n = \frac{1}{q_n} - \frac{1}{q_0} \tag{4}$$

Thus to reduce q from 0.5 ($=q_0$) to 0.33 ($=q_n$), one additional generation is required:

$$n = \frac{1}{0.33} - \frac{1}{0.5} = 3 - 2 = 1$$

With this equation the frequency of a recessive lethal for any number of additional generations can be calculated readily, as shown in Table 14-6.

TABLE 14-6. Reduction in Frequency (q) of a Recessive Lethal by Generations

Generation	q
1	0.500
2	0.333
3	0.250
4	0.200
5	0.167
6	0.143
7	0.125
8	0.111
9	0.100
10	0.091
50	0.020
100	0.010
1,000	0.001

Note that the frequency of the lethal is reduced by half in three generations, but four more are required to reduce it by half again, and so on. Furthermore, application of equation (4) shows that the number of generations required to halve the frequency of such a gene is very large when the initial frequency is quite low. On the other hand, had the lethal been a dominant instead of a recessive, it would have been eliminated in one generation in the absence of complicating influences.

Partial Selection. More often the genotypes involved differ much less in their relative advantage and disadvantage, and the homozygous recessive, for example, will be eliminated much more slowly than in the illustration from tomato. For instance, assume genotype $A-$ produces 100 offspring, all of which reach reproductive maturity in a given environment, whereas genotype aa produces only 80 that do so. We may designate the proportion of the progeny of one genotype surviving to maturity (relative to that of another genotype) as its fitness or *adaptive value, W*. Thus in this case W_{A-} may be set arbitrarily at 1; W_{aa} is then 0.8. The measure of reduced fitness of a given genotype is referred to as its **selection coefficient, s.** We can express the relationship between adaptive value (W) and the selection coefficient (s) as

$$W = 1 - s, \quad \text{or} \quad s = 1 - W$$

In the example we are considering, $s = 0$ for genotype $A-$, and $1 - 0.8 = 0.2$ for aa.

If we let $p =$ the frequency of gene A and $q =$ the frequency of its recessive allele a, selection against the latter is as shown in Table 14-7.

TABLE 14-7. Selection Against a Recessive Allele

	AA	Aa	aa	Total
Initial frequency	p^2	$2pq$	q^2	1
Adaptive value (W)	1	1	$1 - s$	
Frequency after selection	p^2	$2pq$	$q^2(1 - s)$	$p^2 + 2pq + q^2 - sq^2 = 1 - sq^2$

By the same method we have used to calculate frequencies of genes T and t, for example, in the Hardy-Weinberg equilibrium (page 304), the information in Table 14-7 gives us an equation for determining the frequency of gene a when we know the value of W_{aa}:

$$q_1 = \frac{q_0 - sq_0^2}{1 - sq_0^2} \tag{5}$$

where q_0 again is the frequency of a in a given generation, q_1 its frequency one generation later under selection, and s the selection coefficient.

If, for instance, $q = 0.5$ and $s = 0.2$, then substituting in equation (5), $q_1 = 0.4737$. So with *partial selection* against a fully recessive gene, the decrease in its frequency is much less per generation than for complete selection against a recessive lethal. We can represent the change in frequency of gene a under selection (Δq, or $q_1 - q_o$) by the equation

$$\Delta q = \frac{-sq_0^2 p}{1 - sq_0^2} \tag{6}$$

Upon substituting here, $\Delta q = -0.0263$. If q_o were very small, as it would ordinarily be in the case of deleterious mutations, the quantity Δq becomes almost equal to $-sq_o^2$, as we shall see in the next section.

In time a large and randomly mating population attains a genetic equilibrium that is the resultant between selection forces and rate of mutation. This is apparent in many different species of animals, including man, as well as plants. But once a population ceases to be isolated, gene frequencies may begin to change if genetic material is contributed by other populations.

Not all populations, however, are randomly mating. In populations of self-fertilizing plants, for example, given genotypes in effect mate only with like genotypes. This is **complete positive genotypic assortative mating** and results in rapid increase in frequencies of homozygotes. For instance, assume a population where genotypic frequencies in generation 0 are 0.25 AA + 0.5 Aa + 0.25 aa, i.e., $p_o = q_o = 0.5$. With complete positive genotypic assortative mating, progeny frequencies in generation 1 can be calculated as follows:

Mating	Proportions of Matings	Progeny Genotypes		
		AA	Aa	aa
$AA \times AA$	0.25	0.25		
$Aa \times Aa$	0.5	0.125	0.25	0.125
$aa \times aa$	0.25			0.25
Total	1.00	0.375	0.25	0.375

Similarly, progeny in successive generations would occur with the frequencies shown in Table 14-8.

TABLE 14-8. Progeny Genotypic Frequencies Under Complete Positive Genotypic Assortative Mating, Where $p_o = q_o = 0.5$

Generation	Progeny Genotypes		
	AA	Aa	aa
0	0.25	0.5	0.25
1	0.375	0.25	0.375
2	0.425	0.125	0.425
3	0.4961	0.0078	0.4961
4	0.49905	0.00195	0.49905
5	0.4995125	0.000975	0.4995125
n	$P_o - \dfrac{2pq}{2}$	$2p_o q_o (\frac{1}{2})^n$	$q_o - \dfrac{2pq}{2}$

With continued inbreeding, frequencies of homozygotes approach the values of p_o and q_o, respectively. Heterozygotes become so low in frequency that they may easily be eliminated, e.g., by a late freeze or similar minor event. Should this happen, the frequencies of AA individuals would then equal p_o, and the frequency of aa individuals would equal q_o. A similar, but less precipitous, decrease in frequency of heterozygotes will occur in populations where like genotypes, although not "forced" to mate with each other, may mate preferentially (**partial positive genotypic assortative mating**).

On the other hand, mating may occur preferentially (or even only) between *unlike* genotypes (outbreeding). This is **negative genotypic assortative mating,** and may, of course, be either complete or partial. Complete negative genotypic assortative mating results in an increase in the frequency of heterozygotes.

Similarly, *phenotypic* assortative mating may occur. **Positive phenotypic assortative mating** consists in mating of like phenotypes, i.e., $A-$ × $A-$ and aa × $aa;$ **negative phenotypic assortative mating** involves mating of unlike phenotypes, i.e., $A-$ × aa. The effect of the former is to increase homozygosity, of the latter to increase heterozygosity.

Combined Effect of Mutation and Selection. Infants normally exhibit a "startle response," in which they stiffen their arms and legs upon hearing a sudden noise. Continuation of this reaction beyond about the age of four to six months may be symptomatic of Tay-Sachs disorder (infantile amaurotic idiocy), which is due to a recessive autosomal gene. Homozygous recessives fail to produce the enzyme hexosaminidase A, with the result that a lipid, ganglioside GM_2, accumulates in the brain. Mental and motor deterioration follow rapidly, accompanied by paralysis and degeneration of the retina, leading to blindness; the culmination is death, generally before the age of four. This gene has a much higher frequency in Jewish persons of middle and northern Europe (Ashkenazi Jews) than in other groups. Its frequency in the Jewish population of New York City is about 0.015, but approximately 0.0015 in non-Jewish individuals.

The selection coefficient for the Tay-Sachs gene is, of course, 1, inasmuch as homozygotes die in infancy or very early childhood. A reasonable estimate of forward mutation (u) appears to be 1×10^{-6}, although some reports range as high as 1.1×10^{-5}. Because of its lethality, the rate of back mutation (v) is effectively zero. In summary, then, for the Jewish population of New York City

$$q = 0.015 = 1.5 \times 10^{-2}$$
$$p = 0.985 = 9.85 \times 10^{-1}$$
$$u = 0.000001 = 1 \times 10^{-6}$$
$$v = 0$$
$$s = 1$$

What, then, is the equilibrium frequency of this recessive lethal in the Jewish population of New York City under the *combined* effect of mutation and selection?

Recall (page 315) that the rate of change in frequency of a recessive gene under mutation is given by

$$\Delta q_m = up - vq$$

Substituting,

$$\Delta q_m = (1 \times 10^{-6} \times 9.85 \times 10^{-1}) - (0 \times 1.5 \times 10^{-2})$$
$$= (1 \times 10^{-6} \times 9.85 \times 10^{-1}) - 0$$
$$= up = 9.85 \times 10^{-7}$$

This last figure is approximately equal to u, because $0.000000985 \cong 0.000001$. Hence

$$\Delta q_m \cong u \tag{7}$$

Turning for a moment to frequency change under selection, we use equation (6) from page 319:

$$\Delta q_s = \frac{-sq_o^2 p}{1 - sq_o^2}$$

Substituting in this equation,

$$\Delta q_s = \frac{-(0.015)^2 \times 1 \times 0.985}{1 - 1(0.015)^2}$$

$$= \frac{-2.25 \times 10^{-4} \times 9.85 \times 10^{-1}}{1 - 2.25 \times 10^{-4}}$$

$$\equiv \frac{-2.216 \times 10^{-4}}{9.998 \times 10^{-1}}$$

$$= -2.22 \times 10^{-4}$$

So

$$\Delta q_s \cong -sq_o^2 \tag{8}$$

since $-2.22 \times 10^{-4} \cong -2.25 \times 10^{-4}$.

Thus $\Delta q_m \cong u$, $\Delta q_s \cong -sq^2$, and $\Delta q = \Delta q_s + \Delta q_m = -sq^2 + u$. But, by definition, at equilibrium, $\Delta q = 0$. Therefore at equilibrium,

$$u - sq^2 = 0$$

and

$$u = sq^2$$

Rewriting this latter equation to solve for q^2,

$$q^2 = \frac{u}{s}$$

and therefore

$$\hat{q} = \sqrt{\frac{u}{s}} \tag{9}$$

Substituting in equation (9) for the Tay-Sachs problem,

$$\hat{q} = \sqrt{\frac{1 \times 10^{-6}}{1}} = 1 \times 10^{-3}$$

Hence in the population under consideration here the equilibrium frequency of the Tay-Sachs gene under the combined effect of mutation and selection, assuming the values we have used, is 0.001. The frequency of Tay-Sachs births in this group, then, at equilibrium would be 1×10^{-6}, i.e., $(1 \times 10^{-3})^2$, as compared with the present frequency of 2.25×10^{-4}, or $(0.015)^2$.

Random Genetic Drift. From generation to generation the number of individuals carrying a particular allele, either in the homozygous or heterozygous state, may be expected to vary somewhat so that gene frequencies will fluctuate about a mean. The amplitude of this fluctuation is *random genetic drift* and is due to the vagaries of chance mating and to the fact that, even in cases where $p = q = 0.5$, theoretical ratios (e.g., $3:1$, $1:1$, $1:2:1$) are certainly not always produced. If a population is large, drift is nondirectional and of small magnitude; it may be expected to vary within rather narrow limits above and below the mean. But in small breeding populations, all of the progeny might, by chance alone, be of the same genotype with respect to a particular pair of alleles. If this should happen, *fixation* would have occurred at that locus; that is, either $p = 1.00$ or $q = 1.00$. Fixation, if it does occur, may do so in one or any number of generations, but the important fact is that, in moving toward fixation, gene frequencies drift.

The formation of a new population by migration of a sample of individuals may likewise lead to different gene frequencies. Imagine a large population in which $p = 0.4$ and $q = 0.6$. The most probable values of p and q, therefore, in a *sample* from this population are also 0.4 and 0.6, respectively. Expected deviation from these values is, of course, provided by the standard deviation. The standard deviation for a simple proportionality, such as heads versus tails, is given by

$$s = \sqrt{\frac{pq}{n}} \qquad (10)$$

where n is the number of observations. But when gene frequencies are calculated from the frequency of homozygous recessive phenotypes, as we have been doing, the formula for standard deviation becomes

$$s = \sqrt{\frac{pq}{2N}} \qquad (11)$$

where N is the number of (diploid) individuals in the sample.

If, in the large population under consideration, we take a sample of 50,000 persons,

$$s = \sqrt{\frac{0.24}{100,000}} = 0.00155$$

That is, in any sample of 50,000 individuals from this population with $p = 0.4$ and $q = 0.6$, 68 per cent of the time p will lie between 0.39845 and 0.40155, i.e., 0.4 ± 0.00155; 95 per cent of the time it will fall within the range 0.4 ± 0.0031, or 0.3969 to 0.4031. But if the sample were to consist of only 50 persons,

$$s = \sqrt{\frac{0.24}{100}} = 0.049$$

In 68 per cent of samples of this size, p would be expected to fall within the range 0.4 ± 0.049, or between 0.351 and 0.449. Similarly, in 95 per cent of the cases of samples of 50, p would be within the range of 0.302 to 0.498. Formation of a new population by emigration of a sample as small as 50 might be expected to lead purely by chance to a very different gene frequency in the next generation.

A presumed example of just such a case of genetic drift has been reported for the Dunkers of Pennsylvania. These are members of a religious sect who migrated from Germany in the early eighteenth century and have remained relatively isolated. The frequency of blood group A in this small group is almost 0.6, whereas it is between 0.40 and 0.45 in German and American populations, and the I^B allele is nearly absent in the Dunkers, whereas group B persons comprise 10 to 15 per cent of German and American populations.

The Mechanism of Evolution

The principles we have been examining, especially in this and the preceding chapter, constitute the basic mechanism whereby new species evolve, sometimes slowly, sometimes suddenly, from pre-existing ones. A species is more a taxonomic concept than a concrete entity, and its parameters necessarily differ somewhat among various groups of animals and plants. Thus a species in bacteria is quite a different concept from one in birds. Species "boundaries" in viruses and vertebrates have different emphases, as they do even in more closely related units like freely hybridizing, genetically plastic willows versus the more stable maples. However, in very general terms, speciation, or the origin of new species, depends largely on such cytological and genetic mechanisms as these that have been described:

1. Chromosomal aberrations, especially translocations and allopolyploidy (in plants).
2. Addition of new genetic material by mutation (although most are either neutral or disadvantageous in an existing environment).
3. Changes in gene frequencies:
   a. By mutation.
   b. Through random drift.
   c. By migration or the breaking down of isolating barriers.
   d. By environmental selection.

These agencies may sooner or later break up larger populations into smaller units that (1) develop genetically determined, distinctive morphological and/or physiological characters differing from those of other such units, and (2) become reproductively isolated from related groups, developing thereby into "Mendelian populations" having their own gene pools. How, in what direction(s) and to what extent, evolution thus directed will develop depends on an intricate interaction among a variety of influences. But basically only those changes in the genetic material that can be passed on to succeeding generations and do not disappear from the gene pool can serve as agents of evolution in living organisms.

REFERENCES

BLAKESLEE, A. F., and T. N. SALMON, 1935. Genetics of Sensory Thresholds: Individual Taste Reactions for Different Substances. *Proc. Nat. Acad. Sci. (U.S.)*, **21:** 84–90.

FOX, A. L., 1932. The Relationship Between Chemical Composition and Taste. *Proc. Nat. Acad. Sci. (U.S.)*, **18:** 115–120.

GLASS, H. B., M. S. SACKS, E. F. JAHN, and C. HESS, 1952. Genetic Drift in a Religious Isolate: An Analysis of the Causes of Variation in Blood Group and Other Gene Frequencies in a Small Population. *American Naturalist,* **86:** 145–159.

LEWONTIN, R. C., 1967. Population Genetics. In H. L. Roman, ed. *Annual Review of Genetics,* volume 1. Palo Alto, Calif., Annual Reviews, Inc.

LEWONTIN, R. C., 1973. Population Genetics. In H. L. Roman, ed. *Annual Review of Genetics,* volume 7. Palo Alto, Calif., Annual Reviews, Inc.

MORTON, N. E., 1969. Human Population Structure. In H. L. Roman, ed. *Annual Review of Genetics,* volume 3. Palo Alto, Calif., Annual Reviews, Inc.

REED, T. E., 1969. Caucasian Genes in American Negroes. *Science,* **165:** 762–768. Reprinted in L. Levine, ed. *Papers on Genetics.* St. Louis, C. V. Mosby.

SPIESS, E. B., ed., 1962. *Papers on Animal Population Genetics.* Boston, Little, Brown.

SPIESS, E. B., 1968. Experimental Population Genetics. In H. L. Roman, ed. *Annual Review of Genetics,* volume 2. Palo Alto, Calif., Annual Reviews, Inc.

STERN, C., 1943. The Hardy-Weinberg Law. *Science,* **97:** 137–138. Reprinted in L. Levine, ed., 1971. *Papers on Genetics.* St. Louis, C. V. Mosby.

WHITE, M. J. D., 1969. Chromosomal Rearrangements and Speciation in Animals. In H. L. Roman, ed. *Annual Review of Genetics,* volume 3. Palo Alto, Calif., Annual Reviews, Inc.

PROBLEMS

14-1. Considering the data set forth in this chapter for M, MN, and N persons in this country, and assuming you do not know the genotype of either yourself or your parents: (a) What genotype are you most likely to have? (b) What genotype are you least likely to have?

14-2. A sample of 1,000 persons tested for M-N blood antigens was found to be distributed: M, 360; MN, 480; N, 160. What is the frequency of genes *M* and *N?*

14-3. A sample of 1,522 persons living in London disclosed 464 of type M, 733 of type MN, and 325 of type N. Calculate gene frequencies of *M* and *N*.

14-4. A sample of 200 persons from Papua (southeast New Guinea) showed 14 M, 48 MN, and 138 N. Calculate gene frequencies for *M* and *N*.

14-5. A sample of 100 persons disclosed 84 PTC tasters. Calculate gene frequencies for *T* and for *t*.

14-6. (a) How many heterozygotes should there be in the sample of problem 14-5? (b) How many *TT* persons?

14-7. Albinism is the phenotypic expression of a homozygous recessive genotype. One source estimates the frequency of albinos in the American population as 1 in 20,000. What percentage of the population is heterozygous for this gene?

14-8. Alcaptonuria, resulting from the homozygous expression of a recessive autosomal gene, occurs in about 1 in 1 million persons. What is the proportion of heterozygous "carriers" in the population?

14-9. A sample of 1,000 hypothetical persons in the United States showed the following distribution of blood groups: A, 450; B, 130; AB, 60; O, 360. Calculate the frequencies of genes I^A, I^B, and *i*.

14-10. Another sample of 1,000 hypothetical persons had these blood groups: A, 320; B, 150; AB, 40; O, 490. What is the frequency in this sample of each of the following genotypes: $I^A I^A$, $I^A i$, $I^B I^B$, $I^B i$, $I^A I^B$, *ii*?

14-11. Assume the data of Table 14-1 to represent accurately the distribution of A, B, AB, and O blood groups in the United States population. Using the χ^2 test, determine whether the sample of problem 14-8 represents a significant deviation. (Round the frequencies of Table 14-1 to the whole numbers 440, 420, 100, and 40, respectively, to obtain calculated values for computing chi-square.)

14-12. A sample of 429 Puerto Ricans showed the following gene frequencies: I^A, 0.24; I^B, 0.06; *i*, 0.70. Calculate the percentage of persons in this sample with A, B, AB, and O blood.

14-13. What percentage of the sample is (a) homozygous A, (b) heterozygous B?

14-14. If 1 man in 25,000 is a hemophile, what is the frequency of gene *h* in the population?

14-15. One man in 100 exhibits a trait resulting from a certain sex-linked recessive gene. What is the frequency of (a) heterozygous women, (b) homozygous recessive women?

14-16. Data presented in Table 14-4 show a frequency of 0.64 for gene *i*, and 0.61 for the Rh allele *D*. Considering only these two genes, what would be the frequency of O+ persons in the population?

14-17. Considerable progress is being made in the eradication of malaria. If malaria is eventually eliminated, what effect will this be likely to have on the frequency of gene Hb^S?

14-18. Great effort is being made to eliminate the muscular dystrophy caused by gene *d*. If we are someday successful in eliminating the *effect* of this gene in *treated* individuals, will this result in genetic improvement or deterioration in the world population?

14-19. Considering the tabulated data (text) concerning frequency of gene *c* in tomato, what would the frequency of this gene be after one more generation of inbreeding?

14-20. (a) Considering the tabulated data (text) concerning frequency of gene *c* in tomato, what would be the frequency of this gene in the twentieth generation of inbreeding? (b) Considering the parental generation as generation 1, in what generation would the frequency of c be reduced to exactly 0.005?

14-21. Gene f is an autosomal recessive lethal that kills ff individuals before they reach reproductive age. If, in an isolated population, this gene had a frequency of 0.4 in the P generation, what would be its frequency in the F_2?

14-22. For a certain pair of alleles, completely dominant gene A has an initial frequency of 0.7 and an adaptive value of 1. Its recessive allele has a frequency of 0.3 and an adaptive value of 0.5. What is the frequency of a in the next generation?

14-23. Gene A mutates to its recessive allele a four times more frequently than a mutates to A. What will be the equilibrium frequency of a under mutation?

14-24. If gene B has a forward mutation rate of 5×10^{-6} and a back mutation rate of 1×10^{-6}, what is the equilibrium frequency of gene b under mutation?

14-25. In a given population gene C has a frequency of 0.2. If 50 individuals from that population are sampled, there is a 0.68 probability that the frequency of C will be no more than how much above or below 0.2 in that sample?

14-26. Persons with cystic fibrosis of the pancreas occur with a frequency of about 0.0004. This condition is due to a recessive autosomal gene and is fatal in childhood. (a) Give the frequency of this gene in the present population. (b) If the rate of forward mutation is assumed to be 4×10^{-6}, what is the equilibrium frequency of the gene for cystic fibrosis under the combined effect of mutation and selection?

14-27. Based on your answer to 14-26(a), and assuming a human generation to be 30 years, (a) how many years would be required to reduce this frequency to 0.01? (b) Is this likely to occur? Why?

14-28. Let p represent the frequency of dominant autosomal gene A and q the frequency of its recessive allele a. For a randomly mating population, give a mathematical expression for (a) the frequency of an AA individual, (b) the probability of an $AA \times AA$ mating, (c) the frequency of an Aa individual, (d) the probability of an $Aa \times Aa$ mating, (e) the probability of an $Aa \times aa$ mating, (f) the total of all possible matings.

14-29. Give a mathematical expression for the frequency of (a) dominant and (b) recessive progeny phenotypes from a *single* $Aa \times Aa$ mating.

14-30. Considering the *entire* randomly mating population, with all possible matings equally free to occur, give (a) a mathematical expression for the frequency of recessive progeny in the population resulting from such random matings and (b) the numerical value of this expression if $p = q = 0.5$.

14-31. Using the information in Table 14-8, calculate the frequency of heterozygotes in generation 10.

14-32. In view of your answer to the preceding question, what would be the frequency of each of the two homozygous genotypes in generation 10?

14-33. Assuming $p_o = q_o = 0.5$ in generation 0, calculate the frequency of heterozygotes in generation 1 under complete negative genotypic assortative mating.

14-34. Phenylketonuria (PKU) is a condition characterized by such low intelligence that persons not diagnosed at birth and put on immediate treatment never reproduce. This condition is produced by an inherited inability to metabolize the essential amino acid phenylalanine and is caused by a completely recessive autosomal gene. PKUs have an incidence of about 3.6×10^{-5}. (a) Assuming lack of diagnosis in time to treat the condition effectively, what is the frequency of the PKU gene at present? (b) What is the incidence of homozygous normal persons? (c) What is the incidence of heterozygotes? (d) If u is assumed to be 4×10^{-6}, what is this gene's equilibrium frequency under the combined effect of mutation and selection? (e) At this value of u, what is the probability of two homozygous normal persons producing a PKU child?

14-35. Assume a certain dominant gene to have a forward mutation rate of 2×10^{-6}, and its recessive allele to have a frequency now of 0.015 and a selection coefficient of 0.5. What is the equilibrium frequency of the recessive gene under the combined effect of mutation and selection?

14-36. As noted in the text, some reports of the rate of forward mutation for the recessive Tay-Sachs gene are as high as 1.1×10^{-5}. Calculate the equilibrium frequency of this gene under the combined effect of mutation and selection using this higher rate of forward mutation.

14-37. Recall from problem 12-19 that length of index finger relative to that of the fourth finger is thought to be due to a sex-influenced autosomal gene, with gene F (shorter index finger) dominant in males. A class of 85 genetics students consisted of 32 males with short index finger, 8 males with long index finger, 27 females with short index finger, and 18 females with long index finger. Based on the hypothesis of a pair of sex-influenced genes, calculate the frequency of (a) gene F in the females and (b) gene f in the males. (c) How many of the males in this sample should be of genotype FF?

CHAPTER 15

The Identification of the Genetic Material

UP to this point we have made a variety of genetic observations in a wide assortment of organisms. We have seen that all these observations can be explained by theorizing that genes occur at specific loci, in linear order, on chromosomes. But there remains an important body of questions to which we must now seek answers:

1. What is *the* genetic material?
2. How does this genetic material operate to produce detectable phenotypic traits, and can its operation explain such phenomena as dominance and recessiveness?
3. In terms of *the* genetic material, what actually is a gene? Can its molecular configuration be determined?
4. In terms of the nature of the genetic material and of the gene, what is mutation? That is, when a gene changes and produces a different effect, what happens to it at the level of its molecular structure?
5. Differentiation in multicellular organisms suggests that not all genes function all the time. If this is so, how are genes regulated?
6. Is this genetic material (are these genes) all located in chromosomes, or does the cytoplasm play any part in inheritance?
7. What of the future? Can answers to such questions as these be used in any way to mitigate or offset the effect of disadvantageous or lethal genes, or even to replace them?

These questions we shall pursue in the remaining chapters; in this one we shall turn our attention to the problem of identifying the genetic material itself. Basically, there are three avenues of approach, dealing primarily with microorganisms. It is much easier to find answers to these questions in bacteria and viruses than it is in rather complex individuals such as ourselves. Interestingly enough, knowledge gained from studies on these simpler forms is found to be perfectly applicable to all the more highly evolved ones. These three approaches are **transformation, transduction,** and **conjugation.**

Bacterial Transformation

The "Griffith Effect." In 1928 Frederick Griffith published a paper in which he cited a number of what were, at that time, remarkable results for which he had no explanation. His observations involved a particular bacterium, *Diplo-*

coccus pneumoniae, which is associated with certain types of pneumonia. This organism occurs in two major forms: (1) *smooth* (S), whose cells secrete a covering capsule of polysaccharide materials, causing its colonies on agar to be smooth and rather shiny, and which is virulent in that it produces bacterial pneumonia in suitable experimental animals, and (2) *rough* (R), cells of which lack a capsule, whose agar colonies have a rough, rather dull surface, and which is nonvirulent. Smooth and rough characters are directly related to the presence or absence of the capsule, and this trait is known to be genetically determined.

The smooth types (S) can be distinguished by their possession of different capsular polysaccharides (designated as I, II, III, IV); the specific polysaccharide is antigenic and genetically controlled. Mutations from smooth to rough occur spontaneously with a frequency of about one cell in 10^7, though the reverse is much less frequent. Mutation of, say, an S-II to rough may occur, and these may rarely revert to smooth. When that happens the smooth revertants are again S-II.

In the course of his work Griffith injected laboratory mice with live R pneumococci derived from an S-II culture; the mice suffered no ill effects. Injection of mice with a living S-III culture (or smooth cultures of any other antigenic type) was fatal, but logically enough, use of heat-killed suspensions of either S or R bacteria did not produce pneumonia. Inoculating the animals with live R-II bacteria (i.e., a rough form derived from S-II) *plus* dead S-III, however, resulted in a high mortality. This surely was an unexpected turn of events. The heat-killed S-III were checked; all indeed were dead. Yet both living R-II and S-III organisms were subcultured following autopsy of the dead mice!

An explanation was not immediately forthcoming. It was as though the killed S-III individuals were somehow restored to life, but this was patently absurd. The then rather recently understood phenomenon of mutation might be implicated, but the frequency of occurrence in the Griffith experiments was far too high to be compatible with known mutation rates. Scientists were left only with the idea that in some way the heat-killed cells conferred virulence on the previously nonvirulent strain; in short, the living cells were somehow *transformed.* So the Griffith effect gradually became known as *transformation* and turned out to be the first major step in the identification of *the* genetic material.

Identification of the "Transforming Substance."   Sixteen years after Griffith's work, Avery, MacLeod, and McCarty (1944) reported successful repetition of the earlier work, but in vitro, and were able to identify the "transforming substance." They tested fractions of heat-killed cells for transforming ability. Completely negative results were obtained with fractions containing only the polysaccharide capsule, various cell proteins, or ribonucleic acid (RNA); only extracts containing deoxyribonucleic acid (DNA) were effective. Even highly purified fractions containing DNA and less than 2 parts protein per 10,000, for example, retained the transforming ability. DNA, plus even minute amounts of protein, plus proteolytic enzymes, was fully effective. But DNA, plus DNAase, an enzyme that destroys DNA, lost its transforming capability.

Therefore it began to appear that *beyond any reasonable doubt, DNA must be the genetic material.* Moreover, this landmark of genetic research indicated also that *the genetic material, DNA, differs from the end product it determines.* Evidence has been since accumulated to show that transformation is of fairly wide occurrence in bacteria and probably occurs in the blue-green algae as well. We shall examine the intracellular mechanism of this process later in this chapter.

Transduction

The clear implication of DNA as the genetic material that was furnished by transformation experiments was extended and confirmed by a series of experiments begun by Zinder and Lederberg (1952) on the mouse typhoid bacterium (*Salmonella typhimurium*). Their experiments involved the process of **transduction,** in which a bacterium-infecting virus (**phage**) serves as the vector transferring DNA from one bacterial cell to another. To understand this process fully we must first examine the phage structure and "life cycle."

Structure of T-Even Phages. There is no typical virus structure, but there are several modifications of a basic structure. In general terms, viruses consist of an outer, inert, nongenetic protein "shell" and an inner "core" of genetic material. In many cases this genetic material is DNA, but in some it is RNA. Perhaps the best-known viruses from a structural standpoint are the so-called T-even phages (e.g., T2, T4) that infect the colon bacillus (*Escherichia coli*).

The T phages are of a general "tadpole" shape, outwardly differentiated into a *head* and a *tail* region. The former is hexagonal in outline, bearing numerous facets, and contains genetic material, in this case DNA. The DNA is a closed, no-end molecule approximately 68 μm in length. The tail is a hollow cylinder that can contract longitudinally. Six spikes and six tail fibers, the latter bent at an angle about midway in their length, arise from a hexagonal plate at the distal end of the tail. Phage T4 is shown in electron micrograph view in Figure 15-1 and in diagrammatic sectional view in Figure 15-2.

Other phages are quite different in morphology. Phage X174, for example, appears in electron micrographs as a cluster of 12 identical, adherent, but morphologically distinct spherical subparticles. Two general types of phage replication cycles can be recognized: (1) *virulent* phages, in which infection is followed by *lysis* (bursting) of the host cell and the release of new, infective phages, and (2) *temperate* phages, in which infection only rarely causes lysis.

Life Cycle of a Virulent Phage. Infection begins with attachment of the phage by its tail to one of numerous receptor sites on the bacterial host (Fig. 15-3); the sheath then contracts, driving the core through the host cell wall. The phage DNA and a small amount of protein then enter the bacterial cell.

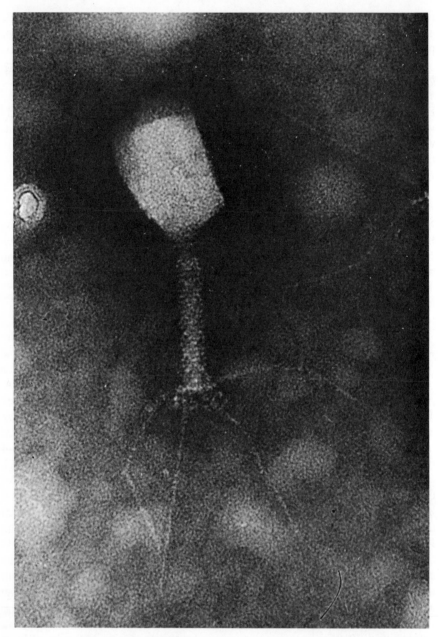

FIGURE 15-1. *Electron micrograph of a T4 bacteriophage,* × *630,000.* (Courtesy Dr. Thomas F. Anderson, Institute for Cancer Research, Philadelphia.)

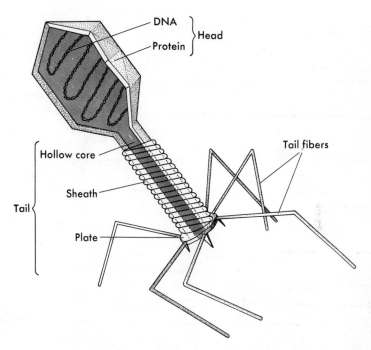

FIGURE 15-2. *Diagrammatic longitudinal section of a T4 bacteriophage. Only a small portion of the total DNA content is shown.*

Determination of immediately succeeding events is facilitated by the fact that the protein outer shell contains sulfur but no phosphorus, whereas DNA, as we shall see shortly in greater detail, contains phosphorus but no sulfur. Using two samples of phage, one containing radioactive ^{35}S and the other radioactive ^{32}P, Hershey and Chase (1952) were able to show that all the phage DNA enters the host cell following attachment; most of the protein remains outside. Details of the Hershey and Chase experiment are diagramed in Figure 15-4. An *eclipse period* ensues, during which the phage DNA replicates numerous times within the bacterial cell. Toward the end of the eclipse period the phage DNA directs the production of protein coats and assembly of some 50 to 200 new, infective viral particles. Within a short time (e.g., 13 minutes for phage T1, and 22 minutes for phage T2), a phage-produced enzyme (lysozyme) brings about lysis of the host cell and release of the mature phages (Fig. 15-3). We can recognize the following steps in the process:

1. *Attachment* of phage tail to specific receptor sites on the bacterial cell wall.
2. *Injection* of phage DNA.
3. *Eclipse period,* in which no infective phage is recoverable if the bacterial

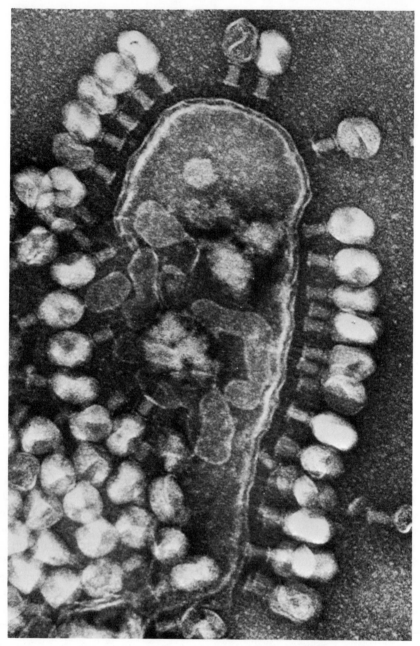

FIGURE 15-3. *Electron micrograph of T4 phage attacking* Escherichia coli. *Not only can phage particles be seen attached by their tail fibers to the cell wall of the bacterium* (top and right), *but new phage particles are shown being released from the lysed bacterial cell* (left). (Photo by Dr. L. D. Simon, courtesy Dr. Thomas F. Anderson, Institute for Cancer Research, Philadelphia.)

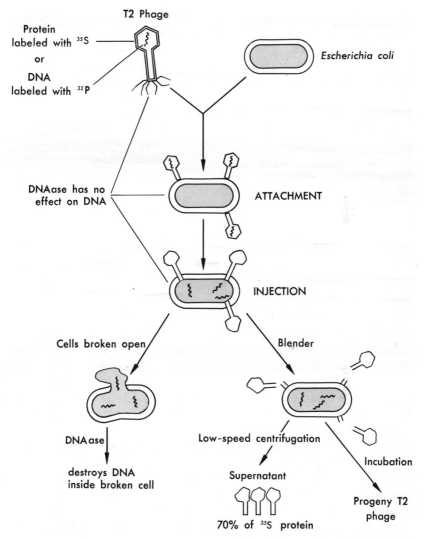

FIGURE 15-4. *Diagram of life cycle of a virulent phage showing details of the Hershey-Chase experiment. Phage particles and bacterial cells are not to scale.* (From Fraser, *Viruses and Molecular Biology*, Macmillan, 1967. By permission.)

cell is artificially lysed and during which synthesis of new phage DNA and protein shells is taking place.

4. *Assembly* of phage DNA into new protein shells.
5. *Lysis* of host cell and release of several hundred infective phage particles.

Life Cycle of a Temperate Phage. Temperate phages do not ordinarily lyse their host; in such cases the phage DNA becomes integrated into the bacterial

DNA as *prophage*. Prophages replicate synchronously with host DNA; progeny bacterial cells then contain this bacterial-plus-phage DNA. Under these conditions both infection by a virulent phage and maturation of infective virus particles are prevented. Infrequently, however, the association between host and phage DNA may be terminated (deintegration); the lytic cycle then follows as in the case of a virulent phage. Because the bacterial hosts in this case are potentially subject to lysis, they are termed *lysogenic.* Phage lambda ($\phi\lambda$) and another, P22, for example, behave as temperate phages. Sometimes deintegration of prophage involves incorporation of a bit of host DNA that can then be transferred to a new bacterial host and be integrated into its DNA, resulting in recombinant progeny cells. This virus-mediated recombination is called **transduction.**

Experiments of Zinder and Lederberg.   In the work of Zinder and Lederberg (1952) several strains of *Salmonella typhimurium* were used. One of these, *met⁻ thr⁺*, was unable to synthesize the necessary amino acid methionine from simple raw materials, that is, it was *auxotrophic* for methionine and required it in the culture medium for growth. It was, however, able to synthesize threonine, another amino acid; that is, it was *prototrophic* for threonine. Another strain was prototrophic for methionine, but auxotrophic for threonine (*met⁺ thr⁻*).

Several of the experiments of Zinder and Lederberg were of critical importance in confirming that DNA is the genetic material and in disclosing a modified method by which it converts a cell from one genotype to another. In each of these experiments, *met⁻ thr⁺* and *met⁺ thr⁻* auxotrophic cultures were first grown separately. In the first experiment the two cultures were simply mixed and plated out on double deficiency medium (i.e., lacking both methionine and threonine); many colonies subsequently appeared (Fig. 15-5). Any colony on such double deficiency agar must be able to produce its own amino acids of both types. The number of such prototrophic colonies was far too great to be accounted for by mutation. But could it be due to conjugation (Appendix B)? Or were these results perhaps to be explained by transformation? Two more experiments answered each of these questions.

Zinder and Lederberg next treated the *met⁺ thr⁻* culture differently before adding it to the *met⁻ thr⁺* cells. The *met⁺ thr⁻* culture was (1) centrifuged to throw down living cells, then (2) the supernatant culture fluid, presumably now cell-free, was heated to kill any cells that might possibly have failed to be sedimented. Upon adding this cell-free, heat-treated supernate to the culture of *met⁻ thr⁺* strain and plating on double deficiency medium, again a very large number of *met⁺ thr⁺* colonies was obtained. Thus because living cells were not required, the origin of the prototrophic colonies is clearly not the result of conjugation (Fig. 15-6).

As a modification of this latter procedure, Zinder and Lederberg added DNAase to the supernate, and again many prototrophs were recovered. Therefore this was not the result of transformation as we saw it in the work of Griffith

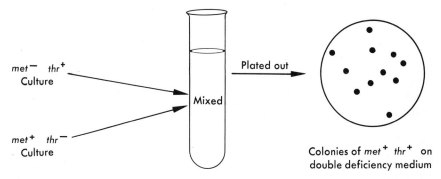

FIGURE 15-5. *Schematic diagram of the first of the Zinder and Lederberg experiments. Two strains, each auxotrophic for a different trait, are grown separately, mixed, then plated out on double deficiency medium. Colonies developing there are prototrophs.*

and of Avery and his colleagues. It was found that the "*met⁺* factor" could pass through filters that hold back bacteria but allow viruses to pass. On the other hand, the reciprocal of this experiment, using filtrate of *met⁻ thr⁺* to convert *met⁺ thr⁻* cells, did not produce prototrophs (Fig. 15-7).

The particular *met⁺ thr⁻* "donor" was found to harbor the temperate phage P22. It was known that the genetic material of *Salmonella* is DNA, so it was considered probable that the *met⁺* "factor" was also composed of this substance. Their use of DNAase convinced Zinder and Lederberg that the *met⁺* "factor" was located inside the virus whose genetic material is, of course, also DNA. Production of recombinant bacterial cells, then, was clearly the result of the process we call transduction.

When we recognize that the material transferred during bacterial conjugation (Chapter 6 and Appendix B) is also DNA, we see that the three processes—conjugation, transformation, and transduction—all implicate DNA as *the* genetic material. Experimental evidence rapidly accumulated to show that in all but the RNA-containing viruses,[1] DNA has this important property, as it does also in all eukaryotes such as man. The essential feature of heredity, then, is the transmission of this "information tape" unchanged from one generation to the next. If we are to identify genes, we must understand the structure of this all-important molecule that has been appropriately called the "thread of life."

DEOXYRIBONUCLEIC ACID

Interestingly enough, recognition of deoxyribonucleic acid as the genetic material was slow in coming. By 1869 Friedrich Miescher, then a 22-year-old Swiss physician, had isolated (by remarkably advanced techniques) from nuclei of pus cells obtained from discarded bandages in the Franco-Prussian

---

[1] For example, influenza, poliomyelitis, and tobacco mosaic viruses.

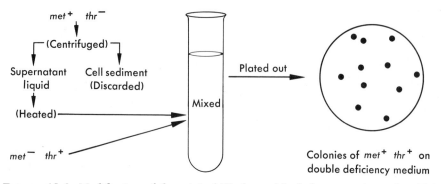

FIGURE 15-6. *Modification of the original Zinder and Lederberg experiment in which only cell-free extract of one auxotroph is added to a culture of the other auxotroph. Again, prototroph colonies develop on double deficiency medium, indicating that their occurrence is not based on conjugation.*

War, and from salmon sperm, a previously unidentified macromolecular substance to which he gave the name *nuclein.* Although he was unaware of the structure and function of nuclein, he submitted his findings for publication. The editor who received the paper was dubious about some aspects of the report and delayed publication for two years while he tried repeating some of what were to him the more questionable aspects of Miescher's work. Finally, in 1871, Miescher's report was published, but made little immediate impact. He continued his careful work up to his death in 1895, recognizing (with the help of his student, Altmann, in 1889) that nuclein was of high molecular weight and was associated in some way with a basic protein, to which he gave the name *protamine.* By 1895 the pioneering cytologist E. B. Wilson suspected that "inheritance . . . may be effected by the physical transmission of a particular chemical compound from parent to offspring."

Nuclein was later renamed *nucleic acid,* the name still used, and work on it continued slowly in several laboratories. Early in this century the biochemist Kossel identified the constitutent nitrogenous bases of nucleic acid, as well as its 5-carbon sugar, and phosphoric acid. Kossel's work and the later investigations of Ascoli, Levene, and Jones during the first quarter of this century disclosed the two kinds of nucleic acid, deoxyribonucleic acid and ribonucleic acid. Development of DNA-specific staining techniques by Feulgen and Rossenbeck in 1924 enabled Feulgen to demonstrate in 1937 that *most* of the DNA content of a cell is located in the nucleus.

Structure.   In 1953 Watson and Crick proposed a molecular model for DNA, based on X-ray diffraction pictures by Wilkins (Fig. 15-8). So completely was their model substantiated by subsequent investigations that this team shared a Nobel Prize in 1962. The basic structure they deduced was that of a very long molecule of high molecular weight and composed of two sugar-phosphate strands oriented in opposite directions and together forming a

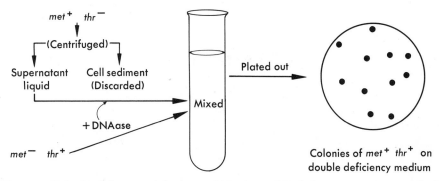

FIGURE 15-7. *Modification of the original Zinder and Lederberg experiment in which DNAase is added to the cell-free extract of one auxotroph. The appearance of prototroph colonies on double deficiency medium suggests that transformation is not occurring.*

double helix making a complete right-handed (clockwise) turn every 34 angstrom units (i.e., 3.4 nm). These sugar-phosphate "backbones" are cross-connected internally by nitrogen-containing bases, two of them of the class of chemicals known as **purines** and two **pyrimidines.** The whole, then, may be likened to a rope ladder that is helically twisted as suggested in Figure 15-9.

The sugar is the 5-carbon **deoxyribose** whose molecular structure is shown in Figure 15-10; these are linked together by **phosphoric acid** bonded to the 3' and 5' carbons of the sugar.[2] The purine and pyrimidine bases are spaced 3.4 angstrom units (0.34 nm) apart, with the result that there are 10 base pairs per complete turn of the sugar-phosphate strands. Each pair of bases is, therefore, oriented 36° clockwise from the preceding pair. The diameter of the molecule is about 20 angstrom units.

Typically, only four different nitrogenous bases occur, the purines **adenine** and **guanine** and the pyrimidines **thymine** and **cytosine.** Their molecular structure, shown in Figure 15-11, is such that adenine is connected by two hydrogen bonds only to thymine, and cytosine by three hydrogen bonds only to guanine:

Purines		Pyrimidines
Adenine	$=$	Thymine
Guanine	$\equiv$	Cytosine

Because pairing occurs in this way, the amounts of adenine and thymine should be equal to each other, as should the amounts of cytosine and guanine. Indeed, Chargaff showed this to be the case (within the limits of experimental

[2] It is customary to prime the positions of carbons in the sugar to distinguish them from carbon positions in the organic bases.

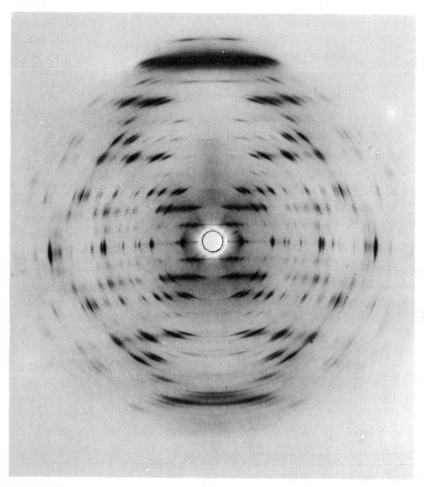

FIGURE 15-8. *X-ray diffraction photograph of the lithium salt of DNA.* (Courtesy Dr. M. H. F. Wilkins and Dr. W. Fuller, Medical Research Council, Biophysics Research Unit, King's College, London.)

error) in 1950. Notice in Table 15-1 that for all sources listed (except phage X174) the ratios A/T and G/C are near 1.

The DNA of phage X174 is single-stranded (except during its replicative phase), hence its values for A/T and G/C depart considerably from unity. In fact, whenever this kind of deviation is found it is indicative of single-strandedness. Note also that values for A + T/C + G vary widely from well below to well above 1; that is, although the relationships A = T and C = G are valid, it is also true that A + T ≠ C + G in most cases. Certainly the structure of DNA does not require equality in that relationship. In fact, *DNA of different species is distinguished by the relative numbers of AT and CG pairs,*

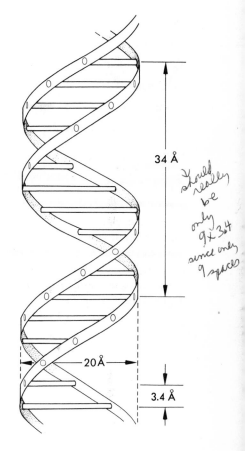

FIGURE 15-9. *The general structure of double-stranded deoxyribonucleic acid. ($1\overset{\circ}{A} = 0.0001 \mu m$ or 0.1 nm)*

34 Å

20Å

3.4 Å

[handwritten margin note: should really be only 9 × 3.4 since only 9 spaces]

TABLE 15-1. Comparison of Nucleotide Composition of DNA

Source	A	T	G	C	$\dfrac{A}{T}$	$\dfrac{G}{C}$	$\dfrac{A + T}{G + C}$
Human sperm	31.0	31.5	19.1	18.4	0.98	1.03	1.67
Salmon sperm	29.7	29.1	20.8	20.4	1.02	1.02	1.43
Euglena nucleus	22.6	24.4	27.7	25.8	0.93	1.07	0.88
Euglena chloroplast	38.2	38.1	12.3	11.3	1.00	1.09	3.23
Escherichia coli	26.1	23.9	24.9	25.1	1.09	0.99	1.00
Mycobacterium tuberculosis	15.1	14.6	34.9	35.4	1.03	0.98	0.42
Phage T2	32.6	32.6	18.2	16.6*	1.00	1.09	1.87
Phage X174	24.7	32.7	24.1	18.5	0.75	1.30	1.35

* 5-hydroxymethyl cytosine.

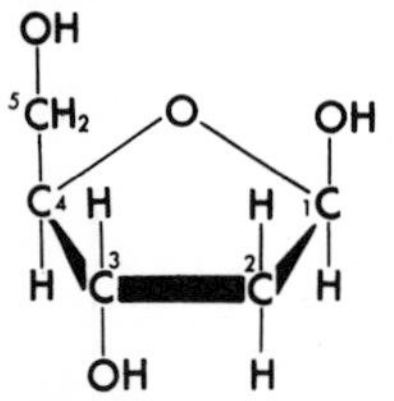

$$^5CH_2 \quad OH$$

FIGURE 15-10. *Molecular structure of deoxyribose, the sugar of DNA.*

their sequence, whether these occur as AT or TA and as CG or GC, and the number of such base pairs (hence the length of the DNA molecules).

Certain useful terms are applied to parts of the DNA molecule. One phosphate and its attached sugar are referred to as **deoxyribose phosphate,** one deoxyribose molecule plus its attached purine or pyrimidine is a **nucleoside** (deoxyribonucleoside or deoxynucleoside) as shown in Figure 15-12, and a purine or a pyrimidine plus one deoxyribose phosphate constitutes a **nucleotide** (deoxyribonucleotide or deoxynucleotide) as shown in Figure 15-13. Deoxyribonucleosides and deoxyribonucleotides have specific names, as shown in Table 15-2.

nucleoside + phosphate = nucleotide

sugar + base = nucleoside

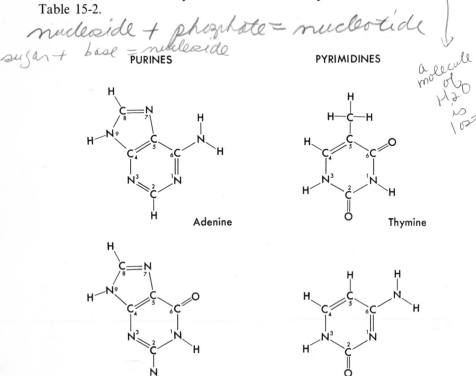

PURINES   PYRIMIDINES

a molecule of H_2O is lost

Adenine

Thymine

Guanine

Cytosine

FIGURE 15-11. *Molecular structure of the four nitrogenous bases of DNA.*

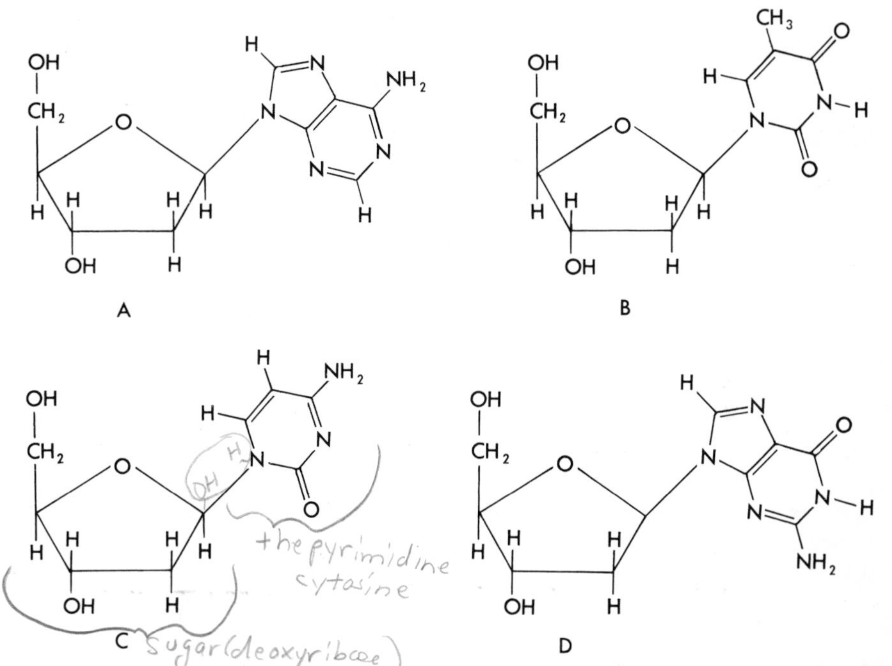

FIGURE 15-12. *The four deoxyribonucleosides.* (*A*) *Deoxyadenosine.* (*B*) *Thymidine.* (*C*) *Deoxycytidine.* (*D*) *Deoxyguanosine.*

[handwritten annotations: "the pyrimidine cytosine", "Sugar (deoxyribose)"]

TABLE 15-2. Deoxyribonucleosides
and Deoxyribonucleotides

Base	Deoxyribonucleoside	Deoxyribonucleotide
Adenine	Deoxyadenosine	Deoxyadenylic acid
Thymine	Thymidine	Thymidylic acid
Cytosine	Deoxycytidine	Deoxycytidylic acid
Guanine	Deoxyguanosine	Deoxyguanylic acid

Techniques involving enzymatic splitting of the DNA molecule have made it possible to determine the exact arrangement of its parts. The nitrogenous bases are linked to deoxyribose at the 1′ carbon (forming nucleosides), and the phosphate is attached to the 5′ carbon (forming nucleotides) as shown in Figures 15-12 and 15-13. Successive nucleotides are linked by 3′, 5′ phosphodiester bonds as shown in the molecular model of a segment of DNA consisting of four deoxyribonucleotide pairs (Fig. 15-14). That is, the 3′ and 5′ hydroxyl groups of two different deoxyribose molecules form a double ester with the PO_4 group. As seen in Figure 15-14, the two sugar-phosphate strands

[handwritten at bottom: "so two OH groups are exist, one from each sugar molecule"]

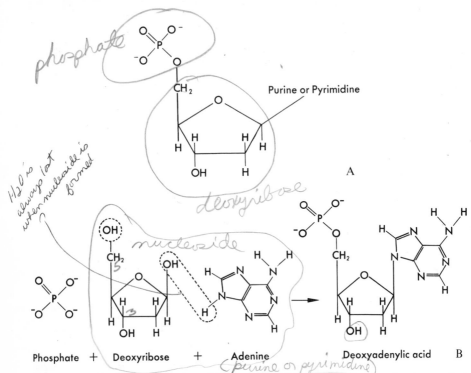

phosphate

1 H₂O is always lost when nucleoside is formed

deoxyribose

nucleoside

Purine or Pyrimidine

A

Phosphate + Deoxyribose + Adenine · Deoxyadenylic acid   B

(purine or pyrimidine)

FIGURE 15-13. (A) *Structural model for a deoxyribonucleotide. A purine is attached to deoxyribose by a bond connecting a ring-nitrogen at the number 9 position to the 1′ position of the sugar; a pyrimidine is attached to deoxyribose by a bond connecting a ring-nitrogen at the number 3 position to the 1′ position of the sugar. (B) Formation of the deoxyribonucleotide deoxyadenylic acid. The other three deoxyribonucleotides are formed in the same way.*

are oriented in opposite directions. In reading the *left* strand from top to bottom the sugar-phosphate linkages are 5′, 3′, whereas in the *right* strand they are 3′, 5′. The two "backbone" strands are thus described as being *antiparallel*. Hydrogen bonding between nucleotide pairs was deduced by Watson to be such as to give closely similar measurements to AT and CG pairs (Fig. 15-15). In a molecule such as this, with only four "usual" bases, the possible kind and number of sequences are virtually infinite, and we might also anticipate that the DNA of different species is distinguished, in part, by length. As we shall see in the next chapter, the DNA of different species is, in fact, distinctive through just such qualitative and quantitative differences in nucleotide pairs.

   On the other hand, not all DNA is built exactly according to the pattern just described. The DNA of some phages (e.g., φX174) is only single-stranded, as we have noted previously. It does, however, become temporarily double-stranded after infection of a host cell as a prelude to replication (see pages

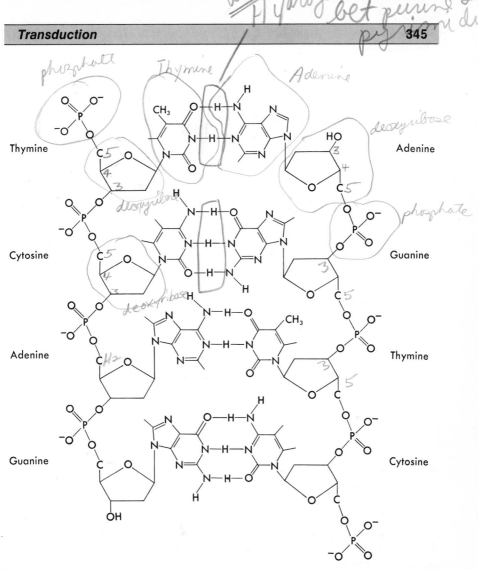

Figure 15-14. *Molecular model of a four nucleotide pair segment of DNA. For simplicity the helical nature of the molecule is not shown. Note antiparallel orientation of the sugar-phosphate strands.*

347–351). In some DNA, other, rarer bases regularly replace some of the four common ones. Phage T2, for instance, contains 5-hydroxymethyl cytosine instead of cytosine, but in an amount equal to the guanine content, thus indicating its pairing qualities. Other similar substitutions are known, but the exact role of these rare bases is only partly understood.

Location of DNA in Cells. As was pointed out earlier, the Feulgen stain technique is specific for DNA; any cellular structure containing DNA retains a purple color. Not only is this process specific for DNA, but the intensity of

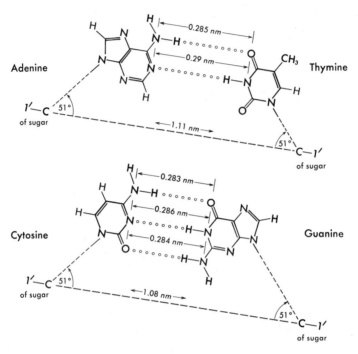

FIGURE 15-15. *Details of hydrogen bonding between deoxyribonucleotide pairs. Note the close similarity of measurements of AT and GC pairs. (0.1 nm = 1 Å.)*

staining is directly proportional to the amount of deoxyribonucleic acid present. Measurements of light absorption by structures colored by the Feulgen process can be used to determine the relative amounts of this nucleic acid present. Such techniques, applied to many different kinds of eukaryotic cells, show DNA to be almost entirely restricted to the chromosomes.[3] Such microspectrophotometric techniques also employ ultraviolet light, the peak absorption being near 260 nm. It is significant that although ultraviolet radiation is a relatively weak mutagenic agent, its most effective wavelengths as such are also near 260 nm.

The physical arrangement of DNA in the chromosomes of eukaryotes is still an unsettled matter of some conjecture. Various experimental techniques of recent development disclose what clearly seem to be extremely long but intricately folded DNA molecules in the chromosome. The total length of the DNA of *Drosophila,* for example, is calculated to exceed by far the length of the entire metaphase chromosome complement. Yet chromosomes replicate, exhibit crossing-over, and mediate the transmission of heritable traits as

[3] The fact that small amounts of DNA are also found in such cytoplasmic organelles as chloroplasts and mitochondria does not weaken the cytological, chemical, and genetic evidence that DNA is genetic material. See also Chapter 20.

though they are each composed of a single DNA molecule running the length of the chromosome. The "packaging problem" has simply not yet been solved (Fig. 15-16).

On the other hand, the bacterial "chromosome" clearly appears to be a single long DNA molecule in the form of a "naked," continuous, closed, no- *circular* end structure. Even here, though, there are some packaging problems because it can be shown that the length of such a "chromosome" in *Escherichia coli,* for instance, is about 1 mm ($= 1,000$ μm) or a little more, whereas the average length of a mature *E. coli* cell is of the order of 1 or 2 μm! Moreover, such a cell may temporarily have more than one of these "chromosomes," depending on its stage in the life cycle (Fig. 15-16).

Cyclic Quantitative Changes in DNA Content. By appropriate quantitative measurements of DNA content of cells, a number of striking parallels with chromosome behavior and our earlier postulates regarding genes can be seen:

1. The quantity of DNA detected in gametes is, within limits of experimental error, half that of diploid meiocytes.
2. Zygotes contain twice the amount of DNA in gametes.
3. The amount of DNA increases during interphase by a factor of 2. Telophase nuclei have half the DNA content of late-interphase (G_2) or early-prophase nuclei.
4. DNA content of polyploid nuclei is proportional to the number of sets of chromosomes present; i.e., tetraploid nuclei can be shown to have twice as much DNA as diploid cells.

Replication. Because DNA is the genetic material, its mode of increase should be such as to *replicate* exactly. Although many details remain to be clarified (especially with respect to the enzymology), recent work of several laboratories appears to have disclosed the major steps in the process, particularly in *Escherichia coli* and some of the DNA phages. As might be expected, the replicative process in eukaryotic cells is less well understood than it is for prokaryotes. The description that follows outlines the general process in bacteria such as *E. coli.*

In *E. coli* the continuous, no-end DNA molecule (the "chromosome") retains its circularity during replication. The process is initiated at an *origin point* where breakage of the weak hydrogen bonds that join the purine-pyrimidine bases commences. Initiation involves the synthesis of a short *RNA* fragment that serves as a primer for the elongation process; it requires three distinctive and highly specific enzyme systems in *E. coli* (Schekman et al., 1974). Breakage of the hydrogen bonds proceeds sequentially along the DNA molecule, which thus begins "unzipping" into a Y-shaped *replication fork.* As this continues, short, discrete primer segments of about 1,000 deoxyribonucleotides are constructed through action of a multisubunit enzyme system, DNA polymerase III. These primer segments are synthesized in the 5' $\rightarrow$ 3' direction and quickly become hydrogen-bonded to the old template

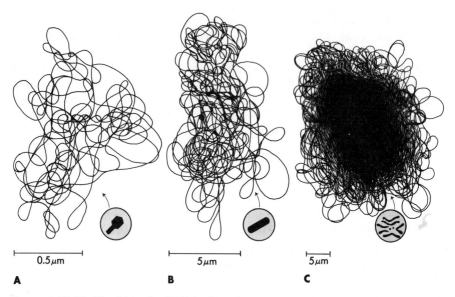

0.5 μm

5 μm

5 μm

A   **B**   **C**

FIGURE 15-16. *Total length of DNA of (A) bacteriophage T-4 (about 68 μm); (B) the bacterium,* E. coli *(about 1,100 μm); (C) the fruit fly.* Drosophila melanogaster *(about 16,000 μm). Compare these lengths to the size of the structures into which the DNA is packed.* (Redrawn from F. W. Stahl, *The Mechanics of Inheritance,* © 1964, Prentice-Hall, Inc., Englewood Cliffs, N.J., by permission of author and publisher.)

strand in complementary fashion in the 3′ → 5′ direction of the template strand (i.e., antiparallel to it) as diagramed in Figure 15-17. The short primer strands are joined sequentially by formation of internucleotide phosphodiester bonds between the 5′-phosphate and 3′-hydroxyl of two adjacent nucleotides by action of another enzyme system, DNA ligase (Lehman, 1974). As a result, inorganic phosphate is split off as pyrophosphate. According to Schekman and his colleagues (1974), termination requires excision of the RNA priming fragment (through action of DNA polymerase I), and closure through action of DNA polymerase I and DNA ligase.

The fact that copying of both strands is discontinuous in the fashion just described, with primer strands being formed in the 5′ → 3′ direction but bonded to the template strand in the latter's 3′ → 5′ direction, poses some problems in view of the antiparallel nature of the template strands and the travel of the replication fork in a single direction. Events at the replication fork have been interpreted somewhat differently by Okazaki et al. (1968, 1969) and by Kornberg (1969). Okazaki has suggested that 5′→3′ synthesis takes place on both template strands, toward the replication point on one and away on the other (Fig. 15-18). Kornberg postulates primer synthesis first on one template strand, then on the other; this process continues, switching rapidly back and forth. The new primer strand is then cleaved at the replication point

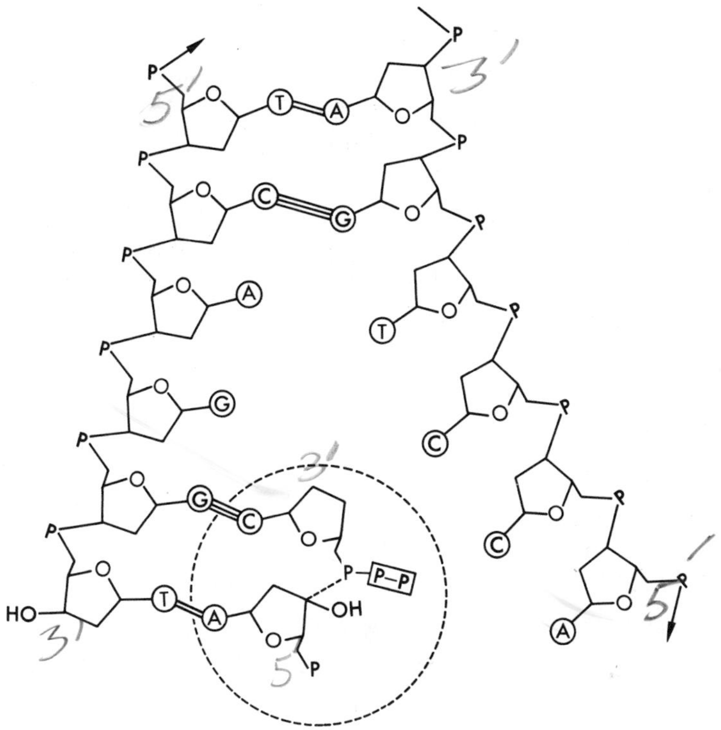

FIGURE 15-17. *Replication of DNA takes place by addition of complementary deoxy-ribonucleotides to the 5'-3' strand from the 3' end toward the 5' end of the template strand. For simplicity, such addition is shown for only one strand. See text for details.*

by an endonuclease. In either case, the process continues until the entire molecule has been replicated.

In addition to our imperfect understanding of the enzymology of the replicative process, there are some physical problems as well. The double helix of DNA must "unwind" as strand separation proceeds, and for the relatively short generation time of bacterial cells, this requires some rather high-speed events. For example, in a bacterium whose nucleoid consists of 4×10^6 deoxyribonucleotide pairs (a reasonable figure), that DNA molecule must make 4×10^5 turns during the unwinding process, inasmuch as one complete turn of the molecule includes 10 nucleotide pairs. For a generation time of 40 minutes, a fairly long period, the DNA molecule would have to make 10,000 revolutions per minute. Moreover, a no-end double helix cannot "unwind" unless one or both template strands are broken. This would require some type of scission enzyme system.

Whatever the precise details of the operation of the replicative mechanism, its accuracy and speed are quite high. Kornberg and his associates (1967)

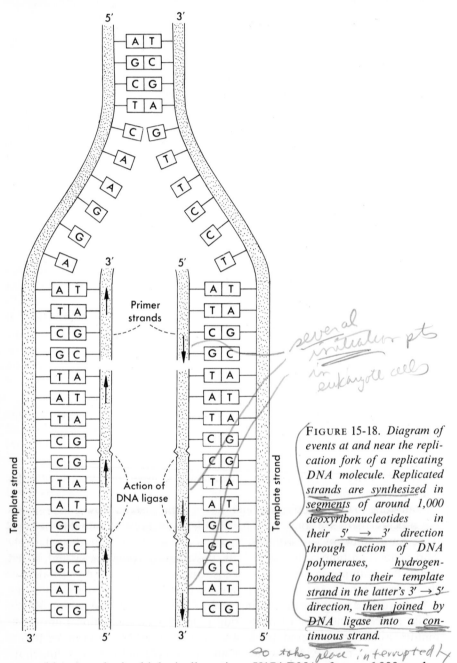

FIGURE 15-18. *Diagram of events at and near the replication fork of a replicating DNA molecule. Replicated strands are synthesized in segments of around 1,000 deoxyribonucleotides in their 5′ → 3′ direction through action of DNA polymerases, hydrogen-bonded to their template strand in the latter's 3′ → 5′ direction, then joined by DNA ligase into a continuous strand.*

were able to synthesize biologically active φX174 DNA of some 6,000 nucleotides (recall that DNA of this phage is single-stranded except in its replicative form). Infection of susceptible *E. coli* cells with this synthetic DNA resulted in replication of φX174 particles identical with natural phage, as well as lysis of host cells in the usual fashion. Speed of synthesis in vitro has been reported as 500 to 1,000 nucleotides per minute, but in vivo speeds as high as 100,000 per minute have been calculated.

Single-stranded viruses such as ϕX174 cannot, of course, follow exactly the procedure outlined for double-stranded bacterial DNA. The single strand present in the infective phage serves as a template in the host cell for synthesis of a complementary strand, forming a double-stranded molecule (the *replicative* form) that replicates many times. Finally, however, the replicative form begins to produce only single (infective) strands, which are then assembled within protein shells as mature phage and released (Sinsheimer et al., 1962).

However, in a pair of papers, Baltimore (1970) and Temin and Mizutani (1970) describe evidence that certain RNA tumor viruses (e.g., Rous sarcoma virus and Rauscher mouse leukemia virus) must either produce or induce an enzyme (RNA-dependent DNA polymerase) that catalyzes synthesis of DNA from an RNA template. Temin and Mizutani point out that "if the present results and Baltimore's results with Rauscher leukemia virus are upheld, they will constitute strong evidence that the DNA provirus hypothesis is correct and that RNA tumor viruses have a DNA genome when they are in cells and an RNA genome when they are in (virus particles)." Thus replication of at least these RNA viruses would take place through a DNA intermediate rather than through an RNA intermediate as described for other RNA viruses. Furthermore, this discovery may be expected to throw considerable light on carcinogenesis by RNA viruses, as well as eventually to clarify the process of formation of RNA from DNA templates.

Semiconservative Nature of DNA Replication. Under the strand separation hypothesis, each newly replicated double helix must consist of one "old" strand and one "new" one. By a series of ingenious experiments, Meselson and Stahl (1958) showed that this does, indeed, occur. Bacteria were cultured for some time in a medium containing only the heavy isotope of nitrogen, ^{15}N, until all the DNA should be labeled with heavy nitrogen. These cells were next removed from the ^{15}N medium, washed, and transferred to a medium all of whose nitrogen was the usual isotope, ^{14}N. After one bacterial generation, a sample of cells was removed, and the DNA extracted. All DNA that replicated once in the ^{14}N medium should be "hybrid"; that is, they should consist of one strand whose nitrogen is all ^{15}N ("old") and one in which the nitrogen should be all ^{14}N ("new"), as suggested in Figure 15-19.

Differentiation between ^{14}N-DNA and ^{15}N-DNA is readily made by *density gradient centrifugation.* The now generally employed technique was developed by Meselson, Stahl, and Vinograd. In it particles of different densities (such as ^{14}N-DNA and ^{15}N-DNA) are suspended in a concentrated solution of the salt of a highly soluble heavy metal such as cesium chloride and subjected to intense centrifugation in an ultracentrifuge. The centrifugal field produces a density gradient in the tube of both the salt and the suspended particles. After several hours the gradient reaches an equilibrium in which densities of the

Original DNA molecule
(all N^{15} labeled)

Once-replicated "hybrid" DNA
(each molecule contains one N^{15}
and one N^{14} strand)

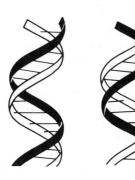

Twice-replicated DNA
(Half the molecules "hybrid"
and half containing only N^{14})

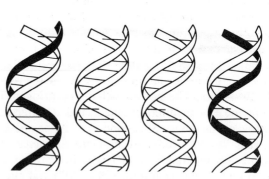

FIGURE 15-19. *The Meselson and Stahl experiment demonstrating the semiconservative replication of DNA. See text for details. Strands labeled with ^{15}N are shaded; ^{14}N-labeled strands are unshaded.*

solution and suspended material match at a level in the tube where all the suspended particles of like density collect in a narrow band (Fig. 15-20). Here the particles will remain, subject only to their diffusibility. It is possible to separate particles differing in density by extremely small amounts with this technique.

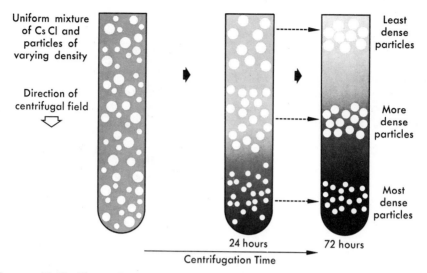

Uniform mixture
of Cs Cl and
particles of
varying density

Direction of
centrifugal field

Least
dense
particles

More
dense
particles

Most
dense
particles

24 hours   72 hours
Centrifugation Time

FIGURE 15-20. *The mechanics of density gradient centrifugation in a cesium chloride solution.* (Adapted from D. Fraser, *Viruses and Molecular Biology,* Macmillan, 1967. By permission.)

When the "hybrid" DNA of Meselson and Stahl was ultracentrifuged in a cesium chloride solution, after some 20 to 50 hours all this DNA was found by ultraviolet absorption (260 nm) to have taken a position exactly intermediate between that previously established for ^{15}N-DNA and ^{14}N-DNA. After two cell generations, in which DNA has replicated a second time, half would be expected to contain only ^{14}N and half should be "hybrid." Therefore there should be two bands of DNA in the ultracentrifuge tube, one at the intermediate position, and one at the "all ^{14}N" position, and this was precisely what was observed. This pattern of replication, in which each of the "old" strands serves as one of the two in each new DNA molecule, is referred to as *semiconservative replication.*

The results of these experiments are entirely consistent with the strand-separation theory (Fig. 15-21), although they do not eliminate such other possibilities as lengthwise extension of the unseparated double strand. But there is good experimental evidence from other sources that this latter type of behavior does not occur. In fact, autoradiographs of DNA from *E. coli* that had been allowed to replicate in radioactive thymidine show that replicating DNA does indeed have a Y-shaped configuration. Cairns (1963, 1966) was able to verify this in *E. coli* by autoradiography that revealed two such Y-shaped regions in the replicating circular DNA molecule (Fig. 15-22). One of these is the replication point at which synthesis of new complementary strands is occurring, and the other is the origin or swivel point that allows for an unwinding of the parent helix. As was pointed out earlier, the question of how

FIGURE 15-21. *DNA replication by the strand separation theory where "old" and "new" duplexes are assumed to unwind and wind simultaneously as nucleotides are assimilated into newly developing strands.* (Redrawn from D. Fraser, *Viruses and Molecular Biology*, Macmillan, 1967. By permission.)

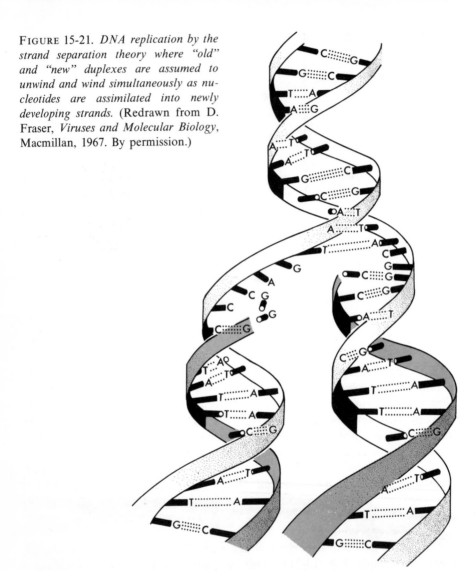

the strands "unwind" to permit this type of replication is unanswered, and the nature of the triggering mechanism that starts it is unclear (Watson, 1970, 1971 and Burdon, 1973). Nevertheless, available evidence as to the mechanics is good, if not quite complete; furthermore, you may be able to discern, in the possibility of an occasional error in the replication process, a mechanism for mutation. This we shall explore in Chapter 18.

Behavior of DNA in Transformation and Transduction. We are now in a position to understand what occurs in a bacterial cell that is about to undergo either transformation or transduction. In both, DNA from another bacterial cell enters the cell that is about to be converted. In the first process, naked

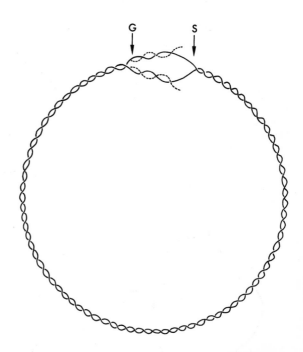

FIGURE 15-22. *Model for replication of circular DNA molecule of* Escherichia coli, *based on work of Cairns (1963, 1966). S, swivel or initiating point at which the template molecule untwists;* R, *replication point at which synthesis of new complementary strands takes place. Parent (template) strands are represented by solid lines, the newly synthesized daughter strands by dashed lines.*

DNA derived from dead cells, and generally a good bit less than their complete DNA complement, penetrates the living cell. In transduction the process differs slightly in that the entering DNA is "injected" from a virus. The events occurring in the living cell are undoubtedly similar in both cases, but the entering DNA differs in one important respect. Transforming DNA is strictly bacterial DNA, whereas transducing DNA is phage material that includes some genetic material of the previous bacterial host. To appreciate this let us look briefly at each process again, as events take place within the living bacterial cell.

If transforming DNA is to bring about transformation, it must be integrated into the host DNA. Chilton (1967) found that each of the two complementary strands of *Bacillus subtilis* DNA has identical transforming capability. She suggests, however, that it is probable that only one is physically integrated into the recipient genome. Both original *(endogenous)* and transforming *(exogenous)* DNA molecules replicate synchronously. Double breaks occur, followed by rejoining so that a double crossover is formed. Early explanations suggested that physical stresses and strains incident to pairing and disjunction produced the breaks, but it is much more likely that breakage is an enzymatically controlled process, though the exact nature of the enzyme action is still unclear. There is, for example, no detectable twisting of bacterial "chromosomes" or parts thereof around each other in transformation, transduction, or conjugation. Repair of the broken sugar-phosphate strands involves DNA ligase.

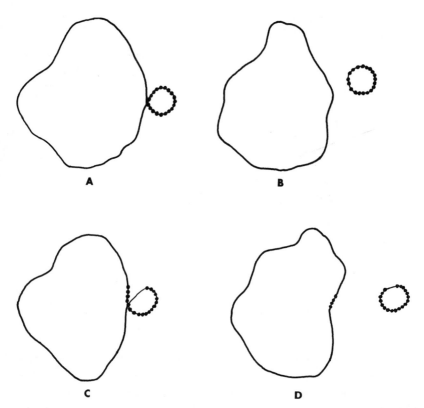

FIGURE 15-23. (*A*) *Normal association of prophage DNA* (*beaded*) *with the bacterial chromosome, and* (*B*) *its separation.* (*C*) *"Faulty" attachment between prophage and bacterial DNA, resulting in* (*D*) *an exchange of DNA between prophage and bacterium. As a consequence of this latter, aberrant association, the phage DNA includes a bit of bacterial DNA and vice versa.* (Redrawn from Werner Braun, *Bacterial Genetics,* W. B. Saunders and Company, 2nd ed., 1965. By permission of author and publisher.)

In a pair of important papers (Meselson and Weigle, 1961, and Kellenberger et al., 1961) it was clearly shown by appropriate labeling techniques that such a break-and-exchange mechanism is substantially correct. Present evidence has been well elaborated by Simon (1965) and by Radding (1973). The sequence of events that follows entry of "foreign" DNA in both transformation and transduction now appears to be

1. Association or pairing of transforming or transducing DNA with corresponding lengths of the recipient's DNA.
2. Synchronous replication of both "foreign" (exogenous) and recipient's (endogenous) DNA.
3. Enzymatic breaking of both exogenous and endogenous DNA.

4. Exchange between exogenous and endogenous DNA by enzymatic repair, resulting in

5. Formation of a recombinant DNA molecule which "breeds true" for the traits introduced by transformation or transduction.

Creation of recombinants by conjugation operates intracellularly in the same manner.

In transduction, the DNA of the temperate phage associates with or attaches to that of the host cell as prophage (Fig. 15-23A) and replicates synchronously with the host DNA. Very occasionally, however, an apparently faulty attachment between prophage and endogenous DNA may occur (Fig. 15-23C), with the result that an exchange of DNA between the two molecules takes place. Subsequent deintegration and infection of another host bacterium may then introduce a bit of exogenous bacterial DNA into the new host, resulting in recombinant bacterial progeny. In the previously cited work of Zinder and Lederberg, the met^+ gene from the met^+ thr^- host is accepted by the phage and later, in the process of transduction, integrated into the DNA of a met^- thr^+ strain, some of whose progeny are then recombinants of the genotype met^+ thr^+.

Transformation, in particular, provides the identification of the genetic material as deoxyribonucleic acid. Our next concern will be to relate the gene to this remarkable substance.

REFERENCES

AVERY, O. T., C. M. MACLEOD, and M. MCCARTY, 1944. Studies on the Chemical Nature of the Substance Inducing Transformation of Pneumococcal Types. *Jour. Exp. Med.,* **79:** 137–158. Reprinted in J. A. Peters, ed., 1959. *Classic Papers in Genetics.* Englewood Cliffs, N.J., Prentice-Hall.

BALTIMORE, D., 1970. Viral RNA-Dependent DNA Polymerase. *Nature,* **226:** 1209–1211.

BURDON, R. H., 1973. Nucleic Acid Biosynthesis and Interactions. In E. E. Bittar, *Cell Biology in Medicine.* New York, Wiley.

CAIRNS, J., 1963. The Bacterial Chromosome and Its Manner of Replication as Seen by Autoradiography. *Jour. Molec. Biol.,* **6:** 208–213.

CAIRNS, J., 1966. The Bacterial Chromosome. *Sci. Amer. Offprints* (1030). San Francisco, W. H. Freeman.

CHARGAFF, E., 1950. Chemical Specificity of Nucleic Acids and Mechanism of Their Enzymatic Degradation. *Experimentia,* **171:** 737–738. Reprinted in J. H. Taylor, ed., 1965. *Selected Papers on Molecular Genetics.* New York, Academic Press.

CHILTON, M. F., 1967. Transforming Activity in Both Complementary Strands of *Bacillus subtilis* DNA. *Science,* **157:** 817–819.

GOULIAN, M., 1971. Biosynthesis of DNA. In E. E. Snell, P. D. Boyer, A. Meister, and R. L. Sinsheimer, eds. *Annual Review of Biochemistry,* volume 40. Palo Alto, Calif., Annual Reviews, Inc.

GOULIAN, M., A. KORNBERG, and R. L. SINSHEIMER, 1967. Enzymatic Synthesis of DNA, XXIV. Synthesis of Infectious Phage φX-174 DNA. *Proc. Nat. Acad. Sci. (U.S.)*, **58**(6):2321–2328.

GRIFFITH, F., 1928. The Significance of Pneumococcal Types. *Jour. Hygiene,* **27:** 113–156.

GROSS, J. D., and L. CARO, 1965. Genetic Transfer in Bacterial Mating. *Science,* **150:** 1679–1684.

HERSHEY, A. D., and M. CHASE, 1952. Independent Functions of Viral Protein and Nucleic Acid in Growth of Bacteriophage. *Jour. Gen. Physiol.,* **36:** 39–56. Reprinted in G. S. Stent, ed., 1965, 2nd ed., *Papers on Bacterial Viruses.* Boston, Little, Brown.

HOTCHKISS, R. D., and M. GABOR, 1970. Bacterial Transformation, with Special Reference to Recombination Process. In H. L. Roman, ed. *Annual Review of Genetics,* volume 4. Palo Alto, Calif., Annual Reviews, Inc.

HOUCK, J. C., H. IRAUSQUIN, and S. LEIKIN, 1971. Lymphocyte DNA Synthesis Inhibition. *Science,* **173:** 1139–1141.

KELLENBERGER, G., J. L. ZICHICHI, and J. J. WEIGLE, 1961. Exchange of DNA in the Recombination of Bacteriophage. *Proc. Natl. Acad. Sci. (U.S.),* **47:** 869–878.

KORNBERG, A., 1960. Biologic Synthesis of Deoxyribonucleic Acid. *Science,* **131:** 1503–1508.

KORNBERG, A., 1969. The Active Site of DNA Polymerase. *Science,* **164:** 1410–1418.

LARK, K. G., 1969. Initiation and Control of DNA Synthesis. In E. E. Snell, P. D. Boyer, A. Meister, and R. L. Sinsheimer, eds. *Annual Review of Biochemistry,* volume 38. Palo Alto, Calif., Annual Reviews, Inc.

LEHMAN, I. R., 1974. DNA Ligase: Structure, Mechanism, and Function. *Science,* **186:** 790–797.

LEWIS, B. J., J. W. ABRELL, R. G. SMITH, and R. C. GALLO, 1974. Human DNA Polymerase III (R-DNA Polymerase): Distinction from DNA Polymerase I and Reverse Transcriptase. *Science,* **183:** 867–869.

MARX, J. L., 1973. Restriction Enzymes: New Tools for Studying DNA. *Science,* **180:** 482–485.

MESELSON, M. S., and F. W. STAHL, 1958. The Replication of DNA in *Escherichia coli. Proc. Nat. Acad. Sci. (U.S.),* **44:** 671–682.

MESELSON, M., and J. J. WEIGLE, 1961. Chromosome Breakage Accompanying Genetic Recombination in Bacteriophage. *Proc. Natl. Acad. Sci. (U.S.),* **47:** 857–868. Reprinted in G. S. Stent, ed., 1965, 2nd ed. *Papers on Bacterial Viruses.* Boston, Little, Brown.

OKAZAKI, T., and R. OKAZAKI, 1969. Mechanism of DNA Chain Growth. IV. Direction of Synthesis of T4 Short DNA Chains as Revealed by Exonucleolytic Degradation. *Proc. Natl. Acad. Sci. (U.S.),* **64:** 1242–1248.

OKAZAKI, R., T. OKAZAKI, K. SAKABE, K. SUZIMOTO, and A. SUGINO, 1968. Mechanism of DNA Chain Growth. I. Possible Discontinuity and Unusual Secondary Structure of Newly Synthesized Chains. *Proc. Natl. Acad. Sci. (U.S.),* **59:** 598–605.

OZEKI, H., and H. IKEDA, 1968. Transduction Mechanisms. In H. L. Roman, ed., 1968. *Annual Review of Genetics,* volume 2. Palo Alto, Calif. Annual Reviews, Inc.

PESSAC, B., and G. CALOTHY, 1974. Transformation of Chick Embryo Neuroretinal Cells by Rous Sarcoma Virus in Vitro: Induction of Cell Proliferation. *Science,* **185:** 709–710.

RADDING, C. M., 1973. Molecular Mechanisms in Genetic Recombination. In H. L. Roman, ed. *Annual Review of Genetics,* volume 7. Palo Alto, Calif., Annual Reviews, Inc.

SCHEKMAN, R., A. WEINER, and A. KORNBERG, 1974. Multienzyme Systems of DNA Replication. *Science,* **186:** 987–993.

SIMON, E., 1965. Recombination in Bacteriophage T-4: A Mechanism. *Science,* **150:** 760–763.

SINSHEIMER, R. L., B. STARMAN, C. NAGLER, and S. GUTHRIE, 1962. The Process of Infection with Bacteriophage φX174, I. Evidence for a Replicative Form. *Jour. Molec. Biol.,* **4:** 142–160.

STOCKDALE, F. E., 1971. DNA Synthesis in Differentiating Skeletal Muscle Cells: Initiation by Ultraviolet Light. *Science,* **171:** 1145–1147.

TEMIN, H. M., and S. MIZUTANI, 1970. RNA-Dependent DNA Polymerase in Virions of Rous Sarcoma Virus. *Nature,* **226:** 1211–1213.

TOMASZ, A., 1969. Some Aspects of the Competent State in Genetic Transformation. In H. L. Roman, ed. *Annual Review of Genetics,* volume 3. Palo Alto, Calif., Annual Reviews, Inc.

WATSON, J. D., 1971. The Regulation of DNA Synthesis in Eukaryotes. In D. M. Prescott, L. Goldstein, and E. McConkey, eds. *Advances in Cell Biology.* New York, Appleton-Century-Crofts.

WATSON, J. D., 1975, 3rd ed. *Molecular Biology of the Gene.* New York, W. A. Benjamin.

WATSON, J. D., and F. H. C. CRICK, 1953. Molecular Structure of Nucleic Acids. A Structure for Deoxyribonucleic Acid. *Nature,* **171:** 737–738. Reprinted in J. H. Taylor, ed., 1965. *Selected Papers on Molecular Genetics.* New York, Academic Press.

ZINDER, N. D., and J. LEDERBERG, 1952. Genetic Exchange in *Salmonella. Jour. Bact.,* **64:** 679–699.

PROBLEMS

dominance, epistasis, sex linkage, crossing-over

15-1. What are some of the genetic phenomena that can be investigated in corn, fruit flies, peas, and human beings that cannot readily be studied in *Neurospora*, bacteria, and viruses? *all are haploid*

15-2. If one strand of DNA is found to have the sequence 5′ A A C G T A C T G C 3′, what is the sequence of nucleotides on the 3′, 5′ strand?

15-3. For a molecule of n deoxyribonucleotide pairs, give a mathematical expression that can be used to calculate the number of possible sequences of those nucleotide pairs if only the "usual" bases are present. *4^n*

15-4. Develop a formula for determining the length in micrometers of a DNA molecule whose number of deoxyribonucleotide pairs is known. (Note: 1 Å = 0.0001 μm.) *$3.4 \times 10^{-4} n$ or $3.4 \times 10^{-4}(n-1)$ for short chains*

15-5. Phage T2 DNA is estimated to consist of about 200,000 deoxyribonucleotide pairs. What is the length in micrometers of its DNA complement? *$3.4 \times 10^{-4}(2 \times 10^5)$*

15-6. Determine the molecular weight of (a) adenine, (b) thymine, (c) cytosine, (d) guanine.

remember base pairs are 3.4 Å apart
so 10 " " " 10 × 3.4 Å apart

15-7. What is the molecular weight of each of these deoxyribonucleosides: (a) deoxyadenosine, (b) thymidine, (c) deoxycytidine, (d) deoxyguanosine? *nucleotides*

15-8. Calculate the molecular weight of each of the following: (a) deoxyadenylic acid, (b) thymidylic acid, (c) deoxycytidylic acid, (d) deoxyguanylic acid.

15-9. From your answers to the preceding problem, give an average molecular weight for any single deoxyribonucleotide pair. (Consider only the "usual" bases.)   *650*

15-10. Based on the answer to the preceding question, and the information given in problem 15-5, what is the molecular weight of phage T2 DNA?

15-11. If the molecular weight of *Escherichia coli* DNA is taken as 2.7×10^9, (a) of how many deoxyribonucleotide pairs would *E. coli* DNA consist, and (b) what would be its length in μm?

15-12. In this chapter the generation time for virulent T2 phage was given as 22 minutes. It appears that there is no replication of DNA for about the first six minutes and that, thereafter, DNA replicates at an exponential rate for five minutes, after which the rate of replication declines to the end of the 22-minute period. In that five-minute period, about 32 phage DNA molecules appear to be formed for each infecting phage particle. At what rate are phage DNA molecules being replicated during this exponential period? *one per minute*   *p.333*

15-13. Based on your answer to the preceding problem, at what rate are deoxyribonucleotide pairs being synthesized per minute in ϕT2? *200,000*

15-14. Assume that you have just determined the adenine content of the DNA of *Bacillus hypotheticus* to be 20 per cent. What is the percentage of each of the other bases?   *$A = T = 20\%$ $\therefore G + C = 60\%$ So $A + T = 40\%$ $G = C = 30\%$*

15-15. If you determined the adenine-guanine content of another species to total 25 per cent, should you believe it?   *NO*   *If $A + G = 25\%$ $A + T = 40$ Then if $A = 20 \& G = 5$ $G + C = \frac{150}{50}$*

15-16. Note the deoxyribonucleotide composition listed in Table 15-1 for ϕX174. Another report in the literature for this phage gives the following deoxyribonucleotide percentages: A, 27.6; T, 27.8; G, 22.3; C, 22.3. (a) What are the A/T and G/C ratios? (b) What does your answer to part (a) suggest regarding the structure of the phage DNA whose composition is given in this problem? (c) How would you account for the considerable difference in these ratios from those given in Table 15-1?   *This must be replicating so double stranded $\therefore A/T = 1$ $G/C = 1$*

15-17. What significance do you think might be attached to the fact that *Euglena* chloroplasts contain DNA and that this chloroplast DNA has a markedly different nucleotide composition from nuclear DNA reported for the same organism as given in Table 15-1?   *not all gene function is nuclear in nature*

15-18. DNA of a species of yeast, a microscopic fungus, is reported to have a thymidylic acid content of 32.6 per cent. The A/T ratio was reported as 0.97. On this basis what is the percentage composition of deoxyadenylic acid?   *nucleotide*

15-19. A culture of *Escherichia coli* was grown in a ^{15}N medium until all of the cells' DNA was so labeled. The culture was then transferred to a ^{14}N medium and allowed to grow for exactly two generations so that all DNA replicated twice in the ^{14}N medium. If the DNA of the latter culture were then subjected to density gradient centrifugation, how much of the DNA should be (a) ^{15}N only, (b) "hybrid," (c) ^{14}N only?   *0*   *1st replication in ^{14}N : $\frac{15}{14}$ $\frac{14}{15}$*

15-20. Differentiate between transformation and transduction.   *2nd replication in ^{14}N : $\frac{15}{14}$ $\frac{14}{14}$ $\frac{15}{14}$ $\frac{14}{14}$*

$\frac{A}{T} = .97$

$A = .97(T) = .97(.326) = .316 = 31.6\%$

CHAPTER 16
Protein Synthesis

WITH its precise replication and transmission, DNA serves to carry genetic information from cell to cell and from generation to generation. We need now to determine the way in which this information is translated into phenotype. Virtually all the phenotypic effects with which we have been dealing are results of biochemical reactions occurring in the cell. All of these reactions require enzymes, and enzymes are proteins, either wholly or in part. Other phenotypic effects are due directly to the kinds and amounts of non-enzymatic proteins present, e.g., hemoglobin, myoglobin, gamma globulin, insulin, or cytochrome c. Proteins are built up from long-chain linear polymers of amino acid residues (polypeptide chains) and are synthesized almost exclusively in the cytoplasm. Because both polypeptide chains and DNA have linear structure, the problem reduces to the question of how DNA, located primarily in nuclear chromosomes, mediates the synthesis of proteins (on ribosomes) in the cytoplasm. In Chapter 4 we saw several cases that appeared to suggest a connection between genes and enzymes. Let us look briefly at a few additional instances that furnish clear proof of this gene-enzyme relationship before going on to an understanding of its operation.

Genes and Enzymes

Phenylalanine Metabolism. The groundwork for a functional relationship between genes and enzymes was laid in 1902, when Bateson reported that a rare human defect, *alkaptonuria,* was inherited as a recessive trait. Then in 1909, the English physician Garrod published a book, *Inborn Errors of Metabolism,* that was far ahead of its time in suggesting a relationship between genes and specific biochemical reactions.[1] Alkaptonuria was among the heritable disorders with which he dealt at length. This condition, manifested by a darkening of cartilaginous regions and a proneness to arthritis, results from a failure to break down alkapton (2,5-dihydroxphenylacetic acid, or homogentisic acid). Alkapton accumulates and is excreted in the urine, which turns black upon exposure to the air. By the 1920s it was discovered that the blood of alkaptonurics is deficient in an enzyme, homogentistic acid oxidase, that catalyzes the oxidation of alkapton.

This defect is but one of a group, all related to the body's metabolism of the essential amino acid phenylalanine. As diagramed in Figure 16-1, man receives his supply of this substance from ingested protein. Once in the body, phen-

---

[1] See also Harris (1975) and Knudson (1969).

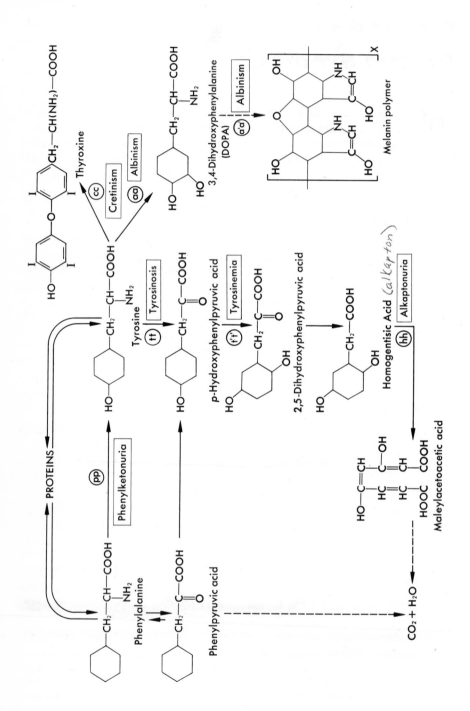

ylalanine may follow any of three paths. It may be (1) incorporated into cellular proteins, (2) converted to phenylpyruvic acid, or (3) converted to tyrosine, another amino acid.

Persons of genotype *pp* (see Fig. 16-1) fail to produce the enzyme phenylalanine hydroxylase, with the result that phenylalanine accumulates in the blood, with excesses up to a gram a day being excreted in the urine. Such persons are said to have *phenylketonuria, or PKU,* which is accompanied by serious mental and physical retardation. About two thirds of all PKUs have an I.Q. below 20, and hence are classed as idiots. However, if symptoms are diagnosed very early and the infant is put on a diet low in phenylalanine (normal development requires some) for at least the first five years, brain development is more nearly normal. PKU can be detected at birth; by the end of 1970 forty-three states had mandatory PKU-screening laws.

Another recessive gene (*t* in Fig. 16-1) blocks conversion of tyrosine to parahydroxyphenylpyruvic acid, probably through failure to produce liver tyrosine transaminase (McKusick and Claiborne, 1973). This leads to accumulation of tyrosine, excesses of which are excreted in the urine. This condition is called *tyrosinosis;* no serious symptoms appear to be associated, although the disorder has been reported for only one human subject. A somewhat less uncommon, but still rare, disorder, *tyrosinemia,* is due to inability of persons who are homozygous recessive for another gene (*t'* in Fig. 16-1) to produce the enzyme parahydroxyphenylpyruvate oxidase. In this condition large quantities of parahydroxyphenylpyruvic acid, as well as lactic and acetic acid derivatives, are excreted in the urine. If not treated with low-tyrosine diets, this condition leads to serious liver problems that may culminate in early death.

Other metabolic blocks associated with faulty utilization of phenylalanine, and all caused by recessive genes, lead to *albinism* or to *cretinism.* Two kinds of albinism have been recognized for some time. The majority of albinos are unable to produce tyrosinase (these are *aa* persons in Fig. 16-1), which catalyzes the conversion of tyrosine to DOPA; hence they do not produce melanin. Some albino persons (*a'a'* in Fig. 16-1), however, can produce DOPA but cannot utilize tyrosinase in oxidizing DOPA to DOPA quinone, thus breaking the melanin production process at a later point. Goitrous cretinism is also of several types, each involving a particular block in the conversion of tyrosine to thyroxine. Cretinism is accompanied by a considerable degree of physical and mental retardation in addition to thyroid defects.

In each case described for phenylalanine metabolism defects, a particular (recessive) gene appears to be associated with nonproduction of a specific enzyme that catalyzes a particular biochemical reaction. Occurrence or non-

FIGURE 16-1. *Metabolic pathways of the amino acid phenylalanine in man. Different genotypes cause blocks at various points because of the failure to produce a given enzyme, resulting in such inherited metabolic disorders as phenylketonuria (PKU), tyrosinosis, tyrosinemia, cretinism, or albinism.*

occurrence of the reaction then determines a related phenotypic effect. Hetero-zygotes do not display the disorder; one "dose" of the normal gene results in enough enzyme to permit the reaction to proceed. Because serious inherited metabolic defects such as PKU are (in the absence of highly uncommon mutation) the result of marriage between heterozygotes, it would be very desirable to be able to detect heterozygosity with certainty in phenotypically normal persons. Considerable progress is being made in this direction, and persons heterozygous for PKU can now be detected by the phenylalanine test. Levels of phenylalanine rise higher and are maintained longer in *Pp* individuals than in homozygotes after administration of a standard dose of this amino acid. Detection of heterozygosity is now possible in more than 70 inherited disorders, although not all tests are completely reliable. Some of the disorders for which carrier tests are available include Tay-Sachs, sickle-cell trait, PKU, galactosemia, cystic fibrosis, glucose-6-phosphate dehydrogenase (G6PD) deficiency, hemophilia A, Huntington's chorea, and Duchenne muscular dystrophy. Depending on the nature of the defect, heterozygosity may be determined even prenatally by amniocentesis. O'Brien et al. (1971), for example, report success in diagnosing the Tay-Sachs disorder by amniocentesis during the sixteenth to twenty-eighth week of pregnancy by testing the sloughed fetal cells present in the amniotic fluid for enzyme level. In this way they were able, in 15 different pregnancies, to detect normal homozygous embryos in seven cases, two heterozygotes, and six with the Tay-Sachs condition. Five of the latter were therapeutically aborted; tests of various organs of these five showed hexosaminidase A to be absent. The remaining one was born and was developing marked symptoms of the Tay-Sachs disorder by the age of nine months.

Neurospora.   Such studies of metabolic defects in man, important as they are to him, do not provide a set of experimental conditions for adequate testing of the gene-enzyme hypothesis. In fact, most other studies up to the 1940s had concerned themselves largely with morphological characters, many of them based on such complex biochemical reactions as to impede analysis, using genetically inefficient subjects such as man. The real breakthrough was furnished by Beadle and Tatum in 1941 and in a series of later reports by them and their colleagues. It occurred to Beadle that an ascomycete fungus, the red bread mold (*Neurospora crassa*), was an ideal organism for this type of investigation. *Neurospora* is easily grown in the laboratory, has a simple life history (Appendix B), in which only the zygote is diploid (hence dominance is not a problem), and it has many easily detected physiological variants. Their aim was to induce metabolic deficiencies by X-irradiation, to identify them by comparing growing ability on minimal and various deficiency media, and so to determine the specific metabolic block. One mutant, for example, was found to be unable to synthesize vitamin B_1 (thiamine). Further tests showed this strain was able to synthesize the pyrimidine half of the thiamine molecule, but not the thiazole portion; the critical chemical difference was one of enzyme-producing ability. So the concept of a one-gene–one-enzyme–one-phenotypic-effect relationship developed.

Protein Structure

As indicated earlier, all enzymes are protein, at least in part, and the catalytic property of an enzyme is conferred by its protein makeup, with or without added cofactors. Proteins are large, heavy, generally complex molecules of great variety and biological significance. Their basic structural unit is the amino acid, numbers of which are linked together to form polypeptide chains. Only 20 amino acids are biologically important, although others are necessary in specific instances. Each may be represented by the general structural formula

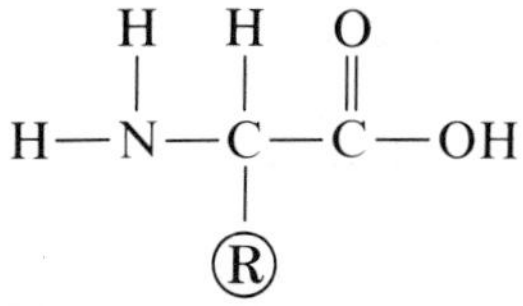

$$
\begin{array}{ccccc}
 & H & H & O & \\
 & | & | & \| & \\
H - & N - & C - & C - & OH \\
 & & | & & \\
 & & Ⓡ & &
\end{array}
\qquad \bigg\} \; \textit{amino acid}
$$

where Ⓡ represents a side chain. Differences in this side chain determine the specific kind of amino acid. Thus in glycine Ⓡ is merely an H atom:

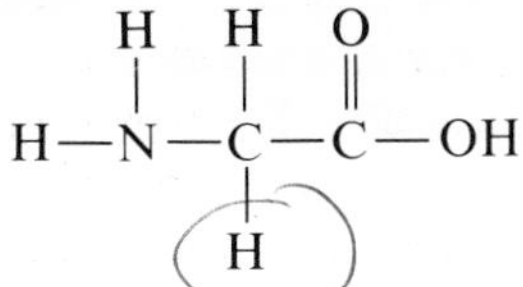

$$
\begin{array}{ccccc}
 & H & H & O & \\
 & | & | & \| & \\
H - & N - & C - & C - & OH \\
 & & | & & \\
 & & \overline{H} & &
\end{array}
\qquad \textit{glycine}
$$

and in alanine Ⓡ is only a little more complex:

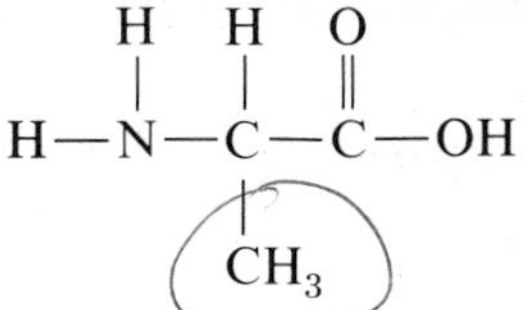

$$
\begin{array}{ccccc}
 & H & H & O & \\
 & | & | & \| & \\
H - & N - & C - & C - & OH \\
 & & | & & \\
 & & \overline{CH_3} & &
\end{array}
\qquad \textit{alanine}
$$

whereas the addition of a benzene ring to the CH_3 side chain of alanine produces phenylalanine:

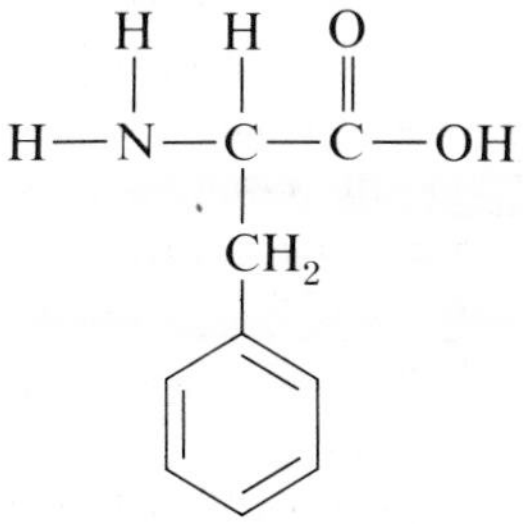

$$
\begin{array}{ccccc}
 & H & H & O & \\
 & | & | & \| & \\
H - & N - & C - & C - & OH \\
 & & | & & \\
 & & CH_2 & &
\end{array}
\qquad \textit{phenylalanine}
$$

and so on. The 20 biologically important amino acids are listed in Appendix C. Amino acids are linked by the **peptide bond** with the loss of the equivalent of a molecule of water. The formation of a dipeptide by condensation of two molecules of alanine may be represented thus:

$$
\begin{array}{cccccccc}
& H & H & O & & CH_3 & & \\
& | & | & \| & & | & & \\
H- & N- & C- & C-(OH & +\ H)- & N- & C- & C-OH \rightarrow \\
& & | & & & | & | & \\
& & CH_3 & & & H & H & O
\end{array}
$$

$$
\begin{array}{cccccccc}
& H & H & O & CH_3 & & \\
& | & | & \| & | & & \\
H- & N- & C- & C- & N- & C- & C-OH\ +\ H_2O \\
& & | & & | & | & \\
& & CH_3 & & H & H & O
\end{array}
$$

in which CONH is the peptide bond ~~in this case~~ A polypeptide is thus a series of **amino acid residues,** joined by peptide bonds and having an amino end (NH_2) and a carboxyl end (COOH).

Protein structural characteristics, then, are based on the number and kind of amino acid residues and, therefore, can occur in infinite variety. In these residues, the side chains confer a number of properties. Some are large and complex, others are small and simple. Some carry a positive charge (lysine, arginine), others a negative charge (glutamic acid, aspartic acid), but most carry none. Differences in bulk and electrical charge along a polypeptide, plus an intricate folding of the protein molecule, impart much of the specificity of enzyme-substrate and antigen-antibody relations.

Polypeptide Chain Synthesis—The Components

THE NUCLEIC ACIDS

By its infinite variety of possible arrangements of groups of the four nucleotides, DNA is admirably suited for carrying the information needed to direct the synthesis of an almost unlimited number of different proteins. But DNA, except for small amounts in chloroplasts and mitochondria (the significance of which we shall examine in Chapter 20), is located in the eukaryote nucleus, whereas protein synthesis occurs almost entirely in the cytoplasm. Moreover, DNA is not degraded or "used up" in performing its function, and it differs from the end product for which it is responsible. It may be thought of as directing, through some intermediary substance or substances, protein synthesis out in the cytoplasm. In this sense we may refer to DNA as being *conserved.* The answer to the problem of how chromosomally located, conserved DNA, which carries the genetic information, mediates the synthesis of proteins in the cytoplasm is provided by a previously mentioned second nucleic acid, ribonucleic acid (RNA), of several kinds.

RIBONUCLEIC ACID

RNA Structure. Ribonucleic acid (RNA) differs from DNA in several important ways. RNA has the following characteristics:

1. Molecules of RNA are single-stranded, linear polymers of mononucleotides, although certain types assume, in part, a secondary double-helix configuration through complementary base pairing.
2. The sugar of the sugar-phosphate strand is **ribose** (Fig. 16-2).
3. Successive **ribonucleotides** are joined by a phosphodiester bond linking the 3′ position of one ribonucleotide to the 5′ position of the next.
4. Nucleotide bases are the purines *adenine* (A) and *guanine* (G), and the pyrimidines *cytosine* (C) and *uracil* (U) (Fig. 16-3), the last replacing the thymine of DNA. However, "unusual" bases occur in certain kinds of RNA; these are chiefly methylated derivatives of A, U, C, and G.

RNA is synthesized in the nucleus of eukaryotes, using one of the strands of DNA as a template. Therefore the sequence of ribonucleotides should be complementary to one of the DNA strands; that it is, except for the occurrence of uracil instead of thymine, and the modification of some of the usual bases after RNA synthesis. The basic function of RNA is the carrying of the genetic message of DNA out into the cytoplasm, where it is responsible for building polypeptide chains—the first step in protein synthesis. Three kinds of RNA, differing in structure and function, occur: *ribosomal, messenger,* and *transfer.*

Ribosomes and Ribosomal RNA (rRNA). The site of polypeptide synthesis is the *ribosome,* a small particle averaging approximately 175 × 225 Angstrom units, composed of protein and ribosomal RNA. Ribosomes are asymmetric aggregates of two subunits that can be distinguished by the rates at which they sediment when centrifuged in an appropriate medium. The rate of sedimentation per unit centrifugal field is referred to as the sedimentation coefficient (s); for most proteins s has value of between 1×10^{-13} and 2×10^{-11} second. An s value of 1×10^{-13} is denoted as one **Svedberg unit** (S). The useful point here is that s (and, therefore, S) for any molecule or small particle such as a ribosome is determined, for any given temperature and solvent, by its shape, mass, and degree of hydration. The larger the S value, the larger the

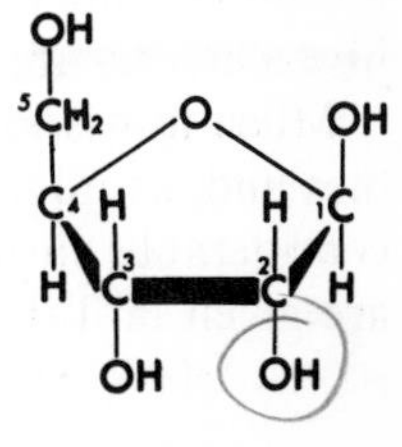

FIGURE 16-2. *Molecular structure of ribose, the sugar of ribonucleic acid. Compare the number 2 carbon with the number 2 carbon of deoxyribose (Fig. 15-10).* p342

its just H

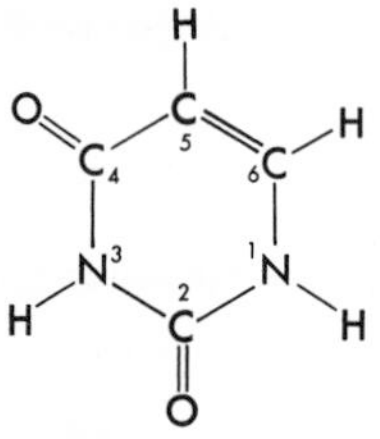

FIGURE 16-3. *Molecular structure of the pyrimidine uracil that replaces thymine in RNA.*

particle, though the relation is not linear. Ribosomes of bacteria have a sedimentation coefficient of 70S and those of eukaryotes 80S. Each of these is composed of two unequal subunits; 50S and 30S in bacteria, and 60S and 40S in eukaryotes. As suggested by these figures, sedimentation velocities are not additive.

Ribosome structure and composition, though not yet fully determined, are best known in *Escherichia coli.* In this organism the 30S subunit is composed of 21 different proteins (Wittmann et al., 1971), and a single 16S rRNA molecule of between 1,600 and 1,700 ribonucleotides, which has a molecular weight of 5.5×10^5. The 50S subunit of *E. coli,* on the other hand, contains about 34 different proteins (Kaltschmidt and Wittmann, 1970) and two rRNA molecules. One of these latter is 5S rRNA and consists of about 120 nucleotides (molecular weight 40,000); the other is 23S rRNA, consisting of some 3,200 ribonucleotides (molecular weight 1.1×10^6). These facts are summarized in Table 16-1.

TABLE 16-1. 70S Ribosome Composition in *Escherichia coli*

Ribosomal Subunit	Proteins	rRNA		
		Sedimentation Coefficient	Molecular Weight	Ribonucleotides
30S	21	16S	5.5×10^5	1,600 to 1,700
50S	34	5S	4×10^4	120
		23S	1.1×10^6	~3,200

Ribosomes functioning in polypeptide synthesis occur in groups connected by messenger RNA (see next section); such groups are called **polysomes.** In *E. coli* and in reticulocytes the ribosomes are scattered through the cell, but in eukaryotes they are situated on the endoplasmic reticulum.

Much less is known about ribosomal structure and composition in eukaryotes and, as might be expected in the diversity of higher plants and animals, considerable variation has been reported. Some data for mammalian rRNA are given in Table 16-2.

TABLE 16-2. Mammalian rRNA

Ribosomal Subunit	rRNA		
	Sedimentation Coefficient	Molecular Weight	Ribonucleotides
40S	18S	7×10^5	~2,100
60S	5S	4×10^4	120
	28S	1.8×10^6	~5,300

The function of rRNA is known only in part; it does play an indispensable structural role in ribosomal subunit assembly, at least. Traub and Nomura (1969) found that *E. coli* 30S subunits are assembled in vitro only in the presence of 16S RNA. That this is a very specific requirement is demonstrated by the fact that assembly cannot proceed in the presence of rRNA from other organisms (e.g., yeast) or by portions of 23S RNA approximating the 16S molecules in size. Some kind of highly specific recognition occurs between six of the 21 proteins of the 30S subunit and the 16S RNA (Mizushima and Nomura, 1970, and Schaup et al., 1970); 5S RNA plays a similar structural part in protein binding in the formation of the 50S ribosomal subunit. What other *functional* role might be performed by rRNA is not yet clear.

Ribosomal RNA in eukaryotes is synthesized from DNA in the nucleolar organizing region, as pointed out in Chapter 3 and by Ritossa and Spiegelman (1965). The major portion of nuclear RNA is to be found in the nucleolus, which led to early speculation that RNA storage occurred in that body. However, the nucleolar region contains deoxyribonucleotide sequences completely complementary to the ribonucleotide sequence of cytoplasmic rRNA (Brown and Dawid, 1968). On this kind of evidence it is now generally accepted that the nucleolar organizing region is the actual site of rRNA *synthesis*. Moreover, each nucleolar organizing region includes many identical, repeated genes responsible for rRNA synthesis. The number of 18S and 28S rRNA genes ranges from 200 (*Drosophila*) to 2,000 in tobacco leaf cells (*Nicotiana*), and an even higher number for 5S RNA. Although there is, of course, no nucleolus in *E. coli,* there are some five or six genes per cell for 16S and 23S RNA.

Ribosomal RNA is nonspecific for the amino acid sequence of polypeptide chains, which as we shall shortly see, are synthesized on ribosomes. Yčas (1969) summarizes three lines of evidence that indicate that rRNA is not specific for the amino acid sequences of polypeptide chains: (1) there is no correlation between the ribonucleotide sequence of rRNA and the type of protein synthesized—that is, although different proteins may be produced characteristically in different cells and in different organisms, there is no significant difference in the rRNA of these cells or organisms; (2) although phage infection results in synthesis of new kinds of protein, ribosomes and rRNA in the bacterial cell remain unchanged; (3) rRNA can be shown to be complementary to only very small sections of DNA (<1 per cent), strongly suggesting that rRNA is too limited in size and variety to carry the information needed for many different proteins.

Messenger RNA (mRNA). Specificity for an amino acid sequence is conferred by messenger RNA (mRNA), as shown, for example, by the fact that phage mRNA utilizes ribosomes made before infection to bring about synthesis of phage proteins. This ribonucleic acid directs assembly of various amino acids from the intracellular pool at the ribosome surface where enzymes effect the peptide linkage, forming polypeptide chains. We shall examine this process in detail in the next section.

The process by which mRNA is synthesized enzymatically from one strand of a length of DNA, to which it is complementary (except that, as we have noted, uracil replaces thymine) is referred to as **transcription.** The enzyme responsible is known as DNA-dependent RNA polymerase. This enzyme from a number of plants and animals has been purified and analyzed. *E. coli* RNA polymerase, for example, consists of several subunits (two α-chains, one β-chain, one β′-chain, and one σ-chain) with a total molecular weight of about 490,000. The β′ subunit is responsible for binding of the enzyme to DNA; sigma (σ) is necessary for initiation of transcription. According to Richardson (1969), DNA includes one binding site for RNA polymerase per 600 to 2,000 deoxyribonucleotides. Once the enzyme binds to DNA, strand separation occurs at the binding site and transcription begins, using one strand as template. When transcription is underway, sigma is released from the RNA polymerase-DNA template and is free to be incorporated into another RNA polymerase molecule. Bautz and colleagues (1969) report that, in the absence of σ, transcription may be initiated anywhere, at random, along DNA.

Roberts (1969) identified a termination, or *release,* protein, which he designated ρ (rho). Rho binds reversibly to DNA at specific sites and directs release of the transcribed mRNA molecule in λ phage in vitro, according to Roberts. On the other hand, work of Pettijohn and his colleagues (1970) indicates that ρ may not be required for in vitro transcription. Once formed, mRNA molecules move into the cytoplasm (although the molecular mechanics are unclear), where they function in polypeptide synthesis.

Synthesis of mRNA takes place in the 5′ → 3′ direction by addition of ribonucleotide triphosphates to the 3′ end of the elongating ribonucleotide chain. The developing mRNA molecule is antiparallel to and its nucleotides complementary to those of the DNA template strand (Fig. 16-4). Messenger RNA chain growth is rapid; reports from several laboratories range from 15 to 55 ribonucleotides *per second* for *E. coli* in vivo, and up to 100 per second in vitro for T7 phage.

The number of ribonucleotides and, therefore, the length of any mRNA molecule depend in large part on the particular polypeptide for whose synthesis it is responsible. In general, lengths of several hundred to several thousand

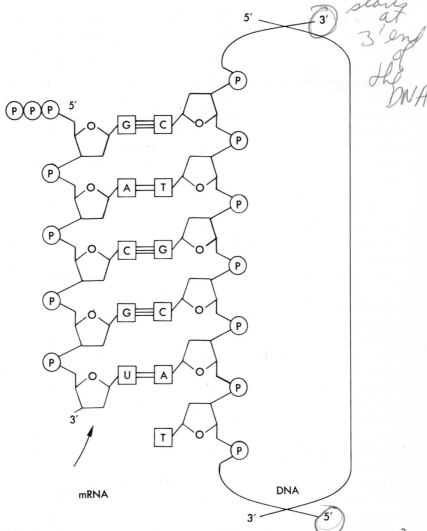

FIGURE 16-4. *Transcription of mRNA from the template strand of DNA showing that mRNA is antiparallel and complementary to the template DNA strand. mRNA is synthesized in the 5′ → 3′ direction by the addition of successive ribonucleotides at the 3′ end of the growing mRNA molecule.*

ribonucleotides have been reported for various organisms. In *E. coli* average mRNA molecular weights fall in the range 300,000 to 500,000. Taking 337 as the mean molecular weight of a ribonucleotide, this would mean an average of 300,000/337 to 500,000/337, or about 900 to 1,500 ribonucleotides per mRNA molecule if the four "usual" bases are equally represented. Of course, the latter assumption may well not be correct in many cases, but this calculation does provide a rough measure of size. Messenger RNA molecules responsible for the α and β polypeptide chains of human hemoglobin consist of more than 400 ribonucleotides each. The matter of size as it relates to specific function is explored in Chapters 17 and 18.

Messenger RNA of *E. coli* is quite short-lived, functioning for only a few minutes. It appears to retain its stability only so long as it is attached to polysomes, with the result that bacterial cells do not become "cluttered" with large amounts of mRNA. For example, the antibiotic actinomycin-D blocks the synthesis of new mRNA in bacteria; studies using actinomycin-treated bacteria indicate that bacterial mRNA is usable only some 10 to 20 times. Therefore if different "species" of mRNA can be selectively produced (see pages 379–381), the cell is able to produce a variety of proteins at different times and under different conditions. On the other hand, mRNA that is associated with hemoglobin production in reticulocytes may persist for a much longer period, even after degeneration of the cell nucleus. In most cases, mRNA appears to be continuously produced, used, and degraded.

In those viruses whose genetic material is RNA instead of DNA—e.g., tobacco mosaic virus (TMV), Rous sarcoma virus (RSV), polio virus, and the phages X174 and Qβ—there can be no transcription of mRNA from a DNA template. The RNA viruses as a group behave in a variety of ways; although research in these viruses is being actively pursued, not all details are yet clear. One of the best understood is phage Qβ. The genome of this virus is single-stranded RNA having a molecular weight of about 1×10^6. Shortly after infection of susceptible bacterial cells, the viral RNA directs the production of (1) a coat protein, (2) a maturation protein, and (3) an RNA replicase subunit. The latter combines with three host proteins to form functional *RNA replicase*. This enzyme attaches to the 3' end of the virus RNA molecule (referred to as the + strand), which it transcribes in the 3' $\rightarrow$ 5' direction to produce anti-parallel polyribonucleotide molecules (the minus strands). Some experimental data suggest that this **replicative form** (**RF**) of RNA is a double helix through extensive base pairing, but other work has been interpreted as indicating that extensive base pairing does not occur. In any event, the minus strands thus produced then serve as templates for synthesis of several new plus RNA strands (Fig. 16-5). The complex of developing viral plus strands and their complementary RNA templates are called **replicative intermediates** (**RI**). In this type of virus, then, the single-stranded RNA genome (+) serves as a template for transcription of antiparallel, complementary minus strands, which, in turn, serve as templates for production of numerous new + strands. Transcription is in the 3' $\rightarrow$ 5' direction of the template, with the new strands thus being synthesized in their 5' $\rightarrow$ 3' direction.

In the *reoviruses*, which are found in the digestive and respiratory tracts of healthy persons, the genome consists of several separate bits of *double-stranded* RNA. Each segment of this RNA is transcribed in infected cells into several "species" of single-stranded mRNA through the action of a viral RNA-

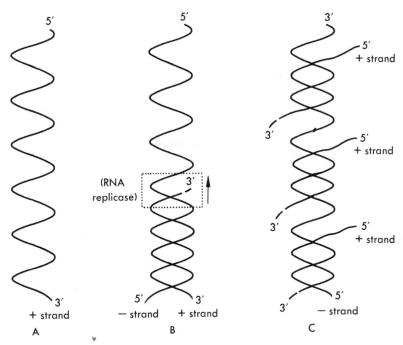

FIGURE 16-5. *Diagrammatic representation of formation of replicative form and replicative intermediates in the phage Qβ. Any base pairing is ignored. See text for details. (A) Single-stranded plus RNA strand of the phage. (B) Beginning of the formation of the replicative form (RF); RNA replicase has attached at the 3' end of the plus strand and has proceeded a short distance in the 3' → 5' direction along the plus strand to begin synthesis of an antiparallel, complementary minus strand (replicative form). (C) Synthesis of new plus strands from the minus RNA template, forming a replicative intermediate (RI). Enzyme attachment is not shown.*

dependent RNA polymerase. The mRNA molecules function directly in synthesis of viral polypeptides. In this process the double-stranded viral RNA is conserved, just as is DNA of bacteria and eukaryotes during transcription.

Perhaps the most interesting variation of RNA transcription occurs in a group of tumor-producing agents, the *leukoviruses*. Baltimore (1970) and Temin and Mizutani (1970) reported that Rous sarcoma virus particles contain RNA-dependent DNA polymerase, or **reverse transcriptase.** In infected cells this enzyme catalyzes the transcription of *DNA* from the viral RNA genome, which it uses as a template. DNA so synthesized is then transcribed into viral RNA, but even more important, such DNA *may also be incorporated into the host cells' DNA.* This, of course, alters the genotype of the host cell and is undoubtedly related to the ability of these viruses to produce tumors.

Transfer RNA (tRNA). The structure of transfer RNA is now well understood. Holley and his colleagues (1965) opened this line of research when, after seven years of intensive work, they reported the complete nucleotide sequence of alanine tRNA of yeast. Nucleotide sequences are now known for more than 60 of the different "species" of tRNA (Holmquist et al., 1973; Kim et al., 1974). Literature on this molecule has now become voluminous and new papers are appearing frequently. Nevertheless, some questions remain to be answered.

Transfer RNA has several unique characteristics: (1) It is a relatively small molecule of 75 to 85 ribonucleotides,[2] and is thus much smaller than either mRNA or any of the rRNAs; (2) a number of "unusual" nucleotides are found in tRNA, many of them methylated derivatives of the common ones (e.g., 1-methylguanylic acid, 2N-methylguanylic acid, 1-methyladenylic acid, ribothymidylic acid, 5-methylcytosine), whose importance lies in the fact that the presence of methyl groups prevents formation of complementary base pairs and thus affects the three-dimensional form of the molecule; (3) hydrogen bonding between usual bases, and its lack between many of the unusual ones, also affects the shape of the tRNA molecule. Research from the late 1960s and early 1970s (Lake and Beeman, 1967; Daykoff and Eck, 1968, Madison et al., 1966; Fuller and Hodgson, 1967; Levitt, 1969; Kim et al., 1974) makes it clear that tRNA molecules may have either three or four lobes, depending on the kind, number, and sequence of nucleotides, giving them a so-called cloverleaf shape. Madison's two-dimensional models are shown in Figure 16-6 and a generalized diagram in Figure 16-7. X-ray diffraction studies are now providing insight into the three-dimensional structure of tRNA (Fig. 16-8).

There are several common characteristics of the cloverleaf configuration: (1) the 3′ end carries a terminal purine-C-C-A sequence; (2) the "stem" or amino acid helix (Fig. 16-7) consists of seven paired bases; (3) the T ψ C stem is composed of five base pairs, the last (i.e., nearest the T ψ C loop) being C-G; (4) the anticodon stem includes five paired bases; (5) the anticodon loop consists of seven bases, with the third, fourth, and fifth (from the 3′ end of the molecule) probably constituting the anticodon, which permits temporary complementary pairing with three bases on mRNA, as will be described shortly in this chapter; (6) the base on the 3′ side of the anticodon is a purine; (7) immediately adjacent to the 5′ side of the anticodon there occurs uracil and another pyrimidine; (8) a purine, often dimethylguanylic acid, is located in the "corner" between the anticodon stem and the DHU stem; and (9) the DHU stem is composed of three or four base pairs (depending on the "species" of tRNA). The extra arm (shown extending toward the lower right in Fig. 16-7) is variable in nucleotide composition, and is lacking entirely in some tRNAs.

Transfer RNA is transcribed from several particular sites on template DNA and comprises about 15 per cent of the RNA present at any one time in an *E. coli* cell.[3]

---

[2] E.g., yeast alanine tRNA, 77 nucleotides; yeast tyrosine tRNA, 78; yeast serine tRNA, 85.

[3] Other approximate quantities are: rRNA, 80 per cent, and mRNA 5 per cent.

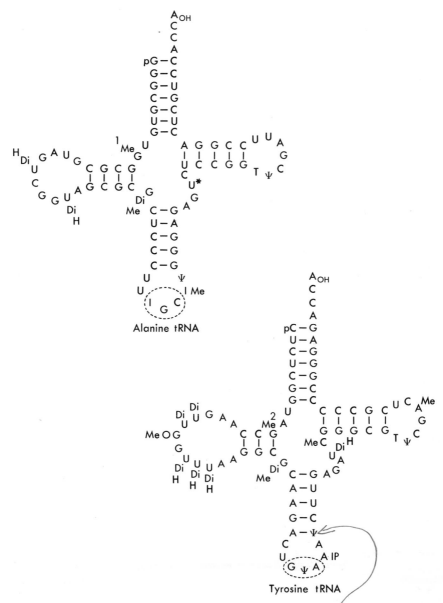

FIGURE 16-6. *"Cloverleaf" model structure for yeast alanine tRNA and tyrosine tRNA. Note that internal pairing is far from complete, and a terminal—A—C—C—A occurs in each molecule. A large number of "unusual" bases, such as ψ, pseudouridine, occurs. The presumed anticodon (see text) is enclosed by the dashed oval.* (Courtesy Dr. J. T. Madison, USDA Agricultural Research Service, Ithaca, N.Y. Reprinted, with author's revisions, from *Science*, **153**:531–534, by permission. Copyright 1966 by the American Association for the Advancement of Science.)

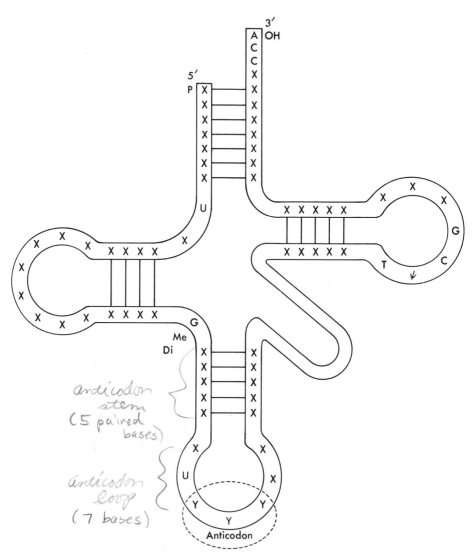

FIGURE 16-7. *Generalized two-dimensional cloverleaf model of tRNA, based on analyses of several yeast tRNA molecules by various investigators. The left lobe includes 8 to 12 unpaired bases, the lower lobe 7 (of which the three central ones are believed to be the anticodon), the upper right lobe also 7, and the lower right several unpaired bases that differ in number from one species of tRNA to another, or may even be entirely absent. Note the common 3' terminal -CCA, the TΨCG in the bottom of the upper right lobe, the "anticodon-U" (here YYYU) in the lower lobe, the DiMeG between the lower and left lobes, and the U as the first unpaired base on the 5' strand. As diagramed here, the anticodon is read from right to left (3' to 5'). The number of paired bases in each lobe shown here appears to be constant for all tRNA species so far reported. (A = adenosine, C = cytidine, G = guanosine, T = ribothymidine, U = uridine, Ψ = pseudouridine, DiMeG = dimethylguanosine, Y = any base of the anticodon.)*

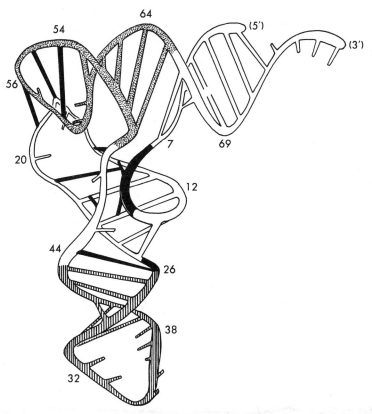

FIGURE 16-8. *Schematic three-dimensional model of yeast phenylalanine tRNA molecule. The ribose-phosphate "backbone" is shown as a continuous cylinder with bars indicating hydrogen-bonded base pairs. Unpaired bases are indicated by shorter "branches." The TΨC arm is shown in heavy strippling; the anticodon arm is indicated by vertical lines. Black portions represent tertiary interactions.* (Redrawn from S. H. Kim, F. L. Suddath, G. J. Quigley, A. McPherson, J. L. Sussman, A. H. J. Wang, N. C. Seeman, and A. Rich, 1974. Three-Dimensional Tertiary Structure of Yeast Phenylalanine Transfer RNA. *Science,* **185**:435–440. Copyright 1974 by the American Association for the Advancement of Science. By permission of Dr. S. H. Kim and the American Association for the Advancement of Science.)

Methylation and formation of other unusual bases take place after transcription, as does the addition of -C-C-A at the 3' end of the molecule. As transcribed, the tRNA *precursors* appear to be slightly larger than the final, functional molecule, the "excess" bases being removed after transcription. Formation and processing of tRNA in eukaryotes follows the same general pattern as has been described for bacteria. Precursors, longer than the final product by 20 or 30 nucleotides, are transcribed from some 0.01 per cent of the nonnucleolar portion of the cell's DNA (Burdon, 1973). Later, in the cytoplasm, these precursors are modified by removal of the excess nucleotides, methylation of others, and addition of the 3′-terminal -C-C-A.

Minimum Necessary Materials

The stage is now set for us to examine the role of each of these components in synthesis of polypeptides. Success in such synthesis in in vitro cell-free systems, to be described, shows that minimal necessary equipment is as follows:

1. Amino acids.
2. Ribosomes containing proteins and rRNA.
3. mRNA.
4. tRNA of several kinds.
5. Enzymes.
   a. Amino-acid-activating system.
   b. Peptide polymerase system.
6. Adenosine triphosphate (ATP) as an energy source.
7. Guanosine triphosphate (GTP) for synthesis of peptide bonds.
8. Soluble protein initiation and transfer factors.
9. Various inorganic cations (e.g., K^+ or NH_4^+, and Mg^{2+}).

Polypeptide Chain Synthesis: The Process

The first critical step in **translation,** i.e., the specification by mRNA of the sequence of amino acid residues in a polypeptide chain, involves the transport of amino acids from an intracellular pool to the ribosomes where they are assembled into polypeptide chains, which are then assembled into proteins elsewhere in the cytoplasm. Some of these amino acids are synthesized in human cells, but others (the *essential* amino acids, Appendix C) cannot be, and must be supplied by the diet. Transfer of amino acids to the ribosome surfaces is accomplished by tRNA and occurs in several stages.

Amino Acid Activation. Each of the 20 kinds of amino acids must be *activated* before it can attach to its tRNA. In activation, an enzyme (amino acyl synthetase) catalyzes the reaction of a specific amino acid with adenosine triphosphate (ATP) to form amino acyl-adenosine monophosphate (amino acyl adenylate) and pyrophosphate:

$$AA + ATP \xrightarrow{\text{amino acyl synthetase}} AA \sim AMP + \text{pyrophosphate}$$

The amino acyl adenylate ($AA \sim AMP$) is referred to as an **activated amino acid,** which is linked to adenosine by a phosphate ester bond.

As many as 20 different amino acids can be involved; hence each cell must have at least 20 different kinds of amino acyl synthetases. By its architecture, each kind of enzyme molecule must "recognize" and be able to fit with a particular amino acid. The specificity of the binding of synthetase to amino acid is dependent upon the structure of the latter's side group, though there is not a great size difference between side groups of, for example, glycine and alanine. In fact, one estimate is that the frequency of wrong insertions may be as high as 1 in 1,000.

The activated amino acid, however, is strongly attached to its enzyme and cannot, in this condition, be joined with another to begin formation of a polypeptide chain. This process requires transfer of the activated amino acid to the ribosomes.

Transfer of Activated Amino Acid to tRNA. The same amino acyl synthetase that catalyzed the activation of the amino acid next attaches to a receptor site at the terminal adenine ribonucleotide of a particular tRNA molecule. That tRNA is then said to be **charged.** The amino acyl adenylate attaches to the free 3'-OH of the ribose of the 3' terminal adenylic acid of tRNA by forming an acyl bond with the α-carboxyl group of the amino acid. Note that the enzyme must be able to bind specifically to both a particular amino acid and also to a particular tRNA molecule. Therefore in addition to the minimum of 20 different kinds of synthetases, each cell must have at least 20 different tRNA species. In fact, there is good experimental evidence that there are many instances in which two or more different species of tRNA have some degree of specificity for the same amino acid; this will be elaborated in the next chapter.

Assembly of Polypeptides. Following amino acid activation and binding to transfer RNA, the charged tRNA diffuses to the polysome, where actual assembly into polypeptide chains takes place. The following description of polypeptide formation applies to *Escherichia coli* in particular and to bacteria in general, organisms in which the process is best known. It apparently differs only in details from the corresponding series of events in eukaryotes.

Recall that mRNA binds to the smaller (i.e., 30S) subparticles of the bacterial ribosome; it displays there at least two groups of three ribonucleotides each, which constitute sites capable of accepting charged tRNA molecules. (Evidence for *three* ribonucleotides, rather than some other number, is explored in the next chapter.) These sites are designated the **aminoacyl (A) site** and the **peptidyl (P) site,** respectively. At or near the 5'-end of the associated mRNA molecule a chain-initiating group of (three) ribonucleotides must be present; the nature of these nucleotides is described in Chapter 17. The first 70S ribosome attaches by its 30S subunit to mRNA at this initiating site, often as the mRNA is still being transcribed (Fig. 16-9). In *E. coli* the first amino acid to be incorporated is N-formylmethionine. This is methionine with a formyl group,

$$\begin{array}{c} O \\ \parallel \\ H-C- \end{array}$$

attached to the amino group:

$$H-\overset{\overset{\displaystyle O}{\|}}{C}-\overset{\overset{\displaystyle H}{|}}{N}-\overset{\overset{\displaystyle H}{|}}{\underset{\underset{\underset{\underset{\displaystyle CH_3}{|}}{\underset{\displaystyle S}{|}}}{\underset{\displaystyle CH_2}{|}}}{\underset{\displaystyle CH_2}{C}}}-\overset{\overset{\displaystyle O}{\|}}{C}-OH$$

The formyl group blocks formation of a peptide bond with the carboxyl group of another amino acid. Depending on the particular polypeptide being synthesized, the N-formylmethionine may or may not later be removed enzymatically before the polypeptide chain is incorporated into a protein.

Steps in polypeptide synthesis on the polysome in *E. coli* are as follows (see also Fig. 16-10):

1. Binding of a tRNA charged with a formylated amono acid directly at the P site of the first ribosome (Thach and Thach, 1971), rather than at its A site as proposed by Kolakofsky and others (1968); this step also requires three protein-initiating factors in the 30S subunit, and GTP.

2. Once the P site of ribosome 1 is thus occupied, a second charged tRNA, whose anticodon must be complementary to the codon of mRNA then at the A site of ribosome 1, arrives and binds to that A site.

3. A peptide bond is synthesized enzymatically between the free α-amino group of the incoming (second) amino acid and the esterified carboxyl group of the (first) amino acid attached to the tRNA that is at the P site. This leaves a peptidyl tRNA carrying a dipeptide at the A site and an uncharged tRNA at the P site.

4. Essentially simultaneously with step 3, there is movement of the ribosome (presumably by a distance of three nucleotides) along the mRNA (translocation) in the 5′ → 3′ direction, permitting release of the now uncharged tRNA (which can participate again later in the process) and resulting in shift of the occupied codon to the P location, thus bringing into position another codon at the new A site.

5. This series of steps (1 through 4) is repeated until the ribosome reaches a chain-terminating signal on mRNA. These termination signals are "recognized" by protein-release factors.

6. Upon reaching the termination signal the ribosome leaves the mRNA as a 70S particle and its completed polypeptide chain is enzymatically released, to be incorporated into the appropriate protein molecule.

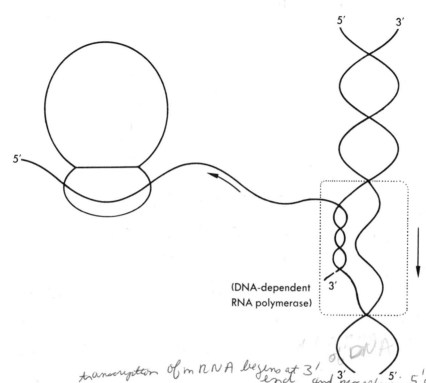

5′   3′

5′

(DNA-dependent RNA polymerase)   3′

transcription of mRNA begins at 3′ end and proceeds in 5′ to 3′ direction of mRNA

FIGURE 16-9. *Transcription of mRNA from DNA template. The enzyme DNA-dependent RNA polymerase attaches to DNA at a binding site, opening up a short section of the DNA double helix by breaking of the hydrogen bonds between deoxyribonucleotide pairs. The enzyme then moves along the DNA molecule in the direction shown here by the arrow at the right (i.e., in the 3′ → 5′ direction of the template strand), bringing about transcription of mRNA, which is antiparallel and complementary to the DNA template strand (except that uracil replaces thymine). New ribonucleotides are added to the mRNA molecule at its 3′ end; its 5′ end may already be engaging ribosomes out in the cytoplasm as shown at the left.*   *since translation begins at 5′ end of mRNA too*

7. After release from mRNA, the 70S ribosome dissociates into its 30S and 50S components, which are then free to combine again with other 50S and 30S subparticles, respectively, and function repeatedly in synthesizing other polypeptides.

As the steps just enumerated take place, additional ribosomes successively engage the initiation site at or near the 5′-end of the mRNA molecule, each being engaged in synthesizing an identical polypeptide chain of its own as it moves along mRNA in the 5′ → 3′ direction. Life of any mRNA molecule is finite, however. It continues to function in the manner described so long as ribosomes attach to its 5′-end, but this terminus is vulnerable to ribonucleases.

so transcription & translation proceed in 5′→3′ direction of mRNA

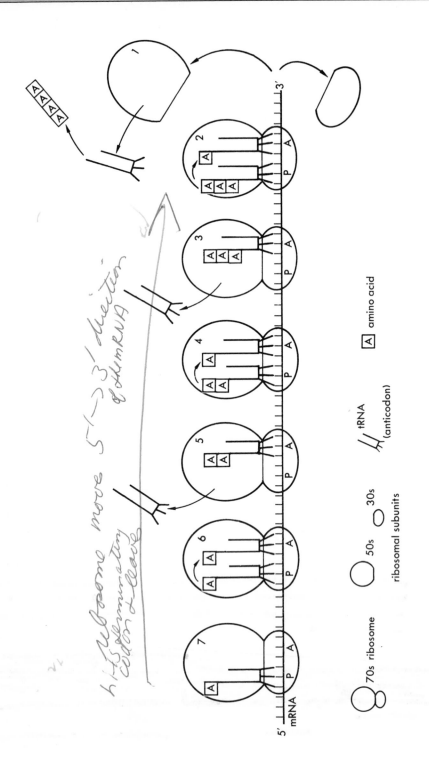

mRNA is translated 10-20 times & then is degraded

These are enzymes that hydrolyze RNA, degrading it into its constituent nucleotides, which may then be used to construct wholly new and different mRNAs. Half-life for much bacterial mRNA appears to be of the order of one to three minutes, but longer in eukaryotes—from a few hours in mouse liver to several or many days in reticulocytes. Relatively little quantitative data are available on the stability of eukaryote mRNA.

Translation occurs with considerable speed. In hemoglobin production, for example, one polypeptide chain of nearly 150 amino acid residues requires about 80 seconds for synthesis. This means addition of one amino acid in a little more than 0.5 second!

SUMMARY

In summary, translation consists of these major steps:

1. Amino acid activation.
2. Charging of tRNA.
3. Polypeptide chain initiation and elongation.
   a. Binding of charged tRNA to the mRNA-ribosome complex by complementarity of the anticodon-codon ribonucleotides.
   b. Synthesis of peptide bonds.
   c. Translocation. *movement of ribosome along mRNA in 5'-3' direction*
4. Chain termination and release.

IN VITRO PROTEIN SYNTHESIS

A superb confirmation of the general outline of protein synthesis is furnished by the work of Von Ehrenstein and Lipmann (1961), who succeeded in synthesizing hemoglobin in a cell-free system. Such a system included

1. Ribosomes in the form of polysomes from rabbit reticulocytes (immature

---

FIGURE 16-10. *Diagrammatic representation of polypeptide synthesis in* Escherichia coli. *The 30S subparticles of the 70S ribosomes successively engage the 5'-end of mRNA during the latter's functional life, displaying there at least two tRNA binding sites:* P (*peptidyl*) *and* A (*aminoacyl*). *A charged tRNA, bearing a formylated amino acid, engages the* P *site of the first ribosome, followed by arrival of a second charged tRNA at the* A site. *A peptide bond is formed between the first and second amino acids, leaving an uncharged tRNA at the* P *site and a tRNA bearing a dipeptide at the* A *site. The uncharged tRNA is released and ribosome 1 moves in the direction of the 3'-end of mRNA (left to right in this diagram) so that the binding site of the second tRNA is shifted to the original* P *location. Another codon is thereby brought into position at the new* A *site. As the first ribosome moves 5' → 3', additional ribosomes successively engage the chain-initiating site at or near the 5' end of mRNA. Ultimately, each ribosome reaches a chain-termination codon on mRNA (at or near the 3'-end) and is released, its 30S and 50S subunits dissociating. See text for details.*

In this figure all participating units are represented by highly stylized diagrams, which are not shown to scale.

genetic from mRNA into the DNA

red blood cells in which hemoglobin is almost the only protein synthesized) that included mRNA for rabbit hemoglobin.
2. An energy-yielding triphosphate.
3. tRNA from the bacterium *Escherichia coli,* previously charged with amino
   acids using *E. coli* activating enzymes.   *has already ?*
                                                        received the

The most exciting aspect of this work is that Von Ehrenstein and Lipmann
were able to synthesize a protein of one species using its mRNA and the tRNA
of a vastly different organism. Not only is universality of this general mechanism of protein synthesis strongly suggested, but the fundamental "kinship"
of all living organisms through their DNA seems inescapable. It is abundantly
clear that DNA is the genetic material of all living organisms (and many viruses, too) and that it is an information vocabulary of four "letters" (adenine,
thymine, guanine, and cytosine). Differences among species thus reside in the
sequences of the nucleotide "letters" to be formed into certain code "words"
(amino acids), which will be combined sequentially into protein "sentences"
and phenotypic "paragraphs."

But there is still another problem in translation that we must explore,
namely, just how the sequence of nucleotides in mRNA is responsible for the
particular sequence of amino acid residues of a given polypeptide and just
how synthesis of such chains starts and stops. In short, we are ready now to
examine the nature of the *genetic code.* This we will do in the next chapter.

REFERENCES

BALTIMORE, D., 1970. Viral RNA-Dependent DNA Polymerase. *Nature,* **226:**
1209–1211.

BALTIMORE, D., 1972. Properties of RNA-Dependent DNA Polymerase. In M.
Sussman, ed. *Molecular Genetics and Developmental Biology.* Englewood Cliffs,
N.J., Prentice-Hall.

BAUTZ, E. K. F., F. A. BAUTZ, and J. J. DUNN, 1969. *E. coli* Sigma Factor: A Positive
Control Element in Phage T4 Development. *Nature, 223:* 1022–1024.

BEADLE, G. W., and E. L. TATUM, 1941. Genetic Control of Biochemical Reactions
in *Neurospora. Proc. Nat. Acad. Sci. (U.S.),* 27: 499–506. Reprinted in J. A. Peters,
ed., 1959. *Classic Papers in Genetics.* Englewood Cliffs, N.J., Prentice-Hall.

BITTAR, E. E., ed., 1973. *Cell Biology in Medicine.* New York, Wiley (Interscience
Division).

BOSCH, L., ed., 1972. *The Mechanism of Protein Synthesis and Its Regulation.* New
York, American Elsevier.

BROWN, D. D., and I. B. DAWID, 1968. Specific Gene Amplification in Oocytes.
Science, **160:** 272.

BURDON, R. H., 1973. Nucleic Acid Biosynthesis and Interactions. In E. E. Bittar, ed.
Cell Biology in Medicine. New York, Wiley (Interscience Division).

BURGESS, R. R., 1971. RNA Polymerase. In E. E. Snell, ed. *Annual Review of Biochemistry,* volume 40. Palo Alto, Calif., Annual Reviews, Inc.

CLARK-WALKER, G. D., 1973. Translation of Messenger RNA. In P. R. Stewart and D. S. Letham, eds. *The Ribonucleic Acids.* New York, Springer-Verlag.

CRAMER, F., and D. H. GAUSS, 1972. Three-Dimensional Structure of tRNA. In L. Bosch, ed. *The Mechanism of Protein Synthesis and Its Regulation.* New York, American Elsevier.

DALGARNO, L., and J. SHINE, 1973. Ribosomal RNA. In P. R. Stewart and D. S. Letham, eds. *The Ribonucleic Acids.* New York, Springer-Verlag.

DAYHOFF, M. O., and R. V. ECK, 1968. *Atlas of Protein Sequence and Structure, 1967–1968.* Silver Springs, Md, National Biochemical Research Foundation.

FULLER, W., and A. HODGSON, 1967. Conformation of the Anticodon Loop in tRNA. *Nature,* **215:** 817–821.

HAENNI, A. -L., 1972. Polypeptide Chain Elongation. In L. Bosch, ed. *The Mechanism of Protein Synthesis and Its Regulation.* New York, American Elsevier.

HARRIS, H., 1975. *The Principles of Human Biochemical Genetics,* 2nd ed. New York, American Elsevier.

HASELKORN, R., and L. B. ROTHMAN-DENES, 1973. Protein Synthesis. In E. E. Snell, ed. *Annual Review of Biochemistry,* volume 42. Palo Alto, Calif., Annual Reviews, Inc.

HOLLEY, R. W., J. APGAR, G. A. EVERETT, J. T. MADISON, M. MARQUISEE, S. H. MERRILL, J. R. PENSWICK, and A. ZAMIR, 1965. Structure of a Ribonucleic Acid. *Science,* **147:** 1462–1465.

HOLMQUIST, R., T. H. JUKES, and S. PANGBURN, 1973. Evolution of Transfer RNA. *Jour. Molec. Biol.,* **78:** 91–116.

HOWELLS, A. J., 1973. Messenger RNA. In P. R. Stewart and D. S. Letham, eds. *The Ribonucleic Acids.* New York, Springer-Verlag.

KALTSCHMIDT, E., and H. G. WITTMANN, 1970. Ribosomal Proteins XII. Number of Proteins in Small and Large Ribosomal Subunits of *Escherichia coli* as Determined by Two-Dimensional Gel Electrophoresis. *Proc. Nat. Acad. Sci. (U.S.),* **67:** 1276–1282.

KAMEYANA, T., ed., 1972. *Selected Papers in Biochemistry,* volume 5, *RNA Synthesis.* Baltimore, University Park Press.

KAZIRO, Y., ed., 1971. *Selected Papers in Biochemistry,* volume 7, *Protein Synthesis.* Baltimore, University Park Press.

KIM, S. H., F. L. SUDDATH, G. J. QUIGLEY, A. MCPHERSON, J. L. SUSSMAN, A. H. J. WANG, N. C. SEEMAN, and A. RICH, 1974. Three-Dimensional Tertiary Structure of Yeast Phenylalanine Transfer RNA. *Science,* **185:** 435–439.

KNUDSON, A. G., JR., 1969. Inborn Errors of Metabolism. In H. L. Roman, ed. *Annual Review of Genetics,* volume 3. Palo Alto, Calif., Annual Reviews, Inc.

KOLAKOFSKY, D., T. OHTA, and R. E. THACH, 1968. Junction of the 50S Ribosomal Subunit with the 30S Initiation Complex. *Nature,* **220:** 244–247.

KURLAND, C. G., 1970. Ribosome Structure and Function Emergent. *Science,* **169:** 1171–1177.

LAKE, J. A., and W. W. BEEMAN, 1967. Yeast Transfer RNA: A Small-Angle X-Ray Study. *Science,* **156:** 1371–1373.

LETHAM, D. S., 1973. Transfer RNA and Cytokinins. In P. R. Stewart and D. S. Letham, eds. *The Ribonucleic Acids.* New York, Springer-Verlag.

LEVITT, M., 1969. Detailed Molecular Model for Transfer Ribonucleic Acid. *Nature,* **224:** 759–763.

LITTAUER, U. Z., and H. INOUYE, 1973. Regulation of tRNA. In E. E. Snell, ed. *Annual Review of Biochemistry*, volume 42. Palo Alto, Calif., Annual Reviews, Inc.

LUCAS-LENARD, J., and F. LIPMANN, 1971. Protein Biosynthesis. In E. E. Snell, ed. *Annual Review of Biochemistry*, volume 40. Palo Alto, Calif., Annual Reviews, Inc.

MADISON, J. T., G. A. EVERETT, and H. KUNG, 1966. Nucleotide Sequence of a Yeast Tyrosine Transfer RNA. *Science,* **153:** 531–534.

MCKUSICK, V. A., and R. CLAIBORNE, eds., 1973. *Medical Genetics.* New York, HP Publishing.

MILLER, O. L. JR., and B. A. HAMKALO, 1972. Visualization of Genetic Transcription. In M. Sussman, ed. *Molecular Genetics and Developmental Biology.* Englewood Cliffs, N. J., Prentice-Hall.

MIZUSHIMA, S., and M. NOMURA, 1970. Assembly Mapping of 30S Ribosomal Proteins from *E. coli. Nature,* **226:** 1214–1218.

MONIER, R., 1972. Structure and Function of Ribosomal RNA. In L. Bosch, ed., *The Mechanism of Protein Synthesis and Its Regulation.* New York, American Elsevier.

NOMURA, M., S. MIZUSHIMA, M. OZAKI, P. TRAUB, and C. V. LOWRY, 1969. Structure and Function of Ribosomes and Their Molecular Components. *Cold Spring Harbor Symposia Quant. Biol.,* **34:** 49–61.

O'BRIEN, J. S., S. OKADA, D. L. FILLERUP, M. L. VEATH, B. ADORNATO, P. H. BRENNER, and J. G. LEROY, 1971. Tay-Sachs Disease: Prenatal Diagnosis. *Science,* **172:** 61–64.

PETTIJOHN, D. E., O. G. STONINGTON, and C. R. KOSSMAN, 1970. Chain Termination of Ribosomal RNA Synthesis in Vitro. *Nature,* **228:** 235–239.

POLYA, G. M., 1973. Transcription. In P. R. Stewart and D. S. Letham, eds., 1973. *The Ribonucleic Acids.* New York, Springer-Verlag.

RICHARDSON, J. P., 1969. RNA Polymerase and the Control of RNA Synthesis. *Prog. Nucleic Acid Res. Molec. Biol.,* **9:** 75.

RITOSSA, F. M., and S. SPIEGELMAN, 1965. Localization of DNA Complementary to Ribosomal RNA in the Nucleolus Organizer Region of *Drosophila melanogaster. Proc. Nat. Acad. Sci. (U.S.),* **53:** 737–745.

ROBERTS, J. W., 1969. Termination Factor for RNA Synthesis. *Nature,* **224:** 1168–1174.

RUDLAND, P. S., and B. F. C. CLARK, 1972. Polypeptide Chain Initiation and the Role of a Methionine tRNA. In L. Bosch, ed. *The Mechanism of Protein Synthesis and Its Regulation.* New York, American Elsevier.

SCHAUP, H. W., M. GREEN, and C. G. KURLAND, 1970. Molecular Interactions of Ribosomal Components. I. Identification of RNA Binding Sites for Individual 30S Ribosomal Proteins. *Molec. Gen. Genet.,* **109:** 193–205.

SILVESTRI, L., 1970. *RNA-Polymerase and Transcription.* New York, Wiley (Interscience Division).

STEWART, P. R., and D. S. LETHAM, eds., 1973. *The Ribonucleic Acids.* New York, Springer-Verlag.

TEMIN, H. M., and S. MIZUTANI, 1970. RNA-Dependent DNA Polymerase in Virions of Rous Sarcoma Virus. *Nature,* **226:** 1211–1213.

THACH, S., and R. E. THACH, 1971. Translocation of Messenger RNA and 'Accommodation' of fMet-tRNA. *Proc. Nat. Acad. Sci. (U.S.),* **68:** 1791–1795.

TRAUB, P., and M. NOMURA, 1969a. Structure and Function of *E. coli* Ribosomes. VII. Mechanism of Assembly of 30S Ribosomes Studied in Vitro. *Jour. Molec. Biol.,* **40:** 391–413.

TRAUB, P., and M. NOMURA, 1969b. Studies in the Assembly of Ribosomes in Vitro. *Cold Spring Harbor Symposia Quant. Biol.,* **34:** 63–67.

WEINBERG, R. A., 1973. Nuclear RNA Metabolism. In E. E. Snell, ed. *Annual Review of Biochemistry,* volume 42. Palo Alto, Calif., Annual Reviews, Inc.

WEISSMANN, C., M. A. BILLETER, H. M. GOODMAN, J. HINDLEY, and H. WEBER, 1973. Structure and Function of Phage RNA. In E. E. Snell, ed. *Annual Review of Biochemistry,* volume 42. Palo Alto, Calif., Annual Reviews, Inc.

WITTMANN, H. G., G. STÖFFLER, C. G. KURLAND, L. RANDALL-HAZELBAUER, E. A. BIRGE, M. NOMURA, E. KALTSCHMIDT, S. MIZUSHIMA, R. R. TRAUT, and T. A. BICKLE, 1971. Correlation of 30S Ribosomal Proteins of *Escherichia coli* Isolated in Different Laboratories. *Molec. Gen. Genet.,* **111:** 327–333.

YČAS, M., 1969. *The Biological Code.* (Neuberger, A., and E. L. Tatum, eds., *Frontiers of Biology,* volume 12.) New York, American Elsevier.

ZACHAU, H. G., 1972. Transfer Ribonucleic Acids. In L. Bosch, ed. *The Mechanism of Protein Synthesis and Its Regulation.* New York, American Elsevier.

Alkapton (Homogentisic acid) accumulates since fail to produce the enzyme homogentisic acid oxidase

Phenylalanine accumulates since fail to produce the phenylalanine hydroxylase

PROBLEMS

16-1. Could PKUs be helped by administration of phenylalanine hydroxylase?

16-2. If alkaptonurics were given large quantities of parahydroxyphenylpyruvic acid, would they excrete larger quantities of alkapton? *Yes*

16-3. Would increased intake of maleylacetoacetic acid increase the excretion of alkapton? *No since this is later on in the system*

16-4. In 1963 P. D. Trevor-Roper reported a family of four normally pigmented children born to a husband and wife, both of whom were albinos. In terms of the gene symbols used in Figure 16-1, give the genotypes of these albino parents and their children. *A A a'a' × A'A'a a → gametes Aa & A'a* *progeny* *AA'aa'*

Use the following information in answering the next four problems. All microorganisms utilize thiamine (vitamin B$_1$) in their metabolism. Final steps in its synthesis involve enzymatic synthesis of a thiazole and of a pyrimidine, followed by the enzymatic combination of these two substances into thiamine. Consider the following mutant strains of *Neurospora:* strain 1 requires only simple inorganic raw materials in order to synthesize thiamine; strain 2 grows only if thiamine or thiazole is supplied; strain 3 requires thiamine or pyrimidine; strain 4 grows if thiamine or both thiazole and pyrimidine are supplied. Call the enzyme responsible for synthesis of thiazole from its precursor *enzyme "a,"* that catalyzing formation of pyrimidine *enzyme "b,"* and the enzyme catalyzing the combining of thiazole and pyrimidine into thiamine *enzyme "c."*

16-5. Which of the strains described above is prototrophic? *can synthesize* *see chap 15 p 336* *strain 1*

16-6. Which enzyme or enzymes is strain 2 incapable of producing? *a*

16-7. Which enzyme or enzymes cannot be produced by strain 4? *a & b*

precursor —a→ thiazole
—b→ pyrimidine } *—c→ thiamine*

strain 1   + + +
2   a + +
3   + b +
4   a b +

16-8. If we assign gene symbol *a* to any strain incapable of producing enzyme "a," symbol *b* to strains not making enzyme "b," and symbol *c* to those not producing enzyme "c," with + signs denoting the ability to produce a given enzyme, each strain can be represented by three gene symbols, plus signs and/or letters in the appropriate grouping. Give the genotype, according to this plan, for each of the four strains listed above.

16-9. A short segment from a long DNA molecule has this sequence of nucleotide pairs:

$$3' \text{ G T C T T T A C G C T A } 5'$$

$$5' \text{ C A G A A A T G C G A T } 3'$$

3′ G U C U U U A C G C U A 5′   mRNA

(a) If the 5′, 3′ ("lower") DNA strand serves as the template for mRNA synthesis, what will be the sequence of ribonucleotides on the latter?

(b) Give the ribonucleotide at the 5′-end of the mRNA molecule thus transcribed. *adenylic acid its fullz name*

16-10. A sedimentation coefficient is calculated as 2×10^{-11} second. Express this value in Svedberg units. 200 $S = 1 \times 10^{-13} s$

16-11. Assume 101 deoxyribonucleotide pairs to be responsible for a particular portion of a certain mRNA molecule. What is the length of that stretch of mRNA in (a) Angstrom units; (b) micrometers? *3.4(100)=340*   *b) 3.4×10⁻⁴ (100)*

use n-1

16-12. Assume a certain RNA molecule to consist of only the "usual" ribonucleotides cytidylic, uridylic, adenylic, and guanylic acids, which occur in equal number. You determine the molecular weight of that RNA molecule to be about 27,000. (a) Of how many ribonucleotides does it consist? (b) How long would this molecule be in micrometers if it were laid out in a straight line? (c) Identify the kind of RNA molecule. *a) 337/27,000 = 80*   *b) 3.4×10⁻⁴(80)*   *c) tRNA (shortest)*

337 is molecular wt of ribonucleotide (single-stranded)

16-13. Recall that the genome of phage Qβ is single-stranded RNA having a molecular weight of approximately 1×10^6. Of about how many ribonucleotides does the genome of this phage consist, if only the four "usual" bases are assumed to occur and in equal number? $\frac{1 \times 10^6}{337}$ ave mol wt of ribonucleotide

16-14. If a portion of the + strand of phage Qβ has the ribonucleotide sequence 3′ A U C G G U U A G . . . 5′, give the ribonucleotide sequence of the − strand. UAG GCC AAUC   p.372   3′

16-15. For what is reverse transcriptase responsible? RNA dependent DNA polymerase   p373

16-16. A certain codon is determined to be AUG. (a) Of what nucleic acid molecule is this codon a part? (b) What is the corresponding anticodon? (c) Of what nucleic acid molecule is this anticodon a part? (d) What is the deoxyribonucleotide sequence responsible for this codon? *Same as for tRNA except T instead of U. so TAC*   mRNA   tRNA UAC

codon on mRNA
anticodon on tRNA

16-17. Differentiate between transcription and translation.

transcription proceeds in 5′→3′ direction of the mRNA (3 to 5 direction of the DNA template) with new ribonucleotides added to the 3′ end of the elongating chain

translation begins proceeds with ribosomes moving along the mRNA in the 5′→3′ direction

CHAPTER 17
The Genetic Code

IN preceding chapters we have established that

1. DNA is *the* genetic material in all but some viruses (in which it is RNA).
2. DNA is responsible for phenotypic expression through transcription of mRNA from DNA templates.
3. Specific nucleotide sequences on mRNA interact with complementary base groups of tRNA to translate mRNA base sequences into polypeptides on ribosomal surfaces.

Studies of phenylalanine metabolism, the Tay-Sachs syndrome, of nutritionally deficient strains of the fungus *Neurospora,* and a host of others, make it clear that occurrence of a given biochemical reaction depends upon the presence of a specific enzyme that, in turn, is due to action of a particular genetic locus. Originally referred to as "one gene–one enzyme," the problem quickly developed into one of determining the precise function of genes in enzyme (and therefore protein) synthesis. Exploration of hemoglobin variants in man (Chapter 18), for example, soon indicated that genes serve to specify the amino acid sequence of proteins (and therefore their precise structure and function), rather than to act merely as "switches" determining whether protein will or will not be produced.

Thus there emerges the concept that the classical particulate gene is really a series of deoxyribonucleotides[1] with the relationship *gene → RNA → polypeptide → phenotype.* But this concept raises two fundamental questions: (1) what sequence of how many mRNA nucleotides *codes* for a particular amino acid and for a given polypeptide chain, i.e., what is the **genetic code,** and (2) in these terms, then, just how many deoxyribonucleotides equal one gene in eukaryotes? The first of these questions we shall explore in this chapter, the second in Chapter 18.

Problems of the Nature of the Code

Investigators of the genetic code faced many questions concerning its nature. Those we shall examine in this chapter include

1. How many nucleotides code for a given amino acid? That is, does a *codon* consist of one, two, three, or more nucleotides?
2. Are codons, whatever their nucleotide number, contiguous, or is there

---

[1] Or ribonucleotides in the RNA viruses.

some sort of spacer "punctuation" separating the codons? In other words, is the code *commaless?*

3. If a codon is composed of two or more nucleotides, is the code *overlapping* or *nonoverlapping?* That is, in the mRNA nucleotide sequence beginning, for example, AUCGUA . . . , how many codons are there, AU, CG, UA (nonoverlapping doublet code), or AUC, GUA (triplet, nonoverlapping), or are the codons AU, UC, CG, GU, UA, A— (doublet, overlapping by one base), or AUC, UCG, CGU, GUA, UA—, A— — (triplet, overlapping by two nucleotides), or some other arrangement?

4. What is the *"coding dictionary"?* That is, precisely which codons code for which amino acids?

5. Is a given amino acid coded for by more than one codon? That is, is the code *degenerate?*

6. Does, on the other hand, one codon code for more than one amino acid? In other words, is the code *ambiguous?*

7. Do the codons of mRNA and the corresponding amino acid residues of polypeptides occur in the same linear order? That is, is the code *colinear?*

8. Are there any codons that serve as "start" or "stop" signals for translation? In other words, are there chain-*initiating* and chain-*terminating* codons?

9. Is the code *universal* for all organisms? Does, in other words, the same codon signal for the same amino acid in phages, bacteria, corn, fruit flies, and mankind?

In succeeding chapters we shall also consider the genetic code as it may relate to the processes of mutation and recombination.

THE BASIC PROBLEM

The basic problem of the genetic code is that there are 20 amino acids that must be coded for by some sequence of four nucleotides in DNA or their complements in mRNA. The mRNA nucleotide or nucleotide sequence that codes for a particular amino acid is called a **codon.** If a codon were to consist of only a single base, the genetic code would have to be quite *ambiguous;* that is, the same codon would have to code for different amino acids under different conditions. In other words, how would four bases code for 20 amino acids? Codons of two bases present the same kind of problem; this provides only $4 \times 4 = 16$ codons for the 20 amino acids. But three-base groups provide 64 codons, *more* than enough for the number of amino acids involved. However, the system will work if the code is *degenerate* in that several codons code for the same amino acid. From the theoretical viewpoint, such a *triplet code* could include both *sense codons* (those that specify particular amino acids) and *nonsense codons* that do not specify any amino acid. Nonsense codons, might, of course, have some other function, such as signaling "start" or "stop" for polypeptide chain synthesis.

Triplet Codons. The first key to the solution was provided in 1955 by Grunberg-Manago and Ochoa, who isolated from bacteria an enzyme, polynucleotide phosphorylase, that catalyzes the polymerization of nucleoside triphosphates into a linear polyribonucleotide sequence, chemically and functionally similar to natural messenger RNA. No template is required by this enzyme and the sequence of ribonucleotides in the resulting polymer is random. Hence polyribonucleotides containing one, two, three, or all four of the different bases can be tested in vitro for their protein-synthesizing ability. To be incorporated into a polypeptide chain, some amino acids require a polyribonucleotide containing only one kind of ribonucleotide (e.g., uridylic acid); others require polyribonucleotides containing two or even three different ones. No amino acid, however, requires all four. This supplies evidence against a four-base code, but does not discriminate among one-, two-, three-, or even five- or six-base codes.

Two other lines of experimental evidence, both dating from 1961, (1) suggested that codons are triplets and (2) permitted the first specific codon assignment. Crick, Barnett, Brenner, and Watts-Tobin (1961) provided a significant clue not only as to the length of the codon but also, as we shall see, to the question of "punctuation" and overlap. To understand the relevance of their experiments we need to digress for a moment to examine one of the important classes of substances that they employed.

Frame Shifts. The acridine dyes (proflavin, acridine orange, and acridine yellow, among others) are substances that bind to DNA and, at least in phages, act as mutagens by causing additions or deletions in the nucleotide sequence during DNA replication. An acridine molecule may become intercalated between previously adjacent nucleotides, doubling the distance between them. This may either allow later insertion of a new nucleotide during replication or result in the deletion of a base, thus altering the sequence. Crick and his colleagues studied a number of acridine-induced mutants in the so-called *rII* region[2] of the DNA of phage T4. In brief, they found that mutant types arose through additions or deletions at any of a large number of sites within the *rII* region but these may revert to the wild type or a very similar one (pseudo-wild type) through deletions or additions elsewhere in the *rII* region. That is, in many cases, one mutation may be suppressed by another, particularly if the second change is located relatively close to the first. To illustrate, assume the code is triplet, nonoverlapping, and commaless, and we are dealing with a repeating deoxyribonucleotide sequence

(transcription →)

$$\overline{\text{TAG}}\ \overline{\text{TAG}}\ \overline{\text{TAG}}\ \overline{\text{TAG}}\ \overline{\text{TAG}} \ldots \tag{1}$$

Now if the second T (the fourth base) in this hypothetical series is deleted,

[2] The rII mutants are described more fully in another context in the next chapter.

[handwritten top margin: i.e. with one more amino acid changed]

[handwritten: missense mutation — the changed codon results in inactive protein (or less likely, a different protein)]

sequence (1) becomes

[handwritten: nonsense mutation — the changed codon codes for no amino acid]

[handwritten: sense mutation — the changed codon still codes for same amino acid / single]

$$\overline{TAG}\ \overline{AGT}\ \overline{AGT}\ \overline{AGT}\ \overline{AG-}\ \ldots \tag{2}$$

A deletion thus alters reading of all following groups; this is termed a *frame shift*. Transcription from the original DNA sequence (1) produces mRNA of repeating AUC codons, determining a peptide consisting only of the amino acid sequence coded for by AUC; and from (2) the codon sequence $\overline{AUC}\ \overline{UCA}\ \overline{UCA}\ \overline{UCA}\ \overline{UC-}$. This could well be expected to result in a change of amino acid composition of the peptide chain beyond the deletion, assuming, of course, that codons AUC and UCA code for different amino acids. We could represent the amino acid of the wild type (1) as aaX aaX aaX aaX . . . , but after deletion the mutant type (2) might produce the amino acid sequence aaX aaY aaY aaY . . . (a *missense* mutant). On the other hand, it might be possible a priori that UCA codes for *no* amino acid (a *nonsense* mutant).

Now, still carrying this deletion, assume a thymidylic acid (T) to be inserted at a different position in sequence (2), say between the sixth and seventh nucleotides. The DNA sequence then becomes

$$\overline{TAG}\ \overline{AGT}\ \overline{TAG}\ \overline{TAG}\ \overline{TAG}\ \ldots \tag{3}$$

and the wild-type reading is restored with the third triplet; only the second produces a misreading. Insertion of any of the other three nucleotides at the same point produces only a slightly longer faulty segment. Suppose deoxycytidylic acid (C) is inserted instead of thymidylic acid at the same point; the sequence then becomes

$$\overline{TAG}\ \overline{AGT}\ \overline{CAG}\ \overline{TAG}\ \overline{TAG}\ \ldots \tag{4}$$

[handwritten: either missense or nonsense]

Wild-type transcription in (4) is restored after two "wrong" triplets instead of one. Thus an insertion farther down the reading sequence corrects for an earlier deletion, regardless of whether the inserted base is the same as the deleted one or not. A *single frame* shift, therefore, may be expected to result in a protein so altered in amino acid sequence that it is nonfunctional. On the other hand, if the altered reading between two opposing events (e.g., a deletion followed by an insertion) is of small enough magnitude, and if the missense mutation is of such a nature as to alter little or not at all the function of the ultimately produced protein, then the second event suppresses the first. The closer the points at which these opposing alterations occur, the higher the probability that this suppression will take place. In essence, this is precisely what the work of Crick and his group showed.

We may summarize the findings of Crick and his coworkers in this way (D = deletion; I = insertion):

[handwritten: single frame shift (single deletion or addition/insertion) → missense mutants / nonsense]

[handwritten: deletion + insertion later (suppression) → function of protein is altered little or not at all]

Pseudo-wild Type	Mutant
D-I	D
I-D	I
D-D-D	D-D
I-I-I	I-I
D-D-D-D-D-D	D-D-D-D
	I-I-I-I
	D-D-D-D-D
	I-I-I-I-I

A Triplet, Commaless, Nonoverlapping Code. Note that several aspects of the genetic code are made clear by these experiments. First, the code is likely to be *triplet*, because a single frame shift results in missense, as do two, four, or five frame shifts, but pairs of opposite kinds of frame shifts restore sense. Likewise, three deletions or three insertions restore sense. Although in a doublet code, for example, one deletion followed by one insertion will also restore sense, such will not always be the case with three (or multiples of three) deletions or insertions. Still other experimental data, to be described shortly, make it clear that the code is, indeed, triplet.

In addition, this kind of restoration of sense sequences clearly suggests that (1) there is no "punctuation" between the codons, that is, each codon is immediately adjacent to the next with no intervening "spacer" bases, and (2) it is *nonoverlapping*. We shall shortly examine some additional grounds against "punctuation." Another line of evidence for nonoverlap arises in the fact that amino acid residues appear to be arranged in completely random sequence when different polypeptide chains are analyzed; no one amino acid always, or even usually, has the same adjacent neighbors. This could be the case only if the code is *nonoverlapping*. If the code did overlap, a given amino acid would always have the same nearest neighbors. An mRNA sequence beginning, for example, A-A-C-C-G-A-G-C-A- . . . consists of three triplets, AAC, CGA, and GCA, which code for asparagine, and arginine, and alanine, respectively (Table 17-1). If the code overlapped by two bases, this sequence of nine bases would consist of the triplets AAC, ACC, CCG, CGA, GAG, AGC, and GCA, coding for asparagine, threonine, proline, arginine, glutamic acid, serine, and alanine. Thus in that particular sequence of ribonucleotides, arginine would always occur between proline and glutamic acid. This is not the case.

Moreover, a replacement that involves a single base pair in DNA would, in an overlapping code, affect more than one amino acid. This does not occur; only single amino acids are changed by such "single site" mutations.

THE CODING DICTIONARY

Once the genetic code was established as a *commaless, nonoverlapping, triplet* code, the question of which triplets code for which amino acids was pursued vigorously. Nirenberg and Matthaei, in the second of the important 1961 papers, pioneered in efforts to crack the code. It is possible to construct short synthetic polyribonucleotides and to test them in cell-free systems for ability to direct incorporation of specific amino acids into polypeptide chains. Using a mixture of amino acids, with a different one radioactively labeled in each run, in a cell-free suspension derived from *Escherichia coli* (tRNA, ribosomes, ATP, GTP, necessary enzymes, and, interestingly enough, mRNA from tobacco mosaic virus), these men were able to bring about polypeptide synthesis in vitro. With polyribouridylic acid, for example, they were able to demonstrate the synthesis of a polypeptide consisting only of phenylalanine, even though other amino acids were present. So the mRNA codon for phenylalanine must be UUU.

Testing copolymers of only the ribonucleotides adenylic and cytidylic acids (poly-AC), Nirenberg and Matthaei showed that proline was coded by at least two triplets, one consisting only of cytidylic acid (CCC) and the other of both cytidylic and adenylic acids. Poly-A (AAA) was soon found to code for lysine. These determinations reinforce the concept of a commaless code for, in a code that includes punctuation, U and A would each have to serve both to code for their respective amino acids as well as to function as commas. From this time on the notion of a commaless code won general acceptance.

Testing random copolymers sheds some additional light on the code. When synthetic mRNA is constructed, it is found that the sequence of nucleotides is random, so that the relative frequency of incorporation of particular nucleotides is mathematically determined. Thus a mixture containing two parts uracil to one part guanine is found to produce triplets in these combinations:

$$UUU \quad \tfrac{2}{3} \times \tfrac{2}{3} \times \tfrac{2}{3} = \tfrac{8}{27}$$
$$GGG \quad \tfrac{1}{3} \times \tfrac{1}{3} \times \tfrac{1}{3} = \tfrac{1}{27}$$
$$UGU \quad \tfrac{2}{3} \times \tfrac{1}{3} \times \tfrac{2}{3} = \tfrac{4}{27}$$
$$UUG \quad \tfrac{2}{3} \times \tfrac{2}{3} \times \tfrac{1}{3} = \tfrac{4}{27}$$
$$GUU \quad \tfrac{1}{3} \times \tfrac{2}{3} \times \tfrac{2}{3} = \tfrac{4}{27}$$
$$GGU \quad \tfrac{1}{3} \times \tfrac{1}{3} \times \tfrac{2}{3} = \tfrac{2}{27}$$
$$UGG \quad \tfrac{2}{3} \times \tfrac{1}{3} \times \tfrac{1}{3} = \tfrac{2}{27}$$
$$GUG \quad \tfrac{1}{3} \times \tfrac{2}{3} \times \tfrac{1}{3} = \tfrac{2}{27}$$

If each of the possible triplets were to code for a different amino acid, then various ones would be incorporated in the proportions shown. Except for the fact that *each* such three-base group does not code a *different* amino acid, this is essentially what occurs. Use of a synthetic poly-UG, in the relative proportions just given, produced a mixture of polypeptides, of which $\tfrac{8}{27}$ were polyphenylalanine.

In this way it was known that poly-UG containing 2 U : 1 G codes for valine, but the sequence of bases, and therefore the exact codon(s) of those possible, cannot be determined from such random copolymers. Nirenberg and Leder

(1965) devised a method by which short-chain polyribonucleotides of known sequence could be obtained. These were introduced into cell-free systems that included ribosomes and a variety of tRNA molecules charged with their amino acids. As in earlier work, one amino acid in each experimental run was labeled with ^{14}C. Although the messengers used were too short for protein synthesis, two important considerations emerged: (1) little or no binding of tRNA took place in the presence of dinucleotide messengers, but occurred preferentially with trinucleotides, and (2) different sequences of the same three bases stimulated binding of different amino acids. Thus the triplet nature of the code was again confirmed, and the way opened to develop a complete coding dictionary. The mRNA triplet code, as established principally in cell-free systems prepared from *E. coli,* is shown in Table 17-1. Kurland (1970) refers to the elucidation of the genetic code as "one of the principal triumphs of molecular biology," and Garen (1968) calls it "a notable milestone in biology," as, indeed, it is.

DEGENERACY

Although the code is extensively degenerate, as seen in Table 17-1, a certain order to this degeneracy can be discerned. In many instances it is the first two bases that are the characteristic and critical parts of the codon for a given amino acid, whereas the third may be read either as a purine only (e.g., glutamine, CAA and CAG), or as a pyrimidine only (e.g., histidine, CAU and CAC). In other cases, however, the third position is read as any base, for example, AC− for threonine, CC− for proline, and so forth.

In 1966 Crick proposed his "**wobble hypothesis**" to account for this lack of specificity in the third base of many codons. He inferred that the first two bases of the triplet codon pair according to the "rules" previously set forth (i.e., A with U, and G with C), but that play, or freedom, in the third position would permit more than one type of pairing there. From a consideration of the molecular structure of the several base molecules, Crick proposed certain pairing rules for the third position (Table 17-2). The wobble hypothesis explains a good bit of the degeneracy in the code, and why most of the "synonymous" codons occur in the same group in Table 17-1 (e.g., CC− for proline, and so forth).

But we also find some latitude in the first two positions. For example, arginine, leucine, and serine each have two different characteristic base pairs in the first two positions (Table 17-1). In addition, pseudouridine (Ψ), which occurs in the second position of the anticodon of yeast tyrosine tRNA, pairs with adenine (A). Methionine and tryptophan, having only one codon each (AUG and UGG, respectively), are exceptions to this general degeneracy.

AMBIGUITY

Essentially, the code is nonambiguous in vivo under natural conditions. Ambiguity is encountered chiefly in cell-free systems under certain conditions.

TABLE 17-1. The mRNA Code as Determined in Vitro for *E. coli*.
(Degeneracies are shown; ambiguities are omitted.)

AA- AAU, AAC } Asparagine AAA, AAG } Lysine	**CA-** CAU, CAC } Histidine CAA, CAG } Glutamine	**GA-** GAU, GAC } Aspartic acid GAA, GAG } Glutamic acid	**UA-** UAU, UAC } Tyrosine UAA End Chain* UAG End Chain†
AC- ACU, ACC, ACA, ACG } Threonine	**CC-** CCU, CCC, CCA, CCG } Proline	**GC-** GCU, GCC, GCA, GCG } Alanine	**UC-** UCU, UCC, UCA, UCG } Serine
AG- AGU, AGC } Serine AGA, AGG } Arginine	**CG-** CGU, CGC, CGA, CGG } Arginine	**GG-** GGU, GGC, GGA, GGG } Glycine	**UG-** UGU, UGC } Cysteine UGA End Chain UGG Tryptophan
AU- AUU, AUC, AUA } Isoleucine AUG Methionine	**CU-** CUU, CUC, CUA, CUG } Leucine	**GU-** GUU, GUC, GUA, GUG } Valine	**UU-** UUU, UUC } Phenylalanine UUA, UUG } Leucine

* Originally called *ochre*.
† Originally called *amber*.

TABLE 17-2. Pairing
Between Codon and
Anticodon at the
Third Position

Anticodon	Codon
A	U
C	G
G	U, C
I (inosine)	U, C, A
U	A, G

Based on the work of Crick,
1966.

In such a system prepared from a streptomycin-sensitive strain of *E. coli* UUU (which ordinarily codes for phenylalanine) may also code for isoleucine, leucine, or serine in the presence of streptomycin. This ambiguity is enhanced at high magnesium-ion concentrations. Poly-U, in a cell-free system from thermophilic bacterial species, has been found to bind leucine at temperatures well below the optimum for growth of living cells of the same species (Friedman and Weinstein, 1964). Changes in pH or the addition of such substances as ethyl alcohol also result in ambiguity. However, ambiguities are not ordinarily encountered in vivo under normal growing conditions for a given species.

COLINEARITY

In a *colinear* code, the sequence of mRNA codons and the corresponding amino acid residues of a polypeptide chain are arranged in the same linear sequence. Now, both DNA and mRNA on the one hand, and polypeptides on the other, are linear, but this, in itself, does not require *colinearity*. That such colinearity does exist, however, is shown by studies of T4 mutants which produce incomplete head protein molecules. These mutants can be shown to map in linear sequence by using recombination techniques. Length of the incomplete head protein molecule made by each mutant is precisely proportional to the map distance involved. Thus the code is determined to be *colinear*. The same conclusions are reached in the work of Yanofsky and his colleagues (1964) on the tryptophan synthetase gene system of the colon bacillus. We shall refer again to Yanofsky's studies in the next chapter. We shall also examine in Chapter 18 similar evidence of colinearity in certain types of human hemoglobin.

CHAIN INITIATION AND TERMINATION

All experimental data clearly indicate that mRNA is synthesized in the $5' \rightarrow 3'$ direction (Chapter 16). The assembly of polypeptide chains is also

see Figure 16-4 p371

sequential, from the amino to the carboxyl termini. The question, then, is what sort of signal (if any) directs the initiation and termination of polypeptide synthesis along mRNA?

Chain Initiation. It appears clear that polypeptide chains in *Escherichia coli* are initiated with N-formylmethionine (Webster et al., 1966; Rudland and Clark, 1972; and Maden, 1973, among many others), which, in some instances, is enzymatically removed before assembly into proteins. To illustrate, the coat protein of the RNA phage R17 begins with the N-terminal sequence alanine-serine-asparagine-phenylalanine-threonine . . . (Adams and Capecchi, 1966). In in vitro systems this sequence is preceded by N-formylmethionine; in such systems it is evident that the necessary enzyme to remove the formylated methionine is lacking.

Only one codon, AUG, exists for methionine (Table 17-1), so the question naturally arises as to how chain-initiating N-formylmethionine and internally located methionine are distinguished in the biosynthesis of polypeptides. The answer in *E. coli*, as well as most if not all organisms, lies in the occurrence of two different tRNAs for methionine (Marcker and Sanger, 1964; Clark and Marcker, 1966). One of these, N-formylmethionyl-tRNA (symbolized as $tRNA_f$ or as fmet-tRNA) is formylated and serves only for initiation of polypeptide synthesis. The other, methionyl-tRNA ($tRNA_m$ or met-tRNA), does not serve as a substrate for the formylating enzyme, and inserts its methionine only into intercalary positions, thus functioning in elongation rather than in initiation. The anticodon for both kinds of tRNA is 3'UAC5'; therefore the same AUG codon codes for both formylated and nonformylated methionine. Fmet-tRNA and met-tRNA do, however, differ in many of their internal nucleotides (Fig. 17-1). Clark-Walker (1973) reports that "fmet-tRNA is required for the initiation of all bacterial proteins," although the formyl group, and in some cases the entire methionine residue, is enzymatically removed from the polypeptide chain at some point prior to its incorporation into protein. There is growing evidence that methionine, but not N-formylmethionine, functions in initiation in mammalian cells. A special met-tRNA is responsible for insertion of methionine into the N-terminal position (Brown and Smith, 1970), although it is removed in some cases, e.g., rabbit globin, before the polypeptide chain is incorporated into hemoglobin (Jackson and Hunter, 1970). Initiation is more than a simple case of an initiation codon alone; at least three protein initiation factors are involved as well (Clark-Walker, 1973; Lengyel and Söll, 1969).

Hartman and Suskind (1969) point out that AUG determines the reading frames of mRNA. The synthetic ribonucleotide AUGGUUUUUUUU . . . is translated only as N-formylmethionine-valine-phenylalanine-phenylalanine . . . ; that is, the reading is AUG-GUU-UUU-UUU . . . and GGU (glycine) is not read as such. Translation, once initiated, continues until a "stop" codon (see next section) is encountered. It would thus appear logical that a chain-initiating AUG should occur adjacent to, or at least fairly close to, a preceding

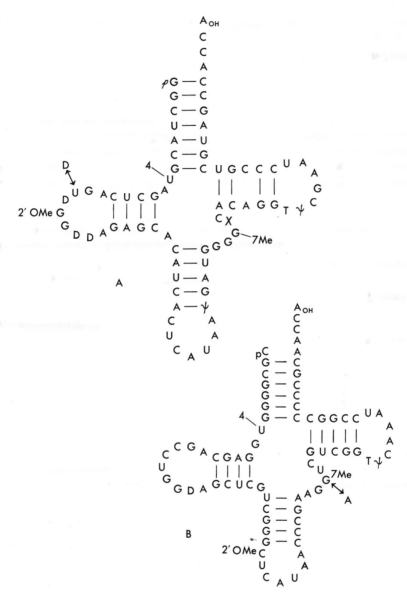

FIGURE 17-1. *Ribonucleotide sequences of (A)* tRNA$_m$, *and (B)* tRNA$_f$. *Unusual ribonucleotides are symbolized as follows: D, dihydrouridylic acid; Ψ, pseudouridylic acid; T, thymidylic acid; 7MeG, 7-methylguanylic acid; 2 OMeC, 2 O-methylcytidylic acid; 2 OMeG, 2 O-methylguanylic acid; 4U, 4-thiouridylic acid.* (Based on work of S. K. Dube, K. A. Marcker, B. F. C. Clark, and S. Cory, 1968. Nucleotide Sequence of N-Formyl-Methionyl-Transfer RNA. *Nature*, **218**:232–233, and S. Cory, K. A. Marcker, S. K. Dube, and B. F. C. Clark, 1968. Primary Structure of a Methionine Transfer RNA from *Escherichia coli. Nature*, **220**:1039–1040.)

chain-terminating codon. Otherwise, AUG would be read as an internal site by tRNA$_m$.

However, Steitz (1969) reports that RNA phage R17 has about 3,300 ribonucleotides, of which some 3,000 code for three proteins (phage coat, replicase enzyme, and a maturation protein involved in assembly of coat and RNA into a mature virus). She was able to isolate the beginning sections of each of the three genes with their initiator regions. Each initiator region began with AUG, but was *not immediately* preceded by any of the chain terminators that are described in the next section. Steitz believes that untranslated sequences intervene between the three structural genes of R17 and that chain termination and initiation codons may, therefore, be rather widely spaced. Various workers with different phages report that AUG codons do not occur until the 60th to 100th position from the 5'-end, depending on the phage analyzed. The role of the untranslated bases has not been fully clarified.

Chain Termination.   Three codons, UAA, UAG, and UGA, do not code for any amino acids and hence are termed *nonsense codons* (Table 17-1). Before their base sequences were determined, UAA was known as "ochre" and UAG as "amber"—terms that are still extensively employed for these two codons as a matter of convenience. Evidence from both in vitro and in vivo experiments demonstrates that all three nonsense codons cause both chain termination and chain release.

Last and others (1967), using short synthetic ribonucleotides (oligonucleotides) showed that UAA is responsible for both termination and release. Among their results, that from using the messenger AUG UUU UAA AAA . . . AAA is highly significant in that only formylmethionyl-phenylalanine *dipeptides* were isolated and recovered. Thus the coding properties of AUG and UUU were verified, the ability of UAA to bring about chain *termination* was demonstrated, and incidentally, the triplet nature of the code was given further proof. It should be remembered, however, that release requires certain protein release factors (Capecchi and Klein, 1970) as well. In highly purified cell-free systems, in which these release factors are absent, no release takes place.

Similarly, Zinder and colleagues (1966) showed that UAG ("amber") also serves in vitro in the same manner. Using natural RNA derived from the RNA phage f2, peptides terminating at the UAG site were recovered. Work of Sarabhai and Brenner (1967a, 1967b) indicates that UGA also serves as a chain terminator.

Stretton and Brenner (1965) used amber mutants of phage T4 (i.e., mutants in which the triplet UAG replaced another by single base substitution) to demonstrate that this codon functions as a chain terminator in vivo. In their work incomplete head proteins were formed, released from ribosomes, and recovered, the length of the protein being directly related to the position of amber. This also suggests that chain termination probably operates in translation rather than in transcription, though some interpretations suggest operation at the transcription level.

UNIVERSALITY

Significantly, the current evidence is that the code is universal for all living organisms and for viruses; the same triplets code similarly in a wide variety of organisms (Table 17-3). Compare codons for isoleucine, as an illustration, for the four very different species listed.

It should be expected that DNA from different species ought to exhibit degrees of similarity in nucleotide sequences in proportion to the closeness of evolutionary relationship of those species. Zoologists have long agreed that man and monkey are more closely related than is either to fish or bacteria, for example. Significant chemical proof was demonstrated in 1964 by Hoyer, McCarthy, and Bolton.

TABLE 17-3. Comparison of mRNA Codons for Twenty Amino Acids in Different Organisms*

Amino Acid	Escherichia coli (a bacterium)	Rat Liver	Wheat Embryo (flowering plant)	Chlamy-domonas (a green alga)
Alanine	GCU GCA GCC GCG			
Arginine	CGU CGA CGC CGG AGA AGG			
Asparagine	AAU AAC			
Aspartic acid	GAU GAC		GAU	
Cysteine	UGU UGC AGU AGC			
Glutamine	CAA CAG			
Glutamic acid	GAA GAG		GAU(?)	
Glycine	GGU GGA GGC GGG	GGU	GGU	
Histidine	CAU CAC			
Isoleucine	AUU AUC AUA	AUU	AUU	AUU
Leucine	CUU CUA CUC CUG UUA UUG		CUU CUC	
Lysine	AAA AAG	AAA	AAU (?)	
Methionine	AUG		AUG	
Phenylalanine	UUU UUC	UUU	UUU	UUU
Proline	CCU CCA CCC CCG		CCU	
Serine	UCU UCA UCC UCG AGU AGC	UCU	UCC	UCU
Threonine	ACU ACA ACC ACG			
Tryptophan	UGG	UGG	UGG	
Tyrosine	UAU UAC	UAU	UAU	UAU
Valine	GUU GUA GUC GUG	GUU	GUU GUG	GUU

* After Groves and Kempner (1967).

Their method was simple in concept but delicate in operation. DNA to be tested was first made to undergo strand separation by heating (denaturation), then cooled quickly to prevent recombining, and immobilized in agar. To this were added from another species short strands of denatured DNA that had previously been made radioactive by incorporation of carbon-14 or phosphorus-32. Pairing of homologous nucleotides from the two species occurred during several hours of incubation. The "hybrid" DNA was then recovered and assayed for radioactivity, yielding a measure of base-pair homology between the two species. Their results are summarized in Table 17-4.

TABLE 17-4. Percentage Recombining of ^{14}C-Labeled Human DNA and ^{32}P-Labeled Mouse DNA with Unlabeled DNA of Other Species*

	Percentage Labeled DNA Bound	
Unlabeled DNA Source	^{14}C Human	^{32}P Mouse
Human	18*	5
Rhesus monkey	14	8
Mouse	6	22*
Cattle	5	4
Rat	4	14
Guinea pig	4	3
Hamster.	4	12
Rabbit	4	3
Salmon	1.5	1.5
Colon bacillus	0.4	0.4

* Human-human and mouse-mouse results serve as a base of comparison.
Data from Hoyer, McCarthy, and Bolton (1961).

The significance of these data lies in the similarity of probably fairly long sequences between man and monkey and between mouse and rat, and the evident dissimilarity of either with such taxonomically distant species as salmon or the colon bacillus. Hoyer and his colleagues conclude,

This observation raises the question of whether there exists among the various animals a particular class of nucleotide sequences which have been retained during the diversification of the vertebrate forms. . . . It is clear . . . that there exist homologies among polynucleotide sequences in the DNAs of such diverse forms as fish and man. These sequences represent genes which have been conserved with relatively little change throughout the long history of vertebrate evolution. Although we have no means yet of relating such genes to particular phenotypic expressions, it is conceivable that they are the determinants of the fundamental conservative characteristics of the vertebrate form.

These latter would include such fundamental traits as skeletal structure and hemoglobin production. Man and mouse are mammals, having many traits in common, but within the primate group, man and monkey are phenotypically similar enough that rather long segments of their genetic material are alike.

Certainly of equal significance in this regard is the fact that, as we noted in Chapter 16, in vitro protein synthesis is successful using rabbit reticulocyte mRNA and *E. coli* tRNA (plus other necessary components). The result was apparently normal rabbit hemoglobin, showing that bacterial tRNA can "recognize" mRNA and polysomes from a taxonomically very different organism. It is true, as has been reported frequently in the literature, that taxonomically divergent organisms do differ in the degree to which a given codon responds to a particular species of tRNA. For example, AAG codes readily for lysine in vertebrates, but does so more weakly in *E. coli*. But this is only a difference in *degree,* not in kind.

Furthermore, such proteins as cytochrome c show uniformity in several sequences of amino acid residues in such very different groups as mammals, fishes, yeasts, and bacteria even though their evolutionary divergence must have taken place many hundreds of millions of years back in geologic time.

The question of *evolution* of the genetic code is, as you might guess, replete with problems. Its very universality complicates the question, as does the paradoxical situation wherein operation of the code requires precise functioning of many enzymes that, it would seem, could not be produced without the translation mechanism for which they are required. Woese (1970) discusses these problems in detail and suggests possible avenues of evolution.

In the next chapter we shall approach the concept of the ultimate structure of the gene through an examination of the process of mutation at the molecular level.

SUMMARY OF CODE CHARACTERISTICS

In summary, the evidence is that the genetic code is *triplet, commaless, nonoverlapping, degenerate, essentially nonambiguous* under natural conditions, *colinear,* and *universal.* In addition, polypeptide chain *initiation* is signaled by certain codons (notably AUG) that bind tRNAs carrying blocked amino acids, although specific protein initiation factors are also required. Chain *termination* is governed by three nonsense codons (UAA, UAG, and UGA), with chain *release* requiring protein release factors as well.

REFERENCES

ADAMS, J. M. and M. R. CAPECCHI, 1965. N-formylmethionine-sRNA as the Initiator of Protein Synthesis. *Proc. Nat. Acad. Sci. (U.S.),* **55:** 147–155.

BEAUDET, A. L., and C. T. CASKEY, 1972. Polypeptide Chain Termination. In L. Bosch, ed. *The Mechanism of Protein Synthesis and Its Regulations.* New York, American Elsevier.

BOREK, E., 1969. *The Code of Life.* New York, Columbia University Press.

BRENNER, S., L. BARNETT, E. R. KATZ, and F. H. C. CRICK, 1967. UGA: A Third Nonsense Triplet in the Genetic Code. *Nature,* **213:** 449–450.

BRENNER, S., A. O. W. STRETTON, and S. KAPLAN, 1965. Genetic Code: The "Nonsense" Triplets for Chain Termination and Their Suppression. *Nature,* **206:** 994–998.

BROWN, J. C., and A. E. SMITH, 1970. Initiator Codons in Eukaryotes. *Nature,* **226:** 610–612.

CAPECCHI, M. R., and H. A. KLEIN, 1970. Release Factors Mediating Termination of Complete Proteins. *Nature,* **226:** 1029–1033.

CLARK, B. F. C., and K. A. MARCKER, 1966. The Role of N-Formyl-Methionyl-sRNA in Protein Biosynthesis. *Jour. Molec. Biol.,* **17:** 394–406.

CLARK-WALKER, G. D., 1973. Translation of Messenger RNA. In P. R. Stewart and D. S. Letham, eds. *The Ribonucleic Acids.* New York, Springer-Verlag.

CRICK, F. H. C., 1963. On the Genetic Code. *Science,* **139:** 461–464.

CRICK, F. H. C., 1966. Codon-Anticodon Pairing: the Wobble Hypothesis. *Jour. Molec. Biol.,* **19:** 548–555.

CRICK, F. H. C., L. BARNETT, S. BRENNER, and R. J. WATTS-TOBIN, 1961. General Nature of the Genetic Code for Proteins. *Nature,* **192:** 1227–1232.

FRIEDMAN, S. M., and I. B. WEINSTEIN, 1964. Lack of Fidelity in the Translation of Synthetic Polyribonucleotides. *Proc. Nat. Acad. Sci (U.S.),* **52:** 988–995.

FRISCH, L., ed., 1967. The Genetic Code. Cold Spring Harbor Symposia Quant. Biol., **31** (1966). Cold Spring Harbor Laboratory of Quantitative Biology, Cold Spring Harbor, Long Island, New York.

GAREN, A., 1968. Sense and Nonsense in the Genetic Code. *Science,* **160:** 149–159.

GOLDBERG, A. L., and R. E. WITTES, 1966. Genetic Code: Aspects of Organization. *Science,* **153:** 420–424.

GROVES, W. E., and E. S. KEMPNER, 1967. Amino Acid Coding in *Sarcina lutea* and *Saccharomyces cerevisiae. Science,* **156:** 387–390.

GRUNBERG-MANAGO, M., and S. OCHOA, 1955. Enzymatic Synthesis and Breakdown of Polynucleotides: Polynucleotide Phosphorylase, *Jour. Amer. Chem. Soc.,* **77:** 3165–3166.

HARTMAN, P. E., and S. R. SUSKIND, 1969. *Gene Action.* Englewood Cliffs, N.J., Prentice-Hall.

HOSMAN, D., D. GILLESPIE, and H. F. LODISH, 1972. Removal of Formyl-methionine Residue from Nascent Bacteriophage f2 Protein. *Jour. Molec. Biol.,* **65:** 163–166.

HOYER, B. H., B. J. McCARTHY, and E. T. BOLTON, 1964. A Molecular Approach in the Systematics of Higher Organisms. *Science,* **144:** 959–967.

JACKSON, R., and T. HUNTER, 1970. Role of Methionine in Initiation of Haemoglobin Synthesis. *Nature,* **227:** 672–676.

LAST, J. A., W. M. STANLEY, M. SALAS, M. B. HILLE, A. J. WAHBA, and S. OCHOA, 1967. Translation of the Genetic Message. IV. UAA as a Chain Terminating Codon. *Proc. Nat. Acad. Sci. (U.S.),* **57:** 1062–1067.

LEDER, P., and M. W. NIRENBERG, 1964. RNA Codewords and Protein Synthesis, III. On the Nucleotide Sequence of a Cysteine and a Leucine RNA Codeword. *Proc. Nat. Acad. Sci. (U.S.),* **52:** 1521–1529.

LENGYEL, P., and D. SÖLL, 1969. Mechanism of Protein Biosynthesis. *Bacteriol. Rev.,* **33:** 264–301.

MADEN, B. E. H., 1973. Protein Synthesis in Animal Cells. In E. E. Bittar, ed. *Cell Biology in Medicine.* New York, Wiley (Interscience Division).

MARCKER, K., and F. SANGER, 1964. N-formylmethionyl-S-RNA. *Jour. Molec. Biol.,* **8:** 835–840.

MARSHALL, R. E., C. T. CASKEY, and M. NIRENBERG, 1967. Fine Structure of RNA Codewords Recognized by Bacterial, Amphibian, and Mammalian Transfer RNA. *Science,* **155:** 820–825.

NIRENBERG, M., and P. LEDER, 1964. RNA Codewords and Protein Synthesis. *Science,* **145:** 1319–1407.

NIRENBERG, M. W., and J. H. MATTHAEI, 1961. The Dependence of Cell-Free Protein Synthesis in *E. coli* upon Naturally Occurring or Synthetic Polyribonucleotides. *Proc. Nat. Acad. Sci. (U.S.),* **47:** 1588–1602.

REVEL, M., 1972. Polypeptide Chain Initiation: the Role of Ribosomal Protein Factors and Ribosomal Subunits. In L. Bosch, ed. *The Mechanism of Protein Synthesis and its Regulation.* New York, American Elsevier.

SARABHAI, A. S., and S. BRENNER, 1967a. Further Evidence that UGA Does Not Code for Tryptophan. *Jour. Molec, Biol.,* **26:** 141–142.

SARABHAI, A. S., and S. BRENNER, 1967b. A Mutant which Reinitiates the Polypeptide Chain after Chain Termination. *Jour. Molec. Biol.,* **27:** 145–162.

SARABHAI, A. S., A. O. W. STRETTON, and S. BRENNER, 1964. Co-linearity of the Gene with the Polypeptide Chain. *Nature,* **201:** 13–17.

STEITZ, J. A., 1969. Polypeptide Chain Initiation: Nucleotide Sequences of the Three Ribosomal Binding Sites in Bacteriophage R17 RNA. *Nature,* **224:** 957–964.

STRETTON, A. O. W., and S. BRENNER, 1965. Molecular Consequences of the *Amber* Mutation and its Suppression. *Jour. Molec. Biol.,* **12:** 456–465.

TAKEDA, M., and R. E. WEBSTER, 1968. Protein Chain Initiation and Deformylation in *B. subtilis* Homogenates. *Proc. Nat. Acad. Sci. (U.S.),* **60:** 1487–1494.

WEBSTER, R. E., D. L. ENGLEHARDT, and N. D. ZINDER, 1966. In Vitro Protein Synthesis: Chain Initiation. *Proc. Nat. Acad. Sci. (U.S.),* **55:** 155–161.

WHITFIELD, H., 1972. Suppression of Nonsense, Frameshift, and Missense Mutations. In L. Bosch, ed. *The Mechanism of Protein Synthesis and its Regulation.* New York, American Elsevier.

WOESE, C. R., 1967. *The Genetic Code—the Molecular Basis for Genetic Expression.* New York, Harper & Row.

WOESE, C. R., 1970. The Problem of Evolving a Genetic Code. *BioScience,* **20:** 471–485.

YANOFSKY, C., B. C. CARLTON, J. R. GUEST, D. R. HELINSKI, and U. HENNING, 1964. On the Colinearity of Gene Structure and Protein Structure. *Proc. Nat. Acad. Sci. (U.S.),* **51:** 266–272.

YČAS, M., 1969. *The Biological Code.* (Neuberger, A., and E. L. Tatum, eds., *Frontiers of Biology,* volume 12.) New York, American Elsevier.

ZINDER, N. D., D. L. ENGLEHARDT, and R. E. WEBSTER, 1966. Punctuation in the Genetic Code. *Cold Spring Harbor Symposia Quant. Biol.,* **31:** 251–256.

ZUBAY, G. L., ed., 1968. *Papers in Biochemical Genetics.* New York, Holt. (A reprinting of some of the most important original papers in this field between 1953 and 1967.)

[handwritten top annotations:]
DNA 3' TAC CGG AAT TGC 5'
mRNA 5' AUG GCC UUA ACG 3'
met ala leu threonine

PROBLEMS

17-1. Assume a length of template <u>DNA</u> with the deoxyribonucleotide sequence 3' T A C C G G A A T T G C 5'. (a) If the code is triplet, nonoverlapping, and commaless, of which amino acid residues (in sequence) will the polypeptide chain for which this stretch of DNA is responsible consist? (b) If the code is triplet, overlapping by two bases, and commaless, how would you answer the preceding question? (c) Of what significance is the TAC triplet in the DNA template? *chain-initiating codon*

17-2. For the DNA template of the preceeding problem assume the second C to be deleted. What now is the sequence of amino acid residues coded for if the code is assumed to be triplet, nonoverlapping, and commaless? *so AUG CCU UAA CG* *Met Pro end*

17-3. In the DNA length of problem 17-1 assume the second C to be deleted and a T to be inserted after the GG sequence so that the DNA strand now reads 3' T A C G G T A A T T G C 5'. (a) How does the amino acid sequence now coded for compare with your answer to part (a) of problem 17-1? (Assume the code to be triplet, nonoverlapping, and commaless.) (b) Does your answer to the preceding part of this problem illustrate a sense, a missense, a nonsense mutation, or none of these? *missense (sense is restored with the third amino acid)*

[handwritten left margin:] mRNA: AUG CCA UUA ACG Met Pro Leu Thr

17-4. Synthetic mRNA is constructed from a mixture of ribonucleotides supplied to a cell-free system in this relative proportion: 3 uracil:2 guanine:1 adenine. What fraction of the resulting triplets would be (a) UGA; (b) UUU?

[handwritten left margin:]
(a) $\frac{3}{6} \cdot \frac{2}{6} \cdot \frac{1}{6} = \frac{6}{216} = \frac{1}{36}$
(b) $\frac{3}{6} \cdot \frac{3}{6} \cdot \frac{3}{6} = \frac{27}{216} = \frac{1}{8}$

17-5. The human hemoglobin molecule includes four polypeptide chains, two α chains of 141 amino acid residues each, and two β chains of 146 amino acid residues each. Neglecting chain-initiating, chain-terminating, and the possibility of untranslated codons, (a) of how many ribonucleotides does the mRNA molecule responsible for the α polypeptide chain consist? (b) What is the length in micrometers of that mRNA molecule?

[handwritten left margin:]
(a) 3(141)=423
(b) 3.4×10⁻⁴(423) = .14

17-6. Assume an alanine tRNA charged with labeled alanine is isolated and the amino acid chemically treated so as to change it to labeled glycine. The treated amino acid-enzyme-tRNA complex is then introduced into a cell-free peptide synthesizing system. At which of two mRNA triplets, say GCU or GGU, would this tRNA now become bound? Why? *still GCU since tRNA anticodon hasn't changed* *ala gly*

17-7. If single-base changes occur in DNA (and therefore in mRNA), which amino acid, tryptophan or arginine, is most likely to be replaced by another in protein synthesis? *only 1 codon codes for tryptophan so it is most likely to be replaced*

17-8. Wittmann-Liebold and Wittman studied a number of mutants in the coat protein of the RNA-containing tobacco mosaic virus. Two of their mutants were

Mutant	Amino Acid Position	Replacement
A-14	129	isoleucine → threonine
Ni-1055	21	isoleucine → methionine

By reference to Table 17-1, explain what has happened in each of these cases.

[handwritten bottom annotations:]
isoleucine { AUU, AUC, AUA } thr { ACU, ACC, ACA } 2nd base changed to C
met AUG 3rd base changed to G

17-9. Yanofsky et al. studied a large number of mutants for the tryptophan synthetase A polypeptide chain of *Escherichia coli.* This polypeptide chain consists of 267 amino acid residues. In the wild-type enzyme a part of the amino acid sequence is: -tyrosine-leucine-threonine-glycine-glycine-glycine-glycine-glycine-serine-. In their mutant A446, cysteine replaces tyrosine; in mutant A187, the third glycine is replaced by valine. By reference to the genetic code, suggest a mechanism for each of these amino acid replacements.

17-10. How many different mRNA nucleotide codon combinations can exist for the internal pentapeptide threonine-proline-tryptophan-leucine-isoleucine?

17-11. From Table 17-1, note that (a) UUU and UUC code for phenylalanine; on the other hand, (b) So and Davie report that, with relatively high concentrations of ethyl alcohol, the incorporation of leucine, and isoleucine to a lesser degree, is sharply increased, whereas incorporation of phenylalanine is decreased, for these same codons. Which of the described situations, (a) or (b), represents ambiguity, and which degeneracy? → a

17-12. Assume a series of different one-base changes in the codon GGA, producing these several new codons: (a) UGA, (b) GAA, (c) GGC, (d) CGA. Which of these represent(s) degeneracy, which missense, and which nonsense?

(handwritten annotations)

17-4 : UGA

$$\frac{3}{6} \cdot \frac{2}{6} \cdot \frac{1}{6} = \frac{1}{36}$$

UUU

$$\frac{3}{6} \cdot \frac{3}{6} \cdot \frac{3}{6} = \frac{27}{6 \cdot 36} = \frac{1}{8}$$

17-9) tyr   cys

UAU   UGU
UAC   UGC   second base change in mRNA from A to G

gly   val

GG-   GU-   second base change

17-12) GGA = glycine → nonsense
UGA = STOP(END chain) → nonsense
GAA = glutamic acid → missense
GGC = glycine → degeneracy
GGC = arginine → missense
CGA = arginine → missense

CHAPTER 18
Molecular Structure of the Gene

IN our discussion of the genetic code we raised an important question: in terms of the operation of the genetic code, just how much of the nucleotide sequence comprises a gene? Even by raising this question, we have come a long way from the early, understandably vague concept of the gene as a "bead on a string" separated from adjacent genes by nongenetic material. To begin to solve this problem at the molecular level, let us first examine the process of mutation.

Mutations are sudden changes in genotype, involving **qualitative or quantitative alterations in the genetic material itself.** Under this concept, recombination is excluded, for this process merely redistributes existing genetic material among different individuals; it makes no *change* in it. Geneticists often distinguish between two kinds of mutation, chromosomal and "point." So-called chromosomal mutations, or, preferably, chromosomal aberrations, may result in alterations in the amount or position of genetic material. These have already been described in Chapter 13. "Point" mutations, however, are *changes within the DNA molecule,* and the term *mutation* is now generally employed by geneticists in this more restricted sense. We shall use the term in this narrower meaning.

Of course, mutations may occur in any living cell that contains genetic material, either a somatic or a reproductive cell, and at any stage in its life cycle. Somatic mutations are perpetuated only in cells descended from the one in which the mutation originally took place. If the mutant trait is clearly detectable, a patch or sector of cells all having this new characteristic will result. For example, Figure 18-1 shows the result of a somatic mutation from "peppermint" to solid color that occurred in a zinnia in the author's garden. Whether or not somatic mutations produce an immediate effect depends on the nature of the change, the dominance relations of the original and mutant genes, and the stage of development of the individual or part at which the mutation occurs. Unless mutation takes place in reproductive cells, or in tissues that will give rise to reproductive cells, and is maintained therein, the mutation will not be passed on to succeeding generations. Even so, whether it is detectable in a later generation depends on many factors, such as dominance, the environment, mating patterns, and so forth.

From our examination of protein synthesis and of the genetic code in the two preceding chapters, it might occur to you that it may well be both possible and useful to think of genes from more than just one viewpoint. Implicit in our discussion of protein synthesis and coding was the suggestion that we may identify a gene as a *functional unit,* having a definite cellular locus and

408

FIGURE 18-1. *Mutant flower heads (inflorescences) of* Zinnia. *The somatic mutation was from "peppermint" (splotched color) to solid color.*

consisting of many nucleotides. But you will recall that a single nucleotide change will, in many instances, code for a different amino acid. Thus base changes of a certain kind and amount (an addition, a deletion, or a replacement) might be expected to result in one or several amino acid substitutions, or even a quite different protein and, therefore, a different (i.e., mutant) phenotype. As we saw in Chapter 17, a single deletion or insertion, for example, produces frame shifts that alter the reading of the remainder of the RNA message. This will often result in production of a very different protein, one that may be reduced in activity, inactive, or even lethal if some vital function is interfered with or prevented. But a replacement of one base by another in certain positions in certain codons will either (1) affect the reading of only the codon in which it occurs, so that a protein with a single amino acid substitution may be produced (a missense mutation), or (2) especially in the third position of the triplet, make no change in the amino acid coded for (a sense mutation).

So, then, we can also think of a gene as a *mutational unit*, which, from what we have seen thus far, can be as small as a single deoxyribonucleotide pair. A functional gene might thus consist of several to very many mutational subunits. Very possibly, too, the genetic *recombinational unit* may be identifiable at the molecular level and therefore constitute a third view of the gene. It is also entirely to be expected that the gene, considered from these three points of view, may well show somewhat different characteristics.

since the first two positions of the triplet are the critical bases

Point mutations, what they consist of and how they come about, shed considerable light on the molecular nature of the gene.

Mutagenic Agents

Although their mode of action is not always fully understood, a wide variety of agents have been implicated in the production of mutations. These mutagens may be broadly grouped into three classes, *radiation, chemicals,* and *temperature shock.* Some examples of the first two groups are the following:

1. Radiation.
   A. Ionizing.
      1. X-rays (wavelength 10^{-8} to 10^{-9} cm).
      2. γ rays (wavelength 10^{-9} to 10^{-10} cm).
      3. Cosmic rays (wavelength about 10^{-11} to 10^{-14} cm).
   B. Nonionizing (e.g., ultraviolet, wavelength about 10^{-4} to 10^{-6} cm).
2. Chemicals.
   A. Reactants and base analogs.
      1. Reactants with purines and pyrimidines.
         a. Nitrous acid.
         b. Formaldehyde.
      2. Base analogs.
         a. 5-bromouracil (a pyrimidine analog).
         b. 2-amino purine (a purine analog).
   B. Acridine dyes (e.g., acridine orange, acridine yellow, proflavin).
   C. Alkylating agents (e.g., mustard gases).
   D. Others.
      1. Carcinogens (e.g., methyl cholanthrene).
      2. Acids (e.g., phenols).

Not all of these are of equal effectiveness. Some are more effective in bacteria, for example, than in mammals, and some are more effective at certain loci than at others in the same organism. Other agents, not listed here, produce chromosomal aberrations (e.g., colchicine). It is not yet possible to induce specific mutations, although the effect of laser microirradiation in damaging small, preselected portions of the DNA of specific chromosomes is being explored (Berns, 1974).

RADIATION

Ionizing Radiation. The spectrum of electromagnetic radiation is a broad one; it extends from the long radio waves, which may have wavelengths as great as several kilometers, down through cosmic rays, which may be as short as 10^{-14} cm. As wavelength becomes progressively shorter, the energy that the radiation particles contain becomes progressively greater, with the result that they penetrate cells and tissues. In this process particles of sufficient energy content may collide with one of the oribiting electrons of an atom, knocking it out of orbit. In this way a previously neutral atom becomes positively charged (an *ion*), because there is then a greater positive charge in the atomic nucleus than there are negative charges in the orbiting electrons. Ionized atoms, and the molecules in which they occur, are chemically much more reactive than neutral ones. In addition, the electrons lost from an atom in this process move off at high speed and cause other atoms to become ionized. Each electron lost from an atom is ultimately gained by another, causing it to become a negatively charged ion. The result is a train of *ion pairs* along the path of the high-energy particles. Mutagenic effects result from the chemical reactions undergone by ions as their charges are neutralized.

One outcome of the ionization process is breakage of the sugar-phosphate strand(s) of DNA. This, of course, can lead to some of the chromosomal structural aberrations (such as deletions) that we explored in Chapter 13. But should strand breakage occur at two or more closely spaced points, the result can be loss of one or more nucleotides or nucleotide pairs, altering the mRNA transcribed. The ultimate outcome, of course, can then be a protein or an enzyme so altered as to be somewhat reduced in capacity to function, or even inoperative. The latter event may well be lethal, particularly in homozygotes. If this occurs in a reproductive cell, a phenotype, new for that particular line of descent, is created; in short, a *mutation* will have occurred.

There is compelling evidence also that some mutagenic effects of irradiation operate indirectly. Irradiated proteins or amino acids within a cell may themselves act as mutagenic agents (Stone et al., 1947). Seeding previously irradiated culture media also results in an increase in mutation rate with bacteria and *Neurospora*. In this situation it appears that hydrogen peroxide (H_2O_2) and other peroxides are produced; these are the actual mutagens.

Radiation applied to tissues is commonly measured in *roentgens* (r), named after the discoverer of X-rays, Wilhelm Roentgen. (It is worth noting that the unit of measurement is not capitalized, even though it represents the discoverer's name.) One roentgen (1r) produces about two ion pairs per cubic micrometer (μm^3), or about 1.6×10^{12} ion pairs per cubic centimeter (cm^3), of tissue. Radiation absorbed dose is expressed in *rads;* one rad is the amount of radiation that liberates 100 ergs of energy in a gram of matter. The rad is slightly larger than the roentgen, inasmuch as a gram of tissue exposed to 1r of γ rays absorbs about 93 ergs.

In many experimental organisms mutation rate is proportional to radiation dose, whether received as a "one-time" massive exposure or gradually over a period of time. Figure 18-2 shows this relationship for *Drosophila*. However, insects are more resistant to the effects of radiation than many organisms. Russell et al. (1958) reported that, for given amounts of radiation applied to mouse spermatogonia, chronic, low-intensity exposure produced a significantly smaller amount of detectable mutations than the same total dose applied acutely. This group found that exposure of mouse spermatogonia to 90r per

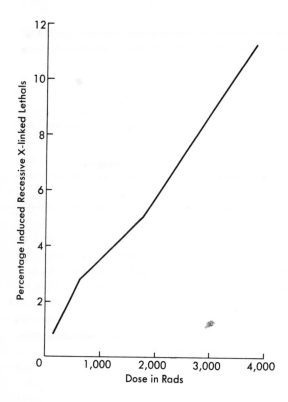

FIGURE 18-2. *Percentage of recessive linked X-linked lethal mutations in Drosophila sperm exposed to different doses of ionizing radiation. There is approximately a 1 per cent increase in (point) mutation frequency for each 380 rad increase in dosage.* (Based on I. H. Herskowitz, H. J. Muller, and J. S. Laughlin, 1959. The Mutability of 18MEV Electrons Applied to *Drosophila* Spermatozoa. *Genetics,* **44**:321–327.)

minute produced about four times the number of mutations as did the same amount of radiation administered at the rate of 90r per week. It is believed that enzymatically controlled repair processes can take place here between successive low-intensity irradiations, whereas at the higher rate of exposure these processes either cannot cope with the larger number of points of simultaneous damage or are themselves impaired.

All of us receive some radiation during our lifetimes from cosmic rays and from radioactive materials in the earth's surface.[1] In addition, medical X-ray examinations add to our accumulation of ionizing irradiations, although only that received by the gonads contributes to the *genetic load* of future generations. The average person in developed countries probably receives less than 50r of radiation to the gonads during a 30-year reproductive period as a result of medical examinations. For example, a single X-ray examination of the teeth is reported to provide about 0.0008r to the gonads; a chest X-ray about 0.0006r in males, but some 0.002r in females; an abdominal X-ray 0.13r in males, and 0.25r in females; and a fluoroscopic examination of the pelvic region about 4 to 6r.

[1] Such unusual factors as the famous radioactive monazite sands of small beach areas in the state of Kerala in extreme southwestern India, which the author has visited, increase this background radiation more than 10 times in that area.

A number of calculations for the amount of radiation required to double the rate of mutation have been accumulated for mice and for *Drosophila*. Applying such figures to human beings requires several assumptions the bases for which are not yet clear. Nevertheless, many authorities presently use the estimate of about 50r as the **doubling dose** for mankind. Although the amount of radiation received by the gonads for a given type of examination will vary among different institutions, it is helpful for each person to keep a record of the number and kind of X-ray and fluoroscopic examinations received.

Fallout of radioactive isotopes following above-ground nuclear explosions adds a small amount of gonadal exposure. In the 1960s, after testing of nuclear weapons in the air was incorrectly assumed to have ceased, it was estimated that such sources would contribute only about 0.1r to the gonads of each person during a reproductive lifetime. But the fission products of this kind of explosion remain in the upper atmosphere for long periods, settling slowly to earth and providing a continuing source of radioactive contamination. By the beginning of the twenty-first century, it has been estimated, some 60 per cent of these fission products will still be present in the atmosphere. Among these are carbon-14, strontium-90, and cesium-137, which contaminate foodstuffs and water, thereby serving as a source of at least somatic mutations in persons now living. So although the total amount of irradiation received per person from testing of nuclear military devices may appear small, experiments already conducted will serve as a source of danger for many years to come and to persons as yet unborn, an unsavory heritage for our children.

Nonionizing Radiation. Ultraviolet light is a fairly effective mutagenic agent, although much less so than the ionizing radiations. Its effects on the pyrimidines cytosine and thymine may produce (1) photoproducts that cause local strand separation in DNA, or (2) dimers that may prevent strand separation and replication, or interfere with normal base-pairing.

Thus ultraviolet may weaken the double bond between the fourth and fifth carbon of cytosine (Fig. 18-3), allowing water to be added at these points. The resulting photoproduct is unable to form hydrogen bonds properly with guanine, and this leads to strand separation. Ultraviolet radiation of about 2,800 Å likewise weakens the 4-5 C double bond, permitting two thymines to link as a dimer (Fig. 18-4). Such dimers may link thymine *between* strands, interfering with replication, or connect adjacent thymines of the same strand, disrupting normal T-A pairing.

Photoproducts and dimers, induced by ultraviolet irradiation, may be "repaired" (in *Escherichia coli* and phage) by (1) photoreactivation or (2) dark reactivation. In photoreactivation an enzyme that has been bound to DNA during ultraviolet irradiation is activated by intense visible light supplied following ultraviolet exposure. The enzyme then catalyzes the cleavage of pyrimidine dimers, restoring their normal structure. Dark reactivation involves instead the enzymatic removal of dimers from the DNA molecule and synthesis of normal replacement segments for those excised, at least in certain genetic strains of bacteria.

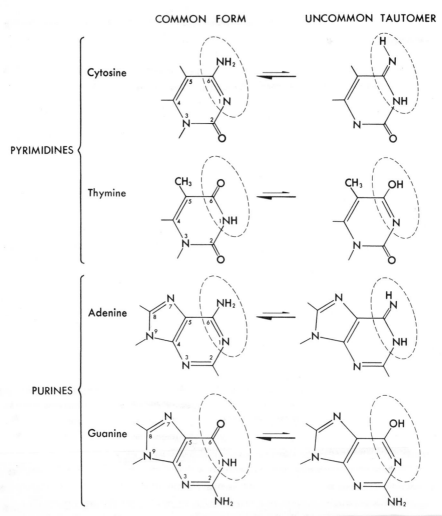

FIGURE 18-3. (A) *Comparison of common forms of DNA bases with their rare tauto-mers. Tautomers, which may occur in several forms, differ from each other by rearrange-ments of protons and electrons. This is symbolized in the structural formula by changes in positions of H atoms and double bonds within the dashed ovals. Only the tautomer thought to be important in each case as a mutagen is shown here; its occurrence is quite rare.*

CHEMICALS AND SUBNUCLEOTIDE CHANGES

Deamination.   It is well known that various chemicals produce mutations; these are especially well documented in bacteria, yeasts, and phages. Nitrous acid (HNO_2) is one such *mutagenic* substance. It brings about changes in DNA bases by replacing the amino group (-NH_2) with an -OH (hydroxyl) group.

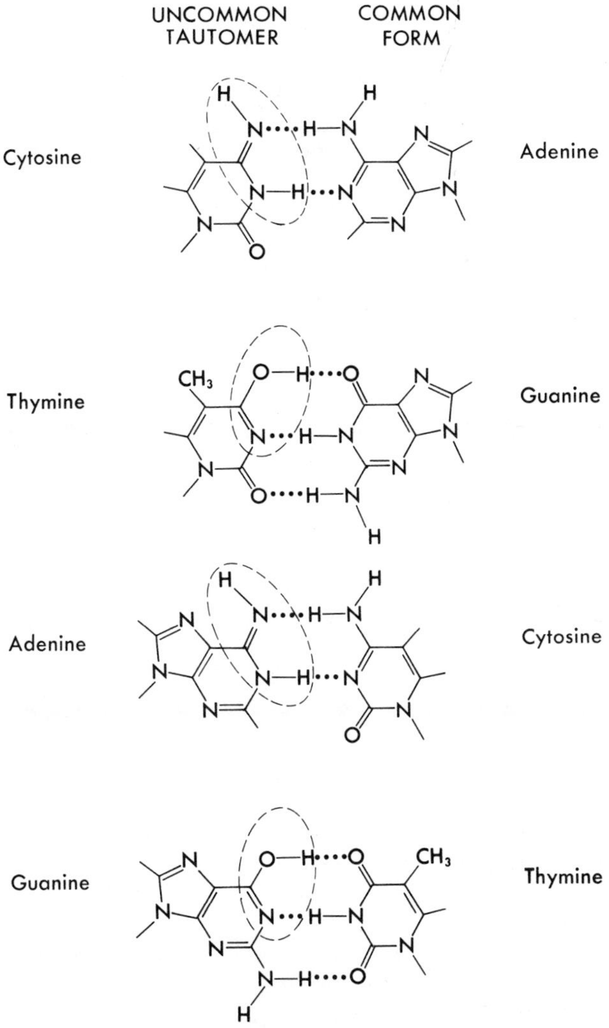

FIGURE 18-3. (B) *Pairing qualities of the rare tautomers of the four bases. Consequences of this "erroneous pairing" are discussed in the text.*

Thus adenine, having an -NH$_2$ at the number 6 carbon (Fig. 18-3A), is deaminated by nitrous acid to hypoxanthine:

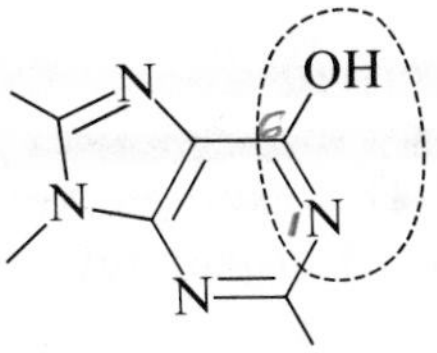

Thymine

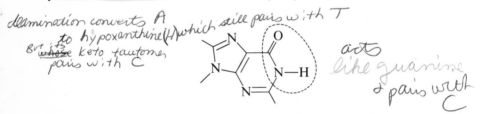

Monomers   Dimer

FIGURE 18-4. *The monomer ⇌ dimer conversion in thymine under ultraviolet light.*

bond between C₄ & C₅ weakened
(no longer double)

By a tautomeric shift, a more common (keto) tautomer is formed

deamination converts A
to hypoxanthine (H) which still pairs with T
But its keto tautomer pairs with C

acts like guanine & pairs with C

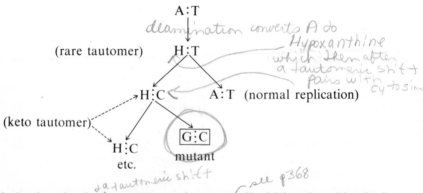

which pairs with cytosine. Thus an A:T pair can be converted to a G:C pair:

A:T

deamination converts A to Hypoxanthine which then after a tautomeric shift pairs with cytosine

(rare tautomer)   H:T

(keto tautomer)   →H:C ←   A:T (normal replication)

H:C   G:C
etc.   mutant

& a tautomeric shift   *see p368*

Similarly, deamination converts cytosine to uracil which pairs with adenine (thus CG becomes TA), and guanine to xanthine which pairs by *two* H bonds with cytosine.

Base Analogs.   Certain substances have molecular structures so similar to the usual bases that, if they are available, such **analogs** may be incorporated into a replicating DNA strand. One example will suffice to indicate the process and consequence. For instance, 5-bromouracil in its usual (keto) form,

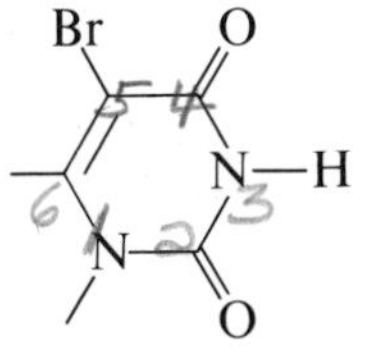

Br O
N—H
N
O

will substitute for thymine, which it closely resembles structurally (Fig. 18-3). Thus an AT pair becomes and remains ABu (Fig. 18-5). There is some in vitro evidence to indicate that Bu immediately adjacent to an adenine in one of the DNA strands causes the latter to pair with guanine. But in its rarer (enol) state,

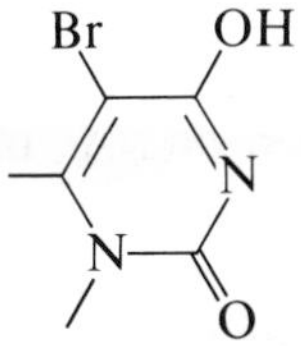

Br OH
N
N
O

5-Bu behaves similarly to the tautomer of thymine (Fig. 18-3) and pairs with guanine (Benzer and Freese, 1958). This converts AT to GC as shown in Figure 18-5. Studies show that 5-Bu increases the mutation rate by a factor of 10^4 in bacteria.

Nitrous acid and base analogs like 5-Bu can produce transitions as well as cause reversion of transition mutants in phages and bacteria to their original state, regardless of what the initial mutagen may have been. Some *rII* mutants of phage T4 (to be discussed shortly) are transitions and can be reversed in this fashion.

Tautomerization. The purines and pyrimidines of DNA and RNA may exist in several alternate forms, or **tautomers.** Tautomerism occurs through rearrangements of electrons and protons in the molecule. Uncommon tautomers of adenine, cytosine, guanine, and thymine, shown in Figure 18-3(A), differ from the common form in the position at which one H atom is attached. As a result, some single bonds become double bonds, and vice versa.

Transitions. The significance of these tautomeric shifts lies in the changed pairing qualities they impart. The normal tautomer of adenine pairs with thymine in DNA; the rare (imino) form pairs with the normal tautomer of cytosine as depicted in Fig. 18-3(B). The rare tautomer is unstable and usually reverts to its common form by the next replication. If tautomerism occurs in an already incorporated adenine, the result is a conversion of an AT pair to GC (Fig. 18-6A). On the other hand, if tautomerism occurs in an adenine about to be incorporated, the result is a conversion of a GC pair to AT (Fig.

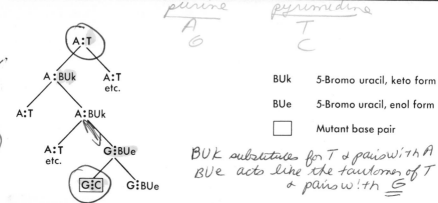

handwritten annotations:
purine   pyrimidine
A   T
G   C

BUk   5-Bromo uracil, keto form

BUe   5-Bromo uracil, enol form

☐   Mutant base pair

handwritten: BUK substitutes for T & pairs with A
BUe acts like the tautomer of T & pairs with G

FIGURE 18-5. *Substitution of the common keto tautomer of 5-bromouracil for thymine; its subsequent tautomerization to the rarer enol form, if it occurs, converts an A: T pair to a G:C pair. Moreover, the presence of 5-BU itself in the DNA sequence may cause imperfections in RNA synthesis.*

18-6B). Such substitution of one purine for another, or of one pyrimidine for another, is termed a **transition.** Transitions may come about in a number of other ways, as described in the following sections.

Acridine Dyes: Whole Nucleotide Changes.   The acridine dyes are effective mutagens. As described in Chapter 17, they act by permitting base additions and/or deletions. If the intercalation of an acridine and consequent stretching of the DNA molecule occurs at the time of crossing-over between DNA molecules, the result can be unequal crossing-over. As a result, one strand may have one more nucleotide than its complement, producing one daughter helix

handwritten: The normal tautomer of A pairs with T
The rare tautomer of A pairs with C
(unstable & reverts back)

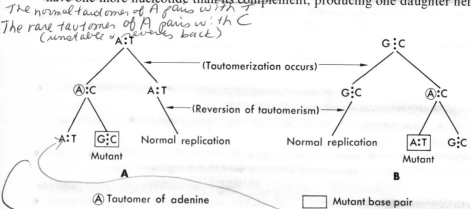

FIGURE 18-6. *(A) Conversion of an A:T pair to a G:C pair by tautomerization of an already incorporated adenine during the first replication of DNA. Reversion of the tautomerism in the second replication leads to the production of one mutant G:C strand. (B) Conversion of a G:C pair to an A:T pair by incorporation of a tautomer of adenine in the first replication of DNA. Assuming reversion of the tautomerism in the second replication, a mutant A:T pair is produced in place of the normal G:C pair. In both figures, the tautomer of adenine is indicated by the circled A; the mutant base pair is boxed.*

with one base pair more than the other. In general, the acridines are mutagenic in bacteria only at the time of recombination.

Mutation Rate in Human Beings

Taking into account the most reasonable estimates of the numerical values involved, some idea of the number of mutations occurring per human generation may be calculated. If one assumes a mean mutation rate per locus of 1×10^{-6} and accepts McKusick's estimate of at least 100,000 genes per person (McKusick, 1975), then the probability that any individual will produce a mutation during his or her reproductive lifetime is $1 \times 10^{-6} \times 1 \times 10^5 = 0.1$. Or it can be stated that one person in 10 may be expected to produce a mutation during reproductive life. For a United States population of 210 million, which is composed of two generations, those in the reproductive age bracket and those not, an average of $(2.1 \times 10^8) \times (1 \times 10^{-1})/2 = 1.05 \times 10^7$, or 10.5 million mutations occur each generation. Because many of these involve nonsense or missense in the code, deletions in DNA, and the like, most of these 10 million mutations per generation may be expected to be deleterious or lethal. Fortunately, however, most mutant genes are recessive, which at least prevents their expression in any but those homozygous for the mutant gene. Each of us, it has been estimated, possesses at least 5 to 10 disadvantageous, deleterious, or lethal genes; because of the relatively large total number of genes that each of us is calculated to have, the probability of two unrelated persons having the *same* such (recessive) gene(s) is quite low. In a randomly mating population the likelihood of homozygous recessives is thereby still further lessened when we are considering low-frequency disadvantageous or lethal genes. However, even rare deleterious genes become very real to a couple who have them in common. To a couple where both persons are normal but heterozygous for the Tay–Sachs gene or for PKU, for example, the problem suddenly becomes much more than a statistical abstraction.

Genetic Polymorphism

The simultaneous occurrence of two or more discontinuously different genotypes in the same interbreeding population is referred to as **genetic polymorphism.** We have already seen an illustration in the Rh + and Rh − genotypes (Chapter 8). These and some other examples that we are about to examine provide evidence for the foregoing mechanisms of change in nucleic acids and indicate that (1) mutation can, and often does, involve a single base change, and (2) at the mutational level a gene may be as little as one nucleotide. We shall look at two illustrations in human beings, glucose-6-phosphate dehydrogenase and hemoglobin variants, and one in bacteria involving the tryptophan synthetase enzyme system.

GENETICS OF GLUCOSE-6-PHOSPHATE
DEHYDROGENASE (G6PD)

[handwritten marginal note: deficiency of this enzyme is caused by recessive sex-linked gene]

The enzyme glucose-6-phosphate dehydrogenase is involved in glucose metabolism. Production of the enzyme, you will recall from Chapter 12, is determined by an X-linked locus. Our interest in it at this point centers in the extreme polymorphism of this locus. McKusick (1971) lists 78 variants of G6PD deficiency, and 11 more possible ones. Activity of so-called normal G6PD (which is determined by the gene Gd^B) may be set arbitrarily at 100 per cent; activity of variant forms ranges from 0 to 400 per cent of that of the normal enzyme (McKusick, 1975). When present, the enzyme occurs in a variety of cells and tissues—red cells, white cells, platelets, spleen, liver, and skin, among others. It also occurs in saliva. G6PD is required for stability of glutathione, a tripeptide capable of alternate oxidation and reduction, which fills an important role in cellular oxidations. Inhalation of the pollen of the broad bean (*Vicia faba*) or ingestion of its seed (which is used as food in some areas of the world) or of such drugs as sulfanilamide, sulfapyridine, aspirin, or primaquine (an antimalarial drug) bring on sudden hemolytic anemia in persons of some, but not all, G6PD deficiency genotypes. In the presence of such agents as these, glutathione levels drop and the red cells are destroyed. Persons so affected recover when the triggering substance is removed. It should be stressed, however, that many variants of G6PD deficiency can be detected only in laboratory tests (e.g., electrophoretic mobility) and these individuals manifest no overt symptoms or discomfort (Table 18-1).

TABLE 18-1. Some Variants of G6PD Deficiency

Variant	Gene Symbol	Red Cell Activity (Per Cent of Normal)	Incidence
Normal (B)	Gd^B	100	Common in all populations
A+	Gd^{A+}	80–100	Common in blacks
A−	Gd^{A-}	8–20	Common in blacks
Mediterranean	$Gd^{Mediterranean}$	0–7	Common in southern Europe
Athens	Gd^{Athens}	25	Common in Greece
Hektoen *	$Gd^{Hektoen}$	400	Rare, primarily in U.S. whites
*Ohio	Gd^{Ohio}	2–16	Rare, primarily in persons of Italian descent
*Chicago	$Gd^{Chicago}$	9–26	Rare, primarily in persons of western European descent

* Associated with hemolytic anemia. More extended data may be found in McKusick (1971), and Harris (1975), Giblett (1969) carries an extensive discussion and tabulation of data.

The important point for our consideration just now is that the many variants of G6PD deficiency are due to a large number of different amino acid substitutions in the enzyme. These have been determined thus far, however, for only two forms, A+ and Hektoen (McKusick, 1971). Yoshida (1967) has identified the A+ substitution as a replacement of asparagine by aspartic acid. The major problem in this work has been securing large enough quantities of the enzyme for the necessary analyses.

GENETICS OF HEMOGLOBIN

Structure of the Hemoglobin Molecule. That mutation may involve as little as a single deoxyribonucleotide pair has been clearly shown by the pioneering work of Ingram (1957) and of Hunt and Ingram (1958, 1960) on the chemical differences between normal and variant hemoglobins. Human hemoglobin is a protein with a molecular weight of about 67,000. The globin consists of four polypeptide chains, two alpha chains and two beta chains, each with iron-containing heme groups. As pointed out in Chapter 17, the α chain includes 141 amino acid residues, the β 146. In one molecule there are thus $(2 \times 141) + (2 \times 146)$, or 574, amino acid residues. Nineteen of the 20 biologically important amino acids are included and their exact sequence in both α and β chains has been determined.

Tryptic Digestion. Before biochemists had analyzed the complete sequence of amino acids in the α and β chains of hemoglobin, Ingram (1956 and 1957) was able to report on chemical differences between hemoglobin of normal persons (hemoglobin A, or Hb-A) and that of individuals suffering sickle-cell anemia (hemoglobin S, or Hb-S). Because the molecule was too large and complex for total analysis at that time, Ingram digested it with trypsin. This enzyme breaks the peptide bonds between the carboxyl group of either arginine or lysine and the amino group of the next amino acid. Because there are about 60 of these amino acids in the hemoglobin molecule, approximately 30 shorter polypeptide segments (in duplicate) are produced in this way. Each of these was then analyzed by Ingram for amino acid content.

"Fingerprinting" Peptides. Ingram placed small samples of the trypsin-digested hemoglobin (A or S) on one edge of a large square of filter paper, then subjected the peptide mixture to an electrical field in the process called electrophoresis. Under these conditions, differently charged portions will migrate characteristically. Next the filter paper with its as yet invisible, spread-out "peptide spots" was dried, turned 90° and placed with one edge in a solvent (normal butyl alcohol, acetic acid, and water). Because of differences in solubility of the peptides in this solvent, additional migration and spreading ensued. The paper was finally sprayed with ninhydrin, which produces a blue color in reaction with amino acids. The resulting chromatogram was called a "fingerprint" by Ingram, an apt description because differences in charge and amino acid content can thereby be picked out (Fig. 18-7) readily and characteristically.

When Ingram fingerprinted the peptides produced from hemoglobin A and S by tryptic digestion, he discovered all "peptide spots," except one which he called "peptide 4," of each to be identical in their locations on filter paper. This means that the long α and β chains of hemoglobins A and S are identical

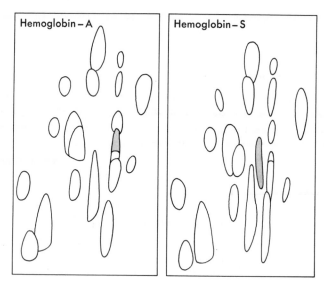

FIGURE 18-7. *Drawings of "fingerprint" chromatogram of the hemoglobins* A (*normal*) *and* S (*sickle cell*). *Each enclosed area represents the final location of a specific peptide; peptide 4 of Ingram is indicated by shading.*

except for one peptide of the 30 kinds. This one carries a positive charge in Hb-S, and no charge in Hb-A, a difference that by 1960, Hunt and Ingram were able to relate directly to a *single amino acid change* in the sequence of 146 composing the β chain.

Analyses of Several Hemoglobins.   With the development of the electro-phoretic-chromatographic technique for peptide analysis, studies of many different hemoglobins progressed rapidly. In addition, the complete amino acid sequence for both polypeptide chains is now known. The amino acid sequence for the β chain of hemoglobin A begins with these eight, starting with the NH$_2$ end of the first: valine, histidine, leucine, threonine, proline, glutamic acid, glutamic acid, and lysine. Table 18-2 compares this series of the first eight amino acids for the β chains of four different hemoglobins.

The electrophoretic behavior of the peptides consisting of these eight amino acids is explainable by their electrical charge differences. Glutamic acid carries a negative charge, valine and glycine none, and lysine a positive charge.

Many other types of hemoglobin have been similarly analyzed. Each differs from Hb-A in the change of at least one amino acid; some of these are in the α chain, others in the β.

Transversions.   Hemoglobin S differs from hemoglobin A, as we have seen (Table 18-2) by the substitution of valine for glutamic acid as the sixth amino acid in the β chain. Now the codons for glutamic acid (Table 17-1) are GAA and GAG, whereas valine is coded for by GUA, GUG, GUC, and GUU. If,

TABLE 18-2. Comparison of the First Eight Amino Acids of the β Chain
of Four Hemoglobins

Hb-A	Hb-S	Hb-C	Hb-G
Valine	Valine	Valine	Valine
Histidine	Histidine	Histidine	Histidine
Leucine	Leucine	Leucine	Leucine
Threonine	Threonine	Threonine	Threonine
Proline	Proline	Proline	Proline
Glutamic acid	VALINE	LYSINE	Glutamic acid
Glutamic acid	Glutamic acid	Glutamic acid	GLYCINE
Lysine	Lysine	Lysine	Lysine

say, a GAA (glutamic acid) codon undergoes a replacement to become GUA,
valine will, of course, be coded for (Fig. 18-8). Such a replacement in mRNA
will occur if the DNA trinucleotide CTT suffers a change to CAT; that is, the
purine adenine replaces the pyrimidine thymine. The substitution of a purine
for a pyrimidine (or vice versa) is called a **transversion.**

So the far-reaching and important phenotypic difference between hemoglobin A and that of persons with either sickle-cell trait or sickle-cell anemia
is the result of transversion, the change of one nucleotide pair in DNA from
AT to TA. A *single nucleotide* in mRNA thus may spell the difference between
life and death.

Transitions Versus Transversions. By definition, transitions and transversions represent different kinds of base substitutions. A comparison of the two
may be represented thus:

The dotted arrows represent transitions, the solid ones transversions.

Inheritance of Hemoglobin Variants. As described in Chapter 2, production
of hemoglobin S is due to a single gene that is codominant with the gene for
hemoglobin A. It is now clear that genes for hemoglobins A, S, C, and G, all
of which affect the β chain, are multiple alleles, and that the α chain is specified
by a different series of multiple alleles. McKusick (1975) lists 65 specific, different amino acid substitutions at 44 positions (plus one deletion) in the alpha
chain, and 111 substitutions at 82 positions (plus seven deletions) for the beta
chain. Mutations producing changes in the α chain do not themselves cause
changes in the β chain, and vice versa.

						c i s t r o n				
DNA	Hemoglobin A	GTA CAT	CAT GTA	CTT GAA	ACT TGA	CCT GGA	GAA CTT	GAA CTT	AAA TTT	... ...
mRNA Codons		GUA	CAU	CUU	ACU	CCU	GAA	GAA	AAA	
Amino Acids		val	his	leu	thr	pro	glu	glu	lys	
DNA	Hemoglobin S						G T A C A T			
mRNA Codons							G U A			
Amino Acid							val			
DNA	Hemoglobin C						A A A T T T			
mRNA Codons							A A A			
Amino Acid							lys			
DNA	Hemoglobin G						G G A C C T			
mRNA Codons							G G A			
Amino Acid							gly			

FIGURE 18-8. *Suggested derivation of three mutant β-chain hemoglobins from hemoglobin* A *by single nucleotide changes. The codon for particular amino acids is arbitrarily chosen when several may code for the same amino acid. Altered nucleotides are enclosed by solid lines.*

THE TRYPTOPHAN SYNTHETASE SYSTEM

In a sense, Ingram's work on hemoglobin really set the stage for the important work of Yanofsky on the tryptophan synthetase enzyme system of *Escherichia coli.* In this bacterium the complete enzyme molecule is a complex of two different subunits, A and B. The A component is a single polypeptide chain of 267 amino acid residues whose exact sequence is now known; the B component is a dimer of two identical polypeptide chains. The enzyme catalyzes three of the steps in tryptophan synthesis (Fig. 18-9). Both the A and B subunits are required for reaction $\overline{AB}$, whereas only the A component is necessary for indole production (reaction (A)), and the B component alone is needed for reaction $\overline{B}$. Both the A and B components are required, then, for synthesis of tryptophan.

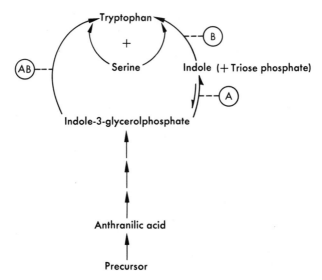

FIGURE 18-9. *Pathways of tryptophan synthesis involving tryptophan synthetase. The enzyme consists of two subunits,* A *and* B, *both of which are required for the reaction indole-3-glycerolphosphate + serine → tryptophan (reaction (AB)). The* A *subunit is required for production of indole and triose phosphate (reaction (A)), and the* B *subunit for the reaction indole + serine → tryptophan (reaction (B)). Thus,* A*-deficient mutants require indole or tryptophan for growth, and* B*-mutants must be supplied with tryptophan.*

Some mutants that are incapable of forming functional tryptophan synthetase do, however, produce a protein that, although enzymatically inactive, does give an immunological reaction with tryptophan synthetase immune serum prepared from rabbit blood. This protein has been named *cross-reacting material* (CRM); both *CRM+* (producing cross-reacting material) and *CRM−* (not producing cross-reacting material) mutants have been detected. It is believed that the difference between *CRM+* and *CRM−* mutants lies in the fact that the latter produce either small fragments of the enzyme molecule or grossly abnormal molecules as the result of nonsense and missense mutations.

Mutants defective in the A polypeptide occur, as do others defective in the B component. Complementation tests, to be described in the next section (pages 426–429), make it clear that two different *functional genes* are involved, one for each of the components (A and B).

Yanofsky and his colleagues (1967) have determined the amino acid substitutions for a number of A mutants (Table 18-3). These not only clearly confirm the colinearity of the genetic code, but also show that so-called point mutations may involve a change in as little as a single ribonucleotide (and therefore in a single deoxyribonucleotide pair). These changes, then, constitute the basis of genetic polymorphism.

TABLE 18-3. Some A Mutants in the Tryptophan Synthetase System of
Escherichia coli

Mutant	Amino Acid Position*	Wild Type Amino Acid Residue	Mutant Amino Acid Residue
A3	48	Glutamic acid	Valine
A33	48	Glutamic acid	Methionine
A446	174	Tyrosine	Cysteine
A487	176	Leucine	Arginine
A223	182	Threonine	Isoleucine
A23	210	Glycine	Arginine
A46	210	Glycine	Glutamic acid
A187	212	Glycine	Valine
A78	233	Glycine	Cysteine
A58	233	Glycine	Aspartic acid
A169	234	Serine	Leucine

* Amino acid residues are numbered consecutively from 1 (amino end) to 267 (carboxyl end).
Based on work of Yanofsky et al., (1967).

Fine Structure of the Gene

The older concept of the gene as a single, indivisible entity can, as predicted in Chapter 8, no longer be valid. On the one hand, there is the series of nucleotides that specifies the sequence of amino acid residues of a polypeptide chain (such as those of the A and B chains of the tryptophan synthetase enzyme, or the α and β chains of hemoglobin). But we have also seen that a change in as little as one nucleotide of the polypeptide-specifying gene may change (mutate) and produce a variant of the wild-type chain that differs in one amino acid residue. So the functional gene is not the same as the mutational gene, but appears to consist of many mutable sites. The gene must also be considered from the standpoint of the nature of the sites at which recombination may occur. The functional gene therefore appears to be composed of many mutational as well as recombinational subunits.

THE CISTRON

Phage T4 is a virus infecting the colon bacillus *Escherichia coli*. Plaques, or clear areas on agar plates of the host, indicate areas of infection and lysis. Those formed by the wild-type phage (*r*⁺) are small and have rough edges. Many phage mutations that alter plaque morphology are known; the *r* (rapid lysis) mutants are easily recognizable because they produce large plaques with sharp edges. The *r* mutants have been shown to involve three major map regions in the phage DNA and are designated *rI*, *rII*, and *rIII*. Both *rI* and *rIII* lyse host strain K; *rII* mutants enter K-type cells but do not lyse them. But all three mutant classes, *rI*, *rII*, and *rIII*, lyse cells of host strain B.

The *rII* mutants fall into two groups, *rIIA* and *rIIB*. Each of these two groups has been found to have an internal linear arrangement and to be adjacent but nonoverlapping on the phage DNA molecule. The *rII* region appears to consist of about 2,000 deoxyribonucleotide pairs. The *A* region transcribes a messenger RNA that translates an *A* polypeptide; the *B* region is similarly responsible for a *B* polypeptide. Both substances are necessary for lysis of K-type host cells. The *r⁺* (wild-type) phage produces both *A* and *B* polypeptides; *A* mutants produce normal *B* polypeptide, but not *A,* and vice versa. So infection only by identical *rIIA* mutants, or by identical *rIIB* mutants alone, fails to result in lysis. Nor does lysis ensue in cases of infection by mutants each of which contains a different *rIIA* (or *rIIB*) mutation; none of the phages can produce *both A* and *B* polypeptides (Fig. 18-10). On the other hand, infection by two different phage strains, one an *rIIA* mutant and the other an *rIIB* mutant, does result in lysis (Fig. 18-10). That is, regions A and B are functionally different and exhibit **complementation** (Benzer, 1955).

Benzer (1955, 1961) found that with infection by two phage strains—one the wild type (*r⁺*) and the other doubly mutant in either the A or the B region, that is, with the mutations in the *cis* position—lysis occurred. But if the A (or B) mutations were in the *trans* configuration, lysis did *not* follow (Fig. 18-11). Thus it was seen that (1) mutations in one functional region (A or B) are complementary only to mutations in the other region, and (2) complementation is detectable by the *cis-trans test.* Each functional region responsible for the production of a given polypeptide chain and defined by the *cis-trans* test was named a **cistron** by Benzer. The cistron therefore may be thought of as the gene at the functional level and as consisting of three times as many deoxyribonucleotide pairs as the polypeptide coded for, neglecting start and stop signals as well as any untranscribed nucleotides. Thus the length of a cistron (either in number of deoxyribonucleotides or in units of length such as the micrometer) is defined by the number of amino acid residues of the polypeptide chain for which it is responsible, plus initiating, terminating, and any untranscribed nucleotides.

THE MUTON

Study of the genetic code (Table 17-1) and the tryptophan synthetase mutants listed in Table 18-3 makes it clear that a change in a single deoxyribonucleotide pair may result in a *missense* codon in transcribed mRNA, for example, AGC → AGA, or *nonsense,* e.g., UGC → UGA. So a cistron may be expected to consist of many mutable sites, or mutons as they have been termed by Benzer. *A muton is thus the smallest length of DNA capable of mutational change;* this may be as little as a single DNA base pair.

We have already seen that mutant hemoglobins differ from normal hemoglobin A by single amino acid substitutions and that these may be accounted for in many instances by single nucleotide changes (both transitions and transversions). By the same kind of reasoning, and based on clear experimental evidence, Yanofsky (1963, 1967, and numerous other papers) has shown how each of the many tryptophan synthetase mutants may occur by nucleotide

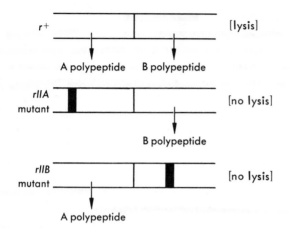

Noncomplementation

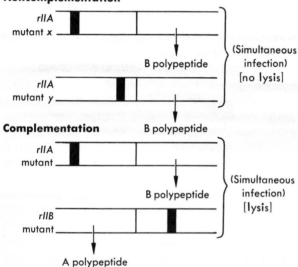

FIGURE 18-10. *Complementation* (see text) *occurs when* Escherichia coli *is infected simultaneously with an* rIIA *and an* rIIB *mutant phage; complementation does not occur if two different A mutants, for example, simultaneoulsy infect the host. For reference,* r+, rIIA, *and* rIIB *mutants are diagramed also. Diagrams are not to any scale.*

substitutions. Two ultraviolet-induced *A* mutants, A23 and A46, have been especially revealing. In one of the A23 mutants, glycine is replaced by arginine, whereas in one of the A46 mutants the *same* glycine (at position 210, as indicated in Table 18-3) is replaced by glutamic acid. The codon changes responsible may be designated, for example, as GGA (glycine, wild-type), *A*GA (arginine, A23 mutant), and G*A*A (glutamic acid, A46 mutant), as diagramed in Figure 18-12. These examples clearly indicate that the muton can be and often is as little as one nucleotide, and that a cistron includes many mutons. Nucleotide changes in these mutable sites may, of course, produce sense, missense, or nonsense reading changes.

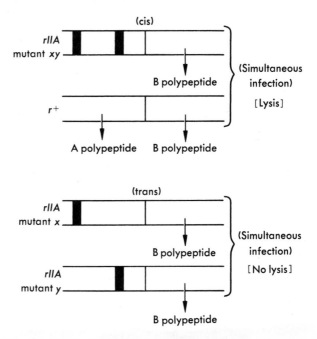

F I G U R E 18-11. *Simultaneous infection of Escherichia coli by* rIIA *(or* rIIB*) double mutants and wild-type* r+ *phages (top diagram), where the cistron defects of the mutant are in the* cis *position (i.e., in the same DNA molecule), results in lysis. When the mutations are in the* trans *position (lower diagram), i.e., in different DNA molecules, no lysis takes place.*

THE RECON

Tryptophan Synthetase System. You will recall that our discussion of pseudoalleles in Chapter 8 suggested that crossing-over may occur within a functional gene or, to use Benzer's terminology, a cistron. The A23 and A46 tryptophan synthetase mutants of *Escherichia coli* confirm this suggestion. These two mutants, involving the same glycine of the A protein have been shown by recombinational studies *not* to map at precisely the same point, but to be 0.001 map unit apart (Yanofsky et al., 1967). In fact, Yanofsky (1967) comes to some interesting conclusions in comparing genetic map length with distances in the polypeptide chain. He finds the map length for 187 amino acid residues (positions 48 through 234) of the A protein in the *E. coli* tryptophan synthetase system to be about three map units, or $\frac{3}{187} = 0.016$ map unit per amino acid residue, and thus also for a group of three adjacent nucleotides. Distances between adjacent nucleotides are thus of the order of 0.001 to 0.005 map unit.

It is therefore evident that, just as mutation may occur *within* a triplet codon, so may recombination. Furthermore, Yanofsky's studies of both mutation and recombination (by means of transduction) clearly show that in each protein with a different amino acid at a given position, the amino acids on either side remain unchanged. To quote Yanofsky (1963), "this is perhaps the most convincing evidence available that excludes overlapping codes." Thus we are provided with still another subdivisional concept of the cistron, namely the

caused by crossing-over

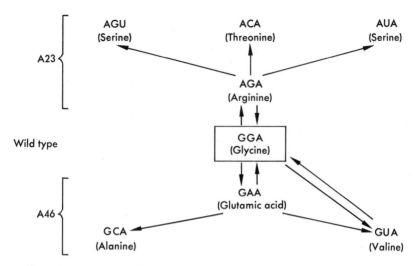

FIGURE 18-12. *Derivation of A23 and A46 tryptophan synthetase mutants in* E. coli *by single nucleotide substitutions. Based on the work of Yanofsky.*

recon. *A recon is the smallest unit of DNA capable of recombination* or (in bacteria) of being integrated by transformation or transduction.

The **rII** *Region of Phage T4.*   Confirmation of this concept of the recon is provided by some of Benzer's work on the *rII* region of phage T4. If two different A mutations or two different B mutations are located sufficiently far apart in their respective cistrons, *recombination* may occur between them. For example, infection of B-type hosts with two *different* A mutants results in lysis (*rII* mutants can lyse B cells) and release of progeny phage particles. These latter are then used to infect K-type hosts. Any lysis occurring in K cells must, therefore, result from recombination between two different A mutations to produce wild type phage (r^+) and their characteristic plaques. The same recombinational effect is observable within the B cistron. Careful study has made it possible to detect recombinants with frequencies as low as 0.00001 per cent (1×10^{-7}). Some of the A $\times$ A and B $\times$ B "crosses" fail, however, to result in recombination; these have been found to be deletions.

From recombination experiments involving both "point mutations" and deletions, Benzer (1962) found over 3,000 mutants, having more than 300 sites identifiable by recombination in the entire *rII* region. Subsequent work has extended these findings to the point where at least 500 recombinational sites have been identified in the *rIIA* region alone. From a variety of evidence, the T4 DNA molecule is calculated to have a molecular weight of about 130,000,000. With 650 as the average molecular weight of a deoxyribonucleotide *pair* (refer back to problems 15-8 and 15-11), the number of nucleotide pairs in the entire genome of T4 is $\frac{130,000,000}{650}$, or about 200,000. Estimates of the number of base pairs in the A cistron range up to 2,500, and generally between 1,000 and 1,500 for the B cistron. With an estimate of 2,000 nucleotide pairs for the *rIIA* region, the unit of recombination (the *recon*) in the A cistron can include no more than an average of $\frac{2,000}{500}$, or four nucleotide pairs. Evidence from the *E. coli* tryptophan synthetase system and the *rII* region of T4 phage indicates that, like the muton, the recon is very small and almost certainly *may consist of as little as one* deoxyribonucleotide pair. It is also quite probable that mutons and recons are structurally indistinguishable.

Dominance and Recessiveness

As a result of our consideration of DNA, the genetic code, polypeptide synthesis, and the molecular nature of the gene, we are now in a position to understand how dominance, incomplete dominance, and codominance operate in terms of the structure and function of the genetic material.

As an illustration of complete dominance, we may return to phenylketonuria. You will recall that PKUs, who are homozygous recessives, are unable to synthesize the enzyme phenylalanine hydroxylase, with the result that phenylalanine accumulates. Heterozygotes are normal, although they have a defective cistron in one set of their chromosomes. Apparently the "wild-type" cistron in the other set of chromosomes is able to mediate synthesis of a sufficient amount of enzyme to convert phenylalanine to tyrosine. The defective cistron may be assumed to contain a gross missense, or even a nonsense, mutation (Fig. 18-13A) so that its product is completely nonfunctional.

Genes for hemoglobins A and S are, as we have learned, codominant. The two hemoglobins differ by a single amino acid substitution in the sixth position of the beta chain (page 423). In this case, clearly a missense mutation is responsible for producing the aberrant hemoglobin S, whereas the other cistron in heterozygotes gives rise to the normally functioning hemoglobin A (Fig. 18-13B).

In Chapter 2 we saw an example of incomplete dominance in the red, pink, and white phenotypes in the flowering plant snapdragon. Here it is quite possible that one cistron, which we can designate a_1, produces the appropriate enzyme needed for synthesis of the red pigment, but, by itself, in an amount insufficient to give the petals the full red color. The result in heterozygotes is the pink phenotype. Thus if we designate the defective cistron a_2, a_1a_1 plants produce sufficient functional enzyme to give enough pigment to result in red petals, the heterozygotes (a_1a_2) produce only enough to give a paler color, and a_2a_2 plants fail to produce any functional enzyme at all so that the petals remain unpigmented (Fig. 18-13C).

Because defects may occur at so many different intracistronic sites, it is not surprising that forward mutations generally have a higher incidence than do back mutations, i.e., why $u > v$ in many cases (Chapter 14).

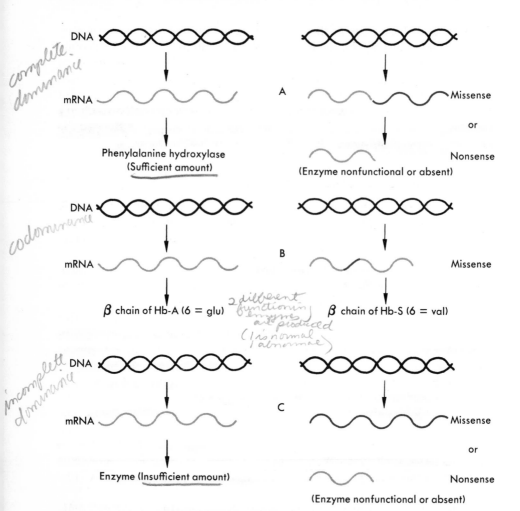

complete dominance

codominance

2 different functioning enzymes are produced (1 is normal 1 abnormal)

incomplete dominance

FIGURE 18-13. *Diagrammatic representation of (A) complete dominance, (B) codominance and (C) incomplete dominance in terms of the nature of DNA and its responsibility for polypeptide translation via transcribed mRNA. See text for details.*

Conclusions

In the last four chapters we have reached some important conclusions regarding the gene:

1. The genetic material is DNA in all organisms except a few viruses, where it is RNA.
2. A gene is a specific linear sequence of nucleotides and is usefully considered at more than one level of organization.

a. The functional unit, the *cistron*, is responsible for specifying a particular polypeptide chain, and consists of three adjacent deoxyribonucleotide pairs for each amino acid residue in the chain, plus a chain-initiating triplet and a chain-terminating triplet or triplets.

b. Each cistron is divisible into

    (1) *Mutons*, the smallest number of nucleotides independently capable of producing a mutant phenotype. This can be as little as one nucleotide.

    (2) *Recons*, the smallest number of nucleotides capable of recombination. This likewise can be as small as one nucleotide. Recons and mutons appear to be structurally identical.

3. Specification of the amino acid sequence of a polypeptide chain by a cistron is made possible through the latter's particular nucleotide sequence.

a. Mediation of protein synthesis operates through an intermediate, messenger RNA, which is ordinarily a single-stranded, base-for-base, complementary transcription of the nucleotide sequence on one DNA strand.

b. Messenger RNA, in conjunction with polysomes, tRNA, energy sources, and a battery of enzymes, is directly responsible through its sequence of ribonucleotides (the genetic code) for protein biosynthesis at ribosomal surfaces, chiefly in the cytoplasm.

c. The genetic code is

    (1) Triplet.

    (2) Commaless.

    (3) Nonoverlapping.

    (4) Degenerate.

    (5) Essentially nonambiguous.

    (6) Colinear.

    (7) Universal.

REFERENCES

AUERBACH, C., and B. J. KILBEY, 1971. Gene Mutation. In H. L. Roman, ed. *Annual Review of Genetics,* volume 5. Palo Alto, Calif., Annual Reviews, Inc.

BENZER, S., 1955. Fine Structure of a Genetic Region in Bacteriophage. *Proc. Nat. Acad. Sci. (U.S.),* **41**: 344–354.

BENZER, S., 1961. On the Topography of the Genetic Fine Structure. *Proc. Nat. Acad. Sci (U.S.),* **47**: 403–415.

BENZER, S., 1962. The Fine Structure of the Gene. *Sci. Amer.,* **206**(1): 70–84.

BENZER, S., and E. FREESE, 1958. Induction of Specific Mutations with 5-Bromouracil. *Proc. Nat. Acad. Sci. (U.S.),* **44**: 112–119. Reprinted in G. S. Stent, ed., 1965, 2nd ed. *Papers on Bacterial Viruses.* Boston, Little, Brown.

BERNS, M. W., 1974. Directed Chromosome Loss by Laser Irradiation. *Science,* **186**: 700–705.

FISHBEIN, L., W. G. FLAMM, and H. L. FALK, 1970. *Chemical Mutagens.* New York, Academic Press.

GIBLETT, E. R., 1969. *Genetic Markers in Human Blood.* Oxford, Blackwell Scientific Publications.

HARRIS, H., 1975. *The Principles of Human Biochemical Genetics,* 2nd ed. New York, American Elsevier.

HUNT, J. A., and V. M. INGRAM, 1958. Allelomorphism and the Chemical Differences of the Human Haemoglobins A, S, C. *Nature,* **181:** 1062.

HUNT, J. A., and V. M. INGRAM, 1960. Abnormal Haemoglobins, IV. The Chemical Difference Between Normal Human Haemoglobin and Haemoglobin C. *Biochim. Biophys. Acta,* **42:** 409–421. Reprinted in J. H. Taylor, ed., 1965. *Selected Papers on Molecular Genetics.* New York, Academic Press.

INGRAM, V. M., 1956. A Specific Chemical Difference Between the Globins of Normal and Sickle-Cell Anaemia Haemoglobin. *Nature,* **178:** 792–794.

INGRAM, V. M., 1957. Gene Mutations in Human Haemoglobin: the Chemical Difference Between Normal and Sickle-Cell Haemoglobin. *Nature,* **180:** 326–328. Reprinted in S. H. Boyer, ed., 1963. *Papers on Human Genetics.* Englewood Cliffs, N.J., Prentice-Hall.

INGRAM, V. M., 1965. *The Biosynthesis of Macromolecules.* New York, W. A. Benjamin.

LUCCHESI, J. C., 1973. Dosage Compensation in *Drosophila.* In H. L. Roman, ed. *Annual Review of Genetics,* volume 7. Palo Alto, Calif., Annual Reviews, Inc.

MCKUSICK, V. A., 1971. *Mendelian Inheritance in Man,* 3rd ed. Baltimore, Johns Hopkins University Press.

MCKUSICK, V. A., 1975. *Mendelian Inheritance in Man,* 4th ed. Baltimore, Johns Hopkins Press.

MULLER, H. J., 1927. Artificial Transmutation of the Gene. *Science,* **66:** 84–87.

RADDING, C. M., 1973. Molecular Mechanisms in Genetic Recombination. In H. L. Roman, ed. *Annual Review of Genetics,* volume 7. Palo Alto, Calif., Annual Reviews, Inc.

RUSSELL, W. L., L. B. RUSSELL, and E. M. KELLY, 1958. Radiation Dose Rate and Mutation Frequency. *Science,* **128:** 1546–1550.

STADLER, D. R., 1973. The Mechanism of Intragenic Recombination. In H. L. Roman, ed. *Annual Review of Genetics,* volume 7. Palo Alto, Calif., Annual Reviews, Inc.

STAMATOYANNOPOULOS, G., 1972. The Molecular Basis of Hemoglobin Disease. In H. L. Roman, ed. *Annual Review of Genetics,* volume 6. Palo Alto, Calif., Annual Reviews, Inc.

STONE, W. S., O. WYSS, and F. HAAS, 1947. The Production of Mutation in *Staphylococcus aureus* by Irradiation of the Substrate. *Proc. Nat. Acad. Sci. (U.S.),* **33:** 59–66.

WELSHONS, W. J., 1965. Analysis of a Gene in *Drosophila. Science,* **150:** 1122–1129.

WOLSTENHOLME, G. E. W., and M. O'CONNOR, eds., 1969. *Mutation as Cellular Process.* London, J. & A. Churchill, Ltd.

YANOFSKY, C., 1963. Amino Acid Replacements Associated with Mutation and Recombination in the A Gene and Their Relationship to in Vitro Coding Data. *Cold Spring Harbor Symposia Quant. Biol.,* **28:** 581–588.

YANOFSKY, C., 1967. Structural Relationships Between Gene and Protein. In H. L.

Roman, ed. *Annual Review of Genetics,* volume 1. Palo Alto, Calif., Annual Reviews, Inc.

YANOFSKY, C., G. R. DRAPEAU, J. R. GUEST, and B. C. CARLTON, 1967. The Complete Amino Acid Sequence of the Tryptophan Synthetase A Protein (α Subunit) and its Colinear Relationship with the Genetic Map of the A Gene. *Proc. Nat. Acad. Sci. (U.S.)* **57:** 296–298.

YOSHIDA, A., 1967. A Single Amino Acid Substitution (Asparagine to Aspartic Acid) Between Normal (B+) and the Common Negro Variant (A+) of Human Glucose-6-Phosphate Dehydrogenase. *Proc. Nat. Acad. Sci. (U.S.),* **57:** 835–840.

YOSHIDA, A., G. STAMATOYANNOPOULOS, and A. G. MOTULSKY, 1967. Negro Variant of Glucose-6-Phosphate Dehydrogenase Deficiency (A−) in Man. *Science,* **155:** 97–99.

PROBLEMS

18-1. From this chapter and any other sources available to you, explain why most mutations are deleterious.

18-2. From this chapter and any other sources available to you, explain why most mutations are recessive.

18-3. From this chapter and any other sources available to you, evaluate the short-term and long-term effects of mass irradiation of the human population.

18-4. Look again at Figure 18-2. Would you say that there is a threshold dose of irradiation below which no mutation is induced?

18-5. X-linked recessive mutations are more easily studied in appropriate organisms than are autosomal ones. Why?

18-6. Which of the following would be likely to suffer the greatest, and which the least, genetic damage from radiation exposure: (a) a monoploid, (b) a diploid, (c) a polyploid?

18-7. Three species of oats (*Avena*) are listed in Table 3-1. Irradiating many samples of these three species with the same amount of X-irradiation resulted in the following rates of induced detectable mutations (given as $\bar{x} \pm s$ per sample):

Species	Irradiated	Control
A. brevis	4.1 ± 1.1	0.2 ± 0.03
A. barbata	2.4 ± 0.6	0.05 ± 0.01
A. sativa	0.0	0.0

Give a reason for this kind of result.

18-8. Suppose samples of the species of wheat (*Triticum*) listed in Table 3-1 were each given the same dose of X-irradiation. Assuming that some radiation-induced mutation occurred in at least some of those species, which species would you expect to show the highest frequency of mutation?

18-9. In an imaginary flowering plant, petal color may be either red, white, or blue. White is due to the presence of a colorless precursor from which red and blue

Handwritten top margin: 16 ths so: 9 blue : 3 red : 4 white so: A → red B → blue / so 3 pairs 2 genes with reduced epistasis / white (colorless)

Handwritten: P. AaBb × AaBb   F₁ blue A-B-; red A--bb, white aa--

pigments may be synthesized in two consecutive, enzyme-controlled processes. The cross of two blue-flowered plants yields an F_1 ratio of nine blue, three red, and four white. (a) How many pairs of genes are involved? (b) Suggest the correct sequence of enzyme-controlled steps and products. (c) Starting with the first letter of the alphabet and using as many more in sequence as needed, give the genotype of (1) the P individuals; (2) the F_1 blue plants. (d) Give genotypes for (1) red and (2) white F_1 plants. (e) According to the system you have worked out, which pair of genes controls the first step of the process you have worked out in part (b) of this problem?

18-10. By deamination a $\begin{array}{c}5'\ T\ T\ T\ 3'\\3'\ A\ A\ A\ 5'\end{array}$ segment of a DNA molecule is changed to a $\begin{array}{c}5'\ T\ T\ G\ 3'\\3'\ A\ A\ C\ 5'\end{array}$ segment. What change in amino acid incorporation does this produce, assuming the 3', 5' ("lower") strand serves as the template for transcription?

Handwritten near 18-10: mRNA 5' U U U 3'   UUU is phenylalanine   mRNA 5' U U G 3'   UUG is leucine

18-11. Look again at Figure 18-5. Does the enol form of 5-bromouracil cause a transition or a transversion? *Handwritten:* transition since causes substitution of G for A, both purines

18-12. Table 18-3 shows two tryptophan synthetase mutants, A78 and A58, that affect amino acid residue 233. In A78 wild-type glycine at this position is replaced by cysteine; in A58 glycine is replaced by aspartic acid. Among the mRNA codons for glycine is GGC; one of the two for cysteine is UGC; and one of the two for aspartic acid is GAC. Which of the mutant strains, A78 or A58, involves a transition and which a transversion in the DNA template?

Handwritten left margin: A58: sub of A for G, so transition / A78: sub of U for G so transversion

18-13. Tryptophan synthetase mutant A446 has cysteine at amino acid position 174 instead of the wild-type tyrosine. This could come about by a change in the DNA template strand from ATG to ACG. Suppose strain A446 were irradiated and the same DNA template triplet were converted from ACG to ACT. (a) What kind of code reading error does this produce? (b) Would you expect a functional polypeptide to be formed or not? (c) Assuming that a polypeptide chain of some length is formed as a result of the change described here, of how many amino acid residues would it consist? *Handwritten:* 173

Handwritten left margin near 18-13: DNA: ACT mRNA: UGA end chain ∴ nonsense / nonsense / Probably not

18-14. Suppose a certain cistron is found to consist of 1,500 deoxyribonucleotides in sequence. (a) What is the maximum number of mutons of which this cistron could consist? (b) Is this number likely to be too high, too low, or about right for an actual organism? Why? *Handwritten:* a) 1500 b) too high since due to degeneracy some nonsense mutations are possible

18-15. For how many codons is the tryptophan synthetase A cistron responsible, if initiating, terminating, and possible untranscribed nucleotides are neglected? *Handwritten left:* 267 amino acids in the polypeptide, so 267 codons

18-16. Neglecting initiating, terminating, and possible untranscribed nucleotides, how many nucleotides are there in the tryptophan synthetase A cistron? *Handwritten:* 267 codons × 3 nucleotides per codon = 801

18-17. One calculation for the molecular weight of the DNA in a single (monoploid) set of human chromosomes is 1.625×10^{12}. (a) Assuming 650 as the average molecular weight of a pair of deoxyribonucleotides, how many nucleotide pairs comprise the DNA of one *diploid* somatic nucleus? (b) What, then, is the maximum number of mutons in each of your somatic cells? (c) What is the total length in micrometers (microns) of the DNA in the nucleus of one of your diploid somatic cells? (d) At these values, for how many mRNA codons is the DNA of *one* of your (monoploid) sets of chromosomes responsible? (e) Assum-

Handwritten left margin: a) $\dfrac{1.625 \times 10^{12}}{650} = .0025 \times 10^{12}$ / $= .005 \times 10^{12}$ / d) same, since

Handwritten bottom: $\dfrac{-4\,(5\times10^9)}{\times 10^5}$ d) $\dfrac{.0025 \times 10^{12}}{3} = .00083 \times 10^{12} = 8.3 \times 10^8$ mRNA codons

[handwritten top margin: 300 | 830 | 2300 | 2100 | 300]

[handwritten: (e) 8.3×10^8 codons means 8.3×10^8 amino acid residues]

[handwritten: $\dfrac{8.3 \times 10^8 \text{ aa residues}}{300 \text{ aa residues per polypeptide}} \rightarrow 2.77 \times 10^6$ polypeptides (i.e. cistrons)]

ing an average of 300 amino acid residues per polypeptide chain, and neglecting start-stop signals, how many cistrons do you have in *one* of your sets of 23 chromosomes?

18-18. Explain why one mutation in a given cistron may be lethal whereas another in the same cistron may produce no adverse phenotypic effect and be detectable only by laboratory tests. *[handwritten annotations: depending on Also, one might be in non-coding DNA strand... the amino acid position involved... one may cause a serious missense (e.g. to STOP) else nonfunctional protein... one may cause amino missense or sense mutation... or silent mutation]*

18-19. Work with phage T4 (among others) discloses many mutational "hot spots" —i.e., certain nucleotides or groups of nucleotides that are more likely to mutate under given conditions than are others. These "hot spots" are not the same for all chemical mutagens tested. Assume two kinds of microorganisms, species *AT* in whose DNA adenine-thymine pairs constitute a large majority of the nucleotide pairs, and species *CG* having much more cytosine-guanine than adenine-thymine in its DNA. Cultures of each of these species are treated with a series of different mutagens. For each of the following indicate which species, *AT, CG,* or both, would be expected to show the higher mutation rate: (a) the enol form of 5-bromouracil, (b) the keto form of hypoxanthine, (c) uracil, (d) acridine orange, (e) ultraviolet radiation.

18-20. How might a multiple-allele series arise at a given locus during the evolution of a particular species? *[handwritten: good question — accumulation of a series of missense mutations affecting different amino acids in the polypeptide chain]*

18-21. Why are not chemical and radiational mutagens used more widely in developing improved varieties of crop plants and domestic animals?

18-22. Distinguish among these three concepts: cistron, muton, and recon.

18-23. How do complementation and recombination differ?

[handwritten worked answers:]

18-12)
	normal	mutant
DNA	CCG	CTG
mRNA	GGC	GAC

sub of T for C in the DNA strand (both pyrimidine)

18-19)a) 5 BUe acts like the tautomer of T & acts like tautomer of T pairs with G instead of A changes pairing of T
So causes sub of G·C for A·T in AT species
b) deamination converts A to hypoxanthine the keto form of hypoxanthine (H) acts like guanine & pairs with C instead of T in AT species changes pairing of A

c) are tautomer of uracil pairs with G instead of A changes pairing of U (won't be incorporated into AT species)

d) both

e) causes pyrimidine dimers (C&T) so both

CHAPTER 19
Regulation of Gene Action

MULTICELLULAR organisms normally develop by an orderly process of differentiation from a single cell, the zygote, to an adult form of many different kinds of cells, tissues, and organs. But because of the behavior of chromosomes in mitosis, all the cells of a complex, multicellular soma may be presumed to have the same genotype. In short, between zygote and adult stages, cells of the organism *differentiate,* both physiologically and physically, yet the genome of all cells should be identical under normal conditions. The problem may be quite simply stated: *how do genetically identical cells become functionally different?*

Up to at least a certain point in development, the cells of a multicellular organism are *totipotent;* that is, they are capable of forming complete bodies if isolated after development has begun. For example, in carrot, excised, mature, differentiated cells from phloem (the food conducting vascular tissue) form an irregular mass of large, thin-walled, undifferentiated cells (called a callus) on a solid nutrient medium. Transfer of the callus to a liquid medium, which is kept gently shaken, separates the cells. Upon subsequent transfer to new cultures, these callus cells develop into embryolike structures, or *embryoids*. Liquid media induce root formation by the embryoid; later transfer to a solid medium results in stem and flower production, in short, a fully differentiated, mature plant. A small section of the body of the flatworm *Planaria* or the mature leaf of a plant like the African violet (*Saintpaulia ionantha*) regenerates the entire organism under proper environmental conditions, though this is by no means true for all animals and plants. Many experiments with very young embryos of sea urchins and frogs indicate that even after the zygote has divided to produce two to eight cells, separation of cells or forcing them to divide in only one plane is followed by normal differentiation and embryo formation. Separation of the cells of two-celled rabbit embryos and then reimplantation into another female similarly produces normal embryos. In the tobacco plant Nitsch and Nitsch (1969) report in vitro development of microspores into monoploid plants that reach flowering stage, although seeds are not produced. (Why?) Yet at later developmental stages, when differentiation has proceeded beyond a critical point, this totipotency is often lost, particularly in many animals.

The key to orderly structural and functional differentiation of multicellular organisms lies in the fact that their cells do not always produce the same proteins all the time. Even unicellular forms such as bacteria display a temporal shift in protein synthesis. Apparently some mechanism exists in the cell to turn genes "on" or "off" at different times and/or in different environments. In

fact, it is now clear from considerable experimental evidence that differentiation is the consequence of orderly, temporal activation and repression of genetic material. This *regulation of gene action* is the problem we need now to
examine.

Evidence of Regulation of Gene Action

CHROMOSOME PUFFS

Giant Chromosomes. In dipteran flies, the chromosomes of larval tissues
such as the salivary glands regularly exhibit unusual behavior that has shed
some light on gene regulation. Although the cells of these tissues do not divide,
their chromosomes replicate repeatedly while permanently synapsed in homologous pairs, producing polytene or many-stranded giant chromosomes. If
these chromosomes are examined at several stages of the individual's development, specific areas (sets of bands) are seen to enlarge into prominent "puffs"
or Balbiani rings (Fig. 19-1). From electron micrographs Beerman and Bahr
(1954) describe the puff as a loosening of the tightly coiled DNA into long,
looped structures not unlike the lampbrush chromosomes described in the next
section. These puffs appear and disappear in a given tissue at certain chromosomal locations as development proceeds, those at particular locations being
correlated with specific developmental stages of the insect. The pattern of puffing varies in a regular and characteristic way with the tissue and its stage of
maturation.

By means of differential staining techniques, biochemical tests, and use of
radioactive isotopes, it has been demonstrated that each puff is an active site
of transcription. If we recall that genes (i.e., cistrons) are associated with particular bands, this temporal puffing clearly indicates changes in gene activity
over time. Tests show that the mRNA synthesized in one puff is characteristic
and differs from that produced by other puffs.

Injection of very small amounts of the hormone ecdysone, which is produced
by the prothoracic glands and which induces molting, causes formation of the
same puffs that occur normally prior to molting in untreated larvae. The prothoracic glands are activated by the flow of brain hormone from neurosecretory
cells. This has been demonstrated in an ingenious experiment reported by
Amabis and Cabral (1970). Tying off the anterior part of the larva of the dipteran *Rhynchosciara* just behind the brain, they found, resulted both in failure
of normal puffing and a decrease in size of puffs already initiated at the time
of ligation (Fig. 19-1). In other experiments, injection of the antibiotic actinomycin D (which inhibits mRNA synthesis) prevents puff formation for several
hours, even when ecdysone is used simultaneously. Radioactive uridine, injected into larvae, accumulates only in the puffs and nucleoli, but fails to do so
if it is preceded by injection of actinomycin D.

All these results clearly point to the puffs as sites of RNA synthesis according
to a pattern closely associated with the development of the individual. Evi-

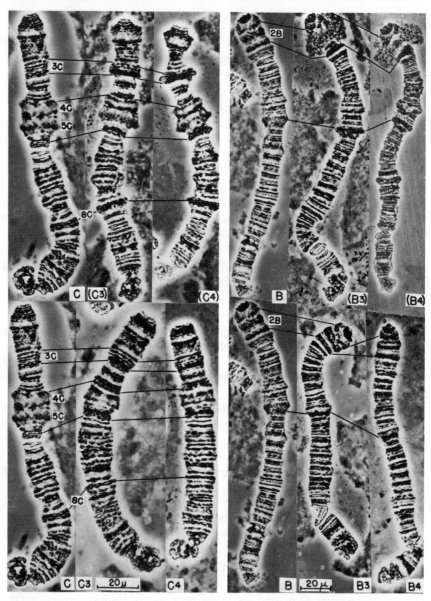

FIGURE 19-1. *Chromosome puffs in polytene chromosomes of* Rhynchosciara, *at 4C and 5C. Top row, chromosomes C and B from controls (no ligation) and their appearance after 3 (C3 and B3) and 4 (C4 and B4) days. Note especially the development of puff 2B (upper right photograph). Bottom row, chromosomes B and C before ligation (C and B); C3 and B3, chromosomes C and B 3 days after ligation; C4 and B4, the same chromosomes 4 days after ligation. Note particularly the failure of puff 2B to develop in the experimental series.* [Courtesy Drs. J. M. Amabis and D. Cabral, Universidade de São Paulo, São Paulo, Brazil. Reprinted from *Science,* **169:**692–694 (14 August 1970) by permission. Copyright 1970 by the American Association for the Advancement of Science.]

dently, then, genes (cistrons) undergo reversible changes in activity (principally mRNA synthesis) that are related to the developmental stage of the organism.

Lampbrush Chromosomes. This "turning on and off" of genes is evident also in the large chromosomes of amphibian oocytes (Fig. 19-2). Here the DNA strands of the long axes form lateral loops. Ribonucleic acid is produced in the loops of these chromosomes in cyclic fashion, very much as in the puffs of dipteran chromosomes. It seems likely that cyclic mRNA production is a general phenomenon, though the morphological details of interphase chromosomes of most eukaryotic cells cannot be studied adequately enough with present techniques to provide visual evidence.

Histones. X-ray diffraction and chemical studies indicate that the DNA of all cells except bacteria and some sperm is intimately associated with a fairly simple class of basic proteins called *histones*. There is some evidence that the nucleolus is the most active stie of histone synthesis (Flamm and Birnstiel, 1964). The exact physical relationship between histones and the helices of DNA is not settled, but it is considered likely that the histone complexing of DNA either causes or enhances coiling, which, in turn, is related to gene repression. Zubay (1964) further suggests that the DNA helices of eukaryote chromosomes are formed into supercoils whose adjacent gyres are held together by histone bridges. Sluyser and Snellen-Jurgens (1970) believe that histones rich in lysine readily form crosslinks between nucleic acid molecules, whereas those rich in arginine do so less readily.

The degree of histone-DNA complexing appears, in at least some organisms, to vary with the body part and with time. In developing pea embryos, for example, the cotyledons (food storage organs of the embryo) develop rather rapidly in embryogeny, then cease to grow further, even in germination. Tests

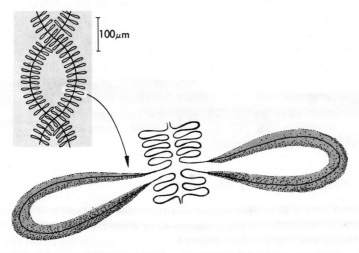

100μm

FIGURE 19-2. *Diagram of portion of a lampbrush chromosome as seen in an amphibian oöcyte.*

indicate that almost all the DNA of mature cotyledons is histone-complexed, whereas the terminal meristems, active growth regions that, at germination, produce all the root and shoot of the seedling, have more noncomplexed DNA.

This suggests an inverse relationship between histone complexing and protein synthesis. Bonner and his colleagues (1963) have presented excellent confirmation of this hypothesis. Pea cotyledons produce a protein, seed reserve globulin, that is not produced in vegetative tissues. DNA from cotyledons can be made to produce the mRNA necessary for in vitro synthesis of this protein. DNA from vegetative parts of the plant does not yield mRNA for globulin unless the histone is chemically removed! Nezgovorova and Borisova (1970) have shown that maize embryos begin growth and development into seedlings only when most of the DNA is not histone complexed. They conclude that histone removal may therefore be involved in the physiological mechanism that initiates germination of the grain. It has been suggested by a number of researchers that histones inhibit DNA strand separation by enhancing coiling; in that case, histone-complexed DNA would be unable to serve as a template for either DNA or RNA synthesis.

However, the way in which histones regulate gene activity must simply be regarded at present as unsettled. It is known that at least certain of the histones do very greatly inhibit DNA-dependent RNA synthesis (transcription); in some way complexing of DNA by histones prevents action of RNA polymerase. It has been suggested (Pogo et al., 1966; Allfrey et al., 1968) that acetylation of histones precedes RNA synthesis by previously inactive cistrons. But Ellgaard (1967) and Clever and Ellgaard (1970) offer experimental evidence that, although this appears to be so, histone acetylation does not necessarily take place at loci being activated in heat-induced puffs in *Drosophila* chromosomes. Georgiev (1969) concludes that "the number of . . . facts is not enough to describe a general scheme of histone participation in the control of gene expression." Of many possibilities he concludes that (1) histones combine with and repress all DNA nonspecifically, whereas other, nonhistone proteins recognize particular deoxyribonucleotide sequences, acting to eliminate histone repression, and/or (2) some histones, especially those rich in lysine, may be specific for certain DNA sequences. Perhaps the most concise statement of the present state of our information is that most histones appear to act as generalized repressors of mRNA transcription, and that various *inducers* (e.g., ecdysone) may bring about histone removal in some way at particular loci that, therefore, become derepressed.

Enzyme Regulatory Mechanisms

Inasmuch as the ultimate product of a cistron is a polypeptide chain, which is frequently incorporated into an enzyme, a simple system of control would operate at the level of enzyme regulation. Evidence accumulated in recent years indicates that there are, in fact, two basic mechanisms of enzyme regulation.

One controls the *activity* of enzymes whose production is not altered, the other regulates the *synthesis* of enzymes. The first of these is generally termed **end-product inhibition,** the second involves the concept of the **operon.** The operon is a group of adjacent cistrons whose function is regulated by other sites, called the operator and the promoter. Operons are common in bacteria, and we shall examine them shortly.

END-PRODUCT INHIBITION

In studies on isoleucine synthesis in *Escherichia coli,* Umbarger (1961) demonstrated that addition of isoleucine (the end product of a five-step conversion of threonine, Fig. 19-3) to a culture of the bacteria resulted in immediate blocking of the threonine $\rightarrow$ isoleucine pathway. In the presence of added isoleucine, the cells preferentially use this *exogenous* end product and cease their own isoleucine synthesis. Moreover, it has been shown that *production* of each of the five enzymes is *not* interfered with, but action of the enzyme responsible for the deamination of threonine to α-ketobutyrate (Fig. 19-3) is *inhibited* by the end product, isoleucine.

Interestingly, this inhibition results from a binding of end product to the enzyme so that the inhibitor appears to compete with the substrate for a site on the enzyme molecule. But it is known that this competition is not for the same site, the enzyme apparently having two specific recognition sites, one for its substrate and another for the inhibiting end product. However, attachment at the inhibitor site affects the substrate site. The term *allosteric interaction* is applied to such changes in enzyme activity produced by binding of a second substance at a different and nonoverlapping site.

THE OPERON: NEGATIVE CONTROL

The Lac *Operon in* Escherichia coli.   Among the many strains of the colon bacillus, two in particular have shed considerable light on another system of gene regulation. Two enzymes are required for lactose metabolism in this organism: *β-galactoside permease,* responsible for transport of lactose into the cell and its concentration there, and *β-galactosidase,* which catalyzes the hydrolysis of lactose to galactose and glucose. The wild-type *E. coli* produces these enzymes only in the presence of lactose; this is an *inductive* strain. One might say the wild-type cells are *induced* to synthesize the enzymes that function in lactose metabolism only when the enzymes' substrate is present. When the source of carbon available to *E. coli* does not include lactose, β-galactoside permease and β-galactosidase are present in the cell in very low amount—about 10 molecules per cell, according to Beckwith and Zipser (1970). When and if lactose is supplied, the rate of synthesis of these enzymes increases as much as 1,000 times according to Beckwith and Zipser (1970).

Other strains of the colon bacillus, derived by mutation from the inductive wild-type, produce these enzymes continuously, whether lactose is present or not. These are *constitutive* strains. After some initial uncertainty over whether

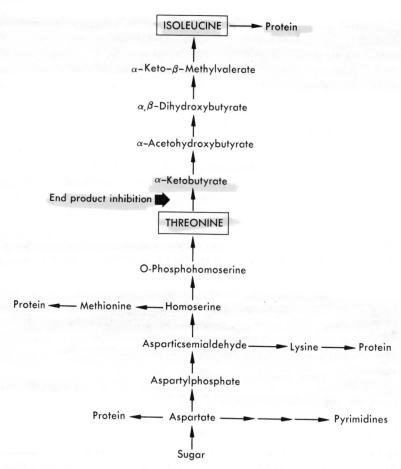

FIGURE 19-3. *Some steps in isoleucine synthesis. In end-product inhibition in* E. coli, *enzyme inhibition occurs in the deamination of threonine to α-ketobutyrate at the point indicated.*

enzyme synthesis or enzyme activity was affected by lactose, it has become clear that the presence of lactose does indeed *induce* synthesis of enzymes in the inductive strain. Lactose, the substrate, is here acting as an enzyme inducer. Enzymes produced only when a substrate is present are designated as *inducible enzymes.* Although inducers are quite specific, structural analogs may often function in the same way.

The *lac* segment of the *E. coli* chromosome is known to include three cistrons, *z* for β-galactosidase, *y* for β-galactoside permease, and *a* for thiogalactoside transacetylase.[1] The three cistrons are closely linked and regulated coordi-

---

[1] The function of thiogalactoside transacetylase in vivo is unknown.

nately. Jacob and Monod and their associates (1959, 1960, 1961) developed a model of a genetic system that regulates lactose metabolism, to which they gave the name **operon.** With minor later refinements, this model appears to explain all observations in this bacterium, and several other similar regulatory systems have subsequently been described for this and other bacteria. Functioning of the *lac* operon is excellently summarized by Beckwith (1967), by Epstein and Beckwith (1968), by Martin (1969), by Beckwith and Zipser (1970), and by Reznikoff (1972), who describe its operation in the following way.

The three cistrons, *z, y,* and *a,* are each responsible for one enzyme (Fig. 19-4); they occupy adjoining positions on the DNA molecule and map in that sequence. The genetic information of these three cistrons is transcribed into a single polycistronic mRNA molecule, which is subsequently translated into the polypeptide chains of the three enzymes. Transcription is initiated at a promoter site, *p,* at which DNA-dependent RNA polymerase binds. Transcription then proceeds in the $5' \rightarrow 3'$ direction in the usual fashion that was described in Chapter 16. An operator site, *o,* lies to the "right" of *p,* i.e., in the direction of the three structural genes (Fig. 19-4) and exercises a control over transcription that we shall examine shortly. Mapping experiments show the sequence of these genetic elements of DNA to be

$$p \quad o \quad z \quad y \quad a$$

The three cistrons, together with the promoter and operator sites comprise the lac *operon.* **An operon, then, consists of a system of cistrons, operator, and promoter sites.** The basis for inductive and constitutive production of the three enzymes of the *lac* operon lies in still another site, the regulator (*i*), and the interaction of its protein product with lactose and with the operator site. The regulator site is not immediately adjacent to the *lac* operon itself; the complete map sequence involved, then, may be represented as

$$i \quad \ldots \quad p \quad o \quad z \quad y \quad a$$

The regulator site produces a repressor protein (molecular weight about 150,000). In the inductive strain of the bacterium, the repressor protein binds to the wild-type operator (o^+), thereby preventing progression of RNA polymerase so that transcription of the three cistrons is prevented. But lactose, if present, combines with the repressor, inactivating it, with the result that the three cistrons are transcribed. Lactose, the substrate for β-galactosidase and β-galactoside permease, thus acts as an effector, or inducer, so that the operator is unrepressed (Fig. 19-4). Miller (1970) has determined that *transcription* is initiated at the promoter site, which is transcribed as part of the overall mRNA product of the operon. Eron and his colleagues (1970) agree, and have further concluded that *translation,* however, begins with the *z* cistron.

Constitutive strains of *E. coli* are characterized by either a defective regulator (i^-) or a mutant operator that is unable to bind the repressor. For example, $i^+ \ldots p^+ o^+ z^+ y^+ a^+$ is an inductive strain, i^+ producing a repressor that

A. INDUCER ABSENT

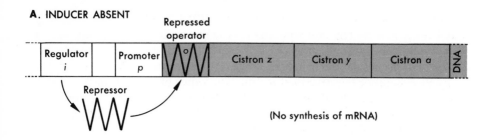

B. INDUCER PRESENT

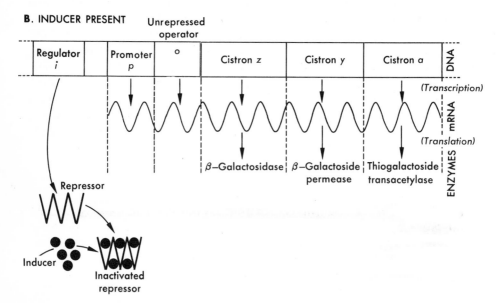

FIGURE 19-4. *Diagrammatic representation of the* lac *operon of* Escherichia coli. *Transcription is initiated at the promoter site* (p) *at which the enzyme DNA-dependent RNA polymerase binds. If the operator site* (o) *is repressed by the binding of the repressor protein to it, not transcription of mRNA occurs but, if the repressor is inactivated by the binding to it of lactose (the effector), the enzyme progresses in the 5′ → 3′ direction, transcribing cistrons* z, y, *and* a *into a single polycistronic mRNA molecule. In inducible systems, the repressor binds to the operator, preventing transcription of the structural genes, but if the effector (lactose in this case) is supplied, it inactivates the repressor so that transcription proceeds and translation ensues. For further details see text.*

binds to o^+, preventing transcription of z^+, y^+, and a^+ (Fig. 19-4). Presence of lactose, of course, inactivates the repressor so that o^+ permits transcription. On the other hand, i^- . . . p^+ o^+ z^+ y^+ a^+ is constitutive because a defective repressor, which will not bind to o^+, is produced, so transcription of the three cistrons is continuous. An i^+ . . . p^+ o^c z^+ y^+ a^+ mutant (an *operator-constitutive* strain) is also constitutive, producing enzymes continuously, but because the repressor binding site (o) rather than the repressor itself, is defective and cannot bind the repressor protein.

Interestingly enough, studies of merozygotes (partial diploids) produced through conjugation show i^+ to be dominant to i^-. In a cell of genotype $i^+ \ldots o^+ z^+ \ldots /i^- \ldots o^+ z^- \ldots$ one might expect normal β-galactosidase to be produced inductively and a modified form of the enzyme (i.e., lacking activity) constitutively. Not so; both types of β-galactosidase are produced inductively. The repressor is able to diffuse from the $i^+ \ldots o^+ z^+$ "chromosome" to the other and repress z^- even though the latter is linked to i^-. The regulator gene thus functions in either the *cis* or *trans* position. On the other hand, the operator (o) functions only in the *cis* position; for example, the diploid $i^+ \ldots o^c z^+ \ldots /i^- \ldots o^+ z^- \ldots$ constitutively produces only normal β-galactosidase and no modified enzyme. This is to be expected if the promoter is the place where mRNA transcription begins. Enzyme production by these several genotypes is summarized in Table 19-1.

? ? should be modified enzyme is produced inductively

TABLE 19-1. Production of Normal and Modified Beta-Galactosidase by Several *Lac* Genotypes in *E. coli*

	Enzyme Production			
	Constitutive		Inductive	
Genotype	Normal	Modified	Normal	Modified
$i^+ o^+ z^+$			+	
$i^- o^+ z^+$	+			
$i^+ o^c z^+$	+			
$i^+ o^+ z^+ / i^- o^+ z^-$			+	+

The *lac* operon illustrates *negative control,* the basis for which is production of a *repressor.* In this case the operator is repressed, and the cistrons turned "off," unless the effector (here, lactose) is present to inactivate the repressor (Fig. 19-5A). (induce)

The His *Operon* in Salmonella Typhimurium. Another type of negative control operates in the histidine (*his*) operon of *Salmonella typhimurium.* Synthesis of histidine occurs from phosphoribosyl pyrophosphate in 10 enzymatically catalyzed steps in this bacterium. Nine enzymes are involved (one functions in two different reactions), resulting from translation of as many cistrons in the *his* operon. As in the case of the *lac* operon, the *his* cistrons transcribe a polycistronic messenger, estimated to be about 13,000 nucleotides in length (Benzinger and Hartman, 1962). The map of the operon, as established by Hartman et al. (1960) and by Ames and his colleagues (1963), is shown in Figure 19-6.

A **Lac operon** (negative control)

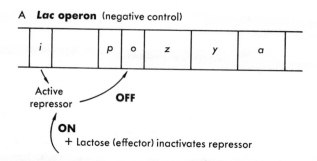

B **His operon** (negative control)

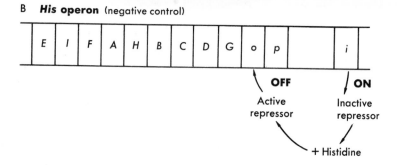

C **Ara operon** (Negative and positive control)

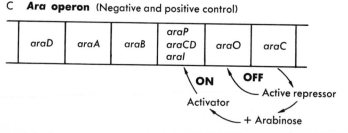

FIGURE 19-5. *Comparison of (A) negative control by a repressor that is inactivated by the effector* (lac *operon), (B) negative control by a repressor that is activated by an effector* (his *operon), and (C) negative and positive control by a repressor that is converted to an activator by the effector* (ara *operon).*

Unlike the *lac* operon, however, the repressor, as produced, is *inactive;* it is activated only by combining with histidine (either exogenously supplied or synthesized intracellularly). When thus activated, the repressor combines with an operator site, with the result that the structural genes are "turned off." In the absence of histidine in the cell, the inactive repressor is unable to repress the structural genes, and production of the enzymes needed for histidine synthesis proceeds. Although the operation of the *his* operon differs somewhat from that of the *lac* operon, it is, nevertheless, a *negative control* because a *repressor* protein is involved (Fig. 19-5B).

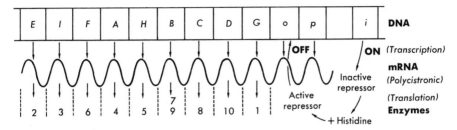

FIGURE 19-6. *The* his *operon of* Salmonella.

THE OPERON: POSITIVE AND NEGATIVE CONTROL

The **Ara** *Operon of* **Escherichia coli.** Regulation of the structural genes responsible for the production of the three enzymes catalyzing the three-step breakdown of the pentose sugar arabinose to xylulose-5-phosphate (another pentose sugar, involved in cellular oxidations) in *Escherichia coli* represents an interesting variation in operon operation. The segment of the *E. coli* molecule involved, shown in Figure 19-7, consists of three structural genes, each responsible for one of the three enzymes, an initiator site (itself consisting of three subsites), an operator, and a regulator. The regulator, *araC,* produces an allosteric protein that serves as both an activator and a repressor of the three cistrons (Cleary and Englesberg, 1974). In the absence of arabinose, this protein serves as a repressor, binding to the operator (*araO,* Fig. 19-5C) and blocking transcription. Wilcox et al. (1974) have shown that the half-life of the *araO*-repressor protein complex is only three minutes; by contrast, the half-life of the *lac* repressor-operator complex is 20 minutes. Arabinose binds to the repressor and/or the repressor-operator complex, producing a conformational change in the *araC* protein, which converts it to an activator. The activator then interacts with the *araI* site, permitting transcription of the structural genes *araB, araA,* and *araD.* So, in the *absence* of the effector (here arabinose, the metabolite), the protein product of the regulator site functions as a *repressor* by binding to the operator site and preventing transcription; this represents, of course, *negative* control. The same protein product of the regulator is converted to an *activator* in the *presence* of the effector by binding to it, allowing transcription to take place; this represents *positive* control.

Present evidence is that gene regulation by operons is widespread in bacteria. Many have been described, and some 100 to 200 have been estimated for *Escherichia coli* alone. For most of the known operons in bacteria the operation has been determined. Other operons have been reported for *Neurospora,* the red bread mold so useful in early biochemical genetics work, and for another fungus, the yeast *Saccharomyces.* However, in these latter two instances, and in reported operonlike systems in other eukaryotic organisms, an operator site appears to be lacking (Fincham and Day, 1971).

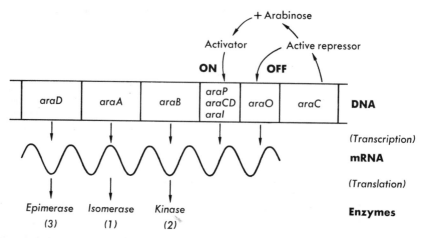

FIGURE 19-7. *The arabinose* (ara) *operon of* Escherichia coli. *The regulator* (araC) *produces an active repressor protein that, in the absence of arabinose, binds to the operator site* (araO) *and thereby blocks transcription. Arabinose serves as an effector and, if present, binds to the repressor and/or the repressor-operator complex, converting the repressor into an activator. The activator then interacts with the* araI *site, permitting transcription of the structural genes* araB, araA, *and* araD *in that sequence.*

Summary of Gene Regulation

Genetic material has turned out to be fairly complicated in its regulation as well as in its action and structure. Yet it is a relatively simple system for the production of an endless variety of phenotypes. Gene action appears to be related to certain proteins (histones) and/or to other genes or even to certain metabolites. So to our summary of gene function and fine structure (Chapter 18) we need to add the following points regarding the regulation of gene action: *e.g. lactose (inducer) (repressor)*

1. The activity of many cistrons is regulated so that their polypeptide product is formed only under certain chemical conditions and/or at certain developmental stages.
2. Although the function of histones is not fully known, it has been suggested that they regulate gene activity in those cells in which they occur by affecting DNA coiling and RNA synthesis. *(transcription of mRNA)*
3. In other organisms, notably bacteria, there is clear evidence for a regulator-operon complex in which the regulator produces a protein that
   a. Represses an operator (active repressor) with the result that transcription does not occur, but that can be inactivated by an effector (inducer), allowing transcripton to take place (negative control, as in the *lac* operon), or

b. Fails to repress the operator (inactive repressor) unless activated by combining with the effector (negative control, as in the *his* operon), or

c. Represses the operator site in the absence of the effector (negative control), or serves as an activator after combining with the effector (positive control), as in the *ara* operon.

In a superb piece of research, Beckwith and his colleagues (Shapiro et al., 1969), using two transducing phages, have been able to isolate and produce electron micrographs óf a segment of the *E. coli lac* operon. This segment consists of only the promoter, operator, and β-galactosidase sites some 1.3 to 1.5 μm in length. Taking an average of 1.4 μm, their calculations show the promoter and operator sites to total approximately 0.14 μm in length (about 410 DNA base pairs) and the *z* cistron 1.26 μm (3,700 base pairs). It is to be expected that the techniques used will make it possible to answer hitherto unexplained problems of gene regulation, such as operation of the repressor, in this and in other operons, as well as to clarify the mechanism of RNA polymerase binding and the control of RNA synthesis. It is also conceivable that the methods used in this remarkable work will eventually have some application in "genetic surgery" (see Chapter 21).

REFERENCES

ALLFREY, V. G., B. G. T. POGO, V. C. LITTAU, E. L. GERSHEY, and A. E. MIRSKY, 1968. Histone Acetylation in Insect Chromosomes. *Science,* **159:** 314–316.

AMABIS, J. M., and D. CABRAL, 1970. RNA and DNA Puffs in Polytene Chromosomes of *Rhynchosciara:* Inhibition by Extirpation of Prothorax. *Science,* **169:** 692–694.

AMES, B. N., and P. E. HARTMAN, 1963. The Histidine Operon. *Cold Spring Harbor Symposia Quant. Biol.,* **28:** 349–356. Reprinted in E. A. Adelberg, ed., 1966. *Papers on Bacterial Genetics.* Boston, Little, Brown.

AMES, B. N., P. E. HARTMAN, and F. JACOB, 1963. Chromosomal Alterations Affecting the Regulation of Histidine Biosynthetic Enzymes in *Salmonella. Jour. Molec. Biol.,* **7:** 23–42.

BECKWITH, J. R. 1967. Regulation of the Lac Operon. *Science,* **156:** 597–604.

BECKWITH, J. R., and D. ZIPSER, eds., 1970. *The Lactose Operon.* Cold Spring Harbor, N.Y., Cold Spring Harbor Laboratory.

BEERMAN, W., and G. F. BAHR, 1954. The Submicroscopic Structure of the Balbiani Ring. *Exptl. Cell Res.,* **6:** 195–201.

BENZINGER, R., and P. E. HARTMAN, 1962. Effects of Ultraviolet Light on Transducing Phage P22. *Virology,* **18:** 614–626.

BONNER, J., R. C. HUANG, and R. V. GILDEN, 1963. Chromosomally Directed Protein Synthesis. *Proc. Nat. Acad. Sci. (U.S.),* **50:** 893–900.

BONNER, J., and P. T'SO, eds., 1964. *The Nucleohistones.* San Francisco, Holden-Day.

CLEARY, P. P., and E. ENGLESBERG, 1974. Transcriptional Control in the L-Arabinose Operon of *Escherichia coli* B/r. *Jour. Bacter.,* **118:** 121–128.

CLEVER, U., and E. G. ELLGAARD, 1970. Puffing and Histone Acetylation in Polytene Chromosomes. *Science,* **169:** 373–374.

ELLGAARD, E. G., 1967. Gene Activation Without Histone Acetylation in *Drosophila melanogaster. Science,* **157:** 1070–1072.

EPSTEIN, W., and J. R. BECKWITH, 1968. Regulation of Gene Expression. *Ann. Rev. Biochem.,* **37:** 411–436.

ERON, L., J. R. BECKWITH, and F. JACOB, 1970. Deletion of Translational Start Signals in the *lac* Operon of *E. coli.* In J. R. Beckwith and D. Zipser, eds. *The Lactose Operon.* Cold Spring Harbor, N.Y., Cold Spring Harbor Laboratory.

FINCHAM, J. R. S., and P. R. DAY, 1971. *Fungal Genetics,* 3d ed. Philadelphia, F. A. Davis.

FLAMM, W. G., and M. L. BIRNSTIEL, 1964. *Studies on the Metabolism of Nuclear Basic Proteins.* In J. Bonner and P. T'so, eds. *The Nucleohistones.* San Francisco, Holden-Day.

GEORGIEV, G. P., 1969. *Histones and the Control of Gene Action.* In H. L. Roman, ed. *Annual Review of Genetics,* volume 3. Palo Alto, Calif., Annual Reviews, Inc.

GROSS, J., and E. ENGLESBERG, 1959. Determination of the Order of Mutational Sites Governing L-Arabinose Utilization in *Escherichia coli* B/r by Transduction with Phage P1bt. *Virology,* **9:** 314–331. Reprinted in E. A. Adelberg, ed., 1966. *Papers on Bacterial Genetics.* Boston, Little, Brown.

HARTMAN, P. E., J. C. LOPER, D. SERMAN, 1960. Fine Structure Mapping by Complete Transduction Between Histidine-Requiring *Salmonella* Mutants. *Jour. Gen. Microbiol.,* **22:** 323–353. Reprinted in E. A. Adelberg, ed., 1966. *Papers on Bacterial Genetics.* Boston, Little, Brown.

JACOB, F., and J. MONOD, 1961. Genetic Regulatory Mechanisms in the Synthesis of Proteins. *Jour. Molec. Biol.,* **3:** 318–356.

JACOB, F., D. PERRIN, C. SANCHEZ, and J. MONOD, 1960. The Operon: A Group of Genes Whose Expression Is Coordinated by an Operator. *Compt. Rend. Acad. Sci.,* **250:** 1727–1729.

LOOMIS, W. F., ed., 1970. *Papers on Regulation of Gene Activity During Development.* New York, Harper & Row.

MARTIN, R. G., 1969. *Control of Gene Expression.* In H. L. Roman, ed. *Annual Review of Genetics,* volume 3. Palo Alto, Calif., Annual Reviews, Inc.

MILLER, J. H., 1970. Transcription Starts and Stops in the *Lac* Operon. In J. R. Beckwith and D. Zipser, eds. *The Lactose Operon.* Cold Spring Harbor, N.Y., Cold Spring Harbor Laboratory.

NEZGOVOROVA, L. A., and N. N. BORISOVA, 1970. On the Trigger Mechanism of Germinating Seeds. V. Histones, Their Relation to Nucleic Acids and Influence of Inhibitors. *Fiziol Rast.,* **17:** 322–329.

NITSCH, J. P., and C. NITSCH, 1969. Haploid Plants from Pollen Grains. *Science,* **163:** 85–87.

PARDEE, A. B., F. JACOB, and J. MONOD, 1959. The Genetic Control and Cytoplasmic Expression of "Inducibility" in the Synthesis of β-galactosidase by *E. coli. Jour. Molec. Biol.,* **1:** 165–178.

POGO, B. G. T., V. G. ALLFREY, and A. E. MIRSKY, 1966. RNA Synthesis and Histone Acetylation During the Course of Gene Activation in Lymphocytes. *Proc. Nat. Acad. Sci. (U.S.),* **55:** 805–812.

REZNIKOFF, W. S., 1972. The Operon Revisited. In H. L. Roman, ed. *Annual Review of Genetics,* volume 6. Palo Alto, Calif., Annual Reviews, inc.

SHAPIRO, J., L. MACHATTIE, L. ERON, G. IHLER, K. IPPEN, and J. BECKWITH, 1969. Isolation of Pure *Lac* Operon DNA. *Nature,* **224:** 768–774.

SLUYSER, M., and N. H. SNELLEN-JURGENS, 1970. Interaction of Histones and Nucleic Acids in vitro. *Biochim. Biophys. Acta,* **199:** 490–499.

UMBARGER, H. E., 1961. Feedback Control by Endproduct Inhibition. *Cold Spring Harbor Symposia Quant. Biol.,* **26:** 301–312.

WILCOX, G., K. J. CLEMETSON, P. CLEARY, and E. ENGLESBERG, 1974. Interaction of the Regulatory Gene Product with the Operator Site in the L-Arabinose Operon of *Escherichia coli. Jour. Molec. Biol.,* **85:** 589–602.

ZIPSER, D., 1967. Orientation of Nonsense Codons on the Genetic Map of the Lac Operon. *Science,* **157:** 1176–1177.

ZUBAY, G. L., 1964. *Nucleohistone Structure and Function.* In J. Bonner, and P. T'so, eds., *The Nucleohistones.* San Francisco, Holden-Day.

PROBLEMS

19-1. In what way or ways are operator and regulator sites similar? Dissimilar?

19-2. Differentiate between repressors and effectors (inducers).

19-3. Distinguish between positive and negative control.

19-4. Synthesis of mRNA by the *lac* operon of *Escherichia coli* increases with addition of the inducer. Is this evidence for action of the inducer at the transcriptional or at at translational level?

19-5. Will production of *normal* beta-galactosidase be constitutive, inductive, or absent for each of the following genotypes: (a) $i^+ p^+ o^+ z^+ y^+ a^+$, (b) $i^- p^+ o^+ z^+ y^+ a^+$, (c) $i^+ p^+ o^c z^+ y^+ a^+$, (d) $i^+ p^+ o^+ z^- y^- a^+$. (e) $i^+ p^+ o^+ z^- y^+ a^+ / i^+ p^+ o^c z^+ y^+ a^+$?

19-6. Would you expect a nonsense mutation in one of the structural genes of the *his* operon to affect transcription or translation?

19-7. Cleary and Englesberg (1974) report an *Escherichia coli* mutant (*araC5*) that they found to carry an amber mutation in the *araC* site. Would this mutant strain produce any of the arabinose enzymes in the presence of arabinose? Explain.

19-8. If a hypothetical *Escherichia coli ara* mutant whose genotype is represented as $D^- A^+ B^+ I^+ O^+ C^+$, where D^- involves a deletion of five deoxyribonucleotide pairs near the beginning of the *D* cistron, is supplied with arabinose, (a) would you expect xylulose-5-phosphate to be produced? (b) What substance(s) in the arabinose metabolic pathway would you expect to accumulate in the presence of arabinose?

19-9. The tryptophan (*trp*) operon of *Escherichia coli* is responsible for coordinate production of the enzymes involved in the synthesis of the amino acid tryptophan. The operon includes five structural genes, an operator, and a promoter. A regulator site produces a repressor that is activated by tryptophan. (a) Is this an illustration of positive or of negative control? (b) In the absence of tryptophan in the cell, are the enzymes of the *trp* operon produced or not?

19-10. State the present concept of the gene.

CHAPTER 20
Cytoplasmic Genetic Systems

THE existence of genes as segments of nucleic acid molecules, located in chromosomes, and controlling phenotypes in known and predictable fashion has been amply demonstrated on sound, observable, verifiable bases. But the firm establishment of such a chromosomal mechanism of inheritance does not necessarily preclude a role by other cell parts. From time to time observations have been reported that suggest a cytoplasmic role in genetics. To evaluate these reports we need first to have a clear understanding of what cytoplasmic inheritance does or should mean.

SUGGESTED CRITERIA FOR CYTOPLASMIC GENES

Before a particular genetic phenomenon is clearly acceptable as having a cytoplasmic basis, the following criteria should be applied:

1. *Is there evidence of genes physically and/or physiologically independent of any nuclear genetic material?*
2. *What is the chemical nature of the suspected cytoplasmic gene?*
   a. *Is it DNA? RNA? Something else?*
   b. *Is it operationally genetic, that is, is it a conserved structure whose activities govern the trait, or is it the trait itself that is being called a cytoplasmic gene?*
   c. *Is it a normal and usual part of the cell's cytoplasm (recognizing that "normal" may not be easy to define)?*
3. *What are its mutational properties and methods?*
4. *Do reciprocal crosses show clear evidence of a cytoplasmic role as opposed to sex (or other) chromosomes?*
5. *Does its pattern of inheritance bear no relation to chromosomal segregation?*
6. *Is it unequivocally not a part of any chromosomal linkage group?*
7. *Are there alternative phenotypic expressions and are they stable through the life cycle?*
8. *Is transmission of the trait unaffected by nuclear transplants?*

Although this list is neither exhaustive nor universally accepted, for our purposes cytoplasmic inheritance will be understood to be based on cytoplasmically located, independent, self-replicating nucleic acids, differing from chromosomal genes only in their location within the cell. Not many of the purported examples of cytoplasmic inheritance fit this concept, and only a few have been unequivocally established. We shall examine some of them in the following sections.

MATERNAL EFFECTS

Shell Coiling in the Snail **Limnaea.** The direction of coiling of the shell in such snails as *Limnaea* illustrates the influence of nuclear genes acting through effects produced in the cytoplasm.

The shells of snails coil either to the right (dextral) or to the left (sinistral), as seen in Figure 20-1. A shell held so that the opening through which the snail's body protrudes is on the right and facing the observer is termed dextral; if the opening is on the left, coiling is sinistral. Direction of coiling is determined by a pair of nuclear genes, dextral ($+$) being dominant to sinistral (s). But expression of the trait depends on the maternal genotype. Because snails may be self-fertilized, the following cross is possible:

$$P \quad + + (♀) \times ss (♂)$$

$$\text{right} \qquad \text{left}$$

$$F_1 \qquad + s$$

$$\text{right}$$
$$\text{(self-fertilized)}$$

$$F_2 \quad \tfrac{1}{4} + + : \tfrac{2}{4} + s : \tfrac{1}{4} ss$$

$$\text{right} \qquad \text{right} \quad \text{right!}$$

When each of the F_2 genotypes is again self-fertilized, progeny of the $+ +$ and $+ s$ animals are dextral, but those of the ss individuals are sinistral.

Additional investigation has disclosed that direction of coiling depends upon the orientation of the spindle in the first mitosis of the zygote. Spindle orientation, in turn, is controlled by the genotype of the oocyte from which the egg develops and appears to be built into the egg before meiosis or syngamy occurs. The exact basis of this rather unusual control is unknown.

Water Fleas and Flour Moths. A closely similar situation occurs in at least two very different invertebrates, the water flea (*Gammarus*) and the flour moth (*Ephestia*). Pigment production in eyes of young individuals depends upon a pair of nuclear genes, *A* and *a*. The dominant gene directs production of kynurenine, a diffusible substance that is involved in pigment synthesis. The cross *Aa* (♀) $\times$ *aa* (♂), for example, produces progeny all of which have dark eyes while young. Upon reaching the adult stage, half the offspring (those of genotype *aa*) become light-eyed. The explanation, of course, is that kynurenine

FIGURE 20-1. *Dextral and sinistral coiling of the shell in the snail,* Limnea.

Dextral **Sinistral**

diffuses from the *Aa* mother into all the young, enabling them to manufacture pigment regardless of their genotype. The *aa* progeny, however, have no means of continuing the supply of kynurenine, with the result that their eyes eventually become light. This is obviously a cytoplasmic effect, but wholly without a cytoplasmic genetic mechanism.

INFECTIVE PARTICLES

***Carbon Dioxide Sensitivity in* Drosophila.**   A certain strain of *Drosophila melanogaster* shows a high degree of sensitivity to carbon dioxide. Whereas the wild type can be exposed for long periods to pure carbon dioxide without permanent damage, the sensitive strain quickly becomes uncoordinated in even brief exposures to low concentrations. This trait is transmitted primarily, but not exclusively, through the maternal parent. Tests have disclosed that sensitivity is dependent upon an infective, viruslike particle, called *sigma,* in the cytoplasm. It is normally transmitted via the eggs' larger amount of cytoplasm but occasionally through the sperms as well. Sensitivity may even be induced by injection of a cell-free extract from sensitives (L'Heritier, 1951).

Sigma contains DNA and is mutable, but is clearly an infective, "foreign" particle. Multiplication is independent of any nuclear gene, but the mechanism of its sensitizing action is unknown.

***Female Production in* Drosophila.**   In another case in *Drosophila,* almost all male offspring die soon after zygote formation. This trait is transmitted by the female and is independent of nuclear genes. Studies have disclosed the trait to be dependent on a spirochete (one of the classes of bacteria) in the hemolymph of the female.

***Killer Trait in* Paramecium.**   Some races ("killers") of the common ciliate *Paramecium aurelia* produce a substance called paramecin that is lethal to other individuals ("sensitives"). Paramecin is water-soluble, diffusible, and depends for its production upon particles called *kappa* in the cytoplasm. Kappa contains DNA and RNA and is mutable, but its presence is dependent on the nuclear gene *K*. Animals of nuclear genotype *kk* are unable to harbor kappa.

K− individuals do not possess kappa unless and until it is introduced through a cytoplasmic bridge during conjugation (Fig. 20-2). Nonkiller *K−* animals may also be derived from killers by decreasing the number of kappa particles. This may be accomplished either by near starvation of the culture or subjecting it to low temperatures or, on the other hand, by causing killers to multiply more rapidly than kappa.

Kappa particles can be seen with the microscope. Electron microscopy reveals them to have minute amounts of cytoplasm and to be bounded by a membrane. They can be transferred to other ciliates by feeding. Far from being the illustration of the cytoplasmic gene it was first surmised to be, kappa must be regarded as an infectious organism that has attained a high degree of symbiosis with its host.

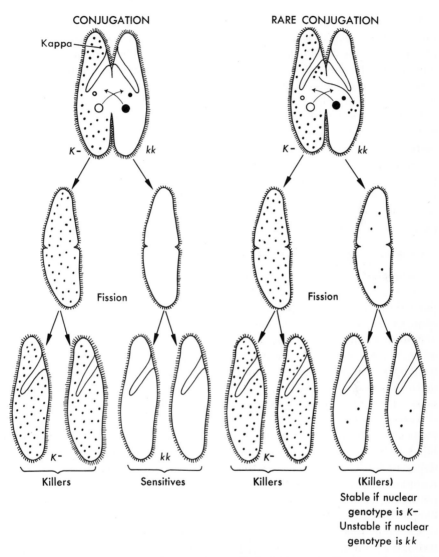

FIGURE 20-2. *Conjugation in* Paramecium *and the killer trait. Progeny of sensitives are killers only in rare situations where conjugation persists for a longer period so that kappa-containing cytoplasm is introduced into the conjugating sensitive. Kappa particles, however, are maintained only in the presence of a K— nuclear genotype.*

Similarly, the mate-killer trait in *Paramecium* is imparted by a *mu* particle that, in turn, exists only in those cells whose micronucleus contains at least one dominant of either of two pairs of chromosomal genes. *Mu* particles, too, appear to be endosymbiotes.

Milk Factor in Mice. The females of certain lines of mice are highly susceptible to mammary cancer. Results of reciprocal crosses between these and
animals of a low-cancer-incidence strain depend on the characteristic of the
female parent. Allowing young mice of a low-incidence strain to be nursed by
susceptible foster mothers produces a high rate of cancer in these low-incidence
young. Apparently this is a case of an infective agent being transmitted in the
milk. This so-called milk factor meets many of the criteria of a virus and has
been discovered to be transmissible also by saliva and semen. Its presence in
body fluids, moreover, is again dependent upon certain nuclear genes.

Situations of the kind just described, often showing a maternal inheritance
pattern, are certainly not examples of cytoplasmic inheritance as we have
defined it. Rather than being components of the individual's genetic mechanism and due to independent, conserved, nucleic acids located in the cytoplasm, they are, in fact, simply acquired infective agents.

NORMAL CELL COMPONENTS AND TRAITS

Chloroplasts. Plastids are cytoplasmic organelles in the cells of many different kinds of plants. Those containing the green photosynthetic pigments,
the chlorophylls, are called chloroplastids or simply *chloroplasts.* Plastids
arise by differentiation from simpler structures, the proplastids. These are
able to divide and so increase in number in a manner only in part under the
control of nuclear DNA. Electron microscopy reveals chloroplasts as fairly
complex structures (Fig. 20-3), and they do contain rather significant amounts
of DNA. Green and Burton (1970) were able to release highly supercoiled
plastid DNA by osmotic shock in the large unicellular green alga *Acetabularia*
(Fig. 20-4). The largest intact segment sufficiently untwisted to permit measurement they report to be 419 μm in length (which would place intact plastid
DNA on the low side of the size range of bacterial DNA), or about 1.23×10^6
deoxyribonucleotide pairs. But the supercoiling of chloroplast DNA, these
authors conclude, does not necessarily imply circularity. In fact, Dr. Green
points out (personal communication) that, "It may well be linear, though quite
long." Raven (1970), Margulis (1970), and other investigators have built a
considerable case for the suggestion that chloroplasts and mitochondria may
well have originated as symbiotic prokaryotes.

It is now clear that chloroplast DNA (cDNA) differs, often greatly, in base
composition from the nuclear DNA of the same species. Ultraviolet microbeam

---

FIGURE 20-3. *Electron micrograph of a thin section of a chloroplast from corn leaf.
The internal membrane system contains the chlorophylls and is the actual site of photo-
synthesis; DNA is contained within the stroma* (S). *GS, grana stacks; GL, grana lamellae;
SL, stroma lamellae; S, stroma containing DNA and ribosomes.* $\times 80,000$. (Photo by
Dr. L. K. Shumway, Program in Genetics, Washington State University. From *Cell
Ultrastructure* by William Jensen and Roderic Park, copyright 1967 by Wadsworth
Publishing Co., Inc., Belmont, California. Reproduced by permission of publisher.)

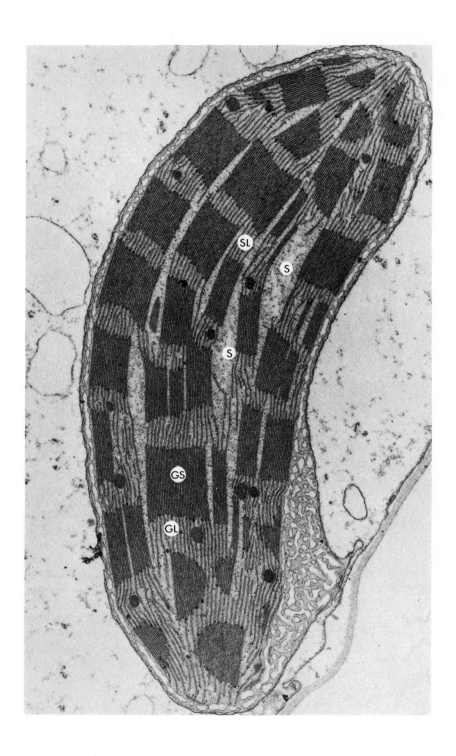

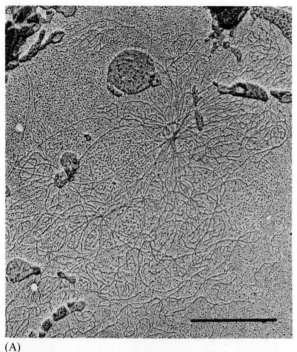

(A)

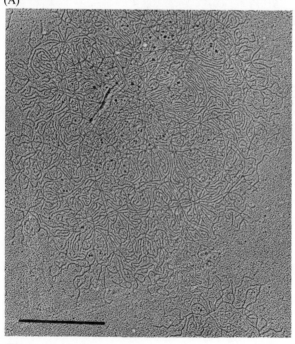

(B)

FIGURE 20-4. *Electron micrographs of chloroplast DNA from the green alga* Acetabularia mediterranea. *(A) Lysed chloroplast showing chloroplast DNA fibrils around a "center." (B) Large array of DNA released from a chloroplast. The bar represents 1 μm.* [Courtesy Dr. Beverley R. Green, University of British Columbia. Reprinted by permission of authors and publisher from Green, B. R., and H. Burton, *Acetabularia* Chloroplast DNA: Electron Microscope Visualization, *Science* (22 May 1970), **168:**981–982. Copyright 1970 by The American Association for the Advancement of Science.]

irradiation of the cytoplasm of the familiar one-celled *Euglena* results in plastid destruction whereas nuclear irradiation does not. This organism's chloroplasts contain a unique DNA component of low guanine-cytosine content that is absent in plastidless mutants. Plastids also contain a DNA polymerase and are able to synthesize their own DNA. Replication of cDNA has been shown to be semiconservative in the green alga *Chlamydomonas,* with chloroplast DNA synthesis occurring at a time quite different from that of nuclear DNA. Sager and Ramanis (1970) have prepared linkage maps for *Chlamydomonas* of eight cytoplasmic genes that they believe to be located in chloroplast DNA.

Chloroplasts also contain RNA and ribosomes. The latter, at least in higher plants, are smaller in mass than those in the cytoplasm of the same cell. They resemble much more closely bacterial ribosomes in size, as demonstrated by determinations of their sedimentation coefficients. They are also inactivated by chloramphenicol (as are bacterial ribosomes), which does not affect cytoplasmic ribosomes. Chloroplasts, then, contain their own machinery for protein synthesis, and it is clear that their DNA does play a part in incorporating amino acids into the characteristic plastid proteins. So here we have self-replicating structures, containing their own unique DNA and an array of protein-synthesizing equipment, although, at the same time, under some control by nuclear DNA as to their existence. However, on the basis of their unique cDNA, their complement of polypeptide-synthesizing apparatus, and their production of proteins, chloroplasts meet many, perhaps all, the criteria listed at the beginning of this chapter for cytoplasmic inheritance.

But chloroplasts themselves also constitute particular phenotypic traits (green color, ability to photosynthesize) of considerable complexity. Actually, we have something of a semantic problem to contend with; that is, we must distinguish between nuclear-controlled presence or absence of plastids per se and their ability to play a part in the synthesis of their own proteins. From the former viewpoint plastids are just another aspect of phenotype, no different from those we have considered thus far; from the latter, they do demonstrate at least a degree of extranuclear genetic control by normally occurring organelles, whatever their ultimate origin and evolution.

Plastid *inheritance* in the four o'clock plant (*Mirabilis jalapa,* Fig. 20-5) is a classic illustration of purported cytoplasmic inheritance. This plant may exist in three forms: normal green, variegated (patches of green and of nongreen tissue), and white (no chlorophyll). Now and then a plant with branches of two or three of these types occurs. Plastids in the white areas have no chlorophyll. Offspring of crosses are phenotypically like that of the pistillate (egg-producing) parent except where the pistillate parent is variegated, in which case the offspring are of all three types, in irregular ratio. Egg cells from green plants carry normal green plastids, those from white plants contain only white plastids, but eggs produced by variegated parents may have both plastid types. Pollen (which produces the sperm and rarely contains any plastids) has no effect on progeny phenotype.

FIGURE 20-5. *The four o'clock plant,* Mirabilis jalapa, *greenleaved variety.* (Photo courtesy Burpee Seeds.)

Note that all we are really dealing with here is the kind of plastid and proplastid present in the egg cytoplasm. If that plastid is defective as to chlorophyll synthesis, the F_1 plant will be nongreen; if not, it will be green. Eggs from variegated parents, containing both normal and defective plastids, give rise to plants some of whose cells fortuitously receive a majority of green plastids, and some that receive larger numbers of white plastids resulting in variegated plants. Recall that division of the cytoplasm following mitosis is not a quantitatively exact process. Yet on the other hand, the situation here could be rationally explained on the basis of an extranuclear determinant that might or might not be located in the chloroplast. In fact, in theory at least, it could be located anywhere in the cytoplasm; but, like some others to be described shortly, no site has ever been demonstrated. Most observational evidence, however, suggests that the chloroplast itself is *not* the actual genetic determiner in this case.

The evening primrose (*Oenothera* spp.) provides an interesting variant of plastid inheritance in which nuclear DNA plays an important part. Many species of *Oenothera* are characterized by their own unique combinations of reciprocal translocations, which result in only two types of functional gametes being produced, those containing only an individual's maternal chromosomes or only its paternal chromosomes. Therefore chromosomes in these species are transmitted as unitary groups.

Both *O. muricata* and *O. hookeri* are normal green plants. Crosses utilizing *O. muricata* as the pistillate parent produce normal green F_1 individuals. The reciprocal cross produces a nongreen F_1, all of which die. Thus *muricata* plastids can develop in the presence of a *muricata-hookeri* nucleus, whereas *hookeri* plastids cannot. Again, *these* latter two illustrations do not appear to provide examples of an *independent* gene system in the cytoplasm.

Mitochondria. All living cells except bacteria, blue-green algae, and mature erythrocytes contain *mitochondria,* small, self-replicating organelles of considerable internal structural complexity that are centers of aerobic respiration. A few are shown in sectional view in Figure 3-2. Although the shape of a given mitochondrion is not constant, mitochondria often appear as elongate, slender rods of rather variable dimensions, averaging some 0.5 μm in diameter and 3 to 5 μm in length. Extremes of 0.2 to 7 μm in width and 0.3 to 40 μm in length have been reported. Their number per cell varies considerably, from one in the unicellular green alga *Micrasterias* to as many as about a half million in the giant amoeba *Chaos chaos.*

Like chloroplasts, mitochondria contain their own unique DNA (mDNA) whose base composition is known to differ from that of both nuclear and plastid DNA. Mitochondrial DNA is replicated independently of nuclear DNA through the action of a DNA polymerase whose chromatographic properties differ from those of the polymerase of the nucleus. In many multicellular animal tissues (e.g., rat liver and mouse fibroblast cells) mitochondrial DNA is present as a double-stranded, circular structure, strongly reminiscent of bacterial DNA, with a molecular weight of 9 to 10 $\times$ 10^6 (Margulis, 1970; Raven, 1970). Satisfy yourself that this would indicate roughly 14,000 base pairs and a length of nearly 5 μm. Circular mitochondrial DNA has also been reported for man. Nass (1969) has described circular DNA molecules several times larger than 5 μm from mouse fibroblast cells (Fig. 20-6). On the other hand, mitochondrial DNA of the higher plants has not yet been shown to have this circular form.

Many of the mitochondrial proteins are synthesized within the organelle itself, utilizing mDNA as a template for unique mitochondrial RNA (Fan and Penman, 1970), but other proteins, notably cytochrome c, may be made elsewhere in the cell. Among the transfer RNAs identified in mitochondria is N-formylmethionyl tRNA ($tRNA_f$), known otherwise only from bacterial cells (Chapter 17). Mitochondrial ribosomes have also been identified (Bernhard, 1969), and Scragg et al. (1971) have obtained a cell-free protein synthesizing system from yeast mitochondria. So, as in the case of choroplasts, we have a semi-independent, DNA-containing organelle that is able to synthesize at least some of its own proteins.

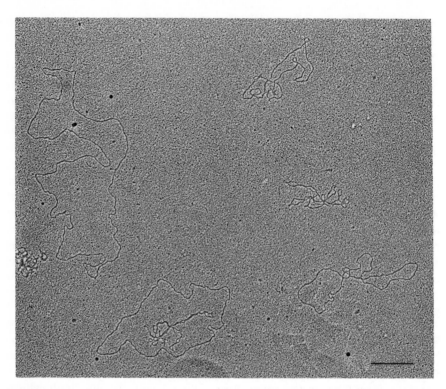

FIGURE 20-6. *Electron micrograph of DNA isolated from mitochondria of mouse fibroblast cells. Note the circularity of these molecules. A highly supercoiled molecule lies inside an open monomer (lower left), and three loosely twisted molecules are located at the right. The bar represents 1 μm.* (Courtesy Dr. Margit M. K. Nass, University of Pennsylvania. Reprinted by permission of author and publisher from Nass, M. M. K., Mitochondrial DNA: Advances, Problems, and Goals, *Science,* **165**:25–35, 4 July 1970. Copyright 1970 by The American Association for the Advancement of Science.)

The studies of Ephrussi (1953, 1955) on respiratory-deficient strains of baker's yeast (*Saccharomyces cerevesiae*) point up the genetic role of mitochondria. Yeasts are unicellular ascomycete fungi. In the life cycle of some species (Appendix B) diploid and monoploid adults alternate, the former reproducing by meiospores called ascospores, the latter by isogametes. Respiratory-deficient strains are able to respire only anaerobically and, on agar, produce characteristically small colonies known as "petites." Ephrussi has shown that both nuclear and cytoplasmic controls affect this trait.

In one type of petite (*segregational petite*) respiratory deficiencies have been clearly shown to be caused by a recessive nuclear gene. Crosses between monoploid petite and normal result in all normal diploid F_1 progeny, but ascospores produced by the latter segregate 1:1 for petite and normal. On the other hand, another strain (*neutral petite*) has defective mitochondrial DNA.

In these the petite character fails to segregate and F_1 and F_2 of the cross petite $\times$ normal are all normal, the cytoplasm that contains normal mitochondria being incorporated into F_1 zygotes and vegetative cells and distributed to ascospores and monoploid vegetative cells of the next generation. In a third type of petite (*suppressive petite*), crosses with normals produce a highly variable fraction of petites in the progeny. Rank (1970) suggests that suppressives have rapidly replicating, abnormal mitochondrial DNA. So petites may possess defective mitochondria because of either mutant nuclear DNA or mutant mitochondrial DNA, or both. Respiratory enzymes (except for cytochrome c) are thus under a double genetic control.

Carnevali, Morpurgo, and Tecce (1969) have shown that mutation of mitochondrial DNA probably involves a relative increase of adenine-thymine pairs and suggest that crossing-over may occur between molecules of cytoplasmic DNA. The action of various mutagens in producing the suppressive petite mutants, they believe, results from incomplete mDNA molecules produced by premature detachment of DNA polymerase. The defective, shorter DNA molecules are replicated more rapidly, resulting in the pseudodominant suppressive trait.

Here, as in the case of mediation of protein synthesis by plastid DNA, we certainly have a degree of cytoplasmic genetic control by DNA of unique and characteristic base sequence not, however, always and wholly independent of nuclear DNA. Whether or not mitochondria (and chloroplasts) have originated as independent, invading organisms that have now become normally occurring organelles in almost all living cells through progressively higher degrees of symbiosis by loss of some of their functions to the nucleus, as suggested by Raven (1970) and others, we clearly have in these two instances cytoplasmic genetic systems quite different from such cases as female production in *Drosophila* or the milk factor in mice.

EXTRANUCLEAR INHERITANCE WITHOUT KNOWN CYTOPLASMIC STRUCTURES

A number of rather perplexing cases of apparent cytoplasmic inheritance that cannot at present be related to any identifiable structures or determinants have been reported from time to time. Examination of one of these will suffice to indicate the scope and nature of the problem.

Uniparental Inheritance in Chlamydomonas. The ubiquitous green alga *Chlamydomonas reinhardi* has a relatively simple life cycle (Appendix B). The motile vegetative cells are monoploid (haploid); the zygote is the only diploid cell. In germination the zygote undergoes meiosis to form four monoploid zoospores that quickly mature into vegetative adults. Sexual reproduction in most species represents morphological isogamy but physiological anisogamy in that the gametes are, as we saw in Chapter 11, differentiated into $+$ and $-$ mating strains that are physically indistinguishable from each other. Mating type depends upon nuclear genes the locus for which has been determined to belong to linkage group 6 (there are 16 linkage groups in *C. reinhardi;* see Table 3-1). Chemical attraction between gametes of opposite mating strain appears to be exerted through the flagella.

Two types of reaction to the antibiotic streptomycin occur in *C. reinhardi.* Vegetative cells of genotype *sr-1* are resistant to 100 µg per milliliter of streptomycin; those of genotype *ss* (sensitives) are killed at this concentration. This pair of genes shows regular Mendelian segregation. Another genotype, *sr-2,* confers resistance to 500 µg per milliliter of streptomycin; sensitives are again designated as *ss.* Reaction to this higher concentration of the drug is *not* transmitted in Mendelian fashion, but streptomycin response of the progeny is that of the + parent, even though the progeny segregate normally for mating type, as seen in this series of crosses:

(a) P $sr\text{-}2^+ \times ss^-$
  F_1 $\frac{1}{2}sr\text{-}2^+ + \frac{1}{2}sr\text{-}2^-$

(b) F_1 $sr\text{-}2^+ \times P\ ss^- \longrightarrow \frac{1}{2}sr\text{-}2^+ + \frac{1}{2}sr\text{-}2^-$
  F_1 $sr\text{-}2^- \times ss^+ \longrightarrow \frac{1}{2}ss^+ + \frac{1}{2}ss^-$

Notice that in each case, mating strain segregates in normal Mendelian fashion, whereas streptomycin response of the progeny follows that of the + parent.

Much of what is known about the genetics of *Chlamydomonas* is due to the work of Sager and her colleagues at Columbia University (1954, 1963, 1970, and numerous other papers). Extensive tests with a variety of mutagens by the Columbia group over a period of years failed to produce any mutation of streptomycin-resistants to sensitives. But Gillham (1965) was able to obtain a high frequency of both *sr-1* and *sr-2* mutants using the mutagen N-methyl-N'-nitro-N'-nitrosoguanidine. Findings up to this point strongly implicate extranuclear DNA, but its location in the cytoplasm remains unknown. That it is not an episome (see next section), transmissible from one individual to another, as in bacteria, is indicated by the work of Gillham and Levine (1962). In their laboratory, a small number of *sr-2* cells were grown in a culture containing an excess of *ss* cells, yet no significant increase in number of *sr-2* cells that could not be accounted for by normal reproduction was found. Had streptomycin response been due to an episome, a considerable increase in the number of resistant cells would be expected. Gillham (1969) has summarized the variety of speculations that have been offered by various investigators to explain the *Chlamydomonas* situation. But although much of the basis for uniparental inheritance in this organism is not yet known, it seems clear that some sort of cytoplasmically based genetic material, deduced to behave like DNA and very probably DNA, is regularly present and operative.

EPISOMES

The F Factor.   Perhaps the most unequivocal instance of a cytoplasmically located, indentified genetic determiner that comes closest to meeting the arbitrary criteria set up at the opening of this chapter is the bacterial episome F, which we have already discussed (Chapter 11).

You will recall that this fertility factor, if present in a cell, may be either located in the cytoplasm or integrated into the single, circular bacterial "chromosome." Cells lacking the F factor (F⁻) are receptor ("female") cells at conjugation; those having F in the cytoplasm (F⁺) are donors ("males"), and cells in which F is a part of the "chromosome" are high-frequency recombinant (Hfr) donors or, again, "males." Current interpretation is that there is but one fertility factor per F⁺ cell (except just prior to fission), and one per "chromosome" in Hfr cells.

The chemical composition of the F factor, its mutability under the effect of acridine dyes, its ability to replicate autonomously, and its ability to be incorporated stably into the "chromosome" show F to be composed of genetic material. In fact, present evidence clearly indicates it to be composed of DNA. Density gradient centrifugation in cesium chloride shows the F episome to consist of about 10^5 nucleotide pairs, as compared to about 4.5×10^6 for the "chromosome" of the colon bacillus. The fertility factor is thus about $\frac{1}{45}$ the size of the entire DNA molecule of *E. coli*. Whatever its evolutionary origin, it cannot now be regarded as an infective particle, but must be considered a normal component of those cells in which it occurs. Furthermore, it is replicated semiconservatively in cell division, autonomously in F⁺ cells, and along with the endogenous DNA molecules into which it has been integrated, in Hfr cells. In conjugation between F⁺ and F⁻ cells, a replicated F factor is the first, and most often the only, DNA element to be transferred to the F⁻ conjugant (which is thereby converted to an F⁺ cell). In conjugation between Hfr and F⁻ cells, the fertility factor, integrated into the Hfr cell's DNA, is the last genetic material to be transferred, but generally conjugation is interrupted by natural forces, so that F in these cases is usually not transferred at all. (See Appendix B-1.) Campbell (1962, 1969) and Falkow et al. (1967) describe the behavior and structure of the episomes at length.

Resistance Transfer Factors. Multiple resistance to a number of drugs, many used in medicine (e.g., penicillin, neomycin, tetracycline, streptomycin, sulfonamide) in several genera of bacteria—among them *Vibrio* (causal agent of cholera), *Salmonella* (responsible for some enteric diseases, among them one type of food poisoning), and *Shigella* (bacterial dysentery)—depends on a *resistance transfer factor* (RTF). Though not fully understood, RTF has some of the characteristics of an episome. It is composed of DNA with a high G-C content and can be transferred during conjugation, even intergenerically between *Shigella* and the related genera *Salmonella* and *Escherichia*. It is also susceptible to modification by acridine dyes and appears to exist largely or entirely as an autonomous element in the cell. The importance of RTF to mankind lies in the multiple drug resistance it imparts to many serious pathogens, making them immune to treatment with such modern drugs as those just listed. RTF is part of a bacterial "sex factor," *R*, which may replicate and be transferred so rapidly that a single R^+ cell can convert an entire culture of R^- cells to R^+ within 24 hours.

Colicins.   Some strains of *Escherichia coli* produce highly specific antibiotic substances consisting of proteins (sometimes along with lipopolysaccharides) that kill other strains of the species. For this reason they have been named *colicins*. An F$^-$ *col$^-$* cell may become *col$^+$* by conjugation; in this case all its asexually derived progeny are also *col$^+$*. In this and other respects, colicins appear to be due to autonomous cytoplasmic particles, referred to as *col* factors. Although interpretations of experimental results are not all in agreement, there is some question that *col* factors (and RTF) may become integrated into a cell's DNA molecule.

CONCLUDING VIEW

In this chapter we have examined the question of the existence of cytoplasmic genetic mechanisms by viewing some of the many purported illustrations in the literature against the backdrop of a working concept of what such mechanisms ought to entail. If we were to accept the broad notion of transmission of heritable traits themselves via the cytoplasm, then, from the examples cited, we would have to acknowledge that *cytoplasmic inheritance* does occur rather widely.

But application of our more rigorous criteria to a heterogeneous collection of cases eliminates many, *but not all,* of these as instances of cytoplasmic inheritance. It is clear that chloroplasts and mitochondria do possess their own DNA genetic mechanisms that operate in the same way as the nuclear genetic material in eukaryotes. In addition, there seems every reason also to consider the F factor, resistance transfer factors, and col factors as examples of true cytoplasmic inheritance.

Drug resistance in *Chlamydomonas,* however, is still an intriguing problem. Every bit of indirect evidence points to a cytoplasmic genetic mechanism, but its nature and identity elude us. Obviously, this is a cause-and-effect relationship, but at present we know only the effect; the cause has not yet been determined. On the other hand, such purely maternal effects as shell coiling in the snail or color in the young water flea can be dismissed rationally as the lingering effects of a parental genotype operating on Mendelian bases. They certainly give no indication of the existence of a genetic system in the cytoplasm even remotely resembling chromosomal DNA.

When we consider infective particles and episomes, it is necessary to distinguish carefully between fortuitous parasites, on the one hand, and "normal" cell components, on the other. At one end of the spectrum is the obvious, identifiable, dispensable parasite (such as the spirochete), on through the viruslike milk factor and sigma to asynchronously multiplying kappa, to apparently nonintegrable factors (such as RTF and col) and the fertility factor at the other. The very fact that we can arrange such a series opens a tempting ground for speculation regarding the possible evolution of such cytoplasmic genetic mechanisms as the F factor, a speculation that, we should clearly note, is as yet wholly without any tangible, experimental proof. Has there been a gradual evolution from virulent parasite through first dispensable then indispensable symbionts to a cytoplasmic genetic system which may even integrate into the chromosomal mechanism? Or has the episomic state derived from originally purely chromosomal genes? If there is a relationship, then why have not episome systems been discovered in any group outside the bacteria? Have virulent and temperate phages had anything to do with the development of extrachromosomal genetic systems? Where in this whole problem do transforming and transducing DNAs fit? At present we can only raise these questions; the answers are still being sought, but will surely one day be in hand.

REFERENCES

BERNHARD, R., 1967. Chromosomes Are Not the Whole Story of Heredity. *Sci. Res.,* **2:** 51–54.

BERNHARD, R., 1969. Mitochondrial "Genes": Some Gambles Pay Off. *Sci. Res.,* **4:** 31–34.

CAMPBELL, A. M., 1962, Episomes. *Advances Genet.,* **11:** 101–145.

CAMPBELL, A. M., 1969. *Episomes.* New York, Harper & Row.

CARNEVALI, F., G. MORPURGO, and G. TECCE, 1969. Cytoplasmic DNA from Petite Colonies of *Saccharomyces cerevesiae:* A Hypothesis on the Nature of the Mutation. *Science,* **163:** 1331–1333.

EPHRUSSI, B., 1953. *Nucleo-Cytoplasmic Relations in Micro-Organisms.* Oxford, Oxford University Press.

EPHRUSSI, B., H. DE MARGERIE-HOTTINGUER, and H. ROMAN, 1955. Suppressiveness: A New Factor in the Genetic Determinism of the Synthesis of Respiratory Enzymes in Yeast. *Proc. Nat. Acad. Sci. (U.S.),* **41:** 1065–1070.

FALKOW, S., E. M. JOHNSON, and L. S. BARON, 1967. *Bacterial Conjugation and Extrachromosomal Elements.* In H. L. Roman, ed. *Annual Review of Genetics,* volume 1. Palo Alto, Annual Reviews, Inc.

FAN, H., and S. PENMAN, 1970. Mitochondrial RNA Synthesis During Mitosis. *Science,* **168:** 135–138.

GILLHAM, N. W., 1965. Induction of Chromosomal and Nonchromosomal Mutations in *Chlamydomonas reinhardi* with N-methyl-N′-nitro-N′-nitrosoguanidine. *Genetics,* **52:** 529–537.

GILLHAM, N. W., 1969. Uniparental Inheritance in *Chlamydomonas reinhardi. Amer. Nat.,* **103:** 355–388.

GILLHAM, N. W., and R. P. LEVINE, 1962. Studies on the Origin of Streptomycin Resistant Mutants in *Chlamydomonas reinhardi. Genetics,* **47:** 1465–1474.

GREEN, B. R., and H. BURTON, 1970. *Acetabularia* Chloroplast DNA: Electron Microscopic Visualization. *Science,* **168:** 981–982.

GREEN, B. R., and M. P. GORDON, 1966. Replication of Chloroplast DNA of Tobacco. *Science,* **152:** 1071–1074.

JINKS, J. L., 1964. *Extrachromosomal Inheritance.* Englewood Cliffs, N.J., Prentice-Hall. (One of the *Foundations of Modern Genetics* series.)

L'HERITIER, P., 1951. The CO_2 Sensitivity Problem in *Drosophila. Cold Spring Harbor Symp. Quant. Biol.,* **16:** 99–112.

MARGULIS, L., 1970. *Origin of Eucaryotic Cells.* New Haven, Yale University Press.

NANNEY, D. L., 1958. Epigenetic Control Systems. *Proc. Nat. Acad. Sci., (U.S.),* **44:** 712–717.

NASS, M. M. K., 1969. Mitochondrial DNA: Advances, Problems, and Goals. *Science,* **165:** 25–35.

PREER, J. R., JR., 1971. *Extrachromosomal Inheritance.* In H. L. Roman, ed. *Annual Review of Genetics,* volume 5. Palo Alto, Calif., Annual Reviews, Inc.

RANK, G. H., 1970. Genetic Evidence for "Darwinian" Selection at the Molecular Level. I. The Effect of the Suppressive Factor on Cytoplasmically Inherited Erythromycin-Resistance in *Saccharomyces cerevesiae. Can. Jour. Genet. Cytol.,* **12:** 129–136.

RAVEN, P. H., 1970. A Multiple Origin for Plastids and Mitochondria. *Science,* **169:** 641–646.

SAGER, R., 1954. Mendelian and Nonmendelian Inheritance of Streptomycin Resistance in *Chlamydomonas. Proc. Nat. Acad. Sci. (U.S.),* **40:** 356–363.

SAGER, R., and Z. RAMANIS, 1963. The Particulate Nature of Nonchromosomal Genes in *Chlamydomonas. Proc. Nat. Acad. Sci. (U.S.),* **50:** 260–268.

SAGER, R., and Z. RAMANIS, 1970. A Genetic Map of Non-Mendelian Genes in *Chlamydomonas. Proc. Nat. Acad. Sci. (U.S.),* **65:** 593–600.

SCRAGG, A. H., H. MORIMOTO, V. VILLA, J. NEKHOROCHEFF, and H. O. HALVORSON, 1971. Cell-Free Protein Synthesizing System from Yeast Mitochondria. *Science,* **171:** 908–910.

PROBLEMS

20-1. You have just discovered a new trait in *Drosophila* and find that reciprocal crosses give different results. How would you determine whether this trait was sex-linked, a purely maternal effect, or the result of an extranuclear genetic system?

20-2. What phenotype would be exhibited by each of the following genotypes in the snail: $+s$, $s+$, ss, $++$?

20-3. What kind(s) of progeny with respect to eye color result from these crosses in the flour moth (*Ephestia*): (a) light ♀ × homozygous dark ♂, (b) homozygous dark ♀ × light ♂?

20-4. A four-o'clock plant having three kinds of branches (green, variegated, and "white") is used in a breeding experiment. What kinds of progeny are to be expected from each of these crosses: (a) green ♀ × white ♂, (b) white ♀ × green ♂, (c) variegated ♀ × green ♂?

20-5. Green and Burton (1970) were able to secure intact segments of *Acetabularia* cDNA as long as 419 μm. As noted in this chapter, such DNA would consist of about 1.23×10^6 deoxyribonucleotide pairs. (a) Assuming all this cDNA is transcribed, for how many codons could it be responsible? (b) If this entire amount of cDNA is transcribed and the resulting codons are all translated, how many polypeptide chains could be produced if each consists of 400 amino acid residues?

20-6. In yeast a neutral petite, although having defective mitochondrial DNA, may have a nuclear gene for normal mitochondrial function. As indicated in the text, a monoploid segregational petite carries a recessive nuclear gene for defective mitochondria, but it may possess normal mitochondrial DNA. If such a neutral is crossed with a segregational petite of the type described here, what is the phenotype of (a) the diploid F_1, (b) the monoploid generation developing from ascospores produced by these diploid cells?

20-7. Employing different substrains of *Escherichia coli* strain K-12 as *Hfr* conjugants produces different results (see linkage map, Fig. 6-6):

K-12 Substrain	Chromosomal genes transferred in conjugation	
	First	Last
C	*lys + met*	*gal*
H	*pil*	*pyr-B*

For each substrain, give (a) the location of F and (b) the second chromosomal gene that would be transferred.

20-8. Defend or oppose the thesis that chloroplast inheritance represents the action of a cytoplasmic genetic system.

20-9. Defend or oppose the thesis that the episome system represents an instance of a cytoplasmic genetic mechanism.

20-10. Defend or oppose the thesis that cytoplasmic genetic systems have not yet been adequately demonstrated.

20-11. Defend or oppose the thesis that mitochondria and/or chloroplasts originated as free-living prokaryotes. (You may wish to read some of the references before answering.)

CHAPTER 21
The New Genetics and the Future

THE exciting, breathtaking progress of the science of genetics following the rediscovery of Mendel's principles at the turn of the century has at no point been more apparent than in recent investigations at the molecular level of the nature and action of genetic material. Yet these studies, fruitful as they have been, have but lifted the veil slightly on many still unresolved problems. Both their solution and their application to the welfare of mankind constitute some of the fascinating, even frightening, challenges confronting not only this and the next generation of the world's scientists, but you and your children as well. In this concluding chapter we shall examine this prologue to the future.

BACKGROUND

Man has made almost exponential progress during his lifetime on this planet in developing better varieties of plants and lower animals and, even more important, in understanding the mechanisms of genetics and genetic change. For example, Fletcher (1974) points out that the doubling time for scientific knowledge in general is 10 years, but five years for the life sciences, and *two years for genetics*. At that rate our genetic knowledge *quadruples* during your college career! Although much remains to be learned, of course, we already know a good deal about such processes as recombination, transformation, transduction, and mutation. But, by and large, application of these processes still entails a considerable degree of chance; that is, they cannot yet be used successfully to bring about only certain desired changes.

Progress in medical science has enabled us to make up, to some extent, for deficiencies resulting from deleterious genes (diabetes is a case in point), but the lethal or disabling consequences of defective genotypes and karyotypes is still fairly considerable. Roberts and his colleagues (1970), in a study of the causes of death of 1,041 children over a seven-year period in the hospitals of Newcastle in Great Britain, found gene and chromosome defects to be responsible for 42 per cent of the deaths in their sample. Single gene defects accounted for 8.5 per cent of these childhood deaths; chromosome aberrations, 2.5 per cent; and those probably resulting from complex genetic causes, 31 per cent. Heller (1969) estimates some type of chromosomal abnormality to be present in almost 0.5 per cent of live-born infants and to occur in nearly one fourth of all spontaneously aborted fetuses. Boué et al. (1967) found polyploidy or aneuploidy in almost two out of three such fetuses. You will recall from Chapter 11 that sex chromosome anomalies occur with frequencies up to 4 per 1,000 live births, and that such aberrations as aneuploidy (trisomy in particular) and even polyploidy are surprisingly frequent (Chapter 13). Moreover, as was pointed out in Chapter 18, each of us carries several deleterious or lethal genes, and roughly 10.5 million mutations occur per human generation in this country, almost none of them neutral or advantageous.

From the personal as well as the humanitarian viewpoint, we would like in some way to minimize these mutations and these chromosomal aberrations and to negate the effect of defective genes. Some would go so far as to raise the question of whether "desirable" genotypes can be made up to order; to put it another way, *can man control his own evolution?* In fact, this query is being raised with increasing frequency; several symposia have addressed themselves solely to this question (Sonneborn, 1965; Roslansky, 1966; Etzioni, 1973a), and a large number of papers and books, some of which are listed at the end of this chapter, are appearing on the same topic. Two major recent developments with lower organisms make it appear that the question as to man's capability to shape in some degree his own evolution must soon be answered in the affirmative. These developments are (1) the construction, in vitro, of biologically active, recombinant ("hybrid") DNA molecules, and (2) the production of exact genotypic copies of individuals (cloning). True, these measures have so far been successfully achieved only with lower organisms, e.g., recombinant DNA using such disparate species as *Escherichia coli* and *Drosophila melanogaster,* and cloning in frogs and carrots. But most geneticists and biochemists believe extension of the technique to mammals, including man, will soon be feasible. As you might well imagine, with the press of population and competition for the world's resources becoming ever more acute, and with the rising hope of eradicating genetic disorders, the question of control of human evolution is increasingly engaging the serious thought of leaders from the natural sciences, the social sciences, philosophy, and religion. It is a problem that will affect you, and one to which, as an educated, intelligent citizen, you must be prepared to give earnest consideration.

There are two broad facets to the question of man's control of his own evolution. Simply stated, they are, "Can he?" and, "Should he?" The second, raising as it does profound moral, legal, and ethical considerations, lies somewhat beyond the immediate scope of an introductory course in genetics. But concern with the first should help us develop a sense of values based on knowledge, as well as help initiate the sober reflection for which there may still be time.

RATIONALE OF HUMAN EVOLUTION CONTROL

Our bank of genetic material is undergoing a slow but inexorable decline in quality. Several dysgenic influences are contributing in different degree to this qualitative dilution and, it is argued, must be counteracted by conscious efforts at quality control. Disagreement lies not so much with the need for genetic improvement as with methods of achieving it. These we shall examine shortly.

There is no doubt that the human population has been increasing at a rate that cannot be ignored (Table 21-1), despite a recent slowing in some Western countries.

TABLE 21-1. Annual Rate of Population Increase
for Selected Countries

Annual Percentage Increase	Country
4.10	Jordan
3.75	Mexico
3.51	Ecuador
3.45	Venezuela
3.24	Israel
3.14	Chile
3.05	Costa Rica
2.78	Brazil
2.75	India
1.75	Australia
1.30	People's Republic of China
1.26	Japan
0.94	U.S.S.R.
0.90	U.S.A.
0.86	France
0.71	Federal Republic of Germany
0.70	Norway
0.52	United Kingdom
0.32	Hungary
0.26	Sweden
−0.10	German Democratic Republic

Data from *Population Index* and other sources.

The overall world average is 2 per cent annually. But when one considers the time required for doubling of populations at these rates (Fig. 21-1), the immediacy of the problem is obvious. The resulting competition for available resources is already strikingly evident in many countries of the world. Moreover, the doubling time for the world's population is decreasing alarmingly fast (Fig. 21-2). But population growth is perhaps more a function of decreased death rate than of increased birth rate. In India, for example, the birth rate has shown a gradual decline for much of this century, yet her population continues to grow. People are simply living longer. Life expectancy did not exceed 25 to 30 years well into the Middle Ages; since then it has risen sharply, especially in the "have" nations, to around 70. A great deal of this increase is due to medical and health advances. But every successful technique that lengthens the life span of persons with inherited defects increases the likelihood that such individuals will reproduce and pass on their defective genes to the genetic load of future generations. As Fleming (1969) puts it, "conventional medicine is

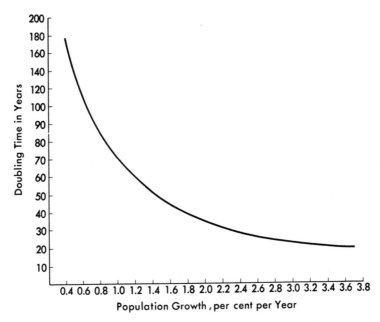

FIGURE 21-1. *Curve showing approximate doubling time in years plotted against rate of population growth in per cent per year. Compare with figures for actual annual growth for selected countries in Table 21-1.*

now seen by the biological revolutionaries as one of the greatest threats to the human race." Selection and survival of the genetically less fit does, indeed, slowly but relentlessly increase the frequencies of these genes and the probability that they will occur in their descendents.

Moreover, the nearly continuous warfare in which "civilized" man has engaged throughout virtually his entire history, now with sophisticated weaponry of terrible efficiency, from which no one is safe any longer, tends to siphon off the physically fit and often the intellectually well endowed. In addition, the danger of accident or misuse of atomic energy is but a hand's motion away from wreaking incalculable genetic damage on a large share of the world's inhabitants. Again, Fleming (1969) has succinctly summed up these grave concerns: "biologists deplore the aggressive instincts of the human animal, now armed with nuclear weapons, his lamentably low average intelligence for coping with increasingly complicated problems, and his terrible prolificity, no longer mitigated by a high enough death rate."

As we have noted, a significant percentage of all births carry some detectable, genetically based defect. Many of the bearers of these defect-producing genes die, some very early, others a little later, and a great many can look ahead only to a life of greater or lesser misery if, indeed, their deficiencies are not such as to impair their mental processes. An additional and apparently

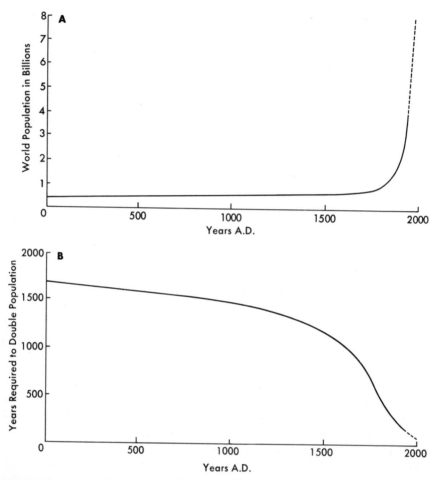

FIGURE 21-2. (A) *The rise of world population from the time of Christ projected to the year 2000, based on present rate of increase.* (B) *The diminishing time required to double the world's population from the beginning of the Christian era to the year 2000.*

rather large number of fetuses, homozygous for recessive lethal genes or carrying some serious chromosome aberration, as we have seen, abort spontaneously. Quite aside from considerations of birth control and death control, those who *are* born ought to have a chance for life free from this type of defect. On the other hand, they should not serve as additional sources of input for deleterious genes.

POSSIBLE SOLUTIONS

Three general avenues of approach, not all presently of equal feasibility or usefulness, and certainly not all equally accepted, are at least theoretically available for gene quality control:

1. *Overriding the expression* of defect-producing genes without trying to control their frequencies.
2. *Directed recombination;* that is, selecting those genes that will be permitted to be passed on to future generations.
3. *Genetic engineering;* that is, modifying or even replacing the defective genes themselves or, as a spinoff of the future, cloning individuals to order.

Overriding Gene Expression.   More than 1,500 hereditary disorders affect human beings. In some of these it is possible to supply affected persons the necessary substance that they are unable to produce, compensating for, or overriding, one or another expression of their genotypes. For example, diabetes mellitus is an inherited disorder that affects several million persons in this country. Although the genetics of this condition are complex and not well understood (autosomal recessive, autosomal dominant, and polygenic bases have been proposed), the physiological basis is failure to produce the hormone insulin. This is one of the simpler proteins (molecular weight about 35,000) synthesized by the human body. Its structure is well known, consisting of two polypeptide chains, one of 21 amino acid residues, the other of 30 (Fig. 21-3). It is formed only in the beta cells of the islets of Langerhans of the pancreas from its precursor protein, proinsulin. Proinsulin is manufactured in the usual biological way, in which the appropriate sequence of deoxyribonucleotides is transcribed into mRNA, which, together with the necessary amino acids, ribosomes, enzymes, tRNA, and so forth, translates this sequence into protein. The evidence indicates that proinsulin is a single polypeptide chain that is cleaved and folded after synthesis to make the two chains of insulin. It is thus presumably the product of a single cistron of some 153 nucleotides, representing only about 0.05 μm of the total length of human DNA.

All somatic cells contain all the genetic information of the individual, yet only a very restricted group of cells is able to manufacture proinsulin. In all other cells of the body the proinsulin cistron is permanently repressed. Knowledge of the mechanism of gene repression in higher organisms is progressing, and a future approach in treating diabetes mellitus may lie in our learning how to derepress the proinsulin cistron, either in the cells where it is normally produced or in other cells. Once such activation has been accomplished, there remains the problem of conversion of proinsulin to insulin and its release from the cells in which its synthesis is effected. By comparison, this may not be very difficult. On the other hand, should the genetic basis of even some cases be nonsense or missense in the codons affecting proinsulin synthesis or the conversion of proinsulin to insulin, or production of an insulin inhibitor, rather than repression, an entirely different dimension arises.

Presently, diabetes mellitus can be largely controlled through injection of the insulin that diabetics fail to synthesize. This, however, permits more affected

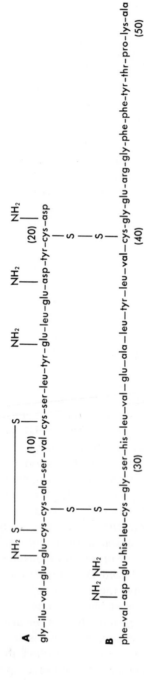

FIGURE 21-3. *Amino acid sequence of the beef insulin molecule, the first protein in which the amino acid sequence was determined. Each molecule consists of an A chain (21 amino acid residues) and a B chain (30 amino acid residues). The A and B chains are joined by disulfide bonds connecting cysteine residues as shown.*

persons to be kept alive, and has no effect in upgrading the human gene pool. Rather, it increases the number of persons who can and do serve as additional sources of deleterious genes for succeeding generations. In terms of the human *species,* as opposed to the *individual,* this control measure is actually counterproductive, as is the case with so many medical advances. So, with reference to mankind, other avenues must be explored; that is, we need to distinguish genetic intervention for purposes of individual therapy from measures designed to alter the course of human evolution. Therein lies the moral and ethical dilemma.

Directed Recombination. In some ways, directed recombination, or gene selection, is mechanically the easiest, though not necessarily the most satisfactory, solution to the problem of quality control in the human gene pool. It requires no sophisticated equipment, no techniques that have not been available, even practiced in other species, for years. Basically, it involves only the selection of parental stocks, those whose genes are to be transmitted to future generations. The oldest technique would merely restrict child-bearing to those best fitted to perpetuate the species. But this raises significant problems. What are the standards? Who determines those standards? Is our knowledge of human genetics sufficiently advanced to permit standards to be set up? For what purposes would those standards actually be devised and applied? Are human societal values sophisticated enough to replace selfish interests by selflesss goals? Racially pernicious, politically unsound, such a "solution" is at best impractical, at worst dangerous.

Heterozygous "carriers" of defect-producing genes can now be identified by relatively simple tests in such varied inherited metabolic disorders as sickle-cell trait, cystic fibrosis, and galactosemia. Could we not limit or prevent reproduction by such persons? The problem of how that control might be accomplished becomes a very personal one when we recognize that each of us carries at least one, and probably several, such lethal, semilethal, or disabling genes. Should control be exercised through advice and counseling? On a distressingly small scale we are doing this now, but only for a small, generally upper, stratum of the human race. Should it be by governmental restraint, by legislation, with penalties for breaking the law—or even by compulsory sterilization? Once again, the question of who should make these judgments is an exceedingly difficult one and may well have no acceptable answer. But, viewed coldly and impersonally, this approach does offer the advantage of being possible now in many instances, for it requires little more in the way of precise genetic knowledge than we already have.

If the concept of selective parenthood seems repugnant or impractical or dangerous, there is the additional problem of whether to allow fetuses that will express defective genes or that have been genetically damaged by such drugs as thalidomide or LSD to come to term. To permit them to do so is regarded by some as the deepest kind of cruelty, both to the child and to his family. By others, *not* to do so is regarded as murder. It is time, as many have pointed out, to re-examine the philosophy underlying selective legalized abortion. Public attitudes toward abortion are changing to the point where it is questionable whether its prohibition under professional medical care is even a proper concern of the state, particularly when it is determined that the fetus carries a gross genetic defect. Accordingly, laws have been markedly liberalized since a 1973 Supreme Court decision declared it unconstitutional for any state to forbid abortion in the first trimester of pregnancy. This decision negated statutes in 46 states.

On the other hand, still another avenue of directed recombination is, even now, open to us. Before long it may become much more practical as our knowledge of the human genotype and linkage groups increases. Rather than select certain *persons* who shall or shall not become parents, or which *embryos* will be allowed to mature, we may select the *reproductive cells* themselves. To a degree this is practiced now. Artificial insemination, so long applied with rather successful results in animal breeding, is currently used to a much more limited extent in mankind. But all too often no account is taken of the genetic constitution of the prospective mother and little more of the sperm donor. Ordinarily the goal is merely the *fact* of child production in families where it would not otherwise be possible and, incidentally, to produce children who could be mistaken for those of the nonbiological father.

The late Herman Muller (1965) proposed setting up sperm and egg banks, because sex cells can be kept alive and functional for considerable periods by freezing. Muller proposed such banks, with the most desirable male and female subjects serving as donors. By whatever standards that might be set up, the best sources could continue so to serve for a considerable period after their deaths and, with suitable precautions, even after a nuclear war. The question of determining genetic desirability here is, of course, no easier than in other systems of restrictive parenthood. But Muller, in his 1965 paper and in many addresses up until his death in 1967, expressed hope in these words: "With the coming of a better understanding of genetics and evolution, the individual's fixation on the attempted perpetuation of just *his* particular genes will be bound to fade. It will be superseded by a more rational view . . . he will condemn as childish conceit the notion that there is any reason for his unessential peculiarities, idiosyncrasies, and foibles to be expressed generation after generation." The background of Muller's philosophy is revealingly described by Sonneborn (1968), a long-time colleague, from a variety of sources that include Muller's unpublished autobiographical notes. Of course, society has not reached the stage of objectivity envisioned by Muller, and one may question whether it ever will.

Use of sperm, frozen in liquid nitrogen at $-321°F$, was first reported to have resulted in successful pregnancies in 1953; since that time, by Fletcher's (1974) estimates, several hundred children have been conceived through use of sperm frozen for as long as two-and-a-half years. Incidence of abnormalities in children so conceived appears not to differ significantly from that of comparable samples of children produced in the customary biological fashion. Of course, use of stored, frozen sperm from selected donors still entails an element of chance in recombination of disadvantageous genes, though the overall probability might be lower than in a system of random mating if donors can be selected with a skill and wisdom that we may not yet possess.

Implantation of selected eggs, fertilized or unfertilized, is already a fact in both laboratory animals and in human beings. Edwards, Steptoe, and Purdy (1970) have brought about fertilization of human ova in vitro and have succeeded in carrying the resulting zygote to at least the 16-cell embryo stage. Fletcher (1974) reports that Shettles, of Columbia University, was able in 1971 to implant an embryo at the blastocyst stage. The purpose of his work was only to determine if it could be accomplished (which it was), inasmuch as the patient was shortly to undergo a hysterectomy because of cervical cancer. A team of British physicians in the early 1970s disclosed that an egg, already fertilized in vitro by sperm from the husband, had been implanted in the womb of a woman who could not conceive because of an obstruction in her fallopian tubes. Although prospects of the birth of the first "test tube conceived" child appeared reasonable, success of this effort has not been reported. Certainly there are difficulties in these procedures, and many presently unanticipated problems will doubtless arise. But work in this direction *is* going on, and success *will* come.

Once the process of implantation has become routine, the prospects are virtually limitless. The possibility has been suggested of a system of volunteer "host" mothers who would bear other people's children for a fee where, perhaps, it may be physically unwise for a woman to undertake the risks and rigors of pregnancy, or even in case she simply does not wish to interrupt a career. Or an infertile woman could bear children who are the product of her husband's sperm and an ovum from an unknown donor; such a fertilized egg might then easily be implanted in her body. Of course, the matter is more complicated than suggested here, but it will be feasible in the relatively near future.

Some writers, only partly with tongue in cheek, have compared human germinal choice to selection of a packet of desirable flower seeds in the supermarket. The container would carry a brief statement of the most probable traits, from intelligence to physical perfection and even sex[1]; selection could be made as simply as deciding on the kind of plants to grow. From the psychological standpoint, there is considerable doubt that man is ready for this, and our genetic knowledge of the human species is not yet up to our understanding of marigolds and petunias.

When it is—and some scientists have estimated that this will be before the end of the present century—perhaps rather than implant an egg in a human female, we could not only select sperm and egg and bring about fertilization

[1] However, predetermination of sex does not appear imminent; for a statement of the problem see Etzioni (1968).

in vitro, but even raise the embryo to term in a glass womb. In fact, although some scientists have expressed doubt, Petrucci in Italy is said to have raised human embryos in this way for as long as two months, and then deliberately terminated the experiment because the embryo was grossly deformed. The moral question "Is it murder?" is reported to have deeply disturbed him as a Catholic. Others have worked and are working along these lines and someday, somewhere, someone is going to be successful in producing such a person. Think of the cognate problems: Who is he? Who is or are his parents? What are his legal rights? Will we mass-produce a new race of slaves? One day these questions, and more, are going to have to be answered.

Further afield, but interesting speculatively, is the possibility of vegetative multiplication or cloning. It has long been possible to maintain cultures of human cells. Will it eventually be possible to induce differentiation and thereby create unlimited numbers of custom-made, identical individuals to certain specific genetic designs? Or will it be useful to have identical clonants to serve as organ donors to each other, at least until effective immunosuppressives are perfected? Sinsheimer is quoted by Fletcher (1974) in these words: "[Cloning would] permit the preservation and perpetuation of the finest genotypes that arise in our species—just as the invention of writing has enabled us to preserve the fruits of their life work." Differentiation in cell cultures, culminating in the formation of complete organisms, has already been accomplished with lower forms of life, and some scientists feel it is more a question of *when* than *if.* Nobel laureate Lederberg is quoted by Toffler (*Future Shock*) as putting "the time scale . . . anywhere from zero to fifteen years from now," i.e., by the mid-1980s.

Another avenue to such asexual reproduction lies in removal of the monoploid nucleus from an egg and replacing it with a diploid nucleus from the female whose egg it was, or from some other selected individual. Theoretically, human eggs so manipulated could either be implanted in a human womb or, with perfection of necessary techniques, brought to term in vitro. Although this technique has been successfully practiced with frogs, the nuclear transplant step has not yet been accomplished with mammals. Individuals produced in this way would be genotypically identical to the one serving as the nuclear donor and, of course, of the same sex. Such immortality of the individual is thought to be fairly remote for human beings, but Signer (1974) believes "a mammal will probably be cloned [in this way] within the next ten years," or by 1984.

Genetic Engineering.   Certainly in some genetic disorders, and very possibly in many, the problem is a cistron whose nucleotide sequence specifies a "wrong" series of amino acids. Even a single base-pair change—for example, a switch from an adenine-thymine pair to one of guanine-cytosine, or even from an adenine-thymine to a thymine-adenine pair, or the deletion or insertion of one nucleotide pair—can, as we have seen, code for an incorrect amino acid at a given position in the polypeptide chain. The result could be the formation of an aberrant or a nonfunctional protein. This is the sort of change, you will recall, that spells the difference between normal hemoglobin A and the hemoglobin S of sickle-cell disease. Could such defective cistrons be replaced by genetic surgery? In brief, the answer today for human beings is "No, not yet." But it can be done in bacteria and in cell cultures, though we cannot yet control very well which genes are replaced. Only the rashest dogmatism would hold that such a possibility could not be brought to bear in human genetics within the lifetime of today's college student.

As we have noted in earlier chapters, transduction is a process involving transfer of DNA from one cell to another through the mediation of a phage. In the infection of a bacterium by a virulent phage, the vast majority of the new virus particles released upon lysis of the host contain DNA identical to that of the original infecting phage. But a very few contain a segment of bacterial DNA from the host cell, which replaces a corresponding bit of viral DNA. On the other hand, recall that the temperate phages do not regularly lyse and destroy the bacterial cell. Infection of a cell by temperate phage *lambda,* for example, results in incorporation of some of the viral DNA into the bacterial cell's chromosome. The bacterial cell survives and multiplies, some of its descendents containing DNA that includes sequences of nucleotides received from the phage.

The question here is whether this same kind of transfer of genetic material could be effected in the cells of higher organisms. There is evidence to suggest an affirmative answer. Aaronson and Todaro (1969), going beyond earlier suggestions of transformation in cultured mammalian cells (Szybalska and Szybalski, 1962), reported that DNA isolated from simian virus 40 can become established in human fibroblast cells in vitro and appears to express itself in protein synthesis in such cells. SV40 is a small virus that produces tumors in appropriate animal hosts. Human cells in which SV40 DNA has been incorporated undergo certain characteristic changes, including loss of sensitivity to contact inhibition of cell division and the production of SV40-specific mRNA. Such human cells also produce a new protein, the so-called T-antigen, which persists in clonal cells. Aaronson and Todaro conclude their report with these words: "There is considerable evidence [to show] that SV40 DNA can become a permanent part of the host cell genome. Most of the SV40 DNA in transformed cells is associated with the chromosomes . . . However, if the viral DNA's ability to integrate into the human cell genome can be separated from its [tumor producing property], it may then be possible to use 'integrating' viral DNA to insert specific information into human cells." The important point here for our purposes is the incorporation and persistence of viral DNA in the human gene complement and its subsequent manifestation in specific new protein synthesis.

Three additional discoveries of tremendous significance (two of which we have already referred to in Chapters 15 and 19) complete the groundwork, although their application to human genetic surgery must still hurdle some problems. But, by comparison, these by no means appear insurmountable. Although it is, of course, risky to speculate, timetables for discoveries and their utilization are frequently much too conservative. To address ourselves to the question of human genetic modification, rather than hope to come upon a convenient transducing virus carrying, for example, a cistron for proinsulin or phenylalanine hydroxylase, it might be more practical to make it to order.

You will recall the isolation of the 4,700 nucleotides of the operator, promotor and z cistron of the *lac* operon of *Escherichia coli* by Beckwith and his colleagues (Shapiro et al., 1969, and Chapter 19) and the synthesis of biologically active, normal DNA of phage X-174 by Kornberg and his associates (Goulian et al., 1967, and Chapter 15). This artificially synthesized single-stranded DNA of some 6,000 nucleotides was found to be infective for the usual host (*E. coli*), and the virus particles resulting from infection with it were completely indistinguishable from wild type phage.

Finally, Osterman, Waddell, and Aposhian (1970) have reported that infection of cultured mouse kidney cells by polyoma virus results in production not only of normal polyoma virus progeny but also of **pseudoviruses.** These consist of *fragments of host cell DNA* contained within normal polyoma virus protein coats. The pseudovirus particles, they found, are adsorbed to host cells of secondary cultures and uncoated by them, *apparently in the host cell* (though this, they point out, remains to be tested further). The Aposhian group is currently at work on the question of whether the genetic information of the pseudovirus is expressed in the host cell.

To summarize, prospects look promising that we may one day have the capability of isolating and synthesizing cistrons coding for almost any trait, of packaging these in protein coats and, we hope, of finding that they will express themselves in the recipient cells by the usual biological mechanism of protein synthesis. Once the original isolation and syntheses have been accomplished, they will be done for all time; following Kornberg's methods, they can be copied accurately forever. If the question of "whether" seems answerable in the affirmative, what about the question of "when"? Fleming (1969), although acknowledging that "gene manipulation and substitution in human beings . . . is the remotest prospect of all" in the biological revolution of which he writes, does suggest its feasiblity "maybe by the year 2000."

If a few small cistrons can one day be synthesized, next may well come the possibility of replicating the entire human DNA complement, literally making human beings to order with almost any set of predetermined characteristics. Combine, if you will, this kind of genetic engineering with the glass womb, and man will, indeed, then control his own evolution, shaping it to suit whatever the need, good or evil. And he *will* be able to do so in the foreseeable future.

CONCLUSIONS

By such techniques and processes as outlined in the last section, then, any faulty or undesirable gene could be changed, first in the somatic cells, and next in the reproductive cells, for there are, of course, two facets to the task here. Modification of the genetic material of somatic cells can have no effect on future generations; to exercise quality control over human evolution, such modification would also have to be made in the sex cells, either directly by genetic engineering or by the more inefficient and problem-laden directed recombination route.

We are still left, however, with several grave questions: (1) To what extent will diminution of negative selection pressures (e.g., keeping alive those whose genotypes are lethal today) "pollute" the human gene pool? (2) How is "desirability" in the genome to be defined; that is, desirability from whose viewpoint? (3) What presently unforeseen dangers lie ahead in genetic engineering experimentation, even with microorganisms? (4) Should research in genetic engineering therefore be halted—either permanently or until our ability to use this new knowledge has matured? (5) What is the responsibility of the scientist in helping the policy makers and the public sort out and understand the complex issues involved? Scientists themselves are devoting more and more thought to these and similar questions, and, as you might surmise, agreement is less than complete.

There is general consensus that medical progress, in keeping alive to reproductive age persons whose genotypes presently doom them to death or incapacitation before reproducing, will have only a small quantitative effect on gene frequencies. These deleterious genes have, for the most part, very low frequencies in the large human gene pool. The number of generations required to double the number of heterozygotes, for example, depends on both the present frequency of the gene under consideration and also the number of births. Dr. Arthur Steinberg, quoted by Etzioni (1973a), has arrived at the figures shown in Table 21-2, where q_0 represents the present frequency of the (recessive) deleterious gene and n the number of children per family.

TABLE 21-2. Number of Generations
Required to Double the Frequency
of Heterozygotes

n	q_0			
	0.1	0.01	0.001	0.0001
2	20	156	1,517	15,127
4	14	110	1,070	10,675

Data of Steinberg as quoted by Etzioni (1973a).

Assuming a human generation of 30 years, and only two children per family, the doubling time for heterozygote frequency ranges from 600 years if $q_0 = 0.1$ to nearly half a million years if $q_0 = 0.0001$. Even without future medical progress in alleviating lethal genetic disorders (which is certainly not to be expected), the effect on the species is far from great. It is only when one focuses on a particular family that concerns take on greater immediacy.

Definitions of "desirability" are simply not possible for at least two reasons. First, man's history has been one of using discoveries for evil as well as for good; atomic fission is a recent illustration and many others doubtless occur to you. Second, future applications of seemingly harmless or even beneficial discoveries cannot always be anticipated. Signer (1974) cites a case in point. The Ph.D. research of Arthur Galston disclosed that 2,3,5-triiodobenzoic acid increases the number of flowers and fruits in soybean, resulting in greater yield per acre—an important discovery in an age of food shortages. However, government chemical warfare research teams found that higher concentrations of this substance produced defoliation. This discovery was followed by development of defoliants widely used in Vietnam. One of these (Agent "Orange," which is 2,4,5-trichlorophenoxyacetic acid plus picloram) produces malformations in fetuses of rats and mice. Unusual increases in stillbirths and in birth defects (e.g., cleft palate and spina bifida—a defect in the walls of the spinal canal resulting in tumor formation) were reported in Tay Ninh Provincial Hospital and in Saigon in 1971 subsequent to heavy and widespread spraying by the United States.

The Committee on Recombinant DNA Molecules of the National Research Council in 1974 proposed voluntary deferral of certain types of experiments dealing with recombinant DNA (Berg et al., 1974). These include "construction of new, autonomously replicating bacterial plasmids (i.e., extrachromosomal genetic determinants) that might result in introduction of genetic determinants for antibiotic resistance or bacterial toxin formation into bacterial strains that do not at present carry such determinants; or construction of new bacterial plasmids containing combinations of resistance to clinically useful antibodies" and linkage of DNA from tumor-producing viruses to autonomously replicating DNA or to other viral DNA. Members of this group were deeply concerned that our ability to construct biologically active recombinant DNA, using *E. coli*, for example, to multiply the number of the recombinant molecules, might permit "new DNA elements introduced into *E. coli* to become widely disseminated among human, bacterial, plant, or animal populations with unpredictable results" (Berg et al., 1974). Dangers cited by this group are very real, for in this kind of research we are moving rapidly toward results whose consequences simply cannot be predicted, and can be only dimly perceived.

Similar concerns had been expressed in 1970 by James Shapiro, the Harvard bacteriologist who, with his colleagues, had only a short time earlier isolated the promoter, operator, and β-galactosidase cistron of the *Escherichia coli lac* operon (Shapiro et al., 1969). His fear of some of the possible evil uses of this kind of work and its eventual ends led him to abandon genetic research.

At a conference in Pacific Grove, California, late in February, 1975, more than 100 of the world's leading biological scientists voted to replace the voluntary ban with a set of stringent safety precautions under which all but the most hazardous experimentation may be resumed. The strict control proposals include one specifying that organisms employed in genetic engineering experiments be incapable of surviving outside the laboratory. Although actions taken at this conference have the power of moral suasion only, the guidelines are expected to be followed closely by regulatory groups in the several countries. Both the earlier moratorium and the recommendations of the Pacific Grove conference represent an unusual, if not unique, effort to impose voluntary safeguards before, rather than after, the occurrence of a hazardous event.

Davis (1974), however, views the situation somewhat differently. He points out that genetic engineering in man and bacteria present quite different biological and moral issues, that our ability to clone in any but a small group of lower organisms is far behind our ability to develop recombinant DNA molecules, and that success in molecular genetics has been confined to single-gene traits as opposed to polygenic characters. He takes the position that "since we cannot predict when a particular kind of manipulation may become feasible, and since moral standards and social needs change with time, it would be presumptuous for us to try to guide future generations by our present wisdom." He does agree that the scientist has an important responsibility to aid in public education, but expresses the fear that public anxiety over unlikely developments "could lead to pruning of valuable major limbs on the tree of knowledge, rather than of branches with dangerous fruit."

The National Research Council group proposed that the National Institutes of Health consider establishment of an advisory committee to evaluate potential hazards of recombinant DNA molecules, to develop procedures to minimize the spread of such molecules, and to draw up guidelines for experimentation with them. Etzioni (1973a) called for a commission of scientists, humanists, and theologians to "lead public awareness and education in the issues" involved in genetic engineering. In 1971 Senator Mondale introduced a bill (S.J. 75) into the 92nd Congress, which he later introduced as a Senate Joint Resolution (S.J. 71) into the 93rd Congress, to establish a National Advisory Commission on Health, Science, and Society, its members drawn from a variety of disciplines.

At the other pole is the view that the compelling duty of scientists is to continue ceaselessly to push back the boundaries of the unknown and to seek out the truth, wherever the search might lead. Under this philosophy, which certainly has some logic, no scientist should ever withhold new facts or techniques because of fear of what their use might hold for mankind. Application of laboratory findings, it is argued, are made by society, by politicians, and by the business community, for example, rather than by scientists themselves. Wilson (1970), a scientist-novelist put it this way:

because of the scientist's inability to look over the walls of history and foresee what subsequent generations will do with the fruits of his discovery, society today blames the scientist for what it wrenched from his hand and turned into engines of evil. . . .

The scientist is bewildered to find himself considered the villain. . . . The very scientists who are being considered bogeymen are the ones who must still be called upon to use their ingenuity to help undo the damage which society has done to itself.

But Etzioni rebuts this view by pointing out that, although scientists discovered the structure of LSD, and how to bring about atomic fission, they are not without responsibility in pointing out the consequences of the use or misuse of new discoveries. Both LSD and atomic fission can be and are put to both "good" and "bad" uses. Similarly, the automobile has given us many benefits, but 56,000 persons died in traffic accidents in 1972 alone, and present models still pollute the air.

Although the successful application of genetic engineering still lies largely in the future, the future quickly becomes the present, and it is not at all unlikely that it will one day be possible in human beings, and sooner rather than later. Man's capacity for good or ill will thereby be immeasurably increased. How wisely he uses this new power depends on how carefully he plans now. As Rosenfeld (1965, 1969) has said so well,

It may be comforting to know that the statesmen and the philosophers and the scientists are worrying about all these things, but we cannot let them do all the worrying for us. The time ahead is wild and uncharted. No one has been there, so there are no experts. Each of us, whose body or brain may be modified or whose descendants' characteristics may be predetermined, has a vast personal stake in the outcome. We can guarantee that good will be done only by looking to it ourselves.

REFERENCES

AARONSON, S. A., and G. J. TODARO, 1969. Human Diploid Cell Transformation by DNA Extracted from the Tumor Virus SV40. *Science,* **166:** 390–391.

AUGENSTEIN, L., 1969. *Come, Let Us Play God.* New York, Harper & Row.

BAER, A. S., ed., 1973. *Heredity and Society: Readings in Social Genetics.* New York, Macmillan.

BEHRMAN, S. J., and D. R. ACKERMAN, 1973. Freeze Preservation of Human Sperm. In J. B. Bresler, ed. *Genetics and Society.* Reading, Mass., Addison-Wesley.

BERG, P., D. BALTIMORE, H. W. BOYER, S. N. COHEN, R. W. DAVIS, D. S. HOGNESS, D. NATHANS, R. ROBLIN, J. D. WATSON, S. WEISSMAN, and N. D. ZINDER, 1974. Potential Biohazards of Recombinant DNA Molecules. *Science,* **185:** 303.

BLESSING, L. C., 1970. Some Social Problems Caused by Scientific Advances. In A. B. Grobman, ed. *Social Implications of Biological Education.* Princeton, N.J., The Darwin Press.

BOREK, E., 1973. *The Sculpture of Life.* New York, Columbia University Press.

BOUÉ, J. G., A. BOUÉ, and P. LAZAR, 1967. Les Aberrations Chromosomique dans les Avortements. *Ann. Génétique,* **10:** 179–187.

BRESLER, J. B., 1973. *Genetics and Society.* Reading, Mass., Addison-Wesley.

BURNS, G. W., 1970. Tomorrow—or the Day After. *Ohio Jour. Sci.,* **70:** 193–198.

DAVIS, B. D., 1970. Prospects for Genetic Intervention in Man. *Science,* **170:** 1279–1283.

DAVIS, B. D., 1974. Genetic Engineering: How Great Is the Danger? *Science,* **186:** 309.

EDWARDS, R. G., P. C. STEPTOE, and J. M. PURDY, 1970. Fertilization and Cleavage *in vitro* of Preovulator Human Oocytes. *Nature,* **227:** 1307–1309.

ETZIONI, A., 1968. Sex Control, Science, and Society. *Science,* **161:** 1107–1112.

ETZIONI, A., 1973a. *Genetic Fix.* New York, Macmillan.

ETZIONI, A., 1973b. Sex Control, Science, and Society. In A. S. Baer, ed. *Heredity and Society, Readings in Social Genetics.* New York, Macmillan.

FLEMING, D., 1969. On Living in a Biological Revolution. *The Atlantic Monthly,* **223:** 64–70.

FLETCHER, J., 1974. *The Ethics of Genetic Control.* Garden City, N.Y., Anchor Press/Doubleday.

GERMAN, J., 1970. Studying Human Chromosomes Today. *Amer. Scientist,* **58:** 182–201.

GOULIAN, M. A., A. KORNBERG, and R. L. SINSHEIMER, 1967. Enzymatic Synthesis of DNA, XXIV. Synthesis of Infectious Phage ϕX-174 DNA. *Proc. Nat. Acad. Sci. (U.S.),* **58(6):** 2321–2328.

GROBMAN, A. B., ed., 1970. *Social Implications of Biological Education.* Princeton, N.J., The Darwin Press.

HAMILTON, M., ed., 1972. *The New Genetics and the Future of Man.* Grand Rapids, Mich. Eerdmans.

HELLER, J. H., 1969. Human Chromosomal Abnormalities as Related to Physical and Mental Dysfunction. *Jour. Hered.,* **60:** 239–248.

HILTON, B., D. CALLAHAN, M. HARRIS, P. CONDLIFFE, and B. BERKLEY, 1973. *Ethical Issues in Human Genetics.* New York, Plenum Press.

HUISINGH, D., Should Man Control His Genetic Future? In J. B. Bresler, ed. *Genetics and Society.* Reading, Mass., Addison-Wesley.

JAROFF, L., ed., 1971. Man into Superman; the Promise and Peril of the New Genetics. *Time,* April 19, 1971, 33–52.

LURIA, S. E., 1973. *Life—The Unfinished Experiment.* New York, Scribner.

MERTENS, T. R., ed., 1975. *Human Genetics—Readings on the Implications of Genetic Engineering.* New York, Wiley.

MULLER, H. J., 1965. Means and Aims in Human Genetic Betterment. In T. M. Sonneborn, ed. *The Control of Human Heredity.* New York, Macmillan.

OSTERMAN, J. V., A. WADDELL, and H. V. APOSHIAN, 1970. DNA and Gene Therapy: Uncoating of Polyoma Pseudovirus in Mouse Embryo Cells. *Proc. Nat. Acad. Sci (U.S.),* **67:** 37–40.

RAMSEY, P., 1973. The Moral and Religious Implications of Genetic Control. In A. S. Baer, ed. *Heredity and Society: Readings in Social Genetics.* New York, Macmillan.

ROBERTS, D. F., J. CHAVEZ, and S. D. M. COURT, 1970. The Genetic Component in Child Mortality. *Arch. Dis. Childhood,* **45:** 33–38.

ROSENFELD, A., 1965. Will Man Direct His Own Evolution? *Life Magazine,* **59(14),** Oct. 1, 1965.

ROSENFELD, A., 1969. *The Second Genesis.* Englewood Cliffs, N.J., Prentice-Hall.

ROSLANSKY, J. D., ed., 1966. *Genetics and the Future of Man.* New York, Appleton-Century-Crofts.

SHAPIRO, J., L. MACHATTIE, L. ERON, G. IHLER, K. IPPEN, and J. BECKWITH, 1969. Isolation of Pure *Lac* Operon DNA. *Nature,* **224:** 768–774.

SIGNER, E., 1974. Gene Manipulation: Progress and Prospects. In E. D. Hay, T. J. King, and J. Papaconstantinou, eds. *Macromolecules Regulating Growth and Development.* New York, Academic Press.

SINSHEIMER, R. L., 1973. The Prospect of Designed Genetic Change. In A. S. Baer, ed., 1973. *Heredity and Society, Readings in Social Genetics.* New York, Macmillan.

SINSHEIMER, R. L., 1973. Prospects for Future Scientific Developments: Ambush or Opportunity. In B. Hilton, D. Callahan, M. Harris, P. Condliffe, and B. Berkley, eds., *Ethical Issues in Human Genetics.* New York, Plenum Press.

SONNEBORN, T. M., ed., 1965. *The Control of Human Heredity and Evolution.* New York, Macmillan.

SONNEBORN, T. M., 1968. H. J. Muller, Crusader for Human Betterment. *Science,* **162:** 772–776.

SONNEBORN, T. M., 1973. Ethical Issues Arising from the Possible Uses of Genetic Knowledge. In B. Hilton, D. Callahan, M. Harris, P. Condliffe, and B. Berkley, eds., *Ethical Issues in Human Genetics.* New York, Plenum Press.

SZYBALSKA, E. H., and W. SZYBALSKI, 1962. Genetics of Human Cell Lines, IV. DNA-Mediated Heritable Transformation of a Biochemical Trait. *Proc. Nat. Acad. Sci. (U.S.),* **48:** 2026–2034.

TOFFLER, A., 1970. *Future Shock.* New York, Random House.

WADE, N., 1974. Genetic Manipulation: Temporary Embargo Proposed on Research. *Science,* **185:** 332–334.

WALLACE, B., 1970. Genetics and Genetic Manipulation. In A. B. Grobman, ed. *Social Implications of Biological Education.* Princeton, N.J., The Darwin Press.

WATSON, J. D., 1971. Moving Toward the Clonal Man—Is This What We Want? *Congressional Record, Senate,* April 29, 1971: 12751–12752.

WEYL, N., 1973. Some Possible Genetic Implications of Carthaginian Child Sacrifice. In J. B. Bresler, ed. *Genetics and Society.* Reading, Mass., Addison-Wesley.

WILSON, M., 1970. On Being a Scientist. *The Atlantic Monthly,* **226:** 101–106.

QUESTIONS FOR REFLECTION

(No claim is made for originality in the following questions—many writers have raised similar ones, implicitly or explicitly. Moreover, absolute answers can be given to none of them, because most are opinion questions to which, it is hoped, your genetics course will have contributed some knowledge as well as a sense of priorities and values. Therefore no answers will be found in Appendix A. However, all these questions are important to you, or will be one day.)

21-1. In a recent case of child abuse, the child was so severely beaten by the parents that it suffered permanent brain damage and was crippled for life. Do you recommend taking children away from such parents?

21-2. A couple has a child who develops the Tay-Sachs syndrome; tests confirm that both husband and wife are heterozygous. Considering the probability existing in this case, should they "try again"?

21-3. The couple in the preceding question does "try again"; amniocentesis discloses that the second child will also have the Tay-Sachs syndrome. Should they abort?

21-4. How would you answer the two preceding questions if one member of the couple is yourself?

21-5. Suppose a child of yours is born with the Tay-Sachs syndrome, but needs a respirator to sustain life during the period shortly after birth. Would you want the respirator used?

21-6. You are one of approximately three in 100 who is heterozygous for cystic fibrosis. In this condition, which is due to a recessive autosomal gene, affected persons are unable to digest food properly and are highly subject to infection; their ligaments do not form properly, and the lungs fill with fluid that has to be removed (sometimes painfully) almost daily. Death often occurs in the teens. (a) Would you marry? (b) If so, would you elect to have children of your own? (c) If you did, and anmiocentesis disclosed the fetus to be homozygous recessive (which, of course, discloses your spouse also to be heterozygous), would you opt for an abortion?

21-7. You are heterozygous for several lethal genes. Would you like to subject yourself to genetic engineering by transducing viruses so as to change your genotype with respect to those genes? Explain the bases for your choice.

21-8. Some people are unable to synthesize arginase and consequently have high blood levels of the amino acid arginine. As a result, they suffer spastic paraplegia, epileptic seizures, and severe mental retardation. This condition, called arginemia, is due to a recessive autosomal gene. Infection with the Shope virus, which carries a cistron for arginase synthesis, has resulted in elevated levels of the enzyme in both normal and affected persons. The virus produces skin cancer in rabbits, though its carcinogenic effect on humans has not been established. If you had a child born with arginemia, would you want it treated with the Shope virus?

21-9. Should parents in general have freedom to choose whether or not to have children of their own (a) if both are heterozygous for several different lethal or disabling genes, (b) if both are heterozygous for the same lethal or disabling gene?

21-10. If you knew you and your spouse were both heterozygous for the same recessive lethal gene (which, when homozygous, causes intense physical suffering, then death, between the ages of 5 and 10), (a) would you want the freedom to make your own choice as to whether or not to have children? (b) Would you want the decision to be made for you, say by a government commission?

21-11. A woman is carrying fraternal (dizygotic) twins; anmiocentesis discloses one of the fetuses to be homozygous for a lethal gene that will kill the child some time between the ages of five and 10 after intense physical suffering. Assume the other fetus to be normal. Knowing that an abortion cannot be selective in aborting only the defective fetus, what option would you accept if (a) the woman were unknown to you, (b) the woman were your sister, (c) the woman were yourself or your spouse?

21-12. Would you like to be able to choose the sex of your own offspring?

21-13. If it were possible to choose the sex of one's offspring with near certainty, can you foresee any possible disadvantages for the human species?

21-14. Would you like to be able to choose the intelligence range of your children within, say, about 10 I.Q. points?

21-15. Would you like to have one or more clonants? Give the bases for your answer.

21-16. Do you feel that we should try to alter human genotypes (a) now or (b) in the future? Give the rationale for your viewpoint.

21-17. Do you feel that research on recombinant DNA should be (a) carried out without interruption but with all possible precaution against contamination of the human gene pool, (b) temporarily and voluntarily halted until risks and advantages can be evaluated, (c) permanently and voluntarily halted, or (d) permanently halted by statutory fiat? Explain your view in the light of your present genetic knowledge and your assessment of the likelihood of future progress and its potential uses.

21-18. Assuming that this planet can no longer sustain its population, that agriculture-related remedies have failed, and that time has run out, do you recommend (a) mass starvation, (b) starvation of selected populations, or (c) the elimination of "substandard" or noncontributing members of the human race? Do you have any other options to suggest?

APPENDIX A
Answers to Problems

Chapter 2

2-2. *cc.*

2-3. (a) 1. *aa;* 2. *Aa;* 3. *Aa;* 4. *aa;* 5. *A—.*
    (b) 1:1.

2-4. (a) 25%.
    (b) 50%.
    (c) Probably zero; evidence *suggests* girl is homozygous normal.

2-5. (a) Incompletely dominant as relates to chloride excretion.
    (b) Recessive lethal.

2-6. (a) None.
    (b) Yes; this couple's children have a 50% probability of being heterozygotes. If one of such heterozygotes marries another, *their* children (grandchildren of the original couple) have a 25% chance of having the disease.

2-7. (a) Purple is heterozygous; blue is homozygous.
    (b) Purple.

2-8. Hornless is the dominant character. Hornless animals producing horned offspring are heterozygotes; any that do so should not be bred. Because he needs to get rid of a recessive, and cattle usually produce but one offspring per year, the problem is not going to be solved quickly. On the other hand, red animals are homozygous, so roans and whites can be excluded from the breeding program.

2-9. (a) *hh.*   (e) *h.*
    (b) *HH.*   (f) *Hh.*
    (c) *H.*   (g) *Hhh.*
    (d) *h.*   (h) *hh.*

2-10. Curly is the heterozygous expression of a recessive lethal; 341:162 is a close approximation of a 2:1 ratio.

2-11. *Ff; Ff; ff; F—; Ff; ff; Ff; ff.*

2-12. Testcross.

2-13. *Aa; Aa; aa.*

2-14. On the basis of her daughter, III-4, who has to be *aa,* inheriting one *a* from each parent.

2-15. Rh positive.

2-16. Incomplete dominance, but with the gene for thalassemia major recessive as to lethality.

2-17. (a) $\frac{1}{2}$. (b) $\frac{1}{4}$.

2-18. No; the tranfused blood does not affect the recipient's genotype.

2-19. (a) Recessive.
    (b) It is maintained in and transmitted by heterozygotes, some of whose children are PKUs. Mutation is another factor, but of lesser magnitude.

2-20. (a) $\frac{2}{3}$ (*not* $\frac{1}{2}$, because his normal phenotype eliminates the possibility that he might be homozygous recessive for PKU).

(b) Probably none; it appears highly likely that the girl is homozygous dominant for normal.

(c) That any heterozygous child of theirs (which could occur if the husband is heterozygous) has 1 chance in 4 of having a PKU child if he or she marries another heterozygote. This eventuality is more likely if such heterozygous child marries a relative such as a cousin.

2-21. (a) $(\frac{2}{3})^2 = \frac{4}{9}$.

(b) $\frac{1}{9}$.

(c) $\frac{1}{4}$.

Chapter 3

3-1. (a) 48.   (e) 48.
     (b) 24.   (f) 24.
     (c) 48.   (g) 12.
     (d) None. (h) 12.

3-2. (a) 20.
     (b) 40.
     (c) 40.
     (d) None.
     (e) 40.

3-3. (a) 20.
     (b) 10.
     (c) 30.
     (d) 10.
     (e) 10.
     (f) 10.
     (g) 20.

3-4. (a) 40.
     (b) 40.
     (c) 40.

3-5. 160.

3-6. (a) 80.
     (b) 160.

3-7. (a) 1.
     (b) 2.
     (c) 2.
     (d) 4.
     (e) 2.
     (f) 32.

3-8. (a) $(\frac{1}{2})^{30}$.
     (b) $[(\frac{1}{2})^{30} \times (\frac{1}{2})^{30}]$.

3-9. (a) 33.
     (b) Irregularities of pairing at synapsis lead to defective gametes having more or less than one complete set of chromosomes.

3-10. All Aa.

3-11. Two A and two a.

3-12. Prophase longest, next telophase, next metaphase, with anaphase shortest; why?

3-13. 7.

3-14. $(\frac{1}{2})^7$.

3-15. $+ + + + y y y y$ indicates segregation of the alleles at the first meiotic division, hence no crossing-over; $+ + y y + + y y$ indicates second-division segregation, hence crossing-over.

3-16. (a) Yes.

    (b) Crossing-over between the two pairs of genes.

3-17. 0.005.

3-18. (a) 0.000,005 m.

    (b) 0.005 mm.

    (c) 5,000 nm.

    (d) 50,000 Å.

3-19. Increases variability by recombining characters from two parents.

3-20. May recombine two or more deleterious genes in a given progeny individual.

Chapter 4

4-1. 6 (i.e., $AA \times AA$; $AA \times Aa$; $AA \times aa$; $Aa \times Aa$; $Aa \times aa$; $aa \times aa$).

4-2. (a) $\frac{3}{16}$.

    (b) $\frac{3}{16}$.

    (c) $\frac{2}{16}$.

    (d) $\frac{1}{16}$.

4-3. (a) 8.

    (b) 2^{12}.

4-4. (a) $1:1$.

    (b) $1:1:1:1:1:1:1:1$.

4-5. 16.

4-6. (a) 16.

    (b) 81.

4-7. 256. → why?

4-8. (a) 4.

    (b) 8.

    (c) 16.

    (d) 2^n.

4-9. (a) 24.

    (b) $\frac{1}{256}$.

    (c) $\frac{1}{128}$.

4-10. Letting Y represent red, and y yellow, the parental genotypes are $Yyh_1h_2 \times Yyh_1h_2$ (red, scattered hairs).

4-11. (a) Cream.

    (b) $\frac{2}{16}$.

4-12. $3:6:3:1:2:1$.

4-13. (a) Two pairs, both incompletely dominant.

    (b) Broad red, narrow red, broad white, narrow white.

4-14. 9 red : 3 pink : 4 white. aren't they assuming white to be wwpp?

4-15. 9 normal : 7 deaf.

4-16. (a) 9:7.

    (b) *A−B−*.

    (c) *AaBb*.

    (d) *AAbb × aaBB*.

4-17. (a) *aabb*.

    (b) *AaBb*.

    (c) anything *except aabb.* why ?

4-18. 9 black : 3 brown : 4 albino.

4-19. (a) 2.

    (b) *AaBb*.

    (c) purple, *A−B−;* red *A−bb;* white *aa−−*, if one assumes gene *A* is responsible for the enzyme converting colorless precursor to cyanidin, and *B* for the enzyme converting cyanidin to dephinidin.

4-20. 9 both enzymes : 3 enzyme number 1 only : 3 enzyme number 2 only : 1 neither enzyme.

    these comprise the aa — class of 4

4-21. *aaBB × AAbb; aaBb × Aabb.*

4-22. Four : red long, red round, white long, and white round.

4-23. (a) 3:6:3:1:2:1.

    (b) 1:2:1.

4-24. (a) Let *A* represent a color inhibitor gene, *a* the gene for color, *B* yellow, *b* green. Then *A−−−* is white, *aaB−* yellow, and *aabb* green.

    (b) P: *AAbb* (white) × *aaBB* (yellow)

      F$_1$: *AaBb*

      F$_2$: 9 *A−B−*$\Big\}$white
          3 *A−bb*

          3 *aaB−* yellow

          1 *aabb* green.

4-25. (a) Disk *C−D−*, sphere *C−dd* and *ccD−*, elongate *ccdd.*

    (b) P: *CCdd × ccDD*

      F$_1$: *CcDd*

      F$_2$: 9 *C−D−* disk

          3 *C−dd*$\Big\}$sphere
          3 *ccD−*

          1 *ccdd* elongate.

4-26. (a) 8.

    (b) 1.

    (c) 24.

4-27. (a) 9.

    (b) $\frac{108}{256}$

4-28. (a) red *R−S−;* sandy *rrS−* and *R−ss;* white *rrss.*

    (b) case 1 *RRSS × RRSS.*

      case 2 *RrSS × RrSS,* or *RRSs × RRSs,* or *RrSs × RrSS,* or *RrSs × RRSs.*

      case 3 *RRSS × rrss.*

      case 4 *rrSS × RRss.*

      case 5 *rrSs × Rrss.*

4-29.

	Genotypic:	*Phenotypic:*
(a)	1:2:1:2:4:2:1:2:1	9:3:3:1
(b)	1:2:1:2:4:2:1:2:1	3:6:3:1:2:1

(c) $1:2:1:2:4:2:1:2:1$ $1:2:1:2:4:2:1:2:1$
(d) $1:2:1:2:4:2$ $3:1$
(e) $1:2:1:2:4:2$ $1:2:1$
(f) $1:2:2:4$ all alike.
4-30. 3 red : 6 purple : 3 blue : 4 white.

Chapter 5

5-1. (a) $(\frac{1}{2})^3$, or $\frac{1}{8}$.
     (b) $\frac{3}{8}$.
5-2. $\frac{1}{2}$.
5-3. $\frac{1}{2}$.
5-4. (a) $\frac{1}{6}$.
     (b) $\frac{1}{36}$.
     (c) $\frac{1}{6}$.
5-5. (a) $\frac{3}{4}$.
     (b) $\frac{1}{4}$.
5-6. (a) $\frac{4}{16}$.   (c) $\frac{9}{16}$.
     (b) $\frac{1}{16}$.   (d) $\frac{3}{16}$.
5-7. (a) $28a^6b^2$.
     (b) $\frac{28}{1024}$.
5-8. (a) $\frac{1}{8}$.
     (b) $\frac{1}{4}$.
     (c) 6.
5-9. $\frac{270}{32,768}$, or about 1 in 121.
5-10. (a) $\frac{243}{32,768}$, or about 1 in 135.
     (b) No, because other phenotypes, not included here, are possible.
5-11. $\frac{24}{81}$, or roughly 3 in 10.
5-12. (a) 0.02.
     (b) 0.0392.
5-13. (a) 1.
     (b) Yes, for $3:1$.
     (c) A recessive lethal.
5-14. P lies between 0.5 and 0.3 ($\chi^2 = 0.993$).
5-15. (a) P is between 0.2 and 0.05 ($\chi^2 = 2.0$).
     (b) P is between 0.8 and 0.7 ($\chi^2 = 0.127$).
     (c) No.
     (d) Either accept results as a better reflection of a $9:7$ expectancy, or obtain a larger sample.
5-16. (a) $P < 0.01$ ($\chi^2 = 20.0$).
     (b) P lies between 0.30 and 0.20 ($\chi^2 = 1.27$).
     (c) Yes; significant for a $1:1$ expectancy.
     (d) The larger the sample the greater the usefulness of the chi-square test.
5-17. (a) $\chi^2 = 0.015$; $P = 0.80 - 0.95$; not significant.
     (b) $\chi^2 = 0.451$; $P = 0.50 - 0.70$; not significant.
     (c) $\chi^2 = 0.563$; $P = 0.30 - 0.50$; not significant.
     (d) $\chi^2 = 0.618$; $P = 0.80 - 0.95$; not significant.

Chapter 6

6-1. (a) *R Ro* and *r ro* each 0.4375; *R ro* and *r Ro* each 0.0625.

(b) 0.1914.

(c) 0.6912.

6-2. (a) *Ed/eD.*

(b) *Trans.*

6-3. (a) *wo dil o aw.*

(b) Double (or any even number of) crossovers are not detected in genes as far apart as *wo* and *aw* are here. The more accurate *wo-aw* distance is 22 map units, i.e., 9 + 6 + 7.

6-4. (a) *jvl fl e.*

(b) Double crossovers are missed in a dihybrid cross involving only *jvl* and *e*.

6-5. It could be either to the "left" of *jvl* or to the "right" of *e*.

6-6. To the "right" of *e*.

6-7. (a) Yes.

(b) 0.77 per cent.

6-8. 4.

6-9. 12.

6-10. (a) 12.

(b) 12.

(c) 23.

(d) 24, because the X and Y chromosomes are only partly homologous.

6-11. Because genes for each of the seven characters happened to be on a different one of the seven pairs of chromosomes of the pea.

6-12. (a) 4.

(b) 2.

6-13. (a) 8 (2 noncrossover, 4 single crossover, 2 double crossover).

(b) 2.

6-14. Yes. Normal beaked plants are doubly homozygous recessive; in this cross, with unlinked genes, about 6 per cent ($= \frac{1}{16}$) of the progeny should have this phenotype. The number actually observed is about 3.68 times greater than is to be expected with unlinked genes.

6-15. (a) Data indicate *cis* linkage (*Cu Bk/cu bk*) in each parent and that the frequency of *cu bk* gametes in each was 0.48 ($= \sqrt{0.2304}$).

(b) Crossover gametes were then produced with a frequency of 0.02 each, so the genes are 4 map units apart.

6-16. 0.4 *Pl Py;* 0.4 *pl py;* 0.1 *Pl py;* 0.1 *pl Py.*

6-17. 16 per cent.

6-18. 9 per cent.

6-19. 34 per cent.

6-20. (a) 88.2 per cent.

(b) 0.2 per cent.

6-21. (a) *h fz eg; h-fz,* 14 map units; *fz-eg,* 6 map units.

(b) 0.238.

6-22. 0.5.

6-23. (a) *d + +* and *+ m p.*

(b) *p d m* (or, of course, *m d p*).

(c) p-d, 4.5 map units; d-m, 4.5 map units.

(d) Yes; coincidence $= 0.5$.

6-24. (a) 0.85.

(b) 0.05.

(c) 0.10.

(d) None.

6-25.

	(a)	(b)
$+\ +\ +$	0.4275	0.42625
$pg_{12}gl_{15}bk_2$	0.4275	0.42625
$+\ gl_{15}bk_2$	0.0225	0.02375
$pg_{12}\ +\ +$	0.0225	0.02375
$+\ +\ bk_2$	0.0475	0.04875
$pg_{12}gl_{15}\ +$	0.0475	0.04875
$+\ gl_{15}\ +$	0.0025	0.00125
$pg_{12}\ +\ bk_2$	0.0025	0.00125

6-26. (a) 0.485.

(b) 0.015.

Chapter 7

7-1. No.

7-2. $c^{ch}c \times c^h c$.

7-3. (a) 3.

(b) Superdouble, double, single.

(c) One parent heterozygous for superdouble and single, the other hetero-zygous for double and single.

7-4. Alexandra, normal, Blue Moon, Primrose Queen.

7-5. 4.

7-6. 1.

7-7. 3.

7-8. All black.

7-9. 3 dark-bellied : 1 black.

7-10. One parent heterozygous for white-bellied and dark-bellied, the other either the same or heterozygous for white-bellied and plain black.

7-11. White-bellied $>$ dark-bellied $>$ black-and-tan $>$ plain black.

7-12. 20.

7-13. 210.

7-14. 8.

7-15. A_1B.

7-16. Yes, if the woman is $I^{A_1}i$ and the man $I^B i$.

7-17. Yes, he is eliminated on the basis of the MNSs test (only). Neither the woman nor the alleged father could have contributed Ns to the child.

linked

Chapter 8

8-1. (a) Lozenge.

(b) Wild.

8-2. The cd pair because of the *cis-trans* effect.

8-3. I-1, *Dd.*
   I-2, *dd.*
   II-1, *Dd.*
   II-2, *Dd.*
   II-3, *Dd.*
   II-4, *dd.*

8-4. II-2 must be Rh+ (*Dd*) because that child evidently sensitized the mother with the result that II-2 and II-3 were erythroblastotic. Had II-1 been *dd,* II-2 should not have been erythroblastotic.

8-5. No cases of erythroblastosis fetalis would have occurred in succeeding children.

8-6. A− ♀ × O+ ♂.

8-7. 16.

8-8. 48.

8-9. 162.

8-10. Yes.

8-11. Yes.

8-12. Yes.

8-13. Yes.

8-14. No. The claimant's M-N type should not be possible with the purported parentage.

8-15. Children numbers 4, 5, and 6 cannot be those of the husband, number 4 on the basis of *NS/Ns*, number 5 on the basis of *dce/dce* and *MS/NS,* number 6 on the basis of all three tests, even allowing for a very low frequency of recombination. Moreover, it appears that more than one additional man was involved; numbers 4 and 5 *could* have been produced by the same man (but not the husband), and number 6 by a third man.

Chapter 9

9-1. Intelligence, height, skin color, eye color.

9-2. Any parental genotypes that can produce at least some F_1 genotypes with a greater number of contributing alleles than they themselves have are possible, e.g., *AaBdCcDd* × *AaBbCcDd, AaBbccdd* × *aabbCcDd,* etc.

9-3. (a) $\frac{1}{4,096}$.
   (b) 13.
   (c) $\frac{924}{4,096}$.

9-4. 8 polygenes (4 pairs); 3 inches per contributing allele.

9-5. 10 polygenes (5 pairs); 4.8 inches per contributing allele.

9-6. $\frac{1}{4,096}$.

9-7. *AaBbCcDd* × *AaBbCcDd.*

9-8. Any in which each parent is homozygous effective for 2 of the 4 pairs, e.g., *AABBccdd* × *aabbCCDD,* etc.

9-9. (a) *AABBCCDD.*
   (b) *aabbccdd.*
   (c) Only green.

9-10. (a) $\frac{1}{256}$.
   (b) $\frac{28}{256}$.
   (c) $\frac{56}{256}$.

9-11. (a) Green.

(b) $\frac{70}{256}$.

9-12. Transgressive variation.

9-13. 769. ($\frac{3}{4}$ of 1,024 which are S————, or 768, plus 1 which is *ssaabbccdd*.)

9-14. 8.

9-15. (a) 25 and 5 cm.

(b) 15 cm.

(c) 1 (25):4 (20):6 (15):4 (10):1 (5).

9-16. (a) 15 and 5 cm.

(b) 15 cm.

(c) 9 (15):6 (10):1 (5).

Chapter 10

10-1. 23.0 (actually, 23.04).

10-2. 20.04.

10-3. (a) 4.477, or approximately 4.5.

(b) That 0.6826 of the sample should lie in the range 23 ± 4.5, and so forth. See text and Appendix D.

10-4. 0.6826, or about $\frac{2}{3}$.

10-5. (a) 0.895.

(b) That there is a 0.6826 probability that the population mean falls in the range 23 ± 0.895, and so forth.

10-6. 0.9544.

10-7. 1.2.

10-8. 0.6826, or about two chances in three.

10-9. No; $S_d =$ only $(\bar{x}_1 - \bar{x}_2)$, and to be significant S_d must exceed $2(\bar{x}_1 - \bar{x}_2)$.

10-10. 8.

Chapter 11

11-1. (a) Metafemale   (d) Intersex.

(b) Metamale.   (e) Female (tetraploid).

(c) Metafemale.

11-2. (a) Female.   (d) Male.

(b) Male.   (e) Male.

(c) Female.   (f) Triploid metamale.

11-3. 9 normal monoecious:3 pistillate (ears terminal and lateral):3 staminate: 1 pistillate (ears terminal only); i.e., 9 monoecious:3 staminate:4 pistillate.

11-4. $\frac{1}{4}$.

11-5. 3 male:1 female.

11-6. (a) bW (or simply b).   (c) BW (or B)

(b) $B - (BB$ or Bb).   (d) bb.

11-7. (a) $\frac{1}{4}$ each of the following: barred ♀, nonbarred ♀, barred ♂, and nonbarred ♂.

(b) All ♀s barred, ♂s 1 barred:1 nonbarred. *other way around ? Yes*

11-8. $\frac{1}{3}$ male, $\frac{2}{3}$ female.

11-9. 0.1. *?*

11-10. 45 per cent (i.e., 20 per cent XX + 25 per cent XXY). *?*

11-11. (a) $\frac{1}{4}$.

(b) $\frac{1}{4}$.

(c) $\frac{1}{3}$.

11-12. (a) White.

(b) Red.

11-13. Fluorescence pattern, especially the bright longer arm, is the best identification criterion; also, the longer arms are close together, and satellites are absent. Its length relative to members of the F and G groups of autosomes is more variable.

11-14. From the theoretical standpoint several alternative explanations are possible. Among the most likely are

(a) nondisjunction of Y in the second meiotic division in spermatogenesis;

(b) nondisjunction in the first meiotic division in spermatogenesis, giving rise to an XY sperm which fertilizes an X egg;

(c) first or second division nondisjunction in either spermatogenesis or oogenesis, producing either O sperm or egg which then fuses with an X gamete from the other sex. The XXY condition could also arise through nondisjunction in the first cleavage division of a normal XY zygote whereby one daughter cell receives XXY, and the other (nonviable) OY.

11-15. Most likely by lagging of an X chromosome in early mitoses of an XX zygote.

11-16. The X chromosome carries a large number of genetic loci, most or all of which appear to be necessary for normal development.

11-17. Man. Single genes (*Asparagus*) are subject to mutation that could result in a sex imbalance in small populations or a lethal condition, but in man there appear to be many genes governing sex on the X and Y chromosomes. Furthermore, the occasional "male × male" crosses that occur in *Asparagus* increase the likelihood of homozygosity of deleterious genes in the progeny. Also, a 1:1 sex ratio in Asparagus occurs only in populations where "males" are heterozygous.

11-18. (a) XX.

(b) XY.

(c) XY.

(d) XX.

Chapter 12

12-1. 1 bent tail female:1 normal male.

12-2. 50 per cent probability that any girls will be heterozygous (slight nystagmus) and 50 per cent chance that any boys will have severe nystagmus.

12-3. Sex-linked recessive.

12-4. 1 barred rose male:1 nonbarred rose female.

12-5. $\frac{6}{16}$; equally divided between male and female.

12-6. 1 barred:1 nonbarred; 3 rose:1 single.

12-7. No chance that any children of the boy will develop the disease as long as he marries a + + girl; each girl has a probability of 0.5 of being heterozygous and, therefore, transmitting the trait to half her sons.

12-8. Sex-linked recessive lethal.

12-9. 2 female:1 male, all normal.

12-10. $\frac{1}{2}$.

12-11. $\frac{3}{64}$.

12-12. (a) 0. (c) $\frac{1}{4}$.

(b) $\frac{1}{8}$. (d) $\frac{1}{2}$.

12-13. (a) early bald.

(b) nonbald.

(c) 9 early bald:3 late bald:4 nonbald.

(d) 12 nonbald:3 late bald:1 early bald.

12-14. Female.

12-15. Male.

12-16. Sex-limited.

12-17. All males yellow; females 3 white : 1 yellow. ?

12-18. Sex-limited.

12-19. (a) All short.

(b) All males short, all females long.

(c) Males: 3 short : 1 long; females: 3 long : 1 short.

(d) All males short; females 1 short : 1 long.

12-20. Sex-linked recessive may appear in either sex, but much more frequently in males; it is often transmitted from father to half the grandsons via a female. Holandric genes appear only in the male sex and are transmitted directly from father to all his sons.

12-21. Traits determined by holandric genes appear only in the heterogametic sex; XY these normally cannot be heterozygous. *ie. since only 1 Y chromosome - only 1 allele*

12-22. Sex-linked dominants may be transmitted directly from an affected mother to her sons; they cannot be transmitted from father to son (why?). *since if for example H Y only the H goes to the son (the H₀ goes to the daughter)*

12-23. All boys will have hairy ears; none of the girls will.

12-24. $HhZW \times hhZZ$, or $hhZW \times HhZZ$.

12-25. I-1, II-3, II-9, III-2, III-4, III-5, III-8.

12-26. $\frac{1}{2}$.

12-27. It would be possible, but very unlikely. It *appears* that all of the persons listed were free of the gene for hemophilia; if this is so, then Elizabeth II is homozygous normal. Under those circumstances, only a mutation, such as apparently occurred with Queen Victoria, could produce the defect in any children of Elizabeth.

12-28. (a) $\frac{3}{16}$.

(b) $\frac{1}{16}$.

(c) All males.

12-29. (a) $Nn \times NY$

(b) $\frac{1}{4}$.

(c) All staminate.

12-30. (a) Hemophilia A.

(b) Sons only.

(c) Yes, all for B, $\frac{1}{2}$ for A and B.

(d) $\frac{1}{4}$

12-31. (a) Maternal.

(b) Paternal.

Chapter 13

13-1. *DDDD*.

13-2. *DD*.

13-3. *DDd*. ~~assuming~~ *(assuming dwarf is homoz. dd)*

13-4. *DDdd*.

13-5. 1 *DD* : 4 *Dd* : *dd*.

13-6. (a) Autotetraploidy.

(b) $\frac{1}{36}$, or $(\frac{1}{6})^2$.

13-7. $\frac{1}{1296}$ or $(\frac{1}{6})^4$.

13-8. $\frac{1}{46,656}$, or $(\frac{1}{6})^6$.

13-9. $(\frac{1}{6})^{2n}$.

(the genes on the chromosomes are subsequently different (e.g. sequence) so can't pair right)

13-10. Reduces it.

13-11. Those having 28 represent diploids; 56, tetraploids; 70, pentaploids; 84, hexaploids.

13-12. A different series of chromosomal aberrations in each species, so that normal pairing is impossible. Translocations are probably the most frequent of these aberrations.

13-13. Yes, by creating an allotetraploid hybrid.

13-14. Euploidy.

13-15. 13.

13-16. 12 (dodecaploid).

13-17. (a) Male.

(b) Female.

13-18. 69, XXY.   *23 × 3*

13-19. (a) 92, XXYY.

(b) No.

13-20. (a) 1 *P* : 2 *p* : 2 *Pp* : 1 *pp*.

(b) 1 *P* : 2 *p*.

(c) 1 *P* : 1 *PP*.

(d) all *P*.

(e) 2 *P* : 1 *p*.

13-21. (a) 15 purple : 3 white (= 5 purple : 1 white).

(b) 11 purple : 1 white.

(c) 12 purple : 6 white (= 2 purple : 1 white).

13-22. (a) 3 normal : 1 eyeless.

(b) 11 normal : 1 eyeless.

(c) 35 normal : 1 eyeless.

13-23.

13-24. *AAA/AA* between heterozygous ultrabar and homozygous ultrabar; *AAAA/AAA* below homozygous ultrabar.

13-25. (a) Deletion (deficiency).

(b) Deletion loop in salivary gland chromosomes.

13-26. See text and references. Although the evidence is presently contradictory, neither drug can yet be completely absolved. Moreover, the time factor must be taken into account. Note that this question and its answer refer only to cytological and genetic consequences; physiological, mental, and other effects are not included here.

13-27. Autosomal monosomy probably constitutes a lethal genic imbalance.

13-28. Trisomy for other chromosomes appears to produce a lethal imbalance.

13-29. Farther from the centromere, which has an interfering effect on crossing-over.

Chapter 14

14-1. (a) *MN*.
   (b) *NN*.

14-2. *M*, 0.6; *N*, 0.4.

14-3. *M*, 0.546; *N*, 0.454.

14-4. *M*, 0.19; *N*, 0.81.

14-5. *T*, 0.6; *t*, 0.4.

14-6. (a) 48.
   (b) 36.

14-7. About 1.4 per cent, or 1 in 70 persons.

14-8. 1 in 500.

14-9. I^A, 0.3; I^B, 0.1; i, 0.6.

14-10. $I^A I^A$, 0.04; $I^A i$, 0.28; $I^B I^B$ 0.01; $I^B i$, 0.14; $I^A I^B$, 0.04; ii, 0.49.

14-11. Deviation is significant (chi-square = 35.69).

14-12. A, 39.36 per cent; B, 8.76 per cent; AB, 2.88 per cent; O, 49.0 per cent.

14-13. (a) 5.76 per cent.
   (b) 8.4 per cent.

14-14. 0.00004.

14-15. (a) 0.0198, or approximately 0.02.
   (b) 0.0001.

14-16. 0.35.

14-17. Decrease the frequency of Hb^S.

14-18. Deterioration.

14-19. 0.25.

14-20. (a) 0.048.
   (b) the 199th generation.

14-21. 0.22.

14-22. 0.267.

14-23. 0.8.

14-24. 0.83.

14-25. 0.04.

14-26. (a) 0.02.
   (b) 2×10^{-3}.

14-27. (a) 1,500 years.
   (b) Probably not, because of mutation from dominant normal to recessive lethal.

14-28. (a) p^2.
   (b) $p^4 (= p^2 \times p^2)$.

(c) $2pq$.

(d) $4p^2q^2\ (= 2pq \times 2pq)$.

(e) $4pq^3\ (= 2(2pq \times q^2))$.

(f) $p^4 + 4p^3q + 6p^2q^2 + 4pq^3 + q^4$, calculated as follows:

$$AA \times AA = p^4, \text{ i.e., } p^2 \times p^2.$$
$$AA \times Aa = 4p^3q, \text{ i.e., } 2(p^2 \times 2pq).$$
$$AA \times aa = 2p^2q^2, \text{ i.e., } 2(p^2 \times q^2).$$
$$Aa \times Aa = 4p^2q^2, \text{ i.e., } 2pq \times 2pq.$$
$$Aa \times aa = 4pq^3, \text{ i.e., } 2(2pq \times q^2).$$
$$aa \times aa = q^4, \text{ i.e., } q^2 \times q^2.$$

14-29. (a) $p^2 \times 2pq$.

(b) q^2.

14-30. (a) $p^2q^2 + 2pq^3 + q^4$.

(b) $0.0625 + 0.125 + 0.0625 = 0.25$.

14-31. $0.0004882\ (= 4.882 \times 10^{-4})$.

14-32. 0.4997559.

14-33. 0.6.

14-34. (a) 0.006.

(b) 0.988

(c) 0.012.

(d) 0.002.

(e) 1.6×10^{-11}.

14-35. 2×10^{-3}.

14-36. 3.317×10^{-3}, or slightly more than 0.003.

14-37. (a) 0.775.

(b) 0.448.

(c) 12.

Chapter 15

15-1. Dominance, epistasis, sex linkage.

15-2. $3'$ T T G C A T G A C G $5'$

15-3. 4^n.

15-4. $L_{\mu m} = 3.4 \times 10^{-4}P$, where $L_{\mu m}$ = length in micrometers and P = the number of pairs of nucleotides.

15-5. 68 μm.

15-6. (a) 135.   (c) 111.

(b) 126.   (d) 151.

15-7. (a) 251.   (c) 227.

(b) 242.   (d) 267.

15-8. (a) 329.   (c) 305.

(b) 320.   (d) 345.

15-9. 649.5, but use **650** for easier calculations.

15-10. About 1.3×10^8.

15-11. (a) 4.15×10^6.

(b) 1,411.

15-12. One per minute.

15-13. 2×10^5.

15-14. Thymine 20 per cent, cytosine and guanine 30 per cent each.

15-15. No. Why?

15-16. (a) A/T = 0.99; G/C = 1.00.

   (b) That it is double-stranded.

   (c) This DNA is the replicative form.

15-17. This suggests that not all gene function is nuclear in nature. The matter is explored in Chapter 20.

15-18. 31.6 per cent.

15-19. (a) None.

   (b) Half.

   (c) Half.

15-20. Both processes may result in recombination, but transformation involves naked DNA from one cell becoming incorporated into another's DNA, whereas in transduction a virus serves as the vector transferring DNA from one cell to another.

Chapter 16

16-1. No. Why?

16-2. Yes.

16-3. No.

16-4. Parents, $aaA'A' \times AAa'a'$; children all $AaA'a'$.

16-5. Strain 1.

16-6. Enzyme "a."

16-7. Enzymes "a" and "b."

16-8. 1, $+ + +$; 2, $a + +$; 3, $+ b +$; 4, $ab+$.

16-9. (a) G U C U U U A C G C U A. 5'

   (b) Adenylic acid (A).

16-10. 200.

16-11. (a) 340.

   (b) 0.034 (= $100 \times 3.4 \times 10^{-4}$; for nucleotides of this rather short length use $n - 1$ internucleotide spaces for greater accuracy). *see 15-4*

16-12. (a) 80.

   (b) 0.027.

   (c) tRNA.

16-13. About 2,967.

16-14. 5' U A G C C A U C . . .3'   *they're missing an A*

16-15. Synthesis of DNA from an RNA template.

16-16. (a) mRNA.

   (b) UAC.

   (c) tRNA.

   (d) TAC.

16-17. Transcription: formation of mRNA from a DNA template. Translation: formation of a particular polypeptide chain consisting of specific amino acid residues in a specific sequence as determined by the sequence of mRNA codons.

Chapter 17

17-1. (a) Methionine, alanine, leucine, threonine.

   (b) Methionine, tryptophan, glycine, alanine, proline, leucine, leucine, end chain.

(c) DNA trinucleotide TAC transcribes into AUG of mRNA which serves as a chain-initiating codon.

17-2. Methionine, proline, end chain.

17-3. (a) It differs by one amino acid residue; the second is now proline instead of alanine. Sense is restored starting with the third amino acid residue.

(b) Missense (in the second amino acid residue).

17-4. (a) $\frac{6}{216}$, or $\frac{1}{36}$.

(b) $\frac{27}{216}$, or $\frac{1}{8}$.

17-5. (a) 423.

(b) 0.14 μm.

17-6. At GCU, an mRNA codon for alanine. tRNA-mRNA pairing is determined by anticodon-codon complementarity, not by the amino acid.

17-7. Tryptophan; it has only one codon, whereas arginine has six.

17-8. A-14 mutant has undergone a base change in the second base of the isoleucine codon (AUU, AUC, or AUA) to one coding for threonine (ACU, ACC, or ACA); Ni-1055 mutant has undergone a change in the third base of AUU, AUC, or AUA (coding for isoleucine) to AUG (coding for methionine).

17-9. In A446, UAU or UAC has been changed to UGU or UGC (second base); in A187, GG– has been changed at the second position to GU–. These represent base changes at position 2 and 17, respectively, of the stretch of mRNA involved.

17-10. 288 ($= 4 \times 4 \times 1 \times 6 \times 3$).

17-11. (a) Represents degeneracy.

(b) Represents ambiguity.

17-12. (a) Nonsense.

(b) Missense.

(c) Degeneracy.

(d) Missense.

Chapter 18

18-1. For the environment of a given species, those mutations that have either a positive or a neutral selection value should be expected to increase in frequency, although deleterious recessive mutations may be expected to persist at a low frequency in heterozygotes. In addition, many (but not all) mutations result in proteins of lowered functional capability; such mutant individuals are usually at a disadvantage in survival and reproduction.

18-2. See Figure 18-13 and accompanying text.

18-3. Short-term effects include radiation sickness, surface and deep tissue burning, loss of hair, and so on. Long-term effects include an increased incidence of leukemia and a variety of mutations.

18-4. No.

18-5. Recessive mutations are detected more readily in the hemizygous males.

18-6. Monoploid greatest, polyploid least. Why?

18-7. *A. brevis* is a diploid, and *A. barbata* is a tetraploid. Because most mutations are recessive, frequency of *detectable* mutations may be expected to be inversely related to ploidy.

18-8. *T. monococcum.* Why?

18-9. (a) 2.

    (b) white → red → blue.

    (c) (1) *AaBb*.

        (2) *A−B−*.

    (d) (1) *A−bb*.

        (2) *aa−−*.

    (e) The *A−a* pair.

18-10. Phenylalanine is coded before deamination; leucine will be coded after deamination.

18-11. A transition, because it pairs with guanine which replaces an adenine, both of which are purines.

18-12. A58: transition; A78: transversion.

18-13. (a) Nonsense.

    (b) No.

    (c) 173.

18-14. (a) 1,500.

    (b) Too high; because of code degeneracy some sense mutations are possible.

18-15. 267.

18-16. 801.

18-17. (a) $5 \times 10^9 \left(= 2 \dfrac{\text{molecular weight}}{650}\right)$.

    (b) 5×10^9.

    (c) 1.7×10^6 (= number nucleotide pairs $\times\ 3.4 \times 10^{-4}$).

    (d) $8.33 \times 10^8 \left(= \dfrac{\text{number of nucleotide pairs}}{3}\right)$.

    (e) $2.78 \times 10^6 \left(= \dfrac{\text{number of codons}}{300}\right)$.

18-18. Depending on the amino acid position involved, some missense mutations result in substitutions that do not materially affect the functioning of the resulting protein but, for example, they may result in peptides with different electrophoretic mobilities because of charge differences.

18-19. (a) *AT*.

    (b) *AT*.

    (c) *CG*.

    (d) Both.

    (e) Both.

18-20. By accumulation of a series of missense mutations affecting different amino acid positions in a given polypeptide; those not lethal or disabling might be expected to be perpetuated and passed on to later generations. This is the basis of genetic polymorphism.

18-21. Because of the present impossibility of developing specific mutations, one simply takes what he gets in induced mutations. Techniques for directed mutation are not yet adequate.

18-22. *Cistron:* a segment of DNA specifying one polypeptide chain. *Muton:* the smallest segment of DNA that can be changed and thereby bring about a

changing amino acid resulting [handwritten annotation]

mutation; can be as small as one deoxyribonucleotide pair. *Recon:* the smallest segment of DNA that is capable of recombination; can be as small as one deoxyribonucleotide pair.

18-23. Complementation: the ability of linearly adjacent segments of DNA to supplement each other in phenotypic effect. Recombination: a new association of genes in a recombinant individual, arising from (1) independent assortment of unlinked genes, (2) crossing-over between linked genes, or (3) intracistronic crossing-over.

Chapter 19

the structural [handwritten annotation]

19-1. They are similar in being composed of a given segment of DNA nucleotides, and each exerts regulatory control over cistrons. On the other hand, no demonstrable product of operators has yet been established.

19-2. The repressor is a protein, the product of a regulator site, which inhibits the action of an operator site so that cistrons of the operon are "turned off." An effector is any substance, often the substrate of the enzyme(s) for which the cistron(s) of the operon are responsible; it binds to the repressor protein, which thereby suffers a change in shape so that it can no longer bind to the operator.

19-3. Both involve control by a regulatory protein which binds to an operator. In positive control the regulatory protein serves as an activator, permitting trans- *transcription* [handwritten annotation] lation of the cistrons of the operon by binding to the operator. In negative control the regulatory protein represses the operator, either in the absence of an effector (*lac* operon) or in the presence of an effector (*his* operon).

19-4. Transcriptional level.

19-5. (a) Inductive.
(b) Constitutive.
(c) Constitutive.
(d) Absent.
(e) Constitutive.

19-6. Translation. Why?

19-7. No, none. No activator is produced; therefore no transcription of the structural genes can occur.

19-8. (a) No.
(b) Ribulose-5-phosphate. ? *Is this arabinose* [handwritten annotation]

19-9. (a) Negative. (b) Produced. *strictly* [handwritten annotation]

19-10. See text, especially Chapters 15–19, and references.

Chapter 20

20-1. Sex-linked recessive traits show a characteristic inheritance sequence from affected father to "carrier" daughter to about half her sons; sex-linked dominants are transmitted by an affected mother ($\times$ normal father) to about half her sons and half her daughters or by an affected father ($\times$ normal mother) to all his daughters and none of his sons. Purely maternal effects are transmitted from mother to all her progeny but do not persist in certain nuclear genotypes. Extranuclear genetic systems would ordinarily operate through the maternal line. If the trait is repeatedly transmitted through backcrosses of F_1 individuals with maternal parent but not with paternal parent, an extra-nuclear genetic system may be involved. It should be identified and located, and such guiding criteria as those listed at the outset of this chapter applied.

20-2. Any genotype will have the phenotype determined by the maternal genotype (a maternal effect).

20-3. (a) Young all light-eyed, adults all dark-eyed.

(b) Both young and adult dark-eyed.

20-4. (a) All green.

(b) All "white."

(c) All three types in irregular ratio.

20-5. (a) 4.1×10^5.

(b) 1,025.

20-6. (a) All normal.

(b) $1:1$ petite : normal.

20-7. (a) In substrain C, between *lys* + *met* and *gal;* in substrain H, between *pil* and *pyr B*.

(b) In C, *muc;* in H, *thr*.

20-8. See text and references.

20-9. See text and references.

20-10. See text and references.

20-11. See references, especially Raven (1970) and Margulis (1970).

APPENDIX B
Selected Life Cycles

1. Bacteria

Bacteria reproduce asexually by cell division, but may also engage in a type of "sexual" reproduction called conjugation. Mating types F^- ("female" or, better, "receptor") and F^+ or Hfr ("male" or, better, "donor") are required. Conjugation of $Hfr \times F^-$ includes the following steps and is diagramed in Figure B-1:

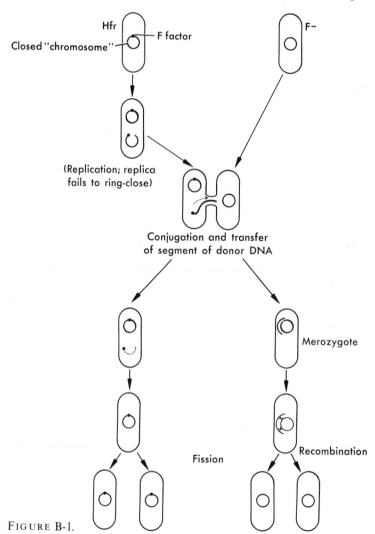

FIGURE B-1.

a. Replication of donor chromosome; replica fails to undergo ring closure.
b. Chance collision of *Hfr* and F^-.
c. Attachment of *Hfr* and F^- cells in pairs by formation of connecting bridge.
d. Transfer of open, replicated, donor "chromosome" to F^- cell; generally only a relatively small segment is so transferred.
e. Partial diploidy of receptor cell (merozygote).
f. Reproduction of both donor and receptor cells by division after separation of the conjugants.
g. Regaining of monoploidy by F^- cells.

In F^+ cells the fertility factor is not integrated into the "chromosome." In $F^+ \times F^-$ conjugation a previously replicated fertility factor is usually the only DNA donated to the F^- cell, which thereby becomes F^+. The original F^+ donor, retaining an F factor, remains F^+. On the other hand, with *Hfr* $\times F^-$ conjugation, the *Hfr* donor remains *Hfr*, but a very few F^- receptors will be converted to donors on rare occasions when all of the *Hfr* open, replicated "chromosome" is transferred.

2. Neurospora

Like most fungi, *Neurospora* produces large numbers of asexual spores (here called conidia), but also reproduces sexually if + and − mating strains come into contact. In *Neurospora*, as in many fungi, pairing of nuclei of sex cells does not result in immediate syngamy. The nuclei that ultimately fuse are daughter nuclei of the original pairing nuclei. The essential steps are diagramed in Figure B-2 and include:

a. Contact between filaments (hyphae) of + and − strains.
b. Pairing (not fusion) of + and − nuclei.
c. Development of the "fruiting body" (ascocarp), called a perithecium, which consists of n^+, n^-, and dikaryon (n^+/n^-) hyphae.
d. Development of large numbers of elongate, saclike sporangia called asci (sing., ascus) in the perithecium.
e. Fusion in the young asci of a + and a − nucleus that were derived through several mitoses from the original pairing nuclei, thus forming a diploid zygote.
f. Meiosis of zygote soon after formation, in the developing ascus, to form four meiospores.
g. Mitosis of the four meiospores to form eight (monoploid) spores called ascospores. Four of these will give rise to + mating strain plants, the other four to − strain plants.
h. Release of ascospores and their germination to form new adults.

3. Saccharomyces (Yeast)

Yeasts are one-celled ascomycete fungi. Multiplication is by "budding," in which the nucleus divides by mitosis. In the full life cycle, however, morphologically identical diploid and monoploid generations alternate as shown in Figure B-3. The life history consists of the following steps:

a. Diploid adults multiply by "budding" but, under certain environmental conditions, undergo meiosis to form four monoploid meiospores within the old cell wall.

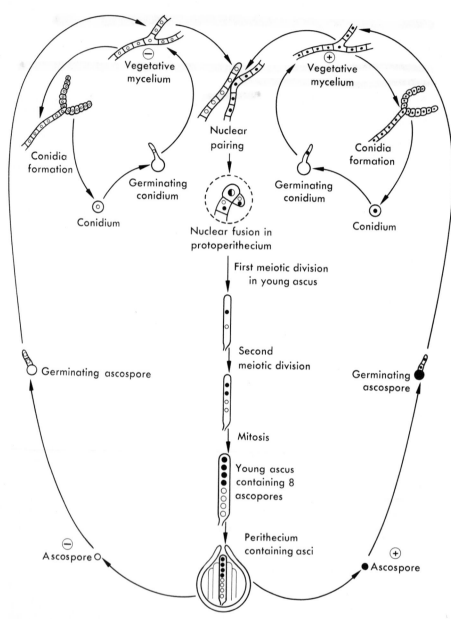

FIGURE B-2.

b. Maturation of the four meiospores to become four ascospores (baker's yeast) or, in some other species, mitosis of each of the four meiospores to form eight ascospores.

c. Liberation of ascospores from ascus (old vegetative cell wall).

d. Germination of ascospores to form monoploid adult cells. Half the ascospores

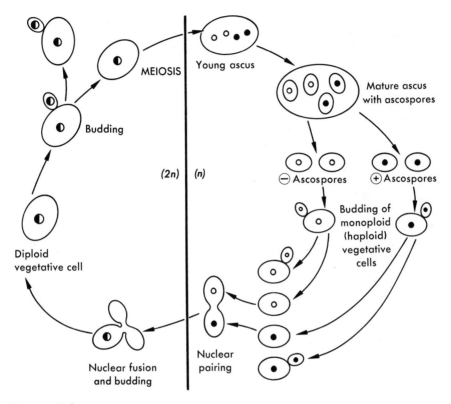

Fɪɢᴜʀᴇ B-3.

from any one ascus give rise to + mating strain adults and half to − mating strain.

e. Multiplication of monoploid + and − adults by "budding."

f. Contact between + and − cells.

g. Formation of intercellular cytoplasmic bridge.

h. Fusion of + and − nuclei (each monoploid cell in the pair furnishes a single gamete).

i. Formation of diploid vegetative cell, often from the cytoplasmic bridge.

4. Chlamydomonas

The small, unicellular, motile, green alga *Chlamydomonas* reproduces freely by cell division (mitosis and cytokinesis). The vegetative cells are monoploid (haploid). In sexual reproduction, the vegetative cell functions as a gametangium, its protoplast dividing mitotically to produce 4, 8, 16, or 32 gametes. These sex cells are morphologically similar to the vegetative cells, but smaller in size. In many species the gametes are identical in appearance, hence may be referred to as isogametes. In other species varying degrees of morphological differentiation of gametes occur. Chemical differences among gametes, and the cells producing them, occur and mating strains are designated as + and −. Gametes of opposite mating strain come into contact at their flagellar ends; the protoplasts fuse to form a four-flagellate zygote. The zygote soon loses its flagella, develops a wall, and becomes dormant. Germination of the zygote begins with meiosis of its diploid nucleus and ends with liberation of biflagellate zoospores from the old zygote wall. In some species only four zoospores are thus formed, but in others meiosis is followed by one or more mitoses so that 8, 16, or more zoospores are produced. Zoospores resemble the vegetative cells into which they will develop, and of the number produced from a single zygote, half are of each mating strain. The process is diagramed in Figure B-4.

5. Sphaerocarpos (Liverwort)

Vegetative plants are small, thin, and lobed; these are monoploid (haploid), unisexual gametophytes (gamete-producing plants). Males produce motile sperms in antheridia; females develop one egg in each of several archegonia. Syngamy occurs when liquid water (from rain or dew) is present, allowing sperms to swim to the archegonia. The resulting zygote develops into a small, multicellular sporophyte (spore-bearing plant), which remains permanently attached to the parent gametophyte. It ultimately protrudes from the remains of the archegonium and produces internally a large number of diploid sporocytes that undergo meiosis to produce four meiospores

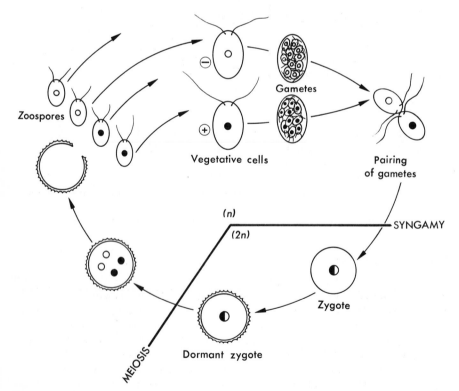

FIGURE B-4.

each. Two of each of these will produce male gametophytes, and two female. The life cycle is diagramed in Figure B-5.

6. Flowering Plants (Class Angiospermae)

The plant we recognize by name in the angiosperms is the diploid sporophyte. The monoploid (haploid) gametophyte is microscopic and contained almost entirely within various floral structures. A (usually) triploid food storage tissue, the endosperm, also occurs and is cytologically unique to the angiosperms. The life cycle of a representative angiosperm is diagramed in Figure B-6 and consists of the following structures and steps:

a. Production of flowers by the sporophyte.
b. Meiosis of microsporocytes (pollen mother cells) in anthers of stamens to form four functional, uninucleate microspores each.
c. Development of young male gametophyte by mitosis of the microspore nucleus within the microspore wall, inside the anther. This two-nucleate structure (tube nucleus and generative nucleus) is sometimes called a pollen grain.
d. Transfer of pollen grains to stigmas of the pistils where each grain produces a tubular outgrowth, the pollen tube, which grows down through structures of the pistil to the ovule, which it enters via the micropyle.
e. Development in each ovule of a megasporocyte, which then undergoes meiosis to form four megaspores, three of which degenerate.
f. Development of the female gametophyte by mitosis from the one functional megaspore. In the classical case the mature female gametophyte consists of eight nuclei (one egg, two polars, two synergids, three antipodals) in a common cytoplasm within each ovule, and contained within the ovary of the pistil.
g. Mitosis of generative nucleus to form two sperms in the pollen tube.
h. Entry of the pollen tube into the embryo sac.

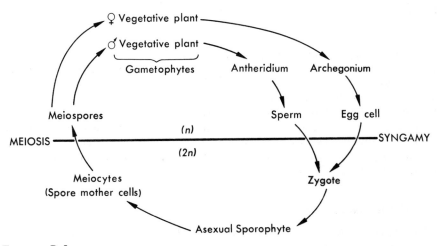

FIGURE B-5.

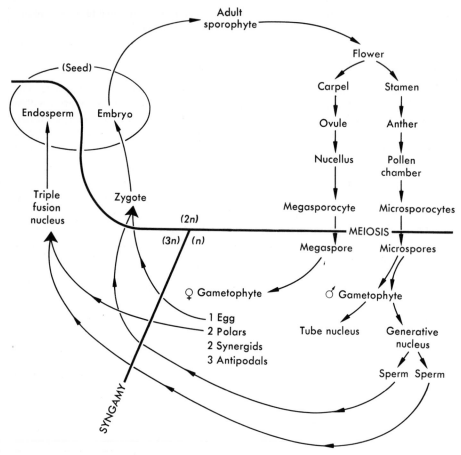

FIGURE B-6.

i. Fusion of egg and sperm to form the zygote.
j. Fusion of the second sperm with the two polars to form the triploid triple fusion nucleus.
k. Degeneration of synergids and antipodals.
l. Development of multicellular embryo from the zygote.
m. Development of endosperm from the triple fusion nucleus. In some plants this tissue is absorbed by the cotyledons of the embryo during the latter's development.
n. Development of seed coat, primarily from the integuments of the ovule.
o. Development of a fruit from the ovary.

7. Paramecium

Paramecia are elongate ciliates of the phylum Protozoa. Each animal contains a large macronucleus, which exerts phenotypic control for that individual, and two micronuclei, which function in the sexual process. The macronucleus is polyploid, the micronucleus diploid in the vegetative animal. Paramecia increase in number only by fission; other processes, important in genetics, also occur and are as follows:

FISSION

a. Mitosis of micronuclei.
b. Constriction of macronucleus to form two.
c. Movement of one macronucleus and one micronucleus to each end of the animal, which then constricts in the middle to form two new individuals.

CONJUGATION (FIG-B-7)

a. Pairing of two animals (conjugants) and formation of intercellular bridge.
b. Disintegration of macronucleus.
c. Meiosis of each of the two micronuclei.
d. Disintegration of seven of the eight products of meiosis.
e. Mitosis of the remaining monoploid nucleus to form two.
f. One of the two monoploid nuclei of each conjugant passes through the connecting bridge to the other animal (reciprocal transfer).
g. Fusion of the two monoploid nuclei in each conjugant, restoring the diploid condition.
h. Two mitoses of the fertilization nuclei, resulting in four diploid nuclei per conjugant.
i. Separation of the conjugants, which are now genetically alike.
j. Two of the four nuclei in each ex-conjugant become macronuclei, two become micronuclei.
k. Distribution of two macronuclei to each daughter cell at next fission.
l. Mitosis of the two micronuclei at next fission, two being distributed to each daughter cell.
m. Conjugation is usually a short-term event with little cytoplasmic transfer. Under certain conditions the intercellular connection may persist for a longer period, allowing exchange of considerable cytoplasm.

AUTOGAMY (FIG-B-8)

This is a type of internal self-fertilization, resembling somewhat the events of conjugation but involving only a single animal. It results in homozygosity of the individual undergoing the process.

a. Macronucleus behaves as in conjugation.
b. Meiosis of the two micronuclei and disintegration of seven of the eight resulting nuclei as in conjugation.
c. Mitosis of the remaining monoploid nucleus to form two.
d. Fusion of the two monoploid nuclei resulting from step (c).
e. Restoration of micronuclei and the macronucleus as for conjugation.

8. Mammals

In mammals, the somatic cells (except for some, such as liver cells, which may be polyploid) are diploid, meiosis immediately preceding the formation of gametes. The

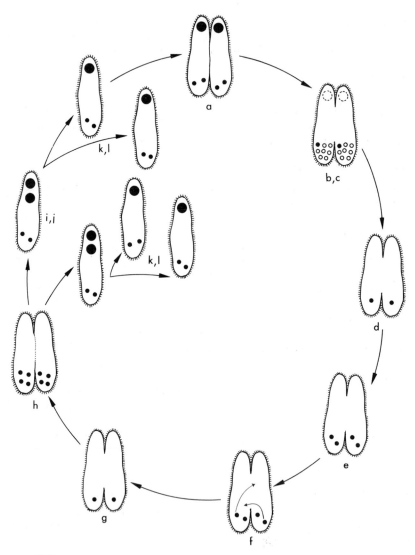

FIGURE B-7.

diploid condition is restored at syngamy (Fig. B-9). The following steps are involved in the male:

a. Development of diploid primary spermatocytes.
b. Meiosis. The two monoploid products of the first meiotic division are called secondary spermatocytes. The four cells resulting from the second meiotic division are called spermatids.
c. Maturation of spermatids into sperms.

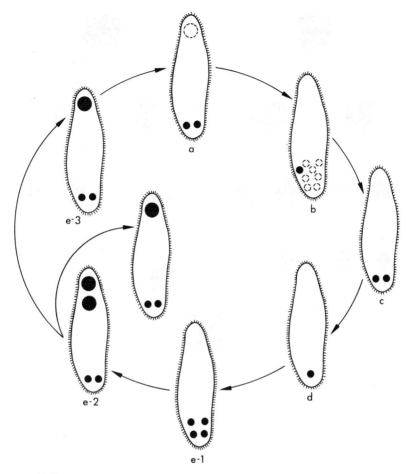

FIGURE B-8.

In the female:

a. Development of diploid primary oocytes.
b. Meiosis. The first division produces two unequal cells, a smaller first polar body and a larger secondary oocyte. The second division produces two second polar bodies from the first polar body and, from the secondary oocyte, a third second polar body and a larger ootid.
c. Maturation of one egg from the ootid; degeneration of the three second polar bodies.

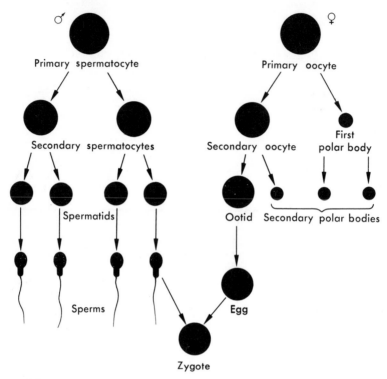

Figure B-9.

The Biologically Important Amino Acids

Alanine (ala)

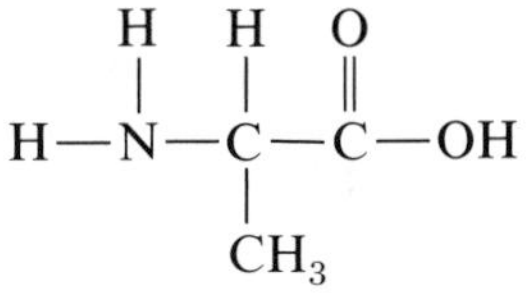

$$H-N-C-C-OH$$

with H, H, O above N, C, C respectively and CH_3 below the central C.

Arginine (arg)

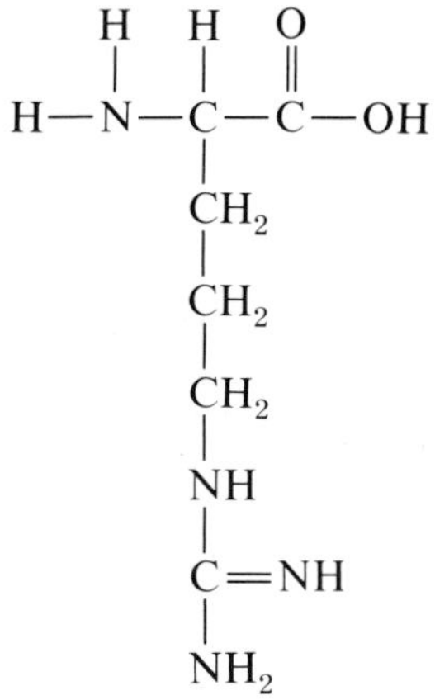

$$H-N-C-C-OH$$

with H, H, O above; side chain: CH_2 — CH_2 — CH_2 — NH — $C=NH$ — NH_2

Asparagine (asn)

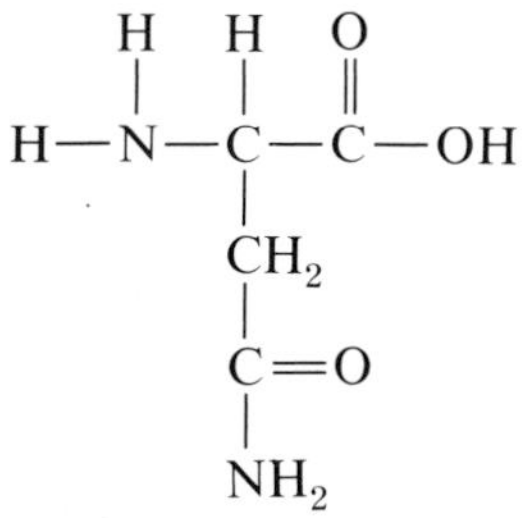

$$H-N-C-C-OH$$

with H, H, O above; side chain: CH_2 — $C=O$ — NH_2

Aspartic Acid (asp)

$$H-N-C-C-OH$$

with H, H, O above; side chain: CH_2 — $C=O$ — OH

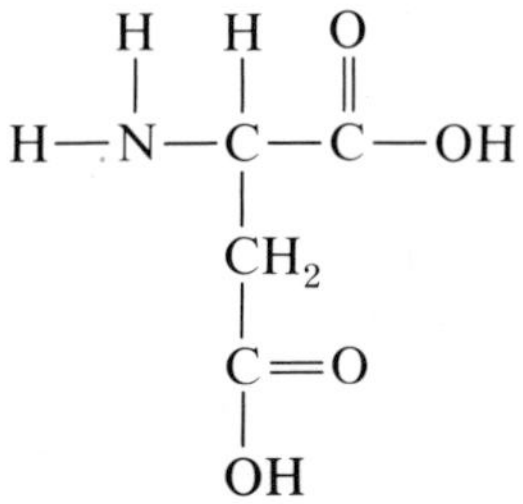

[1] Amino acids marked * are required in the diet of mammals.

Cysteine (cys)

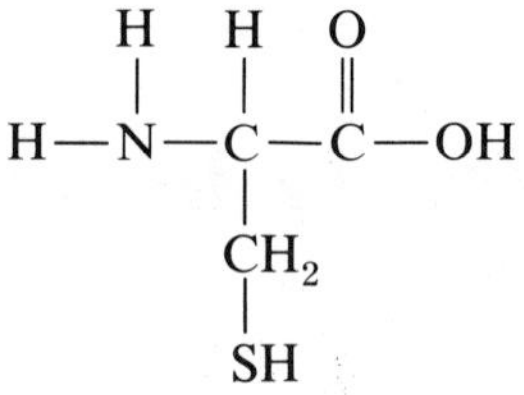

$$\begin{array}{c} H \quad\; H \quad\; O \\ | \qquad | \qquad \parallel \\ H-N-C-C-OH \\ | \\ CH_2 \\ | \\ SH \end{array}$$

Glutamic Acid (glu)

$$\begin{array}{c} H \quad\; H \quad\; O \\ | \qquad | \qquad \parallel \\ H-N-C-C-OH \\ | \\ CH_2 \\ | \\ CH_2 \\ | \\ C=O \\ | \\ OH \end{array}$$

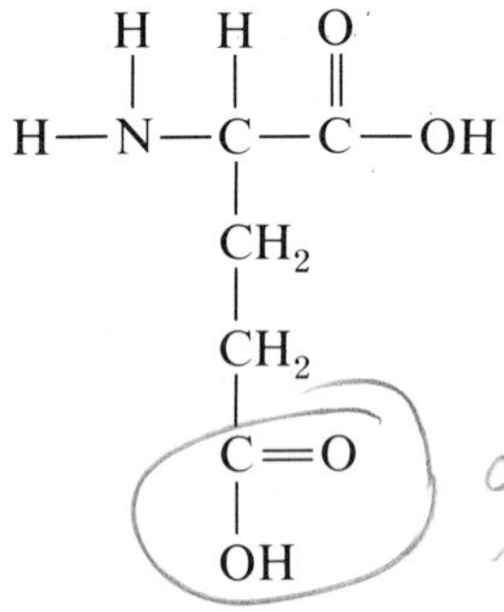

carboxyl at end of side chain

Glutamine (gln)

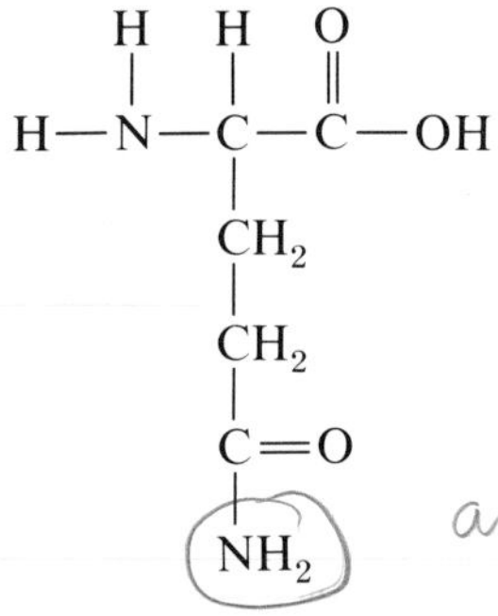

$$\begin{array}{c} H \quad\; H \quad\; O \\ | \qquad | \qquad \parallel \\ H-N-C-C-OH \\ | \\ CH_2 \\ | \\ CH_2 \\ | \\ C=O \\ | \\ NH_2 \end{array}$$

amino at end of side chain

Glycine (gly)

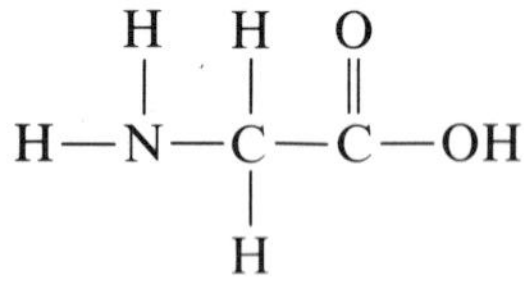

$$\begin{array}{c} H \quad\; H \quad\; O \\ | \qquad | \qquad \parallel \\ H-N-C-C-OH \\ | \\ H \end{array}$$

Histidine (his)

$$\begin{array}{c} H \quad\; H \quad\; O \\ | \qquad | \qquad \parallel \\ H-N-C-C-OH \\ | \\ CH_2 \\ | \\ C=\!=\!=CH \\ | \qquad\quad | \\ H-N \quad\; N \\ \diagdown C \diagup \\ | \\ H \end{array}$$

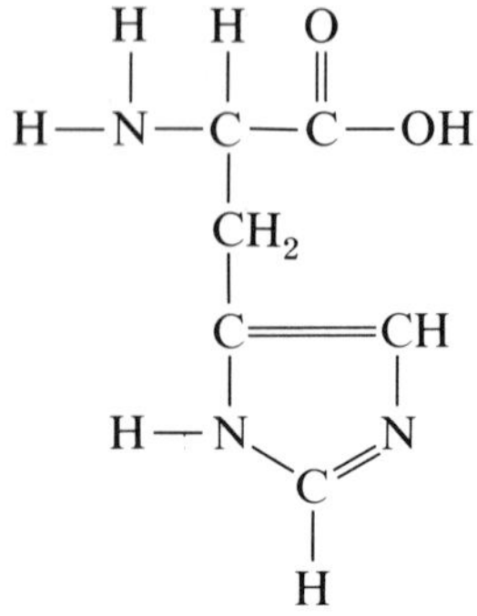

*Isoleucine (ile)

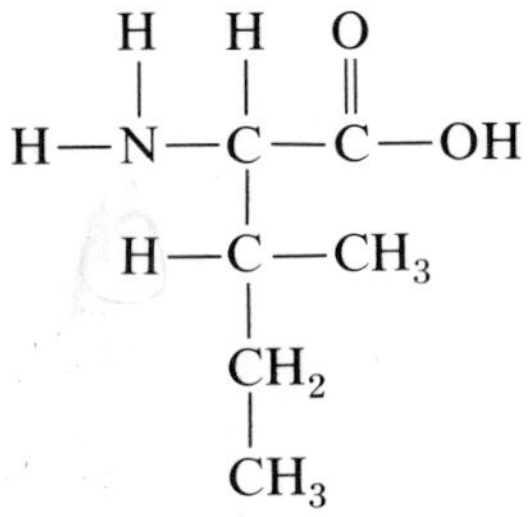

$$
\begin{array}{c}
\text{H} \quad\; \text{H} \quad\;\; \text{O} \\
| \qquad | \qquad\; || \\
\text{H}-\text{N}-\text{C}-\text{C}-\text{OH} \\
| \\
\text{H}-\text{C}-\text{CH}_3 \\
| \\
\text{CH}_2 \\
| \\
\text{CH}_3
\end{array}
$$

*Leucine (leu)

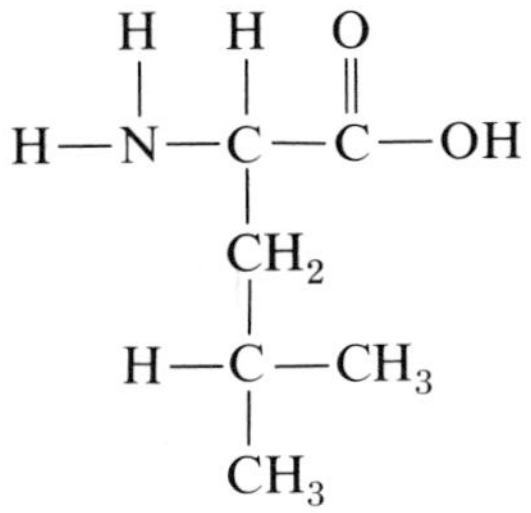

$$
\begin{array}{c}
\text{H} \quad\; \text{H} \quad\;\; \text{O} \\
| \qquad | \qquad\; || \\
\text{H}-\text{N}-\text{C}-\text{C}-\text{OH} \\
| \\
\text{CH}_2 \\
| \\
\text{H}-\text{C}-\text{CH}_3 \\
| \\
\text{CH}_3
\end{array}
$$

*Lysine (lys)

$$
\begin{array}{c}
\text{H} \quad\; \text{H} \quad\;\; \text{O} \\
| \qquad | \qquad\; || \\
\text{H}-\text{N}-\text{C}-\text{C}-\text{OH} \\
| \\
\text{CH}_2 \\
| \\
\text{CH}_2 \\
| \\
\text{CH}_2 \\
| \\
\text{CH}_2 \\
| \\
\text{NH}_2
\end{array}
$$

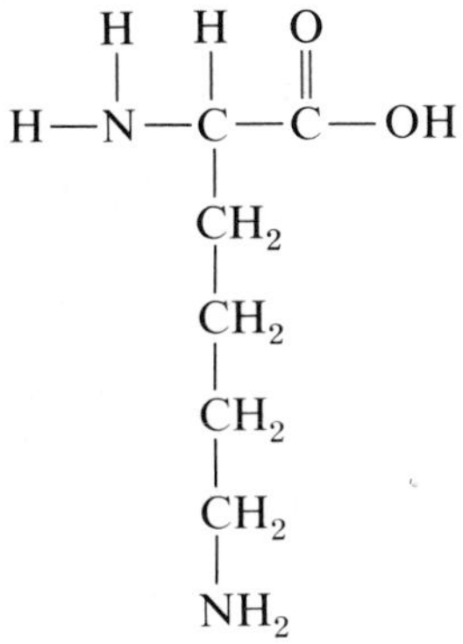

*Methionine (met)

$$
\begin{array}{c}
\text{H} \quad\; \text{H} \quad\;\; \text{O} \\
| \qquad | \qquad\; || \\
\text{H}-\text{N}-\text{C}-\text{C}-\text{OH} \\
| \\
\text{CH}_2 \\
| \\
\text{CH}_2 \\
| \\
\text{S} \\
| \\
\text{CH}_3
\end{array}
$$

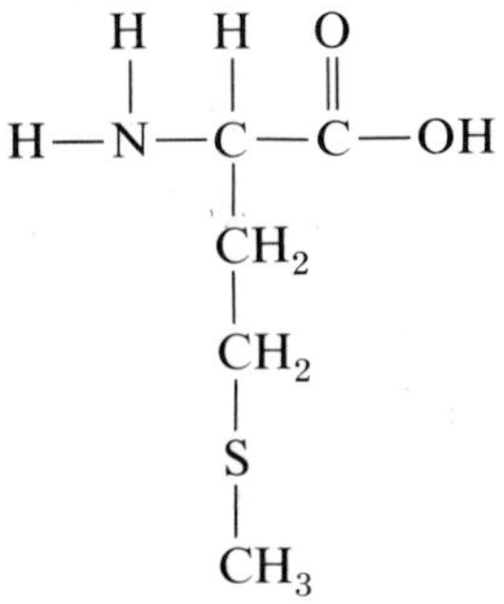

***Phenylalanine (phe)**

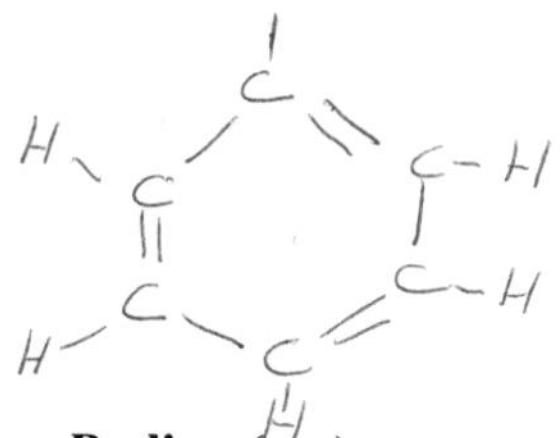

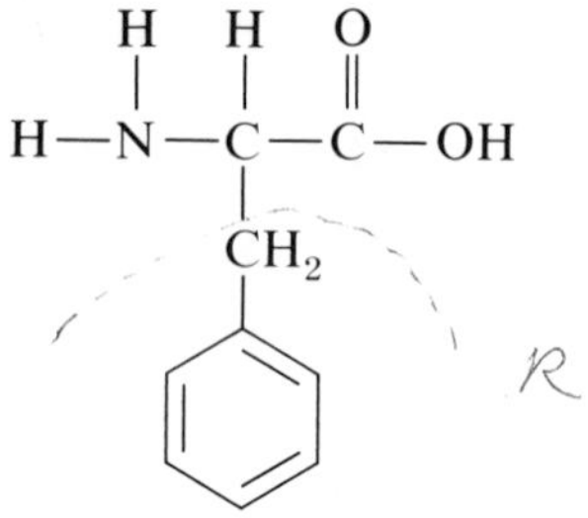

Proline (pro)

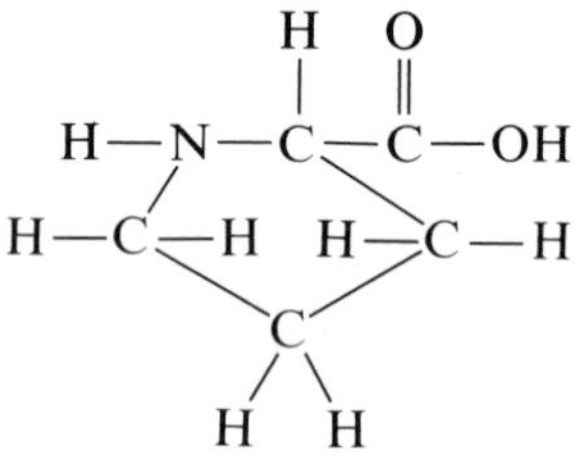

Serine (ser)

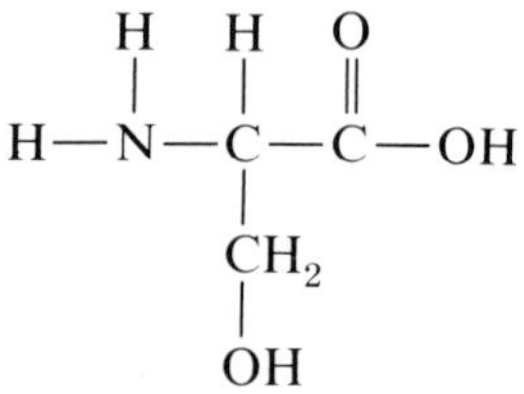

***Threonine (thr)**

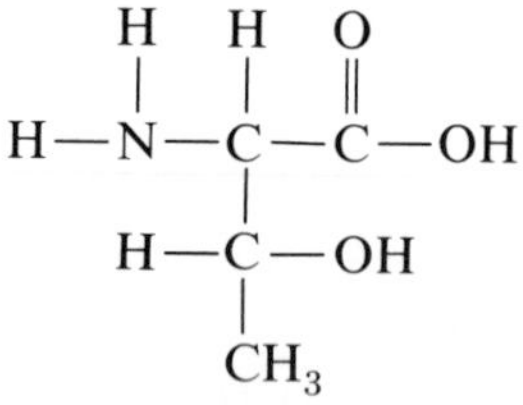

***Tryptophan (try)**

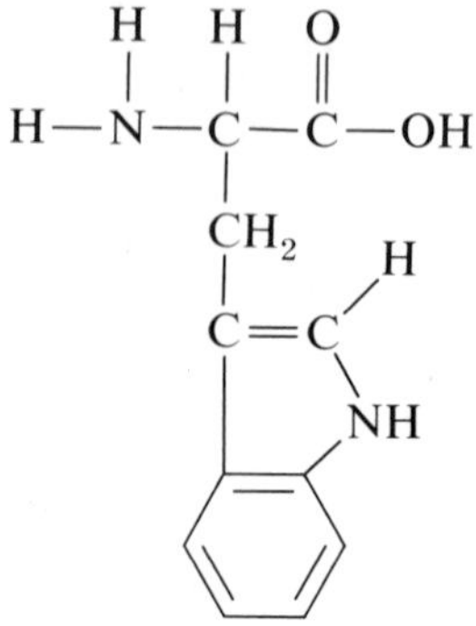

Tyrosine (tyr)

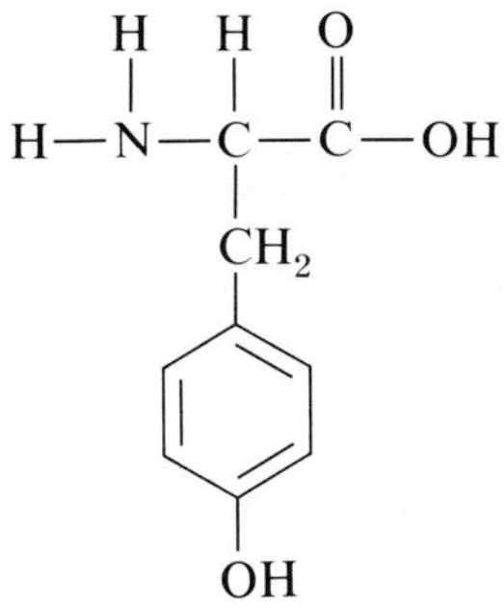

$$
\begin{array}{ccccc}
& H & H & O & \\
& | & | & \parallel & \\
H-N & - & C & - & C-OH \\
& & | & & \\
& & CH_2 & & \\
\end{array}
$$

$$
\begin{array}{c}
\text{(benzene ring)} \\
OH
\end{array}
$$

*Valine (val)

$$
\begin{array}{ccccc}
& H & H & O & \\
& | & | & \parallel & \\
H-N & - & C & - & C-OH \\
& & | & & \\
& H-C & - & CH_3 & \\
& & | & & \\
& & CH_3 & & \\
\end{array}
$$

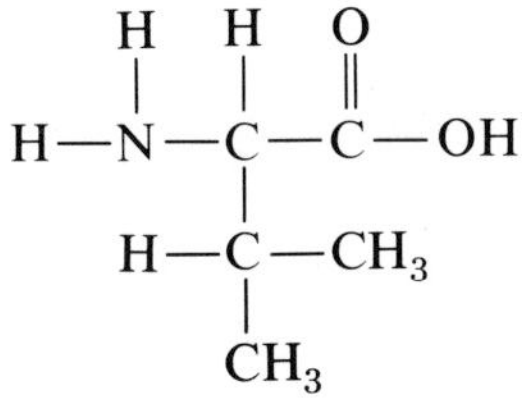

APPENDIX D
Useful Formulas, Ratios, and Statistics

$3:1$	Monohybrid phenotypic ratio produced by $Aa \times Aa$.
$1:2:1$	Monohybrid genotypic ratio produced by $Aa \times Aa$; monohybrid phenotypic and genotypic ratio produced by $a_1a_2 \times a_1a_2$.
$1:1$	Monohybrid testcross phenotypic and genotypic ratio produced by $Aa \times aa$.
$2:1$	Monohybrid lethal genotypic and phenotypic ratio produced by $a_1a_2 \times a_1a_2$ where either a_1a_1 or a_2a_2 is lethal; also sex ratio produced by $Aa \times AY$ where a is a sex-linked recessive lethal.
"$1:0$"	Monohybrid lethal phenotypic ratio produced by $Aa \times Aa$ where aa or $A-$ is lethal; also monohybrid testcross phenotypic and genotypic ratio produced by $AA \times aa$.
$9:3:3:1$	Dihybrid phenotypic ratio produced by $AaBb \times AaBb$ where phenotypes may be represented as 9 $A-B-$, 3 $A-bb$, 3 $aaB-$, 1 $aabb$; note epistatic possibilities producing "condensations" of this ratio (e.g., $9:7$, $9:6:1$).
$1:1:1:1$	Dihybrid testcross genotypic and phenotypic ratio produced by $AaBb \times aabb$ where there is no linkage.
$3:6:3:1:2:1$	Dihybrid phenotypic ratio produced by $Aab_1b_2 \times Aab_1b_2$. Note that dihybrid and polyhybrid ratios are products of their component monohybrid ratios and may be combined in any way.
2^n	Number of gamete genotypes and progeny phenotypes where $n =$ the number of pairs of heterozygous genes with complete dominance.
3^n	Number of zygote genotypes under the preceding conditions.
4^n	Number of possible zygote combinations under the preceding conditions, and yielding 3^n zygote genotypes.
$\frac{n}{2}(n + 1)$	The chance of selecting at random any two items in pairs (e.g., the number of possible genotypes for n multiple alleles in a series).

528

$(a + b)^n$

Expansion of the binomial provides probability determinations where n = the number of independent events and the choices are two.

$(p + q)^2 = 1$

Binomial whose expansion permits calculation of the frequency of each member of a pair of alleles.

$(p + q + r)^2 = 1$

Trinomial whose expansion permits calculation of the frequency of each of three multiple alleles.

$(\frac{1}{4})^n$

In polygene cases, the fraction of the F_2 like either P is given by this expression where n = the number of pairs of genes in which the parents (P) differ.

$(\frac{1}{2})^n$

In polygene cases, the fraction of the F_2 like either P is given by this expression where n = the number of effective or contributing alleles.

$\chi^2 = \Sigma \left[\dfrac{(o - c)^2}{c} \right]$

Consult tables of chi-square for levels of significance (where degrees of freedom equal 1 less than the number of classes). In general, a value of chi-square $\geq$ than that for $P = 0.05$ is regarded as significant; i.e., there is significant evidence against the hypothesis. A value of chi-square showing a level of $P = 0.05$, for example, does *not* mean that a deviation as large or larger will *not* occur by chance alone under the hypothesis adopted, but it is likely to in only five trials out of 100. This is considered too few; at this level, the chance of rejecting a right hypothesis is only 1 in 20.

$\bar{x} = \dfrac{\Sigma fx}{n}$ or $\dfrac{\Sigma x}{n}$

The sample mean is self-explanatory.

$s^2 = \dfrac{\Sigma f(x - \bar{x})^2}{n - 1}$

The variance (s^2) provides an unbiased estimate of the population variance (σ^2).

$s = \sqrt{\dfrac{\Sigma f(x - \bar{x})^2}{n - 1}}$

The standard deviation measures the variability of the sample; in a normal distribution, 68.26 per cent of the sample will lie in the range $\bar{x} \pm s$, and 95.44 per cent will fall in the range $\bar{x} \pm 2s$. Used with normal distributions.

$s_{\bar{x}} = \dfrac{s}{\sqrt{n}}$

The standard error of the sample mean indicates the degree of correspondence between $\bar{x}$ and μ; there is 68.26 per cent confidence that $\mu = \bar{x} \pm s_{\bar{x}}$ by chance alone, and 95.44 per cent confidence that $\mu = \bar{x} \pm 2s_{\bar{x}}$ by chance alone.

$S_d = \sqrt{(s_{\bar{x}_1})^2 + (s_{\bar{x}_2})^2}$

The standard error of the difference in means is useful in comparing two samples to determine whether the difference in their means is significant. If $(\bar{x}_1 - \bar{x}_2) > 2S_d$, the difference in sample means is considered significant and the two samples to represent two different populations.

$$s = \sqrt{\frac{pq}{n}}$$

The standard deviation (s) of a simple proportionality such as heads (p) versus tails (q) for n trials.

$$s = \sqrt{\frac{pq}{2N}}$$

The standard deviation of gene frequencies where N represents the number of *diploid* individuals, and p and q represent the frequency of each of a pair of alleles.

$$n = \frac{R^2}{8(s^2_{F_2} - s^2_{F_1})}$$

The number of pairs of polygenes (n) is calculated from this equation where R is the maximum quantitative range between phenotypes, $s^2_{F_2}$ is the variance of the F_2, and $s^2_{F_1}$ is the variance of the F_1.

$$q_n = \frac{q_0}{1 - nq_0}$$

The frequency of a recessive lethal (q_n) after n additional generations equals the initial frequency of the gene (q_0) divided by 1 plus the product of the number of additional generations (n) times the initial frequency of the gene.

$$n = \frac{1}{q_n} - \frac{1}{q_0}$$

The number of additional generations (n) required to reduce the frequency of a gene from its initial value (q_0) to any particular value (q_n) is given by this equation.

$$W = 1 - s,$$
$$\text{and } s = 1 - W$$

The adaptive value of a genotype (W) is 1 minus its selection coefficient (s), and the selection coefficient is 1 minus the adaptive value of the genotype.

$$q_1 = \frac{q_0 - s(q_0)^2}{1 - s(q_0)^2}$$

The equation for determining the frequency of a gene after one generation under selection (q_1) when its selection coefficient (s) and its initial frequency (q_0) are known.

$$\Delta q = \frac{-sq_0^2 p}{1 - sq_0^2}$$

Change in frequency of a recessive gene under selection in one generation.

$$L_{\mu m} = 3.4 \times 10^{-4} P$$

Length in micrometers ($L_{\mu m}$) of double-stranded DNA equals 0.00034 times the number of deoxyribonucleotide pairs (P).

$$P = \frac{M}{650}$$

The number of deoxyribonucleotide pairs (P) equals the molecular weight (M) of a DNA molecule divided by 650.

$$M = 650P$$

The molecular weight of a DNA molecule (M) equals 650 times the number of deoxyribonucleotide pairs (P) of which it consists.

$$P = \frac{L_{\mu m}}{3.4 \times 10^{-4}}$$

The number of deoxyribonucleotide pairs (P) in a DNA molecule equals its length in micrometers ($L_{\mu m}$) divided by 0.00034.

APPENDIX E

Useful Metric Values

Name	Numerical Value (m)	Power of 10	Symbol	Synonym
Meter	1.0		m	
Decimeter	0.1	10^{-1}	dm	
Centimeter	0.01	10^{-2}	cm	
Millimeter	0.001	10^{-3}	mm	
Micrometer	0.000 001	10^{-6}	μm	micron (μ)
Nanometer	0.000 000 001	10^{-9}	nm	millimicron (mμ)
	0.000 000 000 1	10^{-10}		Angstrom unit (Å)

APPENDIX F
Journals and Reviews

Selected Journals[1]

Advances in Genetics
American Journal of Human Genetics
Annals of Human Genetics
Annual Review of Biochemistry
Annual Review of Genetics
Biochemical Genetics
Chromosoma
Biochemical Genetics
Cytogenetics
Genetica
Genetical Research
Genetics
Genetics Abstracts
Hereditas
Heredity
Journal of Bacteriology
Journal of Genetics
Journal of Heredity
Journal of Medical Genetics
Journal of Molecular Biology
Journal of Virology
Lancet
Molecular and General Genetics
Nature
New England Journal of Medicine
Proceedings of the National Academy of Science (U.S.)
Science
Scientific American

Symposia and Reviews

Advances in Genetics (M. Demerec, ed.), Academic Press.
Brookhaven Symposia on Biology
    No. 8. Mutation (1955)
    No. 9. Genetics in Plant Breeding (1965)
    No. 12. Structure and Function of Genetic Elements (1959)
Cold Spring Harbor Symposia on Quantitative Biology
    Vol. 9. Genes and Chromosomes (1941)

[1]A list of more than 3,400 journals is carried in one number each year of *Genetics Abstracts*.

Vol. 11. Heredity and Variation in Microorganisms (1946)
Vol. 16. Genes and Mutations (1951)
Vol. 20. Population Genetics (1955)
Vol. 21. Genetic Mechanisms (1956)
Vol. 23. Exchange of Genetic Material (1958)
Vol. 26. Cellular Regulatory Mechanisms (1961)
Vol. 28. Synthesis and Structure of Macromolecules (1963)
Vol. 29. Human Genetics (1964)
Vol. 31. The Genetic Code (1966)
Vol. 33. Replication of DNA in Micro-Organisms (1968)
Vol. 34. The Mechanism of Protein Synthesis (1969)
Vol. 35. Transcription of Genetic Material (1970)
Vol. 36. Structure and Function of Proteins at
        the Three-Dimensional Level (1971)
Vol. 38. Chromosome Structure and Function (1974)
The National Foundation March of Dimes[2]
  Syndrome Identification (a continuing series, 1973–)
  Birth Defects—Original Article Series
    Vol. 4, no. 4. Guide to Human Chromosome Defects (1968)
    Vol. 4, no. 6. Human Genetics (1968)
    Vol. 5, no. 5. The First Conference on the Clinical Delineation of
        Birth Defects. Part V, Phenotypic Aspects of Chromosomal
        Aberrations (1969)
    Vol. 6, no. 1. Genetic Counseling (1970)
    Vol. 8, no. 4. Advances in Human Genetics and Their Impact
        on Society (1972)
    Vol. 8, no. 7. Paris Conference (1971): Standardization in Human
        Cytogenetics (1973)
    Vol. 9, no. 1. Long-Term Lymphocyte Cultures in Human Genetics (1973)
    Vol. 9, no. 4. Contemporary Genetic Counseling (1973)

Birth Defects

BERGSMA, D., ed., 1973. *Birth Defects: Atlas and Compendium.* Baltimore, Williams and Wilkins.

BERGSMA, D., H. T. LYNCH, and R. J. THOMAS, 1974, 4th ed. *International Directory of Genetic Services.* White Plains, N.Y., The National Foundation—March of Dimes.

---

[2] An extensive reprint series is also available.

GLOSSARY

Abortus. An aborted fetus.

Acentric. A chromatid or chromosome that lacks a centromere.

Acquired character. An alteration in function or form resulting from a response to environment; it is not heritable.

Acridine dye. Organic molecules that produce mutations by binding to DNA and causing insertions or deletions of bases.

Acrocentric. A chromosome with a nearly terminal centromere.

Adaptation. Adjustment in some way to the environment.

Adaptive value. The proportion of the progeny of a given genotype surviving to maturity (relative to that of another genotype) is its adaptive value (also called *fitness*). Expressed as a pure number between 0 and 1. Thus if only 60 per cent of the progeny of genotype *aa* survive, whereas 100 per cent of those of genotype *A*− survive, the adaptive value of *aa* is 0.6.

Adenine. A purine base occurring in DNA and RNA. Pairs normally with thymine in DNA.

Agglutinin. An antibody that produces clumping (agglutination) of the antigenic structure.

Albino. An individual characterized by absence of pigment. Ordinarily applied to animals whose skin, hair, and eyes are unpigmented and whose skin and eyes are pinkish because of the blood vessels showing through. Also used to describe plants in which a particular pigment (usually the chlorophylls) is absent.

Aleurone layer. Outermost layer of endosperm in some seeds, rich in aleurone (proteinaceous) grains; typically triploid cells.

Allele (also **allelomorph;** adj., **allelic** or **allelomorphic**). One member of a pair or a series of genes that can occur at a particular locus on homologous chromosomes.

Allopolyploid. A polyploid having whole chromosome sets from different species.

Allosteric. Applied to any enzyme (or protein) having two or more nonoverlapping receptor (or attachment) sites.

Alternation of generations. The regularly occurring alternation of monoploid (gametophyte) and diploid (sporophyte) phases in the life cycle of sexually reproducing plants.

Ambiguity. The coding for more than one amino acid by a given codon. See also *genetic code* and *codon*.

Amino acid. A class of chemicals containing an amino (NH_2) group and a carboxyl (COOH) group, plus a side chain; the basic constructional unit of proteins.

Amino group. The $-NH_2$ chemical group.

Amphidiploid. A tetraploid individual having two sets of chromosomes from each of two known ancestral species. An allotetraploid in which the species source of the two different genomes is clearly known. See also *allopolyploid*.

Anaphase. The stage of nuclear division characterized by movement of chromosomes from spindle equator to spindle poles. It begins with separation of the centromeres and closes with the end of poleward movement of chromosomes.

Aneuploidy. Variation in chromosome number by whole chromosomes, but less than an entire set; e.g., $2n + 1$ (trisomy), $2n - 1$ (monosomy).

Angstrom unit (Å). A measurement of

length or distance, often used in describing intra- or intermolecular dimensions; equal to 1×10^{-10} meter and 1×10^{-1} nanometer (which see).

Anisogamy. Sexual reproduction involving gametes similar in morphology but differing in size.

Anther. The microsporangium of flowering plants; distal part of stamen in which microspores and, later, pollen grains are produced.

Antibiotic. Any chemical substance, elaborated by a living organism (usually a microorganism), that kills or inhibits growth of bacteria.

Antibody. A substance that acts to neutralize a specific antigen in a living organism.

Anticodon. The group of three nucleotides in transfer RNA that pairs complementarily with three nucleotides of messenger RNA during protein biosynthesis.

Antigen. Any substance, usually a protein, that causes antibody production when introduced into a living organism.

Ascospore. One of the asexual, monoploid (haploid) spores contained in the ascus of ascomycete fungi such as *Neurospora* and the yeasts.

Ascus (pl., **asci**). The generally elongate, saclike meiosporangium in which ascospores are produced in ascomycete fungi.

Asexual reproduction. Any process of reproduction that does not involve fusion of cells. Vegetative reproduction, by portions of the vegetative body, is sometimes distinguished as a separate type.

ATP. Adenosine triphosphate, an energy-rich compound participating in energy-storing and energy-using reactions in the cell.

Attached-X. A strain of *Drosophila* in which the two X chromosomes of the female are permanently attached (symbolized $\widehat{XX}$) so that only $\widehat{XX}$ and O eggs are produced.

Autoimmunity. The production of antibodies against one's own tissues.

Autopolyploid. A polyploid all of whose sets of chromosomes are those of the same species.

Autosome. A chromosome not associated with the sex of the individual and therefore possessed in matching pairs by diploid members of both sexes.

Auxotroph. An individual unable to carry on some particular synthesis, hence requiring supplementing of minimal medium by some growth factor.

Backcross. The cross of a progeny individual with one of its parents. See also *testcross*.

Bacteriophage. See *phage*.

Balbiani ring. An RNA-producing puff in the polytene chromosomes (which see) of the dipteran *Chironomus*.

Barr body. The inactive, densely staining, condensed X chromosome, generally found next to the nuclear membrane, in nuclei of somatic cells of XX females. The number of Barr bodies in such nuclei is one less than the total number of X chromosomes.

Base analog. A slightly modified purine or pyrimidine molecule that may substitute for the normal base in nucleic acid molecules.

Bivalent. A pair of synapsed homologous chromosomes.

Carboxyl group. An acidic chemical group, $-COOH$.

Carotenoid. Any of a group of yellow, orange, or red pigments, in plants, associated with chlorophylls, and in animal fat.

Carrier. A heterozygous individual, ordinarily used in cases of complete dominance.

Cell culture. A growth of cells in vitro.

Cell-free extract. A fluid extract of soluble materials of cells obtained by rupturing the cells and discarding particulate materials and any intact cells.

Centimeter (cm). 1×10^{-2} meter.

Centriole. The central granule in the *centrosome* (which see).

Centromere. A specialized, complex region of the chromosome, consisting of one kinetochore (which see) for each sister chromatid.

Centrosome. A self-propagating cytoplasmic body present in animal cells and some lower plant cells, consisting of a centriole and astral rays at each pole of the spindle during nuclear division.

Chiasma, (pl., chiasmata). The visible connection or crossover between two chromatids seen during prophase-I of meiosis.

Chi-square test. A statistical test for determining the probability that a set of experimentally obtained values will be equaled or exceeded by chance alone for a given theoretical expectation.

Chloroplast. Cytoplasmic organelle containing several pigments, particularly the light-absorbing chlorophylls, and also DNA and polysomes (which see).

Chromatid. One of the two identical longitudinal halves of a chromosome, which shares a common centromere with a sister chromatid; results from the replication of chromosomes during a nuclear division.

Chromatin. Nuclear material comprising the chromosomes; the DNA–histone complex.

Chromomere. Small, stainable thickenings arranged linearly along a chromosome.

Chromosome. Nucleoprotein structures, generally more or less rodlike during nuclear division, the physical sites of nuclear genes arranged in linear order. Each species has a characteristic number of chromosomes, although individuals with fewer or more than this characteristic number occur, especially in plants.

***Cis* arrangement.** Linkage of the dominants of two or more pairs of alleles on one chromosome and the recessives on the homologous chromosome.

Cistron. A segment of DNA specifying one polypeptide chain in protein synthesis. Under the concept of a triplet code, one cistron must contain three times as many nucleotide pairs as amino acids in the chain it specifies.

Clone. A group of cells or organisms, derived from a single ancestral cell or individual and all genetically alike.

Code. See *genetic code.*

Codominance. The condition in heterozygotes where both members of an allelic pair contribute to phenotype, which is then a *mixture* of the phenotypic traits produced in either homozygous condition. In cattle the cross of red $\times$ white produces roan offspring whose coat consists of both red hairs and white hairs. Codominance differs from incomplete dominance.

Codon. A set of nucleotides that is specific for a particular amino acid in protein synthesis; generally agreed to consist of three nucleotides, the last of which, in the case of some amino acids, may be any of the four nucleotides.

Coincidence. The observed frequency of double crossovers, divided by their calculated or expected frequency. Expressed as a pure number; a measure of interference. In *positive* interference the coincidence is <1; in *negative* interference the coincidence is >1.

Col factors. Cytoplasmic particles in bacteria conferring ability to produce colicins (antibiotic lipocarbohydrate-proteins).

Colinearity. Said of a genetic code (which see) in which the sequence of nucleotides corresponds to the sequence of amino acid residues in a polypeptide.

Commaless. Said of a genetic code (which see) in which successive codons (which see) are contiguous and not separated by noncoding bases or groups of bases.

Complementation. The ability of linearly adjacent segments of DNA to supplement each other in phenotypic effect. Complementary genes, when present together, interact to produce a different expression of a trait.

Conjugation. Side-by-side association of two bodies, as of synapsed chromosomes in meiosis or of two organisms during sexual reproduction.

Constitutive enzyme. A continuously produced enzyme.

Crossing-over. A process whereby genes are exchanged between non-sister chromatids of homologous chromosomes. *Unequal crossing-over* may occur, with the result that one chromatid receives a given gene twice, whereas the other chromatid lacks that gene entirely. Chiasmata (which see) are visible evidences of crossing-over.

Crossover. Said of a chromatid or gamete resulting from crossing-over.

Crossover unit. A crossover value of 1 per cent between linked genes. See also *map unit.*

Cross-reacting material (CRM). A defective protein produced by a mutant gene, enzymatically inactive but antigenically similar to the wild-type protein.

C-terminus. That end of a peptide chain carrying the free alpha carboxyl group of the last amino acid in the sequence; written at the right end of the structural formula.

Cytogenetics. Study of the cellular structures and mechanisms associated with genetics.

Cytokinesis. The division of the cytoplasm during cell division.

Cytology. The study of the structure and function of cells.

Cytosine. A pyrimidine base occurring in DNA and RNA. Pairs with guanine in DNA.

Deficiency. The loss of a part of a chromosome involving one or more genes.

Deficiency loop. The loop of the non-synapsing portion of an unaltered chromosome caused by a deficiency in the homologous chromosome.

Degeneracy. Said of a genetic code (which see) in which a particular amino acid is coded for by more than one codon (which see).

Deletion. See *deficiency.*

Deoxyribonucleic acid (DNA). A usually double-stranded, helically coiled, nucleic acid molecule, composed of deoxyribose-phosphate "backbones" connected by paired bases attached to the deoxyribose sugar; the genetic material of all living organisms and many viruses.

Deoxyribonucleoside. Portion of a DNA molecule composed of one deoxyribose molecule plus either a purine or a pyrimidine.

Deoxyribonucleotide. Portion of a DNA molecule composed of one deoxyribose phosphate bonded to either a purine or a pyrimidine.

Deoxyribose. The 5-carbon sugar of DNA.

Deviation. A departure from the expected or from the norm.

Dicentric. Said of a chromosome or chromatid having two centromeres.

Dihybrid. An individual heterozygous for two pairs of alleles; also said of a cross between individuals differing in two gene pairs.

Dioecious. Individuals producing either sperm or egg, but not both. In dioecious species, the sexes are separate. Compare with *monoecious*.

Diploid. An individual or cell having two complete sets of chromosomes.

Disjunction. The separation of homologous chromosomes during anaphase-I of meiosis.

DNAase. Any enzyme that hydrolyzes DNA.

DNA polymerase. An enzyme catalyzing the formation of DNA from deoxyribonucleotides, using one strand of DNA as a template.

Dominance. That situation in which one member of a pair of allelic genes expresses itself in whole (complete dominance) or in part (incomplete dominance) over the other member.

Dominant. Pertaining to that member of a pair of alleles that expresses itself in heterozygotes to the complete exclusion of the other member of the pair. Also, the trait produced by a dominant gene.

Drift. See *genetic drift*.

Duplication. A chromosomal aberration in which a segment of the chromosome bearing specific loci is repeated.

Dysgenic. Any effect or situation that is or tends to be harmful to the genetics of future generations.

Effector (molecule). A substance that combines with regulatory proteins, either activating or inactivating them with respect to their ability to bind to an operator site.

Embryo sac. The female gametophyte of a flowering plant. A large, thin-walled cell within the ovule, containing the egg and several other nuclei, within which the embryo develops after fertilization of the egg.

Endoplasmic reticulum. A double membrane system in the cytoplasm, continuous with the nuclear membrane and bearing numerous ribosomes.

Endosperm. A polyploid (in many species, triploid) food storage tissue in many angiosperm seeds formed by fusion of two (or more) female cells and a sperm.

Enzyme. Any substance, protein in whole or in part, that regulates the rate of a specific biochemical reaction in living organisms.

Episome. A genetic element that may be present in a given cell, either on a chromosome or separately in the cytoplasm. The *F*, or fertility factor, in *Escherichia coli* is an example.

Epistasis. The masking of the phenotypic effect of either or both members of one pair of alleles by a gene of a different pair. The masked gene is said to be hypostatic.

Equatorial plate. The figure formed at the spindle equator in nuclear division.

Eukaryote (eucaryote). Any organism or cell with a structurally discrete nucleus. Contrast with *prokaryote*.

Euploidy. Variation in chromosome number by whole sets or exact multiples of the monoploid (haploid) number, e.g., diploid, triploid. Euploids above the diploid level may be referred to collectively as polyploids.

F factor. The fertility factor in the bacterium *Escherichia coli;* it is composed of DNA and must be present for a cell to function as a "male" or donor in conjugation. See also *episome*.

F_1. The first filial generation; the first

generation resulting from a given cross.

F_2. The second filial generation; the generation resulting from interbreeding or selfing members of the F_1.

Fertility factor. See *F factor*.

Fitness. See *adaptive value*.

Fixation. Attainment of a gene frequency of 1.0.

Frame shift (reading frame shift). The shift in code reading that results from addition or deletion of nucleotides in any number other than 3 or multiples thereof.

Gamete. A protoplast that, in the process of sexual reproduction, fuses with another protoplast.

Gametophyte. In plants, the phase of the life cycle reproducing sexually by gametes and characterized by having the reduced (usually monoploid or "haploid") chromosome number.

Gene. The particulate determiner of a hereditary trait; a particular segment of a DNA molecule, generally located in the chromosome. See also *cistron, muton,* and *recon*.

Gene frequency. The proportion of one allele of a pair or series present in the population or a sample thereof; that is, the number of loci at which a gene occurs, divided by the number of loci at which it could occur, expressed as a pure number between 0 and 1.

Gene pool. The total of all genes in a population.

Genetic code. The collection of base triplets of DNA and RNA carrying the genetic information by which proteins are synthesized in the cell.

Genetic drift. A change in gene frequency in a population. *Steady drift* is a directed change in frequency toward either greater or lower values; *random drift* is random fluctuation of

gene frequencies caused by chance in mating patterns or to sampling errors.

Genetic equilibrium. Constancy of a particular gene frequency through successive generations.

Genetic load. The proportional reduction in average fitness, relative to an optimal genotype; the average number of lethals per individual in a population.

Genetic polymorphism. The continued occurrence in a population of two or more discontinuous genetic variants in frequencies that cannot be accounted for by recurrent mutation.

Genome. A complete set of chromosomes, or of chromosomal genes, inherited as a unit from one parent.

Genotype (adj., **genotypic**). The genetic makeup or constitution of an individual, with reference to the traits under consideration, usually expressed by a symbol, e.g., "+," "*D*," "*Dd*," "*str*." Individuals of the same genotype breed alike. See *phenotype*.

Guanine. A purine base occurring in DNA and RNA. Pairs normally with cytosine in DNA.

Gynandromorph; gynander. An individual part of whose body exhibits male sex characters and part female characters.

Haplo-. A prefix before a chromosome number denoting an individual whose somatic cells lack one member of that chromosome pair.

Haploid. An individual or cell having a single complete set of chromosomes. Synonym, *monoploid*.

Hemizygous. An individual having but one of a given gene; or a gene present only once. Designates either a monoploid organism, a sex-linked gene in XY males (or an XY individual with regard to a particular X-linked gene), or an individual heterozygous for a given chromosomal deficiency.

Hemoglobin. An iron–protein pigment of blood functioning in oxygen-carbon dioxide exchange of living cells.

Hemophilia. A metabolic disorder characterized by free bleeding from even slight wounds because of the lack of formation of clotting substances. It is associated with a sex-linked recessive gene.

Heterogametic sex. That sex having either only one or two different sex chromosomes, as XO, XY, or ZW. The heterogametic sex thus produces two kinds of gametes with respect to sex chromosomes, e.g., X and Y sperms in human beings.

Heteroploidy. Change in number of whole chromosomes. See also *aneuploidy* and *euploidy*.

Heterospory. In higher plants the production of two kinds of meiocytes (megasporocytes and microsporocytes), giving rise to two different kinds of meiospores. Compare with *homospory*.

Heterozygote (adj., **heterozygous**). An individual whose chromosomes bear unlike genes of a given allelic pair or series. Heterozygotes produce more than one kind of gamete with respect to a particular locus.

Histone. Any of several proteins that can complex with DNA.

Holandric gene. A gene located only on the Y chromosome in XY species.

Homogametic sex. That sex possessing two identical sex chromosomes (XX or ZZ). Gametes produced by the homogametic sex are all alike with respect to sex chromosome constitution.

Homolog. See *homologous chromosomes*.

Homologous chromosomes. Chromosomes occurring in pairs, one derived from each of two parents, normally (except for chromosomes associated with sex) morphologically alike and bearing the same gene loci. Each member of such a pair is the *homolog* of the other.

Homospory. In plants, the production of but one kind of meiocyte, which gives rise to meiospores morphologically indistinguishable from each other.

Homozygote (adj., **homozygous**). An individual whose chromosomes bear identical genes of a given allelic pair or series. Homozygotes produce only one kind of gamete with respect to a particular locus and therefore "breed true."

Hybrid. An individual resulting from a cross between two genetically unlike parents.

Hypha (pl. **hyphae**). One of the filaments of a mycelium in fungi.

Imperfect flower. One lacking either stamens or pistil.

Incomplete dominance. The condition in heterozygotes where the phenotype is intermediate between the two homozygotes. In some plants the cross of red × white produces pink-flowered progeny. Incomplete dominance differs from codominance.

Incompletely sex-linked genes. Genes located on homologous portions of the X and Y (or Z and W) chromosomes.

Independent segregation. The random or independent behavior of genes on different pairs of chromosomes.

Inducer. See *effector*.

Inducible enzyme. An enzyme synthesized only in the presence of an effector (which see).

Interference. The increase (*negative interference*) or decrease (*positive interference*) in likelihood of a second crossover closely adjacent to another. In most organisms interference increases with decreased distance between crossovers. See *coincidence*.

Interphase. The stage of cell life during which that cell is not dividing.

Intersex. An individual showing secondary sex characters intermediate between male and female or some of each sex.

Inversion. Reversal of the order of a block of genes in a given chromosome. PQ*UTSR*VWX would represent an inversion of genes *RSTU* if the normal order is alphabetical.

Inversion loop. The loop configuration in a pair of synapsed, homologous chromosomes, caused by an inversion in one of them.

In vitro. Experimentally induced biological processes outside the organism (literally, "in glass").

In vivo. Experimentally induced biological processes within the organism.

Isogametes. Gametes not sexually differentiated. Fusing isogametes are physically similar but in some species are chemically differentiated.

Karyokinesis. The division of the nucleus during cell division.

Karyolymph. A clear fluid material within the nuclear membrane; "nuclear sap."

Karyotype. The somatic chromosome complement of an individual, usually as defined at mitotic metaphase by morphology (including centromere location and often by special staining techniques, as fluorescence banding) and number, arranged in a sequence that is standard for that organism.

Kinetochore. The attachment region (within the centromere) of the microtubules of the spindle in cells undergoing mitosis or meiosis.

Lampbrush chromosome. A chromosome having paired loops extending laterally, occurring in primary oocyte nuclei; they represent sites of active RNA synthesis.

Lethal gene. A gene whose phenotypic effect is sufficiently drastic to kill the bearer. Death from different lethal genes may occur at any time from fertilization of the egg to advanced age. Lethal genes may be dominant, incompletely dominant, or recessive.

Life cycle. The entire series of developmental stages undergone by an individual from zygote to maturity and death.

Linkage. The occurrence of different genes on the same chromosome.

Linkage group. All of the genes located physically on a given chromosome.

Linkage map. A scale representation of a chromosome showing the relative positions of all its known genes.

Locus (pl., **loci**). The position or place on a chromosome occupied by a particular gene or one of its alleles.

Lysis (n.) (v.i. or vt., **lyse**). Disintegration or dissolution; usually, the destruction of a bacterial host cell by infecting phage particles.

Lysogenic bacteria. Living bacterial cells harboring temperate phages (viruses).

Map unit. A distance on a linkage map represented by 1 per cent of crossovers (recombinants), that is, by a recombination frequency of 1 per cent.

Mean. The arithmetic average; the sum of all values for a group, divided by the number of individuals. Symbolized by $\bar{x}$ (sample) or by μ (population).

Median. The middle value of a series of readings arranged serially according to magnitude.

Megasporocyte. In plants, the meiocyte that is destined to produce megaspores by meiosis. Synonym, *megaspore mother cell.*

Meiocyte. Any cell that undergoes meiosis.

Meiosis. Nuclear divisions in which the diploid or somatic chromosome number is reduced by half. In the first of the two meiotic divisions, homologous chromosomes first replicate, then pair (synapse), and finally separate to different daughter nuclei, which thus have half as many chromosomes as the parent nucleus, and one of each kind instead of two. A second division, in which chromosomal replicates separate into daughter nuclei, follows so that meiosis produces four monoploid daughter nuclei from one diploid parent nucleus.

Meiospore. In plants, one of the asexual reproductive cells produced by meiosis from a meiocyte.

Meristem. An undifferentiated cellular region in plants characterized by repeated cell division.

Merozygote. A partially diploid receptor bacterial cell that results from conjugation.

Messenger RNA. Ribonucleic acid conferring amino acid specificity on ribosomes; complementary to a given DNA cistron.

Metacentric. A chromosome with a centrally located centromere.

Metafemale (also **superfemale**). Abnormal females in *Drosophila,* usually sterile and weak, with an overbalance of X chromosomes with respect to autosomes; X/A ratio greater than 1.0.

Metamale (also **supermale**). Abnormal males in *Drosophila* with an overabundance of autosomes to X chromosomes; X/A ratio less than 0.5.

Metaphase. That stage of nuclear division in which the chromosomes are located in the equatorial plane of the spindle prior to centromere separation.

Micrometer (μm). A commonly employed unit of measurement in microscopy, it equals 1×10^{-6} meter or 1×10^{-3} millimeter. Synonym, *micron (μ)* in older usage.

Micron. See *micrometer.*

Microspore. In plants, the meiospore that gives rise to the male gametophyte and sperms.

Microsporocyte. In plants, a cell that is destined to produce microspores by meiosis. Synonyms, *microspore mother cell* and *"pollen mother cell."*

Microtubule. A hollow tubular cytoplasmic component, outside diameter about 15 to 30 nm, found in the cytoplasm, especially of motile cells, and forming the spindle of the mitotic apparatus (which see).

Millimeter (mm). A unit of distance measurement equal to 1×10^{-3} meter.

Missense. A mutation by which a particular codon is changed so as to incorporate a different amino acid, often resulting in an inactive protein. A *missense codon* results when one or more bases of a *sense codon* (which see) are changed so that a different amino acid is coded for.

Mitochondrion (pl., **mitochondria**). Small cytoplasmic organelle where cellular respiration occurs.

Mitosis. Nuclear division in which a replication of chromosomes is followed by separation of the products of replication and their incorporation into two daughter nuclei. Daughter nuclei are normally identical with each other and with the original parent nucleus in both kind and number of chromosomes. Mitosis may or may not involve cytoplasmic division.

Mitotic apparatus. The collection of cytoplasmic structures present in mitotic cells and consisting of the asters (if present) surrounding each centrosome and the spindle of microtubules.

Mode. The numerically largest class or group in a series of measurements or values.

Monoecious. Individuals producing both sperm and egg.

Monohybrid. The offspring of two homozygous parents that differ in only one gene locus or in which only one such locus is under consideration.

Monohybrid cross. A cross between two parents that differ in only one heritable character or in which only one such character is under consideration.

Monoploid. An individual having a single complete set of chromosomes. Also the fundamental number of chromosomes comprising a single set. Synonym, *haploid.*

Monosomic. An individual lacking one chromosome of a set $(2n - 1)$.

Morphology. The study of form and structure in organisms.

Mosaic. An individual part of whose body is composed of tissue genetically different from the other part.

Multiple alleles. A series of three or more alternative alleles any one of which may occur at a particular locus on a chromosome.

Multiple genes. See *polygenes.*

Mutagen. Any agent that brings about a mutation.

Mutation. A sudden change in genotype having no relation to the individual's ancestry. Used for changes in a single gene itself ("point mutations") and for chromosomal aberrations.

Muton. The smallest segment of DNA or subunit of a cistron that can be changed and thereby bring about a mutation; probably as small as one nucleotide pair.

Mycelium. The threadlike filamentous vegetative body of many fungi.

Nanometer (nm). A unit of distance equal to 1×10^{-9} meter. Synonym, *millimicron (mμ).*

Negative control. Repression of an operator site (which see) by a regulatory protein that is produced by a regulator site (which see).

Nondisjunction. The failure of homologous chromosomes to separate at anaphase-I of meiosis. *Primary nondisjunction* may occur in an XX female, leading to production of XX or O eggs (in addition to the normal X eggs); or it may occur (first division) in an XY male, resulting in XY and O sperm or (second division) in XX and O or YY and O sperm (in addition to normal Y or X sperms). *Secondary nondisjunction* may occur in an XXY female giving rise to eggs with XX, XY, X, or Y chromosomal combinations.

Nonsense. A codon that does not specify any amino acid in the genetic code.

Normal curve. A smooth, symmetrical, bell-shaped curve of distribution.

N-terminus. The amino ($-NH_2$) end of a peptide chain, by convention written as the left end of the structural formula.

Nucleolus. Deeply staining body containing both RNA and protein within the nucleus.

Nucleoside. Portion of a DNA or RNA molecule composed of one deoxyribose molecule (in DNA), or ribose (in RNA), plus a purine or a pyrimidine.

Nucleotide. Portion of a DNA or RNA molecule composed of one deoxyribose phosphate unit (in DNA), or one ribose phosphate unit (in RNA), plus a purine or a pyrimidine.

Nullisomic. A cell or organism lacking both members of a given chromosome pair $(2n - 2)$.

Oligonucleotide. A linear sequence of a few (generally not over 10) nucleotides.

Ontogeny. The complete development of the individual from zygote, spore, and so on, to adult form.

Oocyte. The diploid cell that will undergo meiosis (oogenesis) to form an egg.

Oogenesis. Egg formation.

Operator site. A segment of DNA in an operon that affects activity or nonactivity of associated cistrons; it may be combined with a repressor and thereby "turn off" the associated cistrons.

Operon. A system of cistrons, operator and promotor sites, by which a given genetically controlled, metabolic activity is regulated.

Ovary. The female gonad in animals or the ovule-containing portion of the pistil of a flower.

Ovule. Structure within the ovary of the pistil of a flower that becomes a seed. It represents a megasporangium, together with some overgrowing tissue (integuments), and ultimately contains the female gametophyte (embryo sac), one of whose nuclei is an egg.

P. The parental generation in a given cross.

Parameter. Actual value of some quantitative character for a population. Compare *statistic.*

Parthenogenesis. Development of a new individual from an unfertilized egg.

Pedigree. The ancestral history of an individual; a chart showing such history.

Peptide bond. A chemical bond (CONH) linking amino acid residues together in a protein.

Perfect flower. One having both stamen(s) and pistil(s).

Petite. A slow-growing strain of yeast (*Saccharomyces*) lacking certain respiratory enzymes and forming unusually small colonies on agar. *Segregational petites* bear mutant nuclear gene(s), whereas *neutral petites* (sometimes called *vegetative petites*) bear mutant mitochondrial DNA.

Phage. A virus that infects bacteria.

Phenotype (adj., **phenotypic**). The appearance or discernable character of an individual, which is dependent upon its genetic makeup, usually expressed in words, e.g., "tall," "dwarf," "wild type," "prolineless." Identical phenotypes may not necessarily breed alike.

Phylogeny. The evolutionary development of a species or other taxonomic group.

Pistil. The entire part of the flower that produces megaspores (which, in turn, produce eggs). Sometimes, and *incorrectly,* referred to as a female floral part.

Plaque. Clear area on culture plate of bacteria where these have been killed by phages.

Plasmagene. A self-replicating, cytoplasmically located gene.

Pleiotropy. The influencing of more than one trait by a single gene.

Polar body. One of the three very small cells produced during meiosis of an oocyte, containing a monoploid nucleus but little cytoplasm. It is nonfunctional in reproduction.

Pollen. The young male gametophyte of a flowering plant, surrounded by the microspore wall.

Polygenes. Two or more different pairs of alleles, with a presumed cumulative effect, governing such quantitative traits as size, pigmentation, intelligence, among others. Those contributing to the trait are termed contributing (effective) alleles; those appearing not to do so are referred to as noncontributing or noneffective alleles.

Polymer. A chemical compound composed of two or more units of the same compound.

Polynucleotide. A linear sequence of many nucleotides.

Polypeptide. A compound containing amino acid residues joined by peptide bonds. A protein may consist of one or more specific polypeptide chains.

Polyploid. An individual having more than two complete sets of chromosomes, e.g., triploid ($3n$), tetraploid ($4n$).

Polyribosome. See *polysome.*

Polysome. A group of ribosomes joined by a molecule of messenger RNA.

Polytene chromosome. Many-stranded giant chromosomes produced by repeated replication during synapsis in certain dipteran larval tissues. Synonym, *giant chromosome.*

Population. An infinite group of individuals, measured for some variable, quantitative character, from which a sample is taken.

Position effect. A phenotypic effect dependent on a change in position on the chromosome of a gene or group of genes.

Positive control. Activation of an operator site (which see) by a regulatory protein that is produced by a regulator site (which see).

Probability. The likelihood of occurrence of a given event. Usually expressed as a number between 0 (complete certainty that the event will *not* occur) and 1 (complete certainty that the event *will* occur).

Progeny. Offspring individuals.

Prokaryote. A cell or organism lacking a discrete nuclear body (also *procaryote*).

Promoter site. Site on DNA at which mRNA synthesis is initiated in an operon. May be combined with a repressor protein to inhibit mRNA synthesis by associated cistrons.

Prophase. The first stage of nuclear division, including all events up to (but not including) arrival of the chromosomes at the equator of the spindle.

Protoplast. A structural unit of protoplasm; all the living (protoplasmic) material of a cell. The two principal parts are nucleus and cytoplasm.

Prototroph. An individual able to carry on a given synthesis; a wild type individual able to grow on minimal medium.

Pseudoalleles. Nonalleles so closely linked as often to be inherited as one gene, but shown to be separable by crossover studies.

Pseudodominance. The expression (apparent dominance) of a recessive gene at a locus opposite a deficiency.

Punnett square. A "checkerboard" grid designed to determine all possible genotypes produced by a given cross. Genotypes of the gametes of one sex are entered across the top, those of the other down one side. Zygote genotypes produced by each possible mating are then entered in the appropriate squares of the grid.

Pure line (pure breeding line). A strain of individuals homozygous for all genes being considered.

Purine. Nitrogenous base occurring in DNA and RNA; these are adenine and guanine.

Pyrimidine. Nitrogenous base occurring in DNA (thymine and cytosine) or RNA (uracil and cytosine).

Rad. Term meaning radiation-absorbed dose; it is the amount of ionizing radiation that liberates 100 ergs of energy in 1 gram of matter.

Recessive. An adjective applied to the member of a pair of genes that fails to express itself in the presence of its dominant allele. The term is also applicable to the trait produced by a recessive gene. Recessive genes express themselves ordinarily only in the homozygous state.

Reciprocal cross. A second cross of the same genotypes in which the sexes of the parental generation are reversed. The cross $AA(♀) \times aa(♂)$ is the reciprocal of the cross $aa(♀) \times AA(♂)$.

Reciprocal translocation. The exchange of segments between two nonhomologous chromosomes.

Recombinant. An individual derived from a crossover gamete.

Recombination. The new association of genes in a recombinant individual, arising from independent assortment of unlinked genes, from crossing-over between linked genes, or from intracistronic crossing-over.

Recon. The smallest segment of DNA or subunit of a cistron that is capable of recombination; may be as small as one deoxyribonucleotide pair.

Reduction division. See *meiosis.*

Regulator site. The specific segment of DNA responsible for production of a regulatory protein that may serve as a repressor in some cases or as an activator in others. See also *negative control* and *positive control.*

Relational coiling. The loose coiling of chromatids about each other.

Replicate. To form replicas from a model or template; applies to synthesis of new DNA from pre-existing DNA as part of nuclear division.

Repressor. A protein produced by a regulator gene that can combine with and repress action of an associated operator gene.

Resistance transfer factor (RTF). Possibly an episome conferring multiple resistance to several drugs, which can be transferred between certain genera of bacteria.

Ribonucleic acid (RNA). A single-stranded nucleic acid molecule, synthesized principally in the nucleus from deoxyribonucleic acid, composed of a ribose-phosphate backbone with purines (adenine and guanine) and pyrimidines (uracil and cytosine) attached to the sugar ribose. RNA is of several kinds and functions to carry the "genetic message" from nuclear DNA to the ribosomes.

Ribonucleoside. Portion of an RNA molecule composed of one ribose molecule plus either a purine or a pyrimidine.

Ribonucleotide. Portion of an RNA molecule composed of one ribose-phosphate unit plus a purine or a pyrimidine.

Ribose. The 5-carbon sugar of ribonucleic acid.

Ribosomal RNA. That ribonucleic acid incorporated into ribosomes; it is non-specific for amino acids.

Ribosome. Cytoplasmic structure, usually adherent to the endoplasmic reticulum, which is the site of protein synthesis.

RNAase. An enzyme that hydrolyzes RNA.

RNA polymerase. An enzyme catalyzing the formation of RNA from ribonucleotides (which see) using one strand of DNA as a template.

Roentgen (r). The amount of ionizing radiation that produces about two ion pairs per cubic micrometer of matter, or about 1.6×10^{12} ion pairs per cubic centimeter.

Sedimentation coefficient. The rate of sedimentation of a solute in an appropriate solvent under centrifugation.

An s value of 1×10^{-13} second is one *Svedberg unit.*

Selection coefficient. The measure of reduced fitness of a given genotype; it equals one minus the adaptive value (which see) of that genotype.

Self-fertilization. Functioning of a single individual as both male and female parent. Plants are "selfed" if sperm and egg are supplied by the same individual.

Semiconservative. Replication of DNA in which the two sugar-phosphate "backbones" become separated, each being conserved as one of the two strands of two new DNA molecules.

Sense codon. A codon (which see) specifying a particular amino acid in protein synthesis.

Sex chromosomes. Heteromorphic chromosomes not occurring in identical pairs in both sexes in diploid organisms; in man and fruit fly these are designated as X and Y chromosome.

Sex-influenced trait. One in which dominance of an allele depends on sex of the bearer; e.g., pattern baldness in humans is dominant in males, recessive in females.

Sex-limited trait. One expressed in only one of the sexes; e.g., cock feathering in fowl is limited to normal males.

Sex-linked gene. A gene located only on the X chromosome in XY species (or on the Z chromosomes in ZW species).

Siblings (also **sibs**). Individuals having the same maternal and paternal parents; brother-sister relationship.

Significance. In statistical treatments, probability values of >0.05 are termed *not significant,* those ≤ 0.05 but >0.01 are *significant,* those ≤ 0.01 but >0.001 are *highly significant,* and those ≤ 0.001 are *very highly significant.* Significant probability values indicate that the results, although possible, deviate too greatly from the expectancy to be acceptable when chance alone is operating.

Soma (adj., **somatic**). The body, cells of which in mammals and flowering plants normally have two sets of chromosomes, one derived from each parent.

Spermatid. The monoploid cells, resulting from meiosis of a primary spermatocyte, that will mature into sperms.

Spermatocyte. The cell that undergoes meiosis to produce four spermatids.

Spermatogenesis. Development of sperms.

Spore. An asexual reproductive protoplast capable of developing into a new individual. See also *meiospore.*

Sporocyte. In plants, a meiocyte.

Sporogenesis. Formation of spores.

Sporophyte. In plants, the phase of the life cycle reproducing asexually by meiospores and characterized by having the double (usually diploid) chromosome number.

Stamen. That part of a flower producing microspores (which, in turn, produce sperms). Sometimes, and *incorrectly,* referred to as a male floral part.

Standard deviation. A measure of the variation in a sample. Symbolized by s.

Standard error of difference in means. Measure of the significance of the difference in two sample means. Symbolized by S_d.

Standard error of sample mean. An estimate of the standard deviation of a series of hypothetical sample means, serving as a measure of the closeness with which a given sample mean approximates the population mean. Symbolized by $s_{\bar{x}}$.

Statistic. Actual value of some quantitative character for a sample from which estimates of parameters may be made.

Stigma. The pollen-receptive portion of the pistil of a flower.

Structural gene. A cistron.

Svedberg unit. See *sedimentation coefficient.*

Synapsis (v.i., **synapse**). The pairing of homologous chromosomes occurring in prophase-I of meiosis.

Syngamy. The union of the nuclei of sex cells (gametes) in reproduction.

Tautomer. An alternate molecular form of a compound, characterized by a different arrangement of its electrons and protons as compared with the common form of the molecule.

Taxon. A taxonomic group of any rank.

Taxonomy. The study of describing, naming, and classifying living organisms, and the bases on which resultant classification systems rest.

Telocentric. A chromosome having a terminal centromere.

Telophase. The concluding stage of nuclear division characterized by the reorganization of interphase nuclei.

Temperate phage. A phage (bacterium-infecting virus) that invades and multiplies in but does not ordinarily lyse its host.

Template. A model, mold, or pattern; DNA acts as a template for RNA synthesis.

Testcross. The cross of an individual (generally of dominant phenotype) with one having the recessive phenotype. Generally used to determine whether an individual of dominant phenotype is homozygous or heterozygous, or to determine the degree of linkage.

Tetrad. The four monoploid (haploid) cells arising from meiosis of a megasporocyte or microsporocyte in plants; also, a group of four associated chromatids during synapsis.

Tetraploid. A polyploid cell tissue, or organism having four sets of chromosomes (4*n*). See also *polyploid, allopolyploid,* and *autopolyploid.*

Three-point cross. A trihybrid testcross (e.g., *ABC/abc* $\times$ *abc/abc*, etc.) used primarily in chromosome mapping.

Thymine. A pyrimidine base occurring in DNA. Pairs normally with adenine.

Totipotency. The property of a cell (or cells) whereby it develops into a complete and differentiated organism.

***Trans* arrangement.** Linkage of the dominant allele of one pair and the recessive of another on the same chromosome.

Transcription. Synthesis of messenger RNA from a DNA template.

Transduction. Recombination in bacteria whereby DNA is transferred by a phage from one cell to another.

Transfer RNA. Amino acid-specific RNA that transfers activated amino acids to mRNA where protein synthesis takes place. Sometimes referred to as *soluble RNA.*

Transformation. Genetic recombination, particularly in bacteria, whereby naked DNA from one individual becomes incorporated into that of another.

Transgressive variation. Appearance in progeny of a more extreme expression of a trait than occurs in the parents. Assumed to result from cumulative action of polygenes, but careful testing of variation in parental lines is necessary for verification.

Transition. The substitution in DNA or RNA of one purine for another or of one pyrimidine for another.

Translation. The process by which a particular messenger RNA nucleotide sequence is responsible for a specific amino acid residue sequence of a polypeptide chain.

Translocation. The shift of a portion of a chromosome to another part of the same chromosome or to an entirely different chromosome. (See also *reciprocal translocation.*)

Transversion. The substitution in DNA or RNA of a purine for a pyrimidine or vice versa.

Trihybrid. An individual heterozygous for three pairs of genes.

Triplet. A group of three successive nucleotides in RNA (or DNA) that, in the genetic code (which see), specifies a particular amino acid in the synthesis of polypeptide chains.

Triplo-. Prefix denoting a trisomic individual where the identity of the extra chromosome is known, e.g., triplo-IV *Drosophila,* or triplo-21 human beings.

Triploid. A polyploid cell, tissue, or organism having three sets of chromosomes ($3n$).

Trisomic. An individual having one extra chromosome of a set ($2n + 1$).

Universal donor. A person with group O blood, whose erythrocytes therefore bear neither A nor B antigens, and whose blood can be donated to members of groups O, A, B, and AB if necessary.

Universal recipient. A person of blood group AB who can receive blood from members of groups AB, A, B, or O if necessary.

Uracil. A pyrimidine base occurring in RNA.

Variance. A statistic providing an unbiased estimate of population variability; it is the square of the standard deviation (which see).

Virulence. The ability to produce disease.

Virulent phage. A phage (virus) that destroys (lyses) its host bacterial cell.

Wild type. The most frequently encountered phenotype in natural breeding populations; the "normal" phenotype.

Wobble hypothesis. The partial or total lack of specificity in the third base of some triplet codons (which see) whereby two, three, or four codons differing only in the third base may code for the same amino acid. See Table 17-1.

Zygote. The protoplast resulting from the fusion of two gametes in sexual reproduction; a fertilized egg.

Index